Mathematics for Computer Science

Prentice Hall International Series in Computer Science

C.A.R. Hoare, Series Editor

ARNOLD, A., *Finite Transition Systems*
✗ ARNOLD, A. and GUESSARIAN, I., *Mathematics for Computer Science*
BARR, M. and WELLS, C., *Category Theory for Computing Science* (2nd edn)
BEN-ARI, M., *Principles of Concurrent and Distributed Programming*
BEN-ARI, M., *Mathematical Logic for Computer Science*
BEST, E., *Semantics of Sequential and Parallel Programs*
BIRD, R. and WADLER, P., *Introduction to Functional Programming*
✗ BOVET, D.P. and CRESCENZI, P., *Introduction to the Theory of Complexity*
DE BROCK, B., *Foundations of Semantic Databases*
BRODA, EISENBACH, KHOSHNEVISAN and VICKERS, *Reasoned Programming*
BURKE, E. and FOXLEY, E., *Logic and Its Applications*
DAHL, O.-J., *Verifiable Programming*
DAVIES, J. and WOODCOCK, J., *Using Z: Specification, refinement and proof*
DUNCAN, E., *Microprocessor Programming and Software Development*
ELDER, J., *Compiler Construction*
ELLIOTT, R.J. and HOARE, C.A.R. (eds), *Scientific Applications of Multiprocessors*
FREEMAN, T.L. and PHILLIPS, R.C., *Parallel Numerical Algorithms*
GOLDSCHLAGER, L. and LISTER, A., *Computer Science: A modern introduction* (2nd edn)
GORDON, M.J.C., *Programming Language Theory and Its Implementation*
GRAY, P.M.D., KULKARNI, K.G. and PATON, N.W., *Object-oriented Databases*
HAYES, I. (ed.), *Specification Case Studies* (2nd edn)
HEHNER, E.C.R., *The Logic of Programming*
HINCHEY, M.G. and BOWEN, J.P., *Applications of Formal Methods*
HOARE, C.A.R., *Communicating Sequential Processes*
HOARE, C.A.R. and GORDON, M.J.C. (eds), *Mechanized Reasoning and Hardware Design*
HOARE, C.A.R. and JONES, C.B. (eds), *Essays in Computing Science*
HOARE, C.A.R. and SHEPHERDSON, J.C. (eds), *Mechanical Logic and Programming Languages*
HUGHES, J.G., *Database Technology: A software engineering approach*
HUGHES, J.G., *Object-oriented Databases*
INMOS LTD, *Occam 2 Reference manual*
JACKSON, M.A., *System Development*
JONES, C.B., *Systematic Software Development Using VDM* (2nd edn)
JONES, C.B. and SHAW, R.C.F. (eds), *Case Studies in Systematic Software Development*
JONES, G., *Programming in Occam*
JONES, G. and GOLDSMITH, M., *Programming in Occam 2*
JONES, N.D., GOMARD, C.K. and SESTOFT, P., *Partial Evaluation and Automatic Program Generation*
JOSEPH, M., *Real-time Systems: Specification, verification and analysis*
JOSEPH, M., PRASAD, V.R. and NATARAJAN, N., *A Multiprocessor Operating System*
KALDEWAIJ, A., *Programming: The derivation of algorithms*
KING, P.J.B., *Computer and Communications Systems Performance Modelling*
LALEMENT, R., *Computation as Logic*
McCABE, F.G., *Logic and Objects*
McCABE, F.G., *High-level Programmer's Guide to the 68000*
MEYER, B., *Introduction to the Theory of Programming Languages*
MEYER, B., *Object-oriented Software Construction*
MILNER, R., *Communication and Currency*
MITCHELL, R., *Abstract Data Types and Modula 2*
MORGAN, C., *Programming from Specifications* (2nd edn)
OMONDI, A.R., *Computer Arithmetic Systems*
PATON, COOPER, WILLIAMS and TRINDER, *Database Programming Languages*
PEYTON JONES, S.L., *The Implementation of Functional Programming Languages*

Series listing continues at back of book

Mathematics for Computer Science

A. Arnold

and

I. Guessarian

Prentice Hall
London New York Toronto Sydney Tokyo Singapore
Madrid Mexico City Munich

First published 1996 by
Prentice Hall Europe
Campus 400, Maylands Avenue
Hemel Hempstead
Hertfordshire, HP2 7EZ
A division of
Simon & Schuster International Group

© Prentice Hall Europe/Masson 1996

All rights reserved. No part of this publication may be reproduced,
stored in a retrieval system, or transmitted, in any form, or by any
means, electronic, mechanical, photocopying, recording or otherwise,
without prior permission, in writing, from the publisher.
For permission within the United States of America
contact Prentice Hall Inc., Englewood Cliffs, NJ 07632.

Printed and bound in Great Britain
T. J. Press (Padstow) Ltd

Library of Congress Cataloging-in-Publication Data

Available from the publisher

British Library Cataloguing in Publication Data

A catalogue record for this book is available from
the British Library

ISBN 0-13-234717-2

1 2 3 4 5   00 99 98 97 96

CONTENTS

Preface ix

1 Sets and functions 1
  1.1 Sets 1
  1.2 Functions 4
  1.3 Cardinals 7
  1.4 Operations and relations 9

2 Ordered sets 17
  2.1 Order and preorder relations 17
  2.2 Ordered sets 20
  2.3 Upper and lower bounds 22
  2.4 Well-ordered sets and induction 25
  2.5 Complete sets and lattices 27

3 Recursion and induction 35
  3.1 Reasoning by induction in $\mathbb{N}$ 36
  3.2 Inductive definitions and proofs by structural induction 40
  3.3 Terms 45
  3.4 Closure operations 52

4 Boolean algebras 57
  4.1 Boolean algebras 57
  4.2 Boolean rings 61
  4.3 The Boolean functions 64

5 Logic 67
  5.1 Remarks on mathematical reasoning 68
  5.2 Propositional calculus 70
  5.3 First order predicate calculus 83
  5.4 Herbrand's theorem and consequences 95

6 Combinatorial algebra — 107
6.1 Basics — 107
6.2 Applications: counting techniques for finite sets — 112
6.3 Counting sequences and partitions — 118

7 Recurrences — 121
7.1 Introduction: examples, generalities — 122
7.2 Linear recurrences — 127
7.3 Other recurrence relations — 141
7.4 Complements and examples — 143

8 Generating series — 148
8.1 Generalities — 149
8.2 Applications of generating series to recurrences — 161

9 Asymptotic behaviour — 170
9.1 Generalities — 170
9.2 Criteria of asymptotic behaviour of functions — 174

10 Graphs and trees — 183
10.1 Graphs — 183
10.2 Trees and rooted trees — 197

11 Rational languages and finite automata — 203
11.1 The free monoid — 204
11.2 Regular languages — 206
11.3 Finite automata — 208
11.4 Equation systems — 219

12 Discrete probabilities — 226
12.1 Generalities — 227
12.2 Probability spaces — 229
12.3 Conditional probabilities and independent events — 232
12.4 Random variables — 241
12.5 Generating functions — 253
12.6 Common probability distributions — 255

13 Finite Markov chains — 261
13.1 Introduction — 261
13.2 Generalities — 262
13.3 Classification of states — 268

14 Applications and examples — 280
14.1 Quicksort — 281

14.2 Euclid's algorithm	288
14.3 Proofs of program properties and termination	293

15 Answers to exercises — 301

Chapter 1	301
Chapter 2	309
Chapter 3	314
Chapter 4	321
Chapter 5	326
Chapter 6	334
Chapter 7	340
Chapter 8	345
Chapter 9	353
Chapter 10	356
Chapter 11	362
Chapter 12	370
Chapter 13	383
Chapter 14	393

Index — 397

PREFACE

Like all scientific disciplines, computer science uses mathematics in order to formalize concepts, to model situations and to abstract objects; this enables computer scientists to *reason* and study a *priori* the properties of the entities (machines, programs, systems, etc.) that they use. For instance, we will be able to verify that a program always terminates, to estimate its execution time for certain values of the data, to determine the saturation conditions of a network, to ensure that an operating system can never refuse a user's commands, etc.

Of course, the subset of mathematics used in computer science is different from that used in other disciplines. Differential equations might be less useful for the computer scientist than for the physicist, whilst discrete mathematics might be more useful.

We examine a basic subset of mathematics which underpins computer science, and which we call 'mathematics FOR computer science'. It consists of familiar mathematics which is intensively used by computer scientists. Practically speaking, computer science teachers feel that a certain number of mathematical concepts should be known by students, and hence taught at some point in computer science curricula. The subset of mathematics covered in this handbook can be taught in a one-year course, or divided into two one-semester courses.

Each chapter contains a header section which describes conventional applications of its subject matter.

Z We strongly advise readers to try to work out the exercises while reading, in order to aid understanding of the concepts; we also advise them to try to solve the exercises before checking the solutions, which are given at the end of the book. A diagram of the interdependences of the different chapters concludes this preface; it should aid a non-sequential reading. The most difficult points, which are likely to lead to errors, are indicated by a 'dangerous turn' sign, similar to the one beginning the present paragraph.

Paul Gastin helped in writing Chapters 1 and 3. Many of the exercises of Chapters 12 and 13 were written by Maryse Pelletier-Koskas; Christian Ronse and Guy Melançon, from Bordeaux, also contributed to many of the exercises.

This book has also benefited from the comments of our respective teaching colleagues, and from the advice of Jean Berstel, Paul Gastin, Yuri Gurevich, Jozef Gruska, Colin de la Higueira, Robert Pallu de La Barrière, Dominique Pastre, Laurence Puel, Jean-Claude Raoult, Renaud Rioboo, Michel de Rougemont and Didier Vidal. Finally, we heartily thank Danièle Beauquier, Damian Niwinski, Robert Pallu de La Barrière, Christian Ronse and Violaine Thibau who proofread and corrected this manuscript.

We gratefully acknowledge the help of Guillaume Alexandre, Monique Baron, Patrick Cegielski, Valeria Chiola, Chris Colby, John Fountain, Françoise Gire, Serge Grigorieff, Richard Lassaigne, Jean-Luc Mounier, David Sands, Magnus Steinby and Ben Wagner in preparing the English version of this handbook.

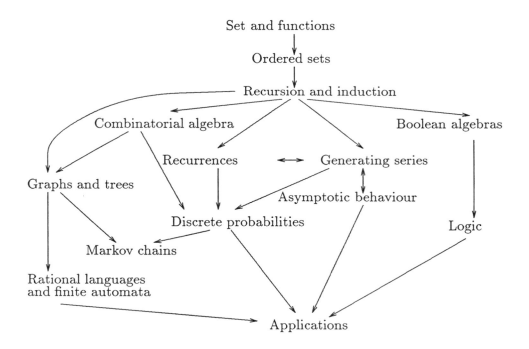

CHAPTER 1

SETS AND FUNCTIONS

In this chapter we review the foundations and notations of naïve set theory. We will define sets and the notions of functions, relations and operations on sets that are required in this text. The reader who is well acquainted with basic mathematics can browse through the present chapter.

1.1 Sets

1.1.1 Set, element, inclusion

Let E be a *set* and e an *element*. $e \in E$ means that e is a member of the set E and is read 'e is in E'. The negation of this relation (i.e. e is not in E) is denoted by $e \notin E$. The empty set is a set containing no elements; it is denoted by $\emptyset$.

Let A, B be two sets. A is said to be a subset of B, or contained in B, if and only if any member of A is a member of B, and this is denoted by $A \subseteq B$. The negation of $A \subseteq B$ is denoted by $A \nsubseteq B$, and this does *not* mean that $B \subseteq A$. We notice that $A = B$ if and only if $A \subseteq B$ and $B \subseteq A$ (i.e. if and only if A and B have the same elements). If $A \subseteq B$, but $A \neq B$, we will write $A \subsetneq B$. Finally, the set of subsets of E is denoted by $\mathcal{P}(E)$; hence $A \subseteq E$ if and only if $A \in \mathcal{P}(E)$. Note that the sets $\emptyset$ and E are elements of $\mathcal{P}(E)$.

EXAMPLE 1.1 Let $E = \{0, 1\}$. $\mathcal{P}(E) = \{\emptyset, \{0\}, \{1\}, \{0, 1\}\}$.

REMARK 1.2 Some authors, mainly French, among them Bourbaki, denote inclusion by $\subset$; so, to avoid ambiguities, strict inclusion is denoted by $\subsetneq$.

The *Cartesian product* of two sets E and F is the set of all ordered pairs consisting of an element of E and an element of F:

$$E \times F = \{(x, y) \, / \, x \in E \text{ and } y \in F\} \, .$$

1

The Cartesian product can be generalized to a finite family of sets:

$$E_1 \times \cdots \times E_n = \{(x_1, \ldots, x_n) \,/\, x_1 \in E_1, \ldots, x_n \in E_n\}.$$

Lastly, the Cartesian product of E by itself n times, for $n \geq 1$, is denoted by $E^n = E \times \cdots \times E$. E^n can be recursively defined by (see Chapter 3): $E^1 = E$ and $E^n = E \times E^{n-1}$.

1.1.2 Union, intersection, difference, complement, partition

Assume that a universal set E is given. Let A, B be two subsets of E. We define:

- the *intersection* of A and B: $A \cap B = \{e \in E \,/\, e \in A \text{ and } e \in B\}$,
- the *union* of A and B: $A \cup B = \{e \in E \,/\, e \in A \text{ or } e \in B\}$,
- the *difference* of A and B: $A \setminus B = \{e \in E \,/\, e \in A \text{ and } e \notin B\}$,
- the *complement* of A with respect to E, denoted by $\overline{A}$ or A^c:
$$\overline{A} = E \setminus A = \{e \in E \,/\, e \notin A\},$$
- the *symmetric difference* of A and B: $A \triangle B = (A \setminus B) \cup (B \setminus A)$.

Two subsets A and B of E are said to be *disjoint* if and only if $A \cap B = \emptyset$.

EXAMPLE 1.3
- $\overline{E} = \emptyset$, and $\overline{\emptyset} = E$.
- $A \cap \emptyset = \emptyset$, $\quad A \cup E = E$, $\quad A \triangle E = \overline{A}$.
- $A \cap A = A \cup A = A \cup \emptyset = A \triangle \emptyset = A \cap E = \overline{\overline{A}} = A$.
- $A \setminus B = A \cap \overline{B}$.
- $A \setminus A = A \triangle A = \emptyset$.

De Morgan's laws enable us to compute the complement of a union and of an intersection: $\overline{A \cup B} = \overline{A} \cap \overline{B}$ and $\overline{A \cap B} = \overline{A} \cup \overline{B}$.

Union and intersection are associative and commutative operations (see Section 1.4.1). Moreover, they distribute over each other (see Section 1.4.1), i.e.

$$A \cap (B \cup C) = (A \cap B) \cup (A \cap C) \quad \text{and}$$
$$A \cup (B \cap C) = (A \cup B) \cap (A \cup C).$$

These properties are to be found again in the study of lattices (see Chapter 2, Section 2.5) and in the study of Boolean algebras (see Chapter 4).

NOTATIONS

$\Longrightarrow$ (resp. $\Longleftrightarrow$) stands for 'implies' (resp. 'if and only if').

$\exists e \in E$ (resp. $\forall e \in E$) stands for 'there exists an e in E' (resp. 'for any e in E').

Sets

EXERCISE 1.1 Let A, B and C be three subsets of E. Show that
$$A \cap \overline{B} = A \cap \overline{C} \iff A \cap B = A \cap C.$$
◇

EXERCISE 1.2 Let A, B and C be three subsets of E. Show that
$$\left(A \cup B \subseteq A \cup C \text{ and } A \cap B \subseteq A \cap C\right) \implies B \subseteq C.$$

When does the equality $B = C$ hold? ◇

EXERCISE 1.3 Let A, B and C be three subsets of E. What is the relationship between
(i) $A \triangle (B \cap C)$ and $(A \triangle B) \cap (A \triangle C)$ and
(ii) $A \triangle (B \cup C)$ and $(A \triangle B) \cup (A \triangle C)$?
When are they equal? ◇

Union and intersection can be generalized to any family of subsets of a set E. A family $(x_i)_{i \in I}$ of elements of a set X is a mapping from I to X (see Section 1.2). I is called the set of indices, and the image by this mapping of the element i of I is denoted by x_i. When we consider a family of subsets of a set E, it means that the elements of the family are subsets of E, i.e. $X = \mathcal{P}(E)$. Let $(A_i)_{i \in I}$ be a family of subsets of E. Recall that $\forall$ (resp. $\exists$) means 'for all' (resp. 'there exists'). We define:

$$\bigcup_{i \in I} A_i = \{e \in E \, / \, \exists i \in I, e \in A_i\},$$

$$\bigcap_{i \in I} A_i = \{e \in E \, / \, \forall i \in I, e \in A_i\}.$$

Note that $\bigcup_{i \in \emptyset} A_i = \emptyset$ and $\bigcap_{i \in \emptyset} A_i = E$. On the other hand, if $I = \{1, \ldots, n\}$, the family $(A_i)_{i \in I}$ is finite, and these two operations are just the finite union and intersection, denoted by $A_1 \cup \cdots \cup A_n$ and $A_1 \cap \cdots \cap A_n$ respectively. This generalization enables us to define the notion of partition of a set E. A finite or infinite family $(A_i)_{i \in I}$ of subsets of E is a *partition* of E if it satisfies

(i) $A_i \neq \emptyset$ for any $i \in I$,
(ii) $A_i \cap A_j = \emptyset$ for all pairwise distinct i, j in I and
(iii) $E = \bigcup_{i \in I} A_i$.

EXAMPLE 1.4 The three sets

- $A_0 = \{n \in \mathbb{N} \, / \, n \text{ is a multiple of } 3\}$,
- $A_1 = \{n + 1 \, / \, n \in A_0\} = \{(\text{multiples of } 3) + 1\}$ and
- $A_2 = \{n + 2 \, / \, n \in A_0\} = \{(\text{multiples of } 3) + 2\}$

form a partition of $\mathbb{N}$.

EXERCISE 1.4 Let A be a subset of E and $(B_i)_{i \in I}$ be a family of subsets of E. Show that

1. $A \cup \left(\bigcap_{i \in I} B_i \right) = \bigcap_{i \in I} \left(A \cup B_i \right),$
2. $A \cap \left(\bigcup_{i \in I} B_i \right) = \bigcup_{i \in I} \left(A \cap B_i \right).$ ◇

1.2 Functions

1.2.1 Definitions

Intuitively, a mapping f from E to F assigns to each element x of E a unique element $f(x)$ of F. The slightly more general notions of correspondence and function will also be of use. Formally, the definitions are as follows.

Let E and F be two sets. A *correspondence* from E to F is a triple $f = (E, F, \Gamma)$, where Γ is a subset of $E \times F$. We say that E is the *pre-domain* of f, F is the *co-domain* of f and Γ is the *graph* of f. The set

$$Dom(f) = \{x \in E \, / \, \exists y \in F, (x,y) \in \Gamma\}$$

is the *domain* of f and the set

$$Im(f) = \{y \in F \, / \, \exists x \in E, (x,y) \in \Gamma\}$$

is the *range* of f. Let $x \in E$. Any element y of F such that $(x,y) \in \Gamma$ is an *image* of x by f. Let $X \subseteq E$. We denote by

$$f(X) = \{y \in F \, / \, \exists x \in X, (x,y) \in \Gamma\}$$

the set of images by f of the elements of X. Similarly, any element $x \in E$ such that $(x,y) \in \Gamma$ is called a *preimage* by f of $y \in F$. For $Y \subseteq F$, we denote by

$$f^{-1}(Y) = \{x \in E \, / \, \exists y \in Y, (x,y) \in \Gamma\}$$

the set of preimages by f of the elements of Y. We have $Dom(f) = f^{-1}(F)$ and $Im(f) = f(E)$.

A correspondence $f = (E, F, \Gamma)$ is said to be a *function* if it assigns at most one image to each element of E; for any element x of E this image is denoted by $f(x)$. The function f is denoted by $f \colon E \longrightarrow F$. A function $f \colon E \longrightarrow F$ is said to be a *mapping* if its domain is the whole set E: $Dom(f) = E$. The set of mappings from E to F is denoted by F^E.

Example 1.5

- The identity mapping on E defined by $f(x) = x$ for any x of E is denoted by $id_E: E \longrightarrow E$.
- The function $\chi_A: E \longrightarrow \{0, 1\}$ defined by:

$$\chi_A(e) = \begin{cases} 1 & \text{if } e \in A, \\ 0 & \text{if } e \notin A, \end{cases}$$

is called the *characteristic function* of the subset A of E.

A mapping $f: E \longrightarrow F$ is said to be *one-to-one* or *injective* (resp. *onto* or *surjective*, *bijective* or a *one-to-one correspondence*) if any element $y \in F$ has at most (resp. at least, exactly) one preimage by f. We thus have:

- f is injective if and only if $\forall x, y \in E, f(x) = f(y) \Longrightarrow x = y$.
- f is surjective if and only if $\forall y \in F, \exists x \in E, y = f(x)$.
- f is bijective if and only if $\forall y \in F, \exists! x \in E, y = f(x)$, i.e. if f is both injective and surjective. (The symbol $\exists!$ means 'there exists a unique'.)

EXERCISE 1.5 Let $f: E \longrightarrow F$ be a mapping and let $(A_i)_{i \in I}$ be a partition of F. Note: $f(E) = \{f(e) / e \in E\}$. Show that $\left(f^{-1}(A_i)\right)_{i \in I}$ is *almost* a partition of E, by which we mean that some $f^{-1}(A_i)$ may be empty, thus preventing $\left(f^{-1}(A_i)\right)_{i \in I}$ from being a partition of E. Give a sufficient condition for all the sets $f^{-1}(A_i)$ to be non-empty. ◇

EXERCISE 1.6 Let $f: E \longrightarrow F$ be a mapping. Find a necessary and sufficient condition for ensuring that the image by f of any partition of E is a partition of F. Assume that if f is injective, then $f(A \cap B) = f(A) \cap f(B)$ (see Exercise 1.8), and recall that the image of a partition $(A_i)_{i \in I}$ of E is defined by $f\big((A_i)_{i \in I}\big) = \big(f(A_i)_{i \in I}\big)$. ◇

Let $f: E \longrightarrow F$ and $g: F \longrightarrow G$ be two mappings. The *composition* of f and g is the mapping $g \circ f: E \longrightarrow G$ defined by $g \circ f(x) = g(f(x))$. For instance, if $f: E \longrightarrow F$ is a mapping, we have $f \circ id_E = id_F \circ f = f$. The composition of mappings is associative; hence, parentheses can be omitted and we can write $h \circ g \circ f$.

1.2.2 Properties

We present some useful properties of sets and mappings.

Let $f: E \longrightarrow F$ be a mapping, let A and B be two subsets of E and let C and D be two subsets of F. We have:

$$f(A \cup B) = f(A) \cup f(B) \quad \text{and} \quad f(A \cap B) \subseteq f(A) \cap f(B),$$
$$f^{-1}(C \cup D) = f^{-1}(C) \cup f^{-1}(D) \quad \text{and} \quad f^{-1}(C \cap D) = f^{-1}(C) \cap f^{-1}(D).$$

If, moreover, f is injective, then $f(A \cap B) = f(A) \cap f(B)$. These properties are easily proved and can be generalized to arbitrary unions or intersections.

EXERCISE 1.7 Let $f: A \longrightarrow B$ be a mapping. Show that

1. f injective $\iff \forall X \subseteq A,\ f^{-1}(f(X)) = X$,
2. f surjective $\iff \forall Y \subseteq B,\ f(f^{-1}(Y)) = Y$. ◇

EXERCISE 1.8 Let $f: A \longrightarrow B$ be a mapping. Show that
$$f \text{ injective} \iff \forall X, Y \subseteq A,\ f(X \cap Y) = f(X) \cap f(Y).$$
◇

EXERCISE 1.9 Let E be a set and let A and B be two subsets of E. Consider the mapping
$$f: \mathcal{P}(E) \longrightarrow \mathcal{P}(A) \times \mathcal{P}(B)$$
$$X \longmapsto (X \cap A, X \cap B),$$

where $X \longmapsto (X \cap A, X \cap B)$ means that f assigns to each X in $\mathcal{P}(E)$ the pair $(X \cap A, X \cap B)$ in $\mathcal{P}(A) \times \mathcal{P}(B)$. Find necessary and sufficient conditions on A and B ensuring that f will be

1. injective,
2. surjective and
3. bijective. ◇

Proposition 1.6 Let $f: E \longrightarrow F$ and $g: F \longrightarrow G$ be two mappings.

(i) f and g injective $\implies g \circ f$ injective.
(ii) f and g surjective $\implies g \circ f$ surjective.
(iii) f and g bijective $\implies g \circ f$ bijective.

EXERCISE 1.10 Prove these properties. ◇

EXERCISE 1.11 Consider two mappings $f: A \longrightarrow B$ and $g: B \longrightarrow C$.

1. Show that
   (i) $g \circ f$ injective $\implies f$ injective,
   (ii) $g \circ f$ surjective $\implies g$ surjective.
2. Find an example for which $g \circ f$ is bijective, f is non-surjective and g is non-injective.
3. Show that
   (i) $g \circ f$ injective and f surjective $\implies g$ injective,
   (ii) $g \circ f$ surjective and g injective $\implies f$ surjective. ◇

EXERCISE 1.12 Consider three mappings $f: A \longrightarrow B$, $g: B \longrightarrow C$ and $h: C \longrightarrow A$.

1. Show that if $h \circ g \circ f$ and $g \circ f \circ h$ are injective and if $f \circ h \circ g$ is surjective then f, g and h are bijective.
2. Show that if $h \circ g \circ f$ and $g \circ f \circ h$ are surjective and if $f \circ h \circ g$ is injective then f, g and h are bijective. ◇

Cardinals

Proposition 1.7 *Let $f: E \longrightarrow F$ be a mapping.*

(i) *If $E \neq \emptyset$ then f is injective if and only if f has a left inverse, i.e. if there exists a mapping $r: F \longrightarrow E$ such that $r \circ f = id_E$. The mapping r is surjective and is called a retraction of f.*

(ii) *f is surjective if and only if f has a right inverse, i.e. if there exists a mapping $s: F \longrightarrow E$ such that $f \circ s = id_F$. The mapping s is injective and is called a section of f.*

(iii) *f is bijective if and only if f has an inverse, i.e. if there exists a mapping $f^{-1}: F \longrightarrow E$ such that $f \circ f^{-1} = id_F$ and $f^{-1} \circ f = id_E$. The mapping f^{-1} is a bijection and is called the inverse bijection of f.*

EXERCISE 1.13 Prove these properties.

1.3 Cardinals

1.3.1 Finite sets

For any integer n, let $[n]$ be the set $\{1, \ldots, n\}$ of integers between 1 and n.

Proposition 1.8 *If $n < m$, there is no injection from $[m]$ to $[n]$.*

EXERCISE 1.14 Prove this proposition by induction (see Chapter 3). ◊

We easily deduce from this proposition that two integers n and m are equal if and only if there exists a bijection from $[n]$ to $[m]$. A set E is *finite* if there exists an integer n and a bijection from E to $[n]$. This integer n is unique and is called the *cardinality* of E; it is denoted by $|E|$.

Let E and F be finite sets and $f: E \longrightarrow F$ be a mapping. With these notations, we have

$$\begin{aligned} f \text{ injective} &\iff \forall y \in F, |f^{-1}(\{y\})| \leq 1 \ , \\ f \text{ surjective} &\iff \forall y \in F, |f^{-1}(\{y\})| \geq 1 \text{ and} \\ f \text{ bijective} &\iff \forall y \in F, |f^{-1}(\{y\})| = 1 \ . \end{aligned}$$

Moreover, if E is finite and $|E| = |F|$ then,

$$f \text{ injective} \iff f \text{ surjective} \iff f \text{ bijective} \ .$$

If E is not finite, these equivalences do not hold; for instance, $f: \mathbb{N} \Longrightarrow \mathbb{N}$ defined by $f(n) = 2n$ is injective, but not surjective.

Finally, the following properties are quite useful.

Proposition 1.9 *Let E and F be two finite sets.*

(i)   *If E and F are disjoint then $|E \cup F| = |E| + |F|$.*
(ii) *If $(A_i)_{i \in [n]}$ is a partition of E then $|E| = |A_1| + \cdots + |A_n|$.*
(iii) *$|E \times F| = |E| \times |F|$.*
(iv) *$|F^E| = |F|^{|E|}$, where F^E denotes the set of mappings from E to F.*
(v)   *$|\mathcal{P}(E)| = 2^{|E|}$.*

EXERCISE 1.15   Prove these properties.   ◇

EXAMPLE 1.10   Let n be the number of strings that can be coded with sixteen bits. Each string is a sequence of 0s and 1s of length 16 and can be viewed as a mapping from $[16]$ to $\mathbb{B} = \{0, 1\}$. Hence, $n = 2^{16} = 65\,536$.

EXERCISE 1.16   A more picturesque way of stating Proposition 1.8 is the 'pigeonhole principle'. If $n < m$ and a flock of m pigeons flies into n pigeonholes then there must be at least one pigeonhole with two or more pigeons in it. More precisely, show that a pigeonhole will contain at least p pigeons, where p is an integer $p \geq m/n$.   ◇

1.3.2 Countable sets

The notion of cardinality can be generalized to infinite sets in such a way that the following properties are verified for any two sets E and F:

(i) $|E| \leq |F|$ $\iff$ there exists an injection from E to F.
(ii) $|E| \geq |F|$ $\iff$ there exists a surjection from E to F.
(iii) $|E| = |F|$ $\iff$ there exists a bijection from E to F.

EXERCISE 1.17   From the above relations, we deduce that if there exists both an injection and a surjection from E to F then there exists a bijection from E to F. Prove this result directly.   ◇

A set E is *countable* if it is either finite or in a one-to-one correspondence with the set $\mathbb{N}$ of natural numbers. We denote by ω the cardinality of $\mathbb{N}$. A union $\bigcup_{i \in I} A_i$ is said to be countable if the set I of indices is countable.

The countable sets satisfy the following properties:

- Any subset of a countable set is countable.
- Any finite Cartesian product of countable sets is countable.
- Any countable union of countable sets is countable.

EXERCISE 1.18   Show that the set $\mathbb{N} \times \mathbb{N}$ is countable.   ◇

Lastly, notice that there exist uncountable sets. This follows from the following result.

Operations and relations

Proposition 1.11 *Let E be a set and let $\mathcal{P}(E)$ be the set of subsets of E. We have*
$$|E| < |\mathcal{P}(E)|.$$

Proof. We show by contradiction that there exists no bijection between E and $\mathcal{P}(E)$. Assume that $f: E \longrightarrow \mathcal{P}(E)$ is a bijective mapping, and let $A = \{x \in E \;/\; x \notin f(x)\}$. Assume that there exists an $a \in E$ such that $A = f(a)$:

- If $a \in A$, we deduce from the definition of A that $a \notin A$: contradiction.
- Similarly, the hypothesis $a \notin A$ leads to a contradiction.

Hence, A has no preimage by f, and this proves that f is not surjective and hence not bijective.

Hence, there exists no bijective mapping from E to $\mathcal{P}(E)$, and the sets E and $\mathcal{P}(E)$ thus have different cardinalities. The result is then deduced from the fact that the mapping $f: E \longrightarrow \mathcal{P}(E)$ defined by $f(x) = \{x\}$ is injective. □

We deduce that $\mathcal{P}(\mathbb{N})$ is uncountable.

EXERCISE 1.19 Consider the set U of sequences $(u_n)_{n \in \mathbb{N}}$ with values ranging over $\{0, 1\}$, i.e. $\forall n \in \mathbb{N}$, $u_n \in \{0, 1\}$. (Sequences are studied in Chapter 7.) Show that U is uncountable. ◇

1.4 Operations and relations

An *operation* Φ on a set E is a mapping $\Phi: E^n \longrightarrow E$. n is called the *arity*, or the *degree*, of Φ. We also say that Φ is an n-ary operation, and this is denoted by $a(\Phi) = n$. We will mainly study operations of arity two, or binary operations.

1.4.1 Binary operations

A *binary operation* or *binary relation* $*$ on a set E is a mapping $*: E \times E \longrightarrow E$. The image of a pair (x, y) by this operation is denoted by $x * y$ (infix notation). We define the following properties:

(i) $*$ is *associative* if $\forall a, b, c \in E$, $a * (b * c) = (a * b) * c$.
(ii) $*$ is *commutative* if $\forall a, b \in E$, $a * b = b * a$.
(iii) $*$ has the element $\mathbb{1}$ for *unit* if $\forall e \in E$, $e * \mathbb{1} = \mathbb{1} * e = e$.

If an operation $*$ is associative, we will denote by $a * b * c$ (without parentheses) the product of the three elements a, b and c that can be computed either as $(a * b) * c$ or as $a * (b * c)$. More generally, we will denote by $e_1 * e_2 * \cdots * e_n$ the product of n elements.

EXERCISE 1.20 Show that if $*$ has a unit, this unit is unique. ◇

Definition 1.12 *A set E together with an associative operation $*$ is a semi-group. If, moreover, E has a unit e for $*$, $(E, *, e)$ is said to be a monoid. If the operation $*$ is commutative, the semi-group (resp. the monoid) is said to be commutative.*

EXAMPLE 1.13 The set $\mathbb{N}$ together with addition $+$ and its unit 0 is a commutative monoid. The set $\mathbb{N}$ together with multiplication $\times$ and its unit 1 is also a commutative monoid.

EXAMPLE 1.14 The set $\mathcal{P}(E)$ equipped with intersection (or union) is a commutative monoid. The set of mappings from E to itself equipped with the composition of mappings is a non-commutative monoid if E has at least two elements. The set of square $n \times n$ matrices with real-valued entries in $\mathbb{R}$ is a monoid, non-commutative if $n > 1$, for the operation of multiplication of matrices.

A special case of monoids, discussed in more detail in Chapter 11 because of their importance in computer science, consists of the *free monoids*.

Definition 1.15 *Let A be a finite set called the alphabet, whose elements are called letters. The free monoid over A, denoted by A^*, is the set of strings written with letters of the alphabet A. A string u is just a finite sequence of letters.*

Given a string u, the number of elements of its sequence is called the length of u and is denoted by $|u|$. If the string u of length n is the sequence $(u_1, u_2, \ldots, u_n)$, it is simply denoted by $u = u_1 u_2 \cdots u_n$. For instance, $aaba, acebdacebd, a, b, aa$ and aaa are strings over the alphabet $A = \{a, b, c, d, e\}$. The empty string, denoted by ε, is a special string of A^* containing no letters (the empty sequence). The operation of the monoid A^* is called *concatenation*. The concatenation of two strings $u = u_1 u_2 \cdots u_n$ and $v = v_1 v_2 \cdots v_m$ is the string $u \cdot v = u_1 u_2 \cdots u_n v_1 v_2 \cdots v_m$. For instance, $aaba \cdot cde = aabacde$. The unit for concatenation is of course the empty string ε.

Definition 1.16 *A set E equipped with an operation $*$ is a group if it is a monoid and every element has an inverse, i.e. $\forall e \in E, \exists e' \in E$ such that $e * e' = e' * e = \mathbb{1}$ (where $\mathbb{1}$ denotes the unit for $*$). If, moreover, $*$ is commutative, the group is said to be commutative.*

EXAMPLE 1.17 The set $\mathbb{Z}$ of integers equipped with the addition operation is a commutative group. If E and operation $*$ form a monoid then the set of invertible elements of E is a group. In particular, the set of invertible $n \times n$ square matrices with real-valued entries in $\mathbb{R}$ is a group, non-commutative if $n > 1$, for the operation of multiplication of matrices.

Operations and relations

Let ⊤ and ⊥ be two operations on a set E. ⊤ is *distributive* over ⊥ if $\forall a, b, c \in E$,
$$a \top (b \bot c) = (a \top b) \bot (a \top c)$$
and
$$(a \bot b) \top c = (a \top c) \bot (b \top c).$$

If the operation ⊤ is commutative, either of the above two conditions alone is of course enough.

EXAMPLE 1.18 In $\mathbb{R}$, multiplication is distributive over addition. In $\mathcal{P}(E)$, intersection and union are distributive over each other.

1.4.2 Relations

Definition 1.19 *A relation on a set E is a subset $\mathcal{R}$ of $E \times E$. To denote that a pair (e, e') of $E \times E$ is in this subset $\mathcal{R}$, we will use one of the following notations: $(e, e') \in \mathcal{R}$, $e \mathcal{R} e'$, $\mathcal{R}(e, e')$.*

EXAMPLE 1.20
1. The following sets define relations on $E = \mathbb{N}$:
   - the set $\{(n, m) \,/\, n \leq m\}$,
   - the set $\{(n, m) \,/\, n \leq m \leq 2n\}$,
   - the set $\{(n, m) \,/\, n \leq m \text{ and } \exists k : n^2 + m^2 = k^2\}$.
2. For any set A, inclusion is a relation on $E = \mathcal{P}(A)$.

1.4.3 Set-theoretic operations on relations

Because a relation is a set, we can easily define:

- the *complement* $\overline{\mathcal{R}}$ of a relation $\mathcal{R}$; $\overline{\mathcal{R}}$ is the complement of $\mathcal{R}$ in $E \times E$:
$$(e, e') \in \overline{\mathcal{R}} \iff (e, e') \notin \mathcal{R},$$

- the *union* $\mathcal{R}_1 \cup \mathcal{R}_2$ of two relations $\mathcal{R}_1$ and $\mathcal{R}_2$:
$$(e, e') \in \mathcal{R}_1 \cup \mathcal{R}_2 \iff (e, e') \in \mathcal{R}_1 \text{ or } (e, e') \in \mathcal{R}_2,$$

- the *intersection* $\mathcal{R}_1 \cap \mathcal{R}_2$ of two relations $\mathcal{R}_1$ and $\mathcal{R}_2$:
$$(e, e') \in \mathcal{R}_1 \cap \mathcal{R}_2 \iff (e, e') \in \mathcal{R}_1 \text{ and } (e, e') \in \mathcal{R}_2.$$

We can also define three special relations:

- the *empty relation*, denoted by $\emptyset_E$: $\forall e, e' \in E, \quad (e, e') \notin \emptyset_E$,

- the *full relation*, denoted by Π_E: $\forall e, e' \in E$, $(e, e') \in \Pi_E$,
- the *identity relation*, denoted by Id_E: $\forall e, e' \in E$, $(e, e') \in Id_E \iff e = e'$.

Because $\mathcal{R}_1$ and $\mathcal{R}_2$ are sets, the property

$$\forall e, e' \in E, \quad (e, e') \in \mathcal{R}_1 \implies (e, e') \in \mathcal{R}_2$$

can be expressed by writing $\mathcal{R}_1 \subseteq \mathcal{R}_2$.

1.4.4 Other operations on relations

Let $\mathcal{R}$ be a binary relation on E. The *inverse relation*, denoted by $\mathcal{R}^{-1}$, is defined by

$$e\, \mathcal{R}^{-1}\, e' \iff e'\, \mathcal{R}\, e.$$

Let $\mathcal{R}_1$ and $\mathcal{R}_2$ be two binary relations on E. Their *product*, denoted by $\mathcal{R}_1.\mathcal{R}_2$, is the relation defined by

$$e\, (\mathcal{R}_1.\mathcal{R}_2)\, e' \iff \exists e'' : e\, \mathcal{R}_1\, e'' \text{ and } e''\, \mathcal{R}_2\, e'.$$

This product is associative; its unit is Id_E.

Let $\mathcal{R}$ be a binary relation. The relation $\mathcal{R}^*$ is equal to

$$Id_E \cup \mathcal{R} \cup (\mathcal{R}.\mathcal{R}) \cup (\mathcal{R}.\mathcal{R}.\mathcal{R}) \cup \cdots$$

or $\bigcup_{i \geq 0} \mathcal{R}^i$ with $\mathcal{R}^0 = Id_E$, $\mathcal{R}^{i+1} = \mathcal{R}.\mathcal{R}^i$, for $i \geq 0$. The relation $\mathcal{R}^+$ is defined by $\mathcal{R}^+ = \bigcup_{i > 0} \mathcal{R}^i$, and hence $\mathcal{R}^* = Id_E \cup \mathcal{R}^+$.

We claim that $\forall i, j \leq 0, \mathcal{R}^{i+j} = \mathcal{R}^i.\mathcal{R}^j$. This result will be proved later (see Exercise 3.5).

EXERCISE 1.21 Let $\mathcal{R}$ be a binary relation on E. Show that
1. $(\mathcal{R}_1 \cup \mathcal{R}_2)^{-1} = \mathcal{R}_1^{-1} \cup \mathcal{R}_2^{-1}$.
2. $(\mathcal{R}_1 \cap \mathcal{R}_2)^{-1} = \mathcal{R}_1^{-1} \cap \mathcal{R}_2^{-1}$.
3. $\left(\overline{\mathcal{R}}\right)^{-1} = \overline{\mathcal{R}^{-1}}$.
4. $\mathcal{R}_1 \subseteq \mathcal{R}_2 \iff \mathcal{R}_1^{-1} \subseteq \mathcal{R}_2^{-1}$.
5. $(\mathcal{R}^{-1})^{-1} = \mathcal{R}$.
6. If $Id_E \subseteq \mathcal{R}$ then $\mathcal{R}^+ = \mathcal{R}^*$. Is the converse true? $\diamond$

EXERCISE 1.22 Let $\mathcal{R}$ be a binary relation on E.
1. Show that
   (a) $(\mathcal{R}_1.\mathcal{R}_2)^{-1} = \mathcal{R}_2^{-1}.\mathcal{R}_1^{-1}$,
   (b) $(\mathcal{R}_1 \cup \mathcal{R}_2).\mathcal{R} = (\mathcal{R}_1.\mathcal{R}) \cup (\mathcal{R}_2.\mathcal{R})$,
   (c) $\mathcal{R}.(\mathcal{R}_1 \cup \mathcal{R}_2) = (\mathcal{R}.\mathcal{R}_1) \cup (\mathcal{R}.\mathcal{R}_2)$.
2. Show that $(\mathcal{R}_1 \cap \mathcal{R}_2).\mathcal{R} = (\mathcal{R}_1.\mathcal{R}) \cap (\mathcal{R}_2.\mathcal{R})$ does not necessarily hold, but that $(\mathcal{R}_1 \cap \mathcal{R}_2).\mathcal{R} \subsetneq (\mathcal{R}_1.\mathcal{R}) \cap (\mathcal{R}_2.\mathcal{R})$ holds. $\diamond$

Operations and relations   13

1.4.5 Some properties of binary relations

A relation $\mathcal{R}$ is said to be

- *left complete*   if $\forall e \in E, \quad \exists e' \in E : e \,\mathcal{R}\, e'$,
- *right complete* if $\forall e' \in E, \quad \exists e \in E : e \,\mathcal{R}\, e'$.

The relation $\mathcal{R}$ is said to be

- *reflexive*   if $\forall e \in E,$ $e \,\mathcal{R}\, e$,
- *irreflexive*   if $\forall e, e' \in E,$ $e \,\mathcal{R}\, e' \implies e \neq e'$,
- *symmetric*   if $\forall e, e' \in E,$ $e \,\mathcal{R}\, e' \implies e' \,\mathcal{R}\, e$,
- *antisymmetric*   if $\forall e, e' \in E,$ $e \,\mathcal{R}\, e'$ and $e' \,\mathcal{R}\, e \implies e = e'$,
- *transitive*   if $\forall e, e', e'' \in E,$ $e \,\mathcal{R}\, e'$ and $e' \,\mathcal{R}\, e'' \implies e \,\mathcal{R}\, e''$.

EXERCISE 1.23
1. Show that if $\mathcal{R}$ is left (resp. right) complete then $\mathcal{R}^{-1}$ is right (resp. left) complete.
2. Show that
   (i) $Id_E \subseteq \mathcal{R}.\mathcal{R}^{-1}$ if and only if $\mathcal{R}$ is left complete ,
   (ii) $Id_E \subseteq \mathcal{R}^{-1}.\mathcal{R}$ if and only if $\mathcal{R}$ is right complete .
3. Show that $\mathcal{R} \cap \mathcal{R}^{-1}$ and $\mathcal{R} \cup \mathcal{R}^{-1}$ are symmetric.
4. Show that if $\mathcal{R}$ and $\mathcal{R}'$ are transitive, then $\mathcal{R} \cap \mathcal{R}'$ is transitive, but $\mathcal{R} \cup \mathcal{R}'$ is not necessarily transitive.
5. Show that $\mathcal{R}^+$ is transitive.
6. Show that if $\mathcal{R}$ is transitive then $\mathcal{R} = \mathcal{R}^+$, and that if $\mathcal{R}$ is reflexive and transitive then $\mathcal{R} = \mathcal{R}^*$. $\diamond$

EXERCISE 1.24   Is the relation $\mathcal{R}$ defined on $\mathbb{N}$ by $n \,\mathcal{R}\, m$ if and only if $m = n+1$ symmetric? reflexive? transitive? What are the relations $\mathcal{R}^+$ and $\mathcal{R}^*$? $\diamond$

EXERCISE 1.25   Is the relation '$n \,\mathcal{R}\, m$ if and only if n and m have a common divisor different from 1' transitive? $\diamond$

EXERCISE 1.26   Let E be the finite set $\{e_1, ..., e_n\}$, and let $\mathcal{R}$ be a binary relation on E. We represent $\mathcal{R}$ by an $n \times n$ matrix, $M_\mathcal{R}$, with entries ranging over $\{0, 1\}$ as follows:

$$m_{i,j} = \begin{cases} 1 & \text{if } e_i \,\mathcal{R}\, e_j, \\ 0 & \text{otherwise.} \end{cases}$$

1. What property of $M_\mathcal{R}$ characterizes the fact that the relation $\mathcal{R}$ is symmetric? reflexive? irreflexive? antisymmetric?
2. Assuming $M_\mathcal{R}$ and $M_{\mathcal{R}'}$ are known, how can we compute $M_{\mathcal{R}^{-1}}$, $M_{\overline{\mathcal{R}}}$ and $M_{\mathcal{R}.\mathcal{R}'}$? $\diamond$

1.4.6 Equivalence relations

Definition 1.21 *An equivalence relation is a reflexive, symmetric and transitive relation.*

EXAMPLE 1.22
(i) The equality on a set E is an equivalence relation, denoted by both $=$ and Id_E (see Section 1.4.3).
(ii) Let n be an integer greater than or equal to 2. The relation on $\mathbb{Z}$: 'x and y have the same remainder when divided by n' is an equivalence relation. It is denoted by both $x \equiv y[n]$ and $x \equiv y \bmod n$, and we say that 'x and y are congruent modulo n'.

The intersection $\mathcal{R} \cap \mathcal{R}'$ of two equivalence relations is also an equivalence relation, but the union $\mathcal{R} \cup \mathcal{R}'$ and the product $\mathcal{R}.\mathcal{R}'$ need not be equivalence relations.

Proposition 1.23 *If $\mathcal{R}$ is an arbitrary relation, $(\mathcal{R} \cup \mathcal{R}^{-1})^*$ is an equivalence relation, and it is the least equivalence relation containing $\mathcal{R}$.*

Proof. By definition, $(\mathcal{R} \cup \mathcal{R}^{-1})^*$ is reflexive and transitive. Since $\mathcal{R} \cup \mathcal{R}^{-1}$ is symmetric, $(\mathcal{R} \cup \mathcal{R}^{-1})^*$ also is symmetric.

Clearly, $(\mathcal{R} \cup \mathcal{R}^{-1})^*$ contains $\mathcal{R}$. Let $\mathcal{R}'$ be an equivalence relation containing $\mathcal{R}$. Since $\mathcal{R}'$ is symmetric, it also contains $\mathcal{R} \cup \mathcal{R}^{-1}$, and since it is reflexive and transitive, it contains $(\mathcal{R} \cup \mathcal{R}^{-1})^*$. □

EXERCISE 1.27 Show that if $\mathcal{R}$ and $\mathcal{R}'$ are two equivalence relations, the least equivalence relation containing $\mathcal{R}$ and $\mathcal{R}'$ is $(\mathcal{R} \cup \mathcal{R}')^+$. ◊

Definition 1.24 *Let $\mathcal{R}$ be an equivalence relation on E and e be an element of E. The set $\{e' \in E \mid e \, \mathcal{R} \, e'\}$, denoted by $[e]_\mathcal{R}$, is called the equivalence class of e.*

Proposition 1.25
1. $\forall e \in E$, $e \in [e]_\mathcal{R}$,
2. $\forall e, e' \in E$, $e \, \mathcal{R} \, e' \Longrightarrow [e]_\mathcal{R} = [e']_\mathcal{R}$,
3. If $[e]_\mathcal{R} \cap [e']_\mathcal{R} \neq \emptyset$, then $[e]_\mathcal{R} = [e']_\mathcal{R}$.

Proof. The first point is obvious because $e \, \mathcal{R} \, e$. To show the second point, consider $e'' \in [e']_\mathcal{R}$; then $e' \, \mathcal{R} \, e''$ and, as $e \, \mathcal{R} \, e'$, we also have $e \, \mathcal{R} \, e''$ and $e'' \in [e]_\mathcal{R}$. Hence, $[e']_\mathcal{R} \subseteq [e]_\mathcal{R}$. Conversely, $[e]_\mathcal{R} \subseteq [e']_\mathcal{R}$ for the same reasons. Lastly, if $e'' \in [e]_\mathcal{R} \cap [e']_\mathcal{R}$, then $e \, \mathcal{R} \, e''$ and $e'' \, \mathcal{R} \, e'$, and hence $e \, \mathcal{R} \, e'$ and $[e]_\mathcal{R} = [e']_\mathcal{R}$. □

Operations and relations

The set $\{[e]_\mathcal{R} \,/\, e \in E\}$ of subsets of E, called the *factor set* of E by $\mathcal{R}$ and denoted by $E/\mathcal{R}$, is hence a *partition* of E. Conversely, if $A \subseteq \mathcal{P}(E)$ is a partition of E (i.e. $\forall E_1, E_2 \in A, E_1 \neq E_2 \Longrightarrow E_1 \cap E_2 = \emptyset$ and $\forall e \in E, \exists E_e \in A : e \in E_e$), we can define an equivalence relation $\mathcal{R}_A$ by $e \,\mathcal{R}_A\, e'$ if and only if e and e' are in the same subset of the partition. It is easy to see that this is indeed an equivalence relation.

EXERCISE 1.28 Let P and P' be two partitions of a set E. P is said to be a refinement of P' if $\forall p \in P, \exists p' \in P': p \subseteq p'$.

Let $\mathcal{R}$ and $\mathcal{R}'$ be two equivalence relations on E. Show that $\mathcal{R} \subseteq \mathcal{R}'$ if and only if $E/\mathcal{R}$ is a refinement of $E/\mathcal{R}'$. ◇

EXERCISE 1.29 Let E be a set and let $\mathcal{F}$ be a set of subsets of E, i.e. $\mathcal{F} \subseteq \mathcal{P}(E)$. For $x \in E$, denote by $\mathcal{F}_x$ the set $\{X \in \mathcal{F} \,/\, x \in X\}$. Let $\mathcal{R}$ be the relation defined on E by

$$x \,\mathcal{R}\, y \iff \mathcal{F}_x \subseteq \mathcal{F}_y.$$

1. Show that $\mathcal{F}$ is reflexive and transitive.
2. Show that $\mathcal{R}$ is antisymmetric if and only if: $\forall x, y \in E$, if $x \neq y$ then there exists $X \in \mathcal{F}$ such that $|X \cap \{x, y\}| = 1$.
3. Show that if $\forall X \in \mathcal{F}$, $\overline{X} \in \mathcal{F}$, then $\mathcal{R}$ is symmetric.
4. Assuming that the union and the intersection of any family of elements of $\mathcal{F}$ are elements of $\mathcal{F}$, prove the converse of 3. ◇

1.4.7 Congruences

Definition 1.26 *An equivalence relation $\mathcal{R}$ on a set endowed with an operation $*$ is said to be a* congruence *if it is compatible with the operation $*$, or, in other words if*

$$\forall e, e', d, d' \in E, \ (e \,\mathcal{R}\, e' \text{ and } d \,\mathcal{R}\, d') \implies (e * d) \,\mathcal{R}\, (e' * d').$$

EXAMPLE 1.27 Let n be an integer greater than or equal to 2. The relation on $\mathbb{Z}$: '$x \equiv y[n]$' is a congruence for addition and for multiplication.

If $\mathcal{R}$ is a congruence on the set E equipped with the operation $*$, then the operation $*$ 'can be factored through' $\mathcal{R}$, i.e. the factor set $E/\mathcal{R}$ can be endowed with an operation $[*]$ defined by $[e][*][e'] = [e * e']$, and this operation $[*]$ is well defined. (It does not depend on the chosen representatives of the equivalence classes.) The operation $[*]$ will simply be denoted by $*$.

Proposition 1.28 *Let $\mathcal{R}$ be a congruence on a monoid (resp. group) $(E, *)$. The factor set $E/\mathcal{R}$ equipped with $*$ is a monoid (resp. group).*

EXAMPLE 1.29 Let n be an integer greater than or equal to 2. The factor of $\mathbb{Z}$ by the relation $x \equiv y[n]$ is denoted by $\mathbb{Z}/n\mathbb{Z}$. It is a group for addition and a monoid for multiplication.

EXERCISE 1.30 On the set $E = \mathbb{Z} \times (\mathbb{Z} \setminus \{0\})$ we define the operations $+$ and $\times$ by

$$(a,b) + (c,d) = (ad + bc, bd),$$
$$(a,b) \times (c,d) = (ac, bd).$$

1. Show that $(E, +)$ and $(E, \times)$ are commutative monoids.

We define on E a relation $\sim$ by: $(a,b) \sim (c,d) \iff ad = bc$.

2. Show that $\sim$ is an equivalence relation on E and that it is a congruence for $+$ and $\times$.

3. Show that $(E/\sim, [+])$ is a group, show that $[\times]$ is distributive over $[+]$ and characterize the elements of $E/\sim$ having an inverse for $[\times]$.

Note: $(E/\sim, [+], [\times])$ is just the field $\mathbb{Q}$ of rational numbers. ◇

CHAPTER 2

ORDERED SETS

In this chapter we study binary relations which will be used extensively in the remainder of the book.

Ordered sets play a fundamental role, similar to the role of metric spaces, because they allow the *comparison* of two or more objects. Ordered sets will be used in most of the subsequent chapters (Chapter 3, Chapter 4, Chapter 9, etc.).

We define order relations and ordered sets, mappings between ordered sets and special elements such as minimal and maximal elements, upper and lower bounds. We study well-founded sets which form the general framework in which we can use proofs by induction. Finally, we study complete sets and monotone functions on complete sets, which form the basis of the semantics of programming languages.

The following book is very complete and at the same time quite readable: Garrett Birkhoff, *Lattice theory*, AMS, 3rd edition, Rhode Island (1979).

2.1 Order and preorder relations

2.1.1 Orders and strict orders

Definition 2.1 *An order relation or ordering is a reflexive, antisymmetric and transitive relation. A strict ordering is an irreflexive and transitive relation.*

REMARK 2.2
1. If $\mathcal{R}$ is a strict ordering on a set E, then the relation $\mathcal{R} \cup Id_E$ is an ordering on E. Conversely, if $\mathcal{R}'$ is an ordering, $\mathcal{R}' \setminus Id_E$ is a strict ordering.
2. If $\mathcal{R}$ is an antisymmetric and transitive relation, then the relation $\mathcal{R} \cup Id_E$ is an ordering and the relation $\mathcal{R} \setminus Id_E$ is a strict ordering.

Orderings are usually denoted by $\leq$, and strict orderings are denoted by $<$. By the preceding remark, it is easy to derive an ordering from the corresponding strict ordering and the converse, i.e.

- $x \leq y$ is equivalent to $x < y$ or $x = y$,
- $x < y$ is equivalent to $x \leq y$ and $x \neq y$.

2.1.2 Total orderings and partial orderings

If ordering $\mathcal{R}$ verifies $\forall e, e' \in E, e \neq e' \implies \left(e \mathcal{R} e' \text{ or } e' \mathcal{R} e \right)$, then $\mathcal{R}$ is called a *total ordering*. Otherwise $\mathcal{R}$ is called a *partial ordering*.

EXAMPLE 2.3
(i) The usual ordering on real numbers is a total ordering.
(ii) The divisibility relation on integers is a partial ordering ($a \leq_{\text{div}} b$ if and only if there exists c such that $b = ac$).
(iii) Inclusion on $\mathcal{P}(E)$ is a partial ordering if $|E| > 1$ and a total ordering if $|E| \leq 1$.

In the next example, we define common orderings on the free monoid A^* (see Definition 1.15).

EXAMPLE 2.4
(i) The *prefix* ordering is a partial ordering on the monoid A^* that is defined as follows. String $u = u_1 \ldots u_n$ is a prefix of string $v = v_1 \ldots v_m$ if $n \leq m$ and $\forall i \leq n, u_i = v_i$.
(ii) We assume that alphabet A has a total ordering $\leq$. The *lexicographic* ordering $\preceq$ is a total ordering on the monoid A^* that is defined as follows. Let $u = u_1 \ldots u_n$ and $v = v_1 \ldots v_m$ be two strings. $u \prec v$ if
- either u is a prefix of v,
- or u and v coincide up to letter k, $u_{k+1} \neq v_{k+1}$ and $u_{k+1} \leq v_{k+1}$, with $0 \leq k < \inf(n, m)$.

2.1.3 Preorders

Definition 2.5 *A preorder relation is a transitive relation.*

EXAMPLE 2.6 Let $E = \mathcal{P}_f(\mathbb{N})$ be the set of finite subsets of $\mathbb{N}$. Each subset X contains a least element, denoted by $\inf(X)$, and a greatest element, denoted by $\sup(X)$. We define the relation $\mathcal{R}$ on E by: $X \mathcal{R} X'$ if and only if $\inf(X) \leq \inf(X')$ and $\sup(X) \leq \sup(X')$. It is easy to see that this relation is transitive and reflexive. But it is not antisymmetric because $(X \mathcal{R} X' \text{ and } X' \mathcal{R} X)$ implies $\inf(X) = \inf(X')$ and $\sup(X) = \sup(X')$, but not necessarily $X = X'$.

Order and preorder relations 19

Proposition 2.7 *If $\mathcal{R}$ is a preorder relation on E then $Id_E \cup (\mathcal{R} \cap \mathcal{R}^{-1})$ is an equivalence relation.*

Note that the relation $\equiv_\mathcal{R} = Id_E \cup (\mathcal{R} \cap \mathcal{R}^{-1})$ can be translated by $e \equiv_\mathcal{R} e'$ if and only if either $e = e'$ or $(e \,\mathcal{R}\, e'$ and $e' \,\mathcal{R}\, e)$.
Proof. The relation $Id_E \cup (\mathcal{R} \cap \mathcal{R}^{-1})$ is obviously reflexive and symmetric. We show that it is transitive, i.e. that
$$(Id_E \cup (\mathcal{R} \cap \mathcal{R}^{-1}))^2 \subseteq Id_E \cup (\mathcal{R} \cap \mathcal{R}^{-1}).$$
We see that $(Id_E \cup (\mathcal{R} \cap \mathcal{R}^{-1}))^2$ is equal to $Id_E \cup (\mathcal{R} \cap \mathcal{R}^{-1}) \cup (\mathcal{R} \cap \mathcal{R}^{-1})^2$. Because $\mathcal{R} \cap \mathcal{R}'$ is the intersection of two transitive relations, it is also transitive (see Exercise 3.5) and thus $(\mathcal{R} \cap \mathcal{R}^{-1})^2 \subseteq (\mathcal{R} \cap \mathcal{R}^{-1})$. □

EXAMPLE 2.8 If we consider the preorder $\mathcal{R}$ of Example 2.6, the equivalence relation defined in Proposition 2.7 is
$$X \left(Id_E \cup (\mathcal{R} \cap \mathcal{R}^{-1}) \right) X'$$
if and only if $\inf(X) = \inf(X')$ and $\sup(X) = \sup(X')$.

Let $\mathcal{R}$ be a preorder relation on E and let $\mathcal{E}$ be the associated equivalence. On the factor set $E/\mathcal{E}$ of E by $\mathcal{E}$, we can define the relation $\mathcal{R}'$ by $[e]_\mathcal{E} \,\mathcal{R}'\, [e']_\mathcal{E}$ if and only if $e \,\mathcal{R}\, e'$. This definition does not depend on the choice of the elements e and e' within their equivalence class, because if $e \,\mathcal{E}\, e_1$ and $e' \,\mathcal{E}\, e'_1$, then $e \,\mathcal{R}\, e'$ if and only if $e_1 \,\mathcal{R}\, e'_1$.

EXERCISE 2.1 Prove that $\mathcal{R}'$ does not depend on the choice of the elements e and e' within their equivalence class. ◇

Proposition 2.9 *The relation $\mathcal{R}'$ is antisymmetric and transitive; $\mathcal{R}'$ is an ordering if $\mathcal{R}$ is reflexive and $\mathcal{R}'$ is a strict ordering if $\mathcal{R}$ is irreflexive.*

Proof. $\mathcal{R}'$ is transitive, because $[e]_\mathcal{E} \,\mathcal{R}'\, [e']_\mathcal{E}$ and $[e']_\mathcal{E} \,\mathcal{R}'\, [e'']_\mathcal{E}$ imply $e \,\mathcal{R}\, e'$ and $e' \,\mathcal{R}\, e''$, and hence $e \,\mathcal{R}\, e''$. It is antisymmetric because $([e]_\mathcal{E} \,\mathcal{R}'\, [e']_\mathcal{E}$ and $[e']_\mathcal{E} \,\mathcal{R}'\, [e]_\mathcal{E})$ implies $(e \,\mathcal{R}\, e'$ and $e' \,\mathcal{R}\, e)$. Thus $(e, e') \in (\mathcal{R} \cap \mathcal{R}^{-1}) \subseteq \mathcal{E}$, and hence $[e]_\mathcal{E} = [e']_\mathcal{E}$.
$\mathcal{R}'$ is reflexive (resp. irreflexive) if $\mathcal{R}$ is reflexive (resp. irreflexive). □

This ordering $\mathcal{R}'$ will be called the *factor ordering* of the preorder $\mathcal{R}$.

EXAMPLE 2.10 Again with the preorder of Example 2.6, the factor set of $E = \mathcal{P}_f(\mathbb{N})$ can be identified with the set of pairs (a, b) of integers such that $a \leq b$. On this set the ordering $\mathcal{R}'$ is defined by $(a, b) \,\mathcal{R}'\, (a', b')$ if and only if $a \leq a'$ and $b \leq b'$.

EXERCISE 2.2 Let $\mathcal{R}$ be a preorder relation. Show that the relation $\mathcal{R}^\dagger$ defined by $x \,\mathcal{R}^\dagger\, y$ if and only if $x = y$ or $(x \,\mathcal{R}\, y$ and $y \,\overline{\mathcal{R}}\, x)$, where $\overline{\mathcal{R}}$ denotes the complementary of relation $\mathcal{R}$ (see Section 1.4.3), is an ordering. ◇

2.2 Ordered sets

Definition 2.11 *An ordered set $(E, \leq)$ is a set E together with an ordering $\leq$.*

The same set E can be equipped with different orderings. We then have different ordered sets.

EXAMPLE 2.12 The set of integers $\mathbb{N}$ can be equipped with the usual ordering or with the divisibility ordering of Example 2.3.

2.2.1 Monotonic mappings

Definition 2.13 *Let $(E_1, \leq_1)$ and $(E_2, \leq_2)$ be two ordered sets. A mapping f from E_1 to E_2 is said to be monotonic, or monotone, if*

$$\forall x, y \in E_1, \quad x \leq_1 y \implies f(x) \leq_2 f(y).$$

f is also said to be a homomorphism from the ordered set $(E_1, \leq_1)$ to the ordered set $(E_2, \leq_2)$.

$(E_1, \leq_1)$ and $(E_2, \leq_2)$ are said to be isomorphic if there is a bijection b between E_1 and E_2 with the property that both b and b^{-1} are monotone.

EXAMPLE 2.14
1. If two ordered sets $(E_1, \leq_1)$ and $(E_2, \leq_2)$ have the same underlying set, namely, if $E_1 = E_2$, then the inclusion $\leq_1 \subseteq \leq_2$, i.e. $\forall x, y, \quad x \leq_1 y \implies x \leq_2 y$, holds if and only if the identity mapping from E_1 to E_2 is monotone.
2. In order for a bijection to be an isomorphism, monotonicity is not sufficient; for instance, the identity mapping from $(\mathbb{N}, \leq_{\text{div}})$ to $(\mathbb{N}, \leq)$ is a monotone bijection but it is not an isomorphism.

2.2.2 Totally ordered sets

An ordered set $(E, \leq)$ is said to be *totally ordered* if $\leq$ is a total ordering, i.e. if $\forall x, y, x \neq y \implies x \leq y$ or $y \leq x$. Otherwise, i.e. if $\exists x, y, x \neq y, x \not\leq y$ and $y \not\leq x$, it is said to be a *partially ordered* set or *poset*. Let $(E, \leq)$ be a partially ordered set. A *linear extension* of $(E, \leq)$ is a totally ordered set $(E, \leq_t)$ with the same underlying set such that $\leq \subseteq \leq_t$.

Theorem 2.15 *Let $(E, \leq)$ be an ordered set. It has at least one linear extension, and $\leq$ is equal to the intersection of all its linear extensions.*

This theorem will not be proved in the general case.

EXERCISE 2.3 Prove the statement of Theorem 2.15 for the case when E is finite. ◇

Ordered sets 21

2.2.3 Products of ordered sets

Let $(E_1, \leq_1)$ and $(E_2, \leq_2)$ be two ordered sets. The *direct product* of these two ordered sets is $(E_1 \times E_2, \leq)$ with the ordering $\leq$ defined by $(x_1, x_2) \leq (y_1, y_2)$ if and only if $x_1 \leq_1 y_1$ and $x_2 \leq_2 y_2$.

REMARK 2.16
1. The ordering on the direct product is also called the *product ordering*.
2. We can define orderings on $E_1 \times E_2$ other than the product ordering; for instance, we can define $(x_1, x_2) \leq' (y_1, y_2)$ if and only if $y_1 \leq_1 x_1$ and $x_2 \leq_2 y_2$.

EXERCISE 2.4
1. Show that the projections π_i from $E_1 \times E_2$ onto E_i are monotonic.
2. Show that if $|E_1| \geq 2$ and $|E_2| \geq 2$, $E_1 \times E_2$ is not totally ordered even when E_1 and E_2 are.
3. Show that the direct product is associative and commutative up to isomorphism (i.e. the mapping b from $E_1 \times E_2$ to $E_2 \times E_1$ associating (x_2, x_1) with (x_1, x_2) is an isomorphism). ◊

The *lexicographic product* of $(E_1, \leq_1)$ by $(E_2, \leq_2)$ is $(E_1 \times E_2, \leq)$ with $(x_1, x_2) \leq (y_1, y_2)$ if and only if $x_1 < y_1$ or $(x_1 = y_1$ and $x_2 \leq y_2)$.

EXERCISE 2.5
1. Show that the lexicographic product of two ordered sets is an ordered set.
2. Show that this product is not commutative, i.e. $(E_1 \times E_2, \leq)$ is not isomorphic to $(E_2 \times E_1, \leq)$.
3. Show that the lexicographic product of total orderings is a total ordering. ◊

2.2.4 Ordered subsets, chains and antichains

Let $(E, \leq)$ be an ordered set. A *subordered set* of $(E, \leq)$ is an ordered set $(E', \leq')$ such that $E' \subseteq E$ and $\leq' = \leq \cap (E' \times E')$, i.e. $\forall x, y \in E'$, $x \leq' y$ if and only if $x \leq y$.

A *chain* of E is a totally ordered subset of E. A chain is *maximal* if it is not strictly included in another chain.

An *antichain* E' of E is a subset of E such that

$$\leq \cap (E' \times E') = Id_{E'}.$$

In other words, any two elements of an antichain are incomparable, because if they are in the ordering then they must be equal. An antichain is *maximal* if it is not strictly included in any other antichain.

EXERCISE 2.6
1. If $(E, \leq)$ is a totally ordered set then its only antichains are singletons.
2. Show that the intersection of a chain and an antichain has at most one element. ◊

A *left segment* is a subset E' of E such that

$$y \in E' \text{ and } x \leq y \implies x \in E'.$$

An *interval* $[x, y]$, with $x \neq y$ and $x \leq y$, is the subset

$$\{z \mid x \leq z \text{ and } z \leq y\}.$$

An ordered set is *locally finite* if all its intervals are finite.

EXAMPLE 2.17 For the usual ordering on numbers, $\mathbb{N}$ is locally finite and $\mathbb{Q}$ is not.

EXERCISE 2.7 Show that the interval $[x, y]$ is empty if and only if $x \not\leq y$. ◇

We say that x is *covered by* y if interval $[x, y]$ contains only x and y. This relation will be denoted by $x \prec y$.

EXERCISE 2.8 Show that if interval $[x, y]$ is finite then there exists an element of this interval covering x. ◇

Proposition 2.18 *If $(E, \leq)$ is locally finite then $\leq \, = \, \prec^*$.*

EXERCISE 2.9 Prove Proposition 2.18. ◇

2.3 Upper and lower bounds

Definition 2.19 *Let E' be a subset of an ordered set $(E, \leq)$. An element x of E is an upper bound of E' (resp. lower bound) if $\forall y \in E'$, $y \leq x$ (resp. $x \leq y$).*

We denote by $\text{Maj}(E')$ the set of upper bounds of E' and by $\text{Min}(E')$ the set of lower bounds of E'. It is easy to see that $\text{Maj}(\emptyset) = \text{Min}(\emptyset) = E$.

Proposition 2.20 *$\text{Maj}(E') \cap E'$ and $\text{Min}(E') \cap E'$ each have at most one element.*

Proof. Assume that $\text{Maj}(E') \cap E'$ contains two distinct elements x and y. We thus have $x \leq y$ and $y \leq x$, a contradiction.

The proof is similar for Min. □

If $\text{Maj}(E') \cap E'$ is non-empty then the unique element of this set is called the *greatest element* or *maximum* of E'. Similarly, if $\text{Min}(E') \cap E'$ is non-empty then its unique element is called the *least element* or *minimum* of E'.

Proposition 2.21 *Let E' be a subset of E and let $z \in E$. The following three conditions are equivalent:*

(i) z is the greatest element of E'.
(ii) $z \in E'$ and $\forall x \in E'$, $x \leq z$.
(iii) $z \in E'$ and z is the least element of $\mathrm{Maj}(E')$.

The least element of E' has a similar characterization.

Proof.
(i) $\implies$ (ii): If z is the greatest element of E', then $z \in E'$ and $z \in \mathrm{Maj}(E')$, and thus (ii) is true.
(ii) $\implies$ (iii): $\forall x \in E'$, $\forall y \in \mathrm{Maj}(E')$, $x \leq y$, and thus $E' \subseteq \mathrm{Min}(\mathrm{Maj}(E'))$. Hence $z \in E' \cap \mathrm{Maj}(E') \subseteq \mathrm{Min}(\mathrm{Maj}(E')) \cap \mathrm{Maj}(E')$ and z is the least element of $\mathrm{Maj}(E')$.
(iii) $\implies$ (i): The least element of $\mathrm{Maj}(E')$ is in $\mathrm{Maj}(E')$, and thus $z \in E' \cap \mathrm{Maj}(E')$. $\square$

Let E' be a subset of E. An element x of E' is said to be *maximal in E'* if $\forall y \in E'$, $y \geq x \implies y = x$ or, equivalently, $y \neq x \implies y \not\geq x$. If E' has a greatest element, this greatest element is its unique maximal element, but the converse is false (see Exercise 2.10).

We define the *minimal* elements of a subset E' similarly.

EXAMPLE 2.22 $\mathbb{N}$ has a minimal element which is its least element (it is 0), but it has no maximal element.

EXERCISE 2.10
1. Show that if a subset E' of E has a unique maximal element, this element is not necessarily the greatest element of E'.
2. What can you say if E is totally ordered? $\diamond$

Definition 2.23 *An element x is the least upper bound of a subset E' of an ordered set E if*

$$(\forall y \in E', y \leq x) \quad \text{and} \quad (\forall z \in E, ((\forall y \in E', y \leq z) \implies x \leq z)).$$

Similarly, an element x is the greatest lower bound of a subset E' of an ordered set E if

$$(\forall y \in E', x \leq y) \quad \text{and} \quad (\forall z \in E, ((\forall y \in E', z \leq y) \implies z \leq x)).$$

The terminology 'the' least upper bound (resp. greatest lower bound) is justified because there is at most one least upper bound (resp. greatest lower bound).

Indeed, the definition of the least upper bound (resp. greatest lower bound) of a subset E' of E is identical to the definition of the least element of $\text{Maj}(E')$ (resp. the greatest element of $\text{Min}(E')$). The least upper bound of subset E' is thus an upper bound of E' that is less than all other upper bounds of E', i.e. the least upper bound of E' is the least among the upper bounds of E'. Similarly, the greatest lower bound of E' is the greatest among the lower bounds of E'.

We denote by $\sup(E')$ and $\inf(E')$ the least upper bound and greatest lower bound, respectively, when they exist.

Proposition 2.24 *Let E' be a subset of E.*

(i) *If z is the greatest element of E', then $z = \sup(E')$.*
(ii) *If $\sup(E') \in E'$, then $\sup(E')$ is the greatest element of E'.*

We have a similar result for the least element and the greatest lower bound of E'.

Proof. This result is a consequence of Proposition 2.21:

(i) If z is the greatest element of E', then z is the least element of $\text{Maj}(E')$. It is thus the least upper bound of E'.
(ii) The least upper bound of E' is the least element of $\text{Maj}(E')$. If it belongs to E' it is thus the greatest element of E'. □

EXAMPLE 2.25
1. Let $\mathbb{N}$ be ordered by the divisibility relation (see Example 2.3 and Exercise 2.17). For this ordering, the greatest lower bound of a set of two integers always exists and is the greatest common divisor of these two integers. The least upper bound also always exists and is their least common multiple.
2. The least upper bound and the greatest lower bound do not always exist. Consider the set $E = \{a, b, c, d\}$ ordered by : $a \leq c$, $a \leq d$, $b \leq c$, $b \leq d$. See Figure 2.1.

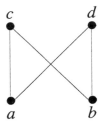

Figure 2.1

Then $\{a, b\}$ has neither a least upper bound nor a greatest lower bound; the same holds for $\{c, d\}$.

EXAMPLE 2.26 Let $\mathcal{P}(E)$ be the set of subsets of E, ordered by inclusion. Let E_i for $i \in I$ be a family of subsets of E. The least upper bound of this family is $\bigcup_{i \in I} E_i$ and its greatest lower bound is $\bigcap_{i \in I} E_i$.

Proposition 2.27 *Let E_i, for $i \in I$, be a family of subsets of an ordered set and let $E' = \bigcup_{i \in I} E_i$ be its union. If each set E_i has a least upper bound (resp. greatest lower bound) e_i, and if the set $\{e_i \,/\, i \in I\}$ has a least upper bound (resp. greatest lower bound) e, then e is the least upper bound (resp. greatest lower bound) of E'.*

Proof. We show this result only in the case of the least upper bound; the other case is completely similar.

First, we show that e is an upper bound of E'. Let x be any element of E'. It thus belongs to some E_i, and so $x \leq e_i \leq e$.

Now we let z be any upper bound of E' and show that $e \leq z$. Because z is an upper bound of E', it also is an upper bound of E_i, for any i in I, and we have: $\forall i \in I,\ e_i \leq z$. Therefore, because e is the least upper bound of $\{e_i \,/\, i \in I\}$, we have that $e \leq z$. □

2.4 Well-ordered sets and induction

Well-founded sets form the general framework in which we can use proofs by induction. All induction principles stated in Chapter 3 are thus justified by the present section.

Definition 2.28 *An ordered set $(E, \leq)$ is said to be well founded if there is no infinite strictly decreasing sequence of elements of E; $\leq$ is then said to have the well-founded ordering property or to be a well founded ordering. A total ordering $\leq$ having the well-founded ordering property is called a well ordering.*

We now prove an important characterization of well-founded ordered sets.

Proposition 2.29 *An ordered set $(E, \leq)$ is well founded if and only if any non-empty subset of E has at least one minimal element.*

Proof. It is equivalent to show the contrapositive of this result, namely, that $(E, \leq)$ has an infinite decreasing sequence if and only if there exists a non-empty subset having no minimal element. Assume that there exists a strictly decreasing infinite sequence $(x_n)_{n \in \mathbb{N}}$ in E. The set $X = \{x_n \,/\, n \in \mathbb{N}\}$ is a non-empty subset having no minimal element.

Conversely, assume that there exists a non-empty subset having no minimal element. Because X has no minimal element, any element x of X is strictly larger than at least one other element y of X. Thus there exists a function f from X to X verifying $\forall x \in X$, $f(x) < x$. (It suffices to choose one among the elements $y < x$ and to let $f(x) = y$.) Let $x_0 \in X$ (where X is non-empty by hypothesis). For any integer n, we define $x_n = f^n(x_0)$. The sequence $(x_n)_{n \in \mathbb{N}}$ is strictly decreasing because $\forall n \in \mathbb{N}$, $x_n = f(x_{n-1}) < x_{n-1}$. □

EXAMPLE 2.30
1. The usual ordering is a well ordering on $\mathbb{N}$ but not on $\mathbb{Z}$.
2. $\mathbb{N}^2$ equipped with the product order $\leq$ (see Section 2.2.3) is well founded. Indeed, any element of $\mathbb{N}^2$ has a finite number of lower bounds. Consequently, there can exist no strictly decreasing infinite sequence. More generally, it is easy to see that the product of two well-founded sets is also well founded.
3. The lexicographic ordering $\preceq$ on $\mathbb{N}^2$ is defined by $(n, m) \prec (n', m')$ if and only if $(n < n')$ or $(n = n'$ and $m < m')$. We note that if $n > 0$ then (n, m) has infinitely many lower bounds. For instance, $\forall p \in \mathbb{N}$, $(n-1, p) \prec (n, m)$. Nevertheless, the lexicographic ordering is a well ordering on $\mathbb{N}^2$. Indeed, let X be a non-empty subset of $\mathbb{N}^2$, and let $n = \min\{p \in \mathbb{N} \,/\, \exists q \in \mathbb{N}, (p, q) \in X\}$ and $m = \min\{q \in \mathbb{N} \,/\, (n, q) \in X\}$. We easily verify that (n, m) is the least element of X.

EXERCISE 2.11   Let $<_1$ be a well ordering on E_1 and let $<_2$ be a well ordering on E_2; we define the *lexicographic product* $\preceq'$ of $<_1$ and $<_2$ on $E_1 \times E_2$ by $(n, m) \prec' (n', m')$ if and only if $(n <_1 n')$ or $(n = n'$ and $m <_2 m')$. Verify that $\preceq'$ is a well ordering on $E_1 \times E_2$. ◇

The induction principle for well-founded sets is stated in the following theorem.

Theorem 2.31   *Let $(E, \leq)$ be a well-founded set and let P be an assertion depending on an element x of E. (P is called a predicate, see Chapter 5.) If the following property is verified:*

$$(\mathrm{I}) \quad \forall x \in E, \quad \Big((\forall y < x, P(y)) \quad \Longrightarrow \quad P(x) \Big),$$

then $\forall x \in E$, $P(x)$.

Proof. Let $X = \{x \in E \,/\, P(x) \text{ is false}\}$. If X is non-empty, X has a minimal element x_0. $\forall y < x_0$, $y \notin X$ and thus $P(y)$ is true. Using (I) we deduce that $P(x_0)$ is true, which contradicts $x_0 \in X$. Thus $X = \emptyset$, which means that $\forall x \in E$, $P(x)$ is true. □

Unfortunately, sets equipped with their natural orderings are not always well founded. We have already seen that $\mathbb{Z}$ with the usual ordering is not well founded. It is of course possible to define well-founded orderings and even well orderings on $\mathbb{Z}$, but these orderings are not very natural. For instance, a well ordering $\preceq$ is defined on $\mathbb{Z}$ by using the usual ordering $\leq$ as follows:

- $\forall n > 0, \forall m > 0, n \prec m \iff n < m$ ($\prec$ coincides with $\leq$ on $\mathbb{N}$).
- $\forall n < 0, \forall m \geq 0, n \prec m$ (negative integers are less than positive ones).
- $\forall n < 0, \forall m < 0, n \prec m \iff m < n$ (the inverse ordering on negative integers).

Below we give yet another example where the usual ordering is not a well-founded ordering.

EXAMPLE 2.32 Let A be an alphabet with at least two letters a and b. The free monoid A^*, together with the lexicographic ordering (see Example 2.4), is not well founded. Indeed, $(a^n b)_{n \in \mathbb{N}}$ is a strictly decreasing infinite sequence. Thus, proofs by induction on A^* equipped with the lexicographic ordering will not be valid.

On the other hand, A^* equipped with the prefix ordering (Example 2.4) is well founded. Finally, a well ordering on A^* is defined by: $x \prec y$ if and only if

$$(|x| < |y|) \quad \text{or} \quad (|x| = |y| \text{ and } x \prec y \text{ in the lexicographic ordering}).$$

Hence proofs by induction on A^* using either the prefix ordering or the ordering $\prec$ will be valid.

2.5 Complete sets and lattices

2.5.1 Complete sets and continuous functions

Definition 2.33 *An ordered set $(E, \leq)$ is said to be a lattice (resp. complete lattice) if any finite subset (resp. any subset) of E has a least upper bound and a greatest lower bound.*

If E is a lattice, then the greatest lower bound of E is less than any element of E; hence a lattice has a least element that is denoted by $\bot$ and pronounced 'bottom'. Similarly, a lattice has a greatest element that is denoted by $\top$ and pronounced 'top'.

EXAMPLE 2.34 $\mathcal{P}(E)$ ordered by inclusion is a complete lattice.

EXERCISE 2.12
1. Show that an ordered set $(E, \leq)$ is a lattice if and only if any two-element subset of E has a least upper bound and a greatest lower bound.
2. Show that an ordered set $(E, \leq)$ is a complete lattice if and only if any subset of E has a least upper bound. $\diamond$

Z If an ordered set is a lattice, its least element $\bot$ is also the least upper bound of the empty set. Because the set $\text{Maj}(\emptyset)$ of upper bounds of the empty set is the whole of E, $\bot$ is the least element of E.

Similarly, the greatest element $\top$ is also the greatest lower bound of the empty set.

Definition 2.35 *A mapping f from an ordered set $(E_1, \leq_1)$ to an ordered set $(E_2, \leq_2)$ is said to be continuous (or, more precisely, sup-continuous) if it preserves the least upper bounds of non-empty subsets. In other words, if the subset $E' \neq \emptyset$ has a least upper bound $e = \sup(E')$, then $f(E') = \{f(x) / x \in E'\}$ also has a least upper bound equal to $f(e)$.*

REMARK 2.36 Since the least upper bound of the empty set is $\bot$, the condition 'f preserves the least upper bound of the empty set' is simply $f(\bot_1) = \bot_2$. This is a very exacting requirement that we will not demand for a continuous function.

Since in a complete lattice least upper bounds always exist, the continuity of a mapping between two complete lattices is then simply expressed by:

$$f(\sup(E)) = \sup(f(E)).$$

EXERCISE 2.13 Show that any continuous function is monotonic. $\diamond$

Let $C(E)$ be the set of left segments of E ordered by inclusion. Let i be the mapping from E to $C(E)$ defined by $i(x) = \{y \in E \,/\, y \leq x\}$, and let $i(E)$ be the image of E by i.

Proposition 2.37 *$C(E)$ is a complete set. The mapping i is monotonic and is an isomorphism between E and $i(E)$.*

Proof. In order for $C(E)$ to be complete for inclusion, it suffices that any union of left segments is a left segment, and this clearly holds.

If $x \leq y$, it is clear that $i(x) \subseteq i(y)$ and thus i is monotonic.

Conversely, if $i(x) \subseteq i(y)$, then because $x \in i(x)$ we have that $x \in i(y)$ and thus $x \leq y$. Hence $i(x) = i(y)$ implies $x \leq y$ and $y \leq x$, and thus $x = y$. □

However, i is not always continuous, as shown by the next example.

EXAMPLE 2.38 Let $E = \mathbb{N}$, together with the usual ordering. For $n \in \mathbb{N}$, $i(n) = \{0, 1, \ldots, n\}$, and the only left segment that is not of this form is the whole of $\mathbb{N}$. We can thus identify $C(\mathbb{N})$ with the complete ordered set $\overline{\mathbb{N}} = \mathbb{N} \cup \{\omega\}$, where $\forall n \in \mathbb{N}$, $n < \omega$.

We may again consider the set $C(\overline{\mathbb{N}})$ of left segments of $\overline{\mathbb{N}}$ which is equal to $\{i(n) \,/\, n \in \mathbb{N}\} \cup \{\mathbb{N}, \overline{\mathbb{N}}\}$. The mapping i' from $\overline{\mathbb{N}}$ to $C(\overline{\mathbb{N}})$ is defined by $\forall n \in \mathbb{N}$, $i'(n) = \{0, 1, \ldots, n\}$ and $i'(\omega) = \overline{\mathbb{N}}$. This mapping is not continuous. Indeed, in $\overline{\mathbb{N}}$ the least upper bound of $\mathbb{N}$ is ω, whilst in $C(\overline{\mathbb{N}})$ the least upper bound of the set $\{i'(n) \,/\, n \in \mathbb{N}\}$ is $\mathbb{N}$.

2.5.2 Fixed points of monotone functions

Let f be a mapping from a set E to itself. A *fixed point* of f is an element x of E such that $f(x) = x$.

If E is an ordered set, the set of fixed points of f is a subordered set of E, possibly empty. If this subset has a least element, this least element is called the *least fixed point* of f, and if it has a greatest element, this greatest element is called the *greatest fixed point* of f.

Theorem 2.39 *If f is a monotone mapping from a complete ordered set to itself, then f has a greatest fixed point.*

Proof. We verify that f has a greatest fixed point. Let

$$X = \{x \in E \; / \; x \leq f(x)\}$$

and let $z = \sup(X)$. By the definition of z, we have $\forall x \in X$, $x \leq z$ and hence, since f is monotonic, $f(x) \leq f(z)$. As $x \leq f(x)$, $f(z)$ is an upper bound of X, and hence $z \leq f(z)$. We deduce that $f(z) \leq f(f(z))$, and hence $f(z) \in X$ and thus $f(z) \leq z$. It follows that z is a fixed point of f. If z' is another fixed point, then $z' \in X$ and thus $z' \leq z$. □

Theorem 2.40 *If f is a continuous mapping from a complete ordered set to itself, then f has a least fixed point. This least fixed point is equal to*

$$\sup(\{f^n(\bot) \; / \; n \in \mathbb{N}\}).$$

Proof. Let $x = \sup(\{f^n(\bot) \; / \; n \in \mathbb{N}\})$. Because f is continuous,

$$f(x) = \sup(\{f^{n+1}(\bot) \; / \; n \in \mathbb{N}\}),$$

and because $\bot = f^0(\bot)$ is the least element of E,

$$\sup(\{f^{n+1}(\bot) \; / \; n \in \mathbb{N}\}) = \sup(\{f^n(\bot) \; / \; n \in \mathbb{N}\}) = x,$$

which is thus a fixed point of f. If y is another fixed point of f, we first show by induction that $\forall n \in \mathbb{N}$, $f^n(\bot) \leq y$; because $\bot$ is the least element of E, $\bot = f^0(\bot) \leq y$; if $f^n(\bot) \leq y$ then $f^{n+1}(\bot) \leq f(y) = y$. Hence, $x = \sup(\{f^n(\bot) \; / \; n \in \mathbb{N}\}) \leq y$. □

Theorem 2.41 *If E is a finite ordered set having a least element $\bot$, then for any monotone function f from E to itself there exists $k \leq \text{card}(E)$ such that the least fixed point of f is $f^k(\bot)$.*

Proof. Consider the sequence

$$\bot, f(\bot), f^2(\bot), \ldots, f^n(\bot), \ldots,$$

which is increasing because f is monotone. If the sequence has two consecutive equal elements then it is stationary:

$$f^i(\bot) = f^{i+1}(\bot) \implies f^{i+1}(\bot) = f^{i+2}(\bot)$$

and thus, by induction, $\forall j \geq i$, $f^i(\bot) = f^j(\bot)$. In the $|E|+1$ first elements of this sequence, there must be two consecutive elements that are equal. We thus have $f^k(\bot) = f^{k+1}(\bot)$ for $k \leq |E|$. The fact that $f^k(\bot)$ is less than any other fixed point of f is proved as in the preceding theorem. □

EXERCISE 2.14
1. What is the value of $f(\bot)$ if f preserves all least upper bounds?
2. If $f(\bot) = \bot$, what is the least fixed point of f?   ◇

EXERCISE 2.15 Consider $\mathcal{P}(E \times E)$ ordered by inclusion. Let $\mathcal{R}$ be a binary relation on E and let f be the mapping of $\mathcal{P}(E \times E)$ to itself defined by $f(X) = Id_E \cup \mathcal{R}.X$.

Show that this mapping is continuous and that its least fixed point is $\mathcal{R}^*$.   ◇

2.5.3 Lattices

Definition 2.42 *An ordered set is a lattice (see Definition 2.33 and Exercise 2.12) if any pair of elements has a least upper bound and a greatest lower bound. We will sometimes denote by $x \sqcup y$, instead of $\sup(\{x, y\})$, the least upper bound of x and y, and by $x \sqcap y$ their greatest lower bound.*

EXAMPLE 2.43 $\mathbb{N}$ equipped with the divisibility ordering is a lattice. The binary operations $\sqcup$ and $\sqcap$ are, respectively, the lcm (least common multiple) and the gcd (greatest common divisor), $\bot$ is 1, and $\top$ is 0. Indeed, 1 divides any number n because $1n = n$, and any number n divides 0 because $n0 = 0$.

EXAMPLE 2.44 $\mathcal{P}(E)$ together with inclusion is a lattice. The binary operations $\sqcup$ and $\sqcap$ are, respectively, $\cup$ and $\cap$.

If E is a lattice, we may thus consider that E is a set equipped with two binary operations $\sqcup$ and $\sqcap$.

Complete sets and lattices

Proposition 2.45 *The operations $\sqcup$ and $\sqcap$ have the following properties:*

- *idempotence:* $x \sqcup x = x$ *and* $x \sqcap x = x$.
- *commutativity:* $x \sqcup y = y \sqcup x$ *and* $x \sqcap y = y \sqcap x$.
- *associativity:* $(x \sqcup y) \sqcup z = x \sqcup (y \sqcup z)$ *and* $(x \sqcap y) \sqcap z = x \sqcap (y \sqcap z)$.
- *absorption:* $x \sqcap (x \sqcup y) = x = (y \sqcap x) \sqcup x$.

Conversely, assume that on a set E there exist two binary operations $\sqcup$ and $\sqcap$ that have the four properties mentioned in the above proposition. Then we can order E in such a way that $x \sqcup y$ and $x \sqcap y$ are, respectively, the least upper bound and the greatest lower bound of x and y. It suffices to let $x \leq y$ if and only if $x \sqcup y = y$, which, because of the absorption property, is equivalent to $x \sqcap y = x$:

$$x \sqcup y = y \Longrightarrow x \sqcap (x \sqcup y) = x \sqcap y \Longrightarrow x = x \sqcap y .$$

Since $\sqcup$ is idempotent, $\leq$ is reflexive: from $x \sqcup x = x$ we deduce that $x \leq x$. From the commutativity of $\sqcup$, we easily deduce that $\leq$ is antisymmetric; $x \leq y$ implies $x \sqcup y = y$; $y \leq x$ implies $y \sqcup x = x$; as $x \sqcup y = y \sqcup x$ we have $x = y$. Finally, the transitivity of $\leq$ is an immediate consequence of the associativity of $\sqcup$: $x \leq y \Longrightarrow x \sqcup y = y$; $y \leq z \Longrightarrow y \sqcup z = z$; hence $z = (x \sqcup y) \sqcup z = x \sqcup (y \sqcup z) = x \sqcup z$ and thus $x \leq z$. Moreover, the least upper bound of x and y is indeed $x \sqcup y$: as $x \sqcap (x \sqcup y) = x$, we have $x \leq (x \sqcup y)$, and for the same reasons, $y \leq (x \sqcup y)$; if $x \leq z$ and $y \leq z$, we have $z = x \sqcup z = y \sqcup z$ and thus $z = z \sqcup z = x \sqcup z \sqcup y$, hence $x \sqcup y \leq z$. The fact that the greatest lower bound of x and y is $x \sqcap y$ is proved similarly.

From the associativity and the commutativity of $\sqcup$ and $\sqcap$, it immediately follows that in a lattice, any *finite non-empty* subset has a least upper bound and a greatest lower bound. This can be proved by induction on the number of elements of the finite subset by writing $\{e_1, e_2, \ldots, e_n, e_{n+1}\}$ as the union of two sets $\{e_1, e_2, \ldots, e_n\}$ and $\{e_{n+1}\}$ and by applying Proposition 2.27.

EXERCISE 2.16 Show that both operations $\sqcap$ and $\sqcup$ are monotone, i.e. if $x \leq x'$ and $y \leq y'$ then $x \sqcap y \leq x' \sqcap y'$ and $x \sqcup y \leq x' \sqcup y'$. ◇

Definition 2.46 *A lattice is said to be* distributive *if $\sqcap$ and $\sqcup$ distribute over each other, i.e. if*

(i) $\forall x, y, z, \quad x \sqcup (y \sqcap z) = (x \sqcup y) \sqcap (x \sqcup z)$ *and*
(ii) $\forall x, y, z, \quad x \sqcap (y \sqcup z) = (x \sqcap y) \sqcup (x \sqcap z)$.

These two conditions are indeed equivalent. If (i) is true, then (ii) is true. To show this, let us compute

$$(x \sqcap y) \sqcup (x \sqcap z).$$

By (i), this can be written
$$((x \sqcap y) \sqcup x) \sqcap ((x \sqcap y) \sqcup z).$$

By using the absorption property, we obtain
$$x \sqcap ((x \sqcap y) \sqcup z)$$

and, by again applying (i),
$$x \sqcap ((x \sqcup z) \sqcap (y \sqcup z)).$$

By the associativity of $\sqcap$, this is equal to
$$(x \sqcap (x \sqcup z)) \sqcap (y \sqcup z)$$

and, by again using the absorption property, this is equal to
$$x \sqcap (y \sqcup z).$$

The converse implication is proved similarly.

EXAMPLE 2.47 $\mathcal{P}(E)$ is a distributive lattice.

EXAMPLE 2.48 Assume E contains three elements a, b and c pairwise incomparable, a least element $\bot$ and a greatest element $\top$, see Figure 2.2.

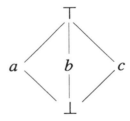

Figure 2.2

It is a lattice because
$$\forall x, y \in \{a, b, c\}, \, x \neq y, \, x \sqcap y = \bot \text{ and } x \sqcup y = \top.$$

It is not distributive, because
$$a \sqcap (b \sqcup c) = a \sqcap \top = a,$$

while
$$(a \sqcup b) \sqcap (a \sqcup c) = \top \sqcap \top = \top.$$

Complete sets and lattices 33

Definition 2.49 *A lattice E is said to be complemented if*

(i) *it has a least element $\bot$ and a greatest element $\top$, with $\bot \neq \top$ and*
(ii) *there exists a mapping ν from E to E such that*
- $\forall x \in E, \quad x \sqcap \nu(x) = \bot$ *and*
- $\forall x \in E, \quad x \sqcup \nu(x) = \top.$

EXAMPLE 2.50
1. The lattice $\mathcal{P}(E)$ is complemented. Its least element is the empty set, its greatest element is E and the mapping ν is the usual complement operation.
2. The lattice of Example 2.48 is complemented. Let $\nu(\bot) = \top$, $\nu(\top) = \bot$, $\nu(a) = b$, $\nu(b) = c$ and $\nu(c) = a$.

EXERCISE 2.17 The set $\mathbb{N}$ equipped with the divisibility ordering is a lattice.
1. Is it distributive?
2. Is it complemented? ◇

EXERCISE 2.18
1. Show that the set of equivalence relations on a set E is a lattice for inclusion.
2. Is it distributive?
3. Is it complemented? ◇

EXERCISE 2.19 Show that, in a complemented lattice,
$$\nu(\top) = \bot \text{ and } \nu(\bot) = \top.$$ ◇

EXERCISE 2.20 Show that, in a complemented lattice, $\forall x, \nu(x) \neq x$. ◇

Proposition 2.51 *If a complemented lattice is distributive, there exists exactly one operation of complement ν. This operation, moreover, verifies*

(i)   *involution:* $\forall x, \quad \nu(\nu(x)) = x$,
(ii) *De Morgan's laws:* $\forall x, y, \quad \nu(x \sqcup y) = \nu(x) \sqcap \nu(y)$ *and* $\nu(x \sqcap y) = \nu(x) \sqcup \nu(y)$ *and*
(iii) *antimonotonicity:* $x \leq y \iff \nu(y) \leq \nu(x)$.

Proof.
(i) We first show that in a distributive lattice with a least element and a greatest element, we have the property

$$\forall x, y, z, \quad x \sqcap y = \bot \text{ and } x \sqcup z = \top \implies y \leq z.$$

Indeed, $z = z \sqcup \bot = z \sqcup (x \sqcap y) = (z \sqcup x) \sqcap (z \sqcup y) = \top \sqcap (z \sqcup y) = z \sqcup y$, and thus $y \leq z$.

Assume now that there exist two mappings ν and μ verifying

$$\forall x \in E, \quad x \sqcap \nu(x) = \bot,$$
$$\forall x \in E, \quad x \sqcup \nu(x) = \top,$$
$$\forall x \in E, \quad x \sqcap \mu(x) = \bot,$$
$$\forall x \in E, \quad x \sqcup \mu(x) = \top.$$

Because $x \sqcap \nu(x) = \bot$ and $x \sqcup \mu(x) = \top$, we have that $\nu(x) \leq \mu(x)$. Similarly, $x \sqcap \mu(x) = \bot$ and $x \sqcup \nu(x) = \top$, and hence $\mu(x) \leq \nu(x)$ and thus $\mu(x) = \nu(x)$. Because $\nu(x) \sqcap x = \bot$ and $\nu(x) \sqcup \nu(\nu(x)) = \top$, we have that $x \leq \nu(\nu(x))$, and, for similar reasons, $\nu(\nu(x)) \leq x$.

(ii) In order to show the De Morgan's laws, it suffices to show, taking into consideration the uniqueness of the complement, that

1. $(x \sqcup y) \sqcap (\nu(x) \sqcap \nu(y)) = \bot$ and $(x \sqcup y) \sqcup (\nu(x) \sqcap \nu(y)) = \top$ and
2. $(x \sqcap y) \sqcap (\nu(x) \sqcup \nu(y)) = \bot$ and $(x \sqcap y) \sqcup (\nu(x) \sqcup \nu(y)) = \top.$

We show only the first identity; the second one can be proved similarly:

$$(x \sqcup y) \sqcap (\nu(x) \sqcap \nu(y)) = (x \sqcap \nu(x) \sqcap \nu(y)) \sqcup (y \sqcap \nu(x) \sqcap \nu(y)) = \bot \sqcup \bot = \bot.$$

$$(x \sqcup y) \sqcup (\nu(x) \sqcap \nu(y)) = (x \sqcup y \sqcup \nu(x)) \sqcap (x \sqcup y \sqcup \nu(y)) = \top \sqcap \top = \top.$$

(iii) To show the last equivalence, notice that

$$x \leq y \iff x = x \sqcap y \iff \nu(x) = \nu(x \sqcap y) = \nu(x) \sqcup \nu(y)$$
$$\iff \nu(y) \leq \nu(x). \qquad \square$$

EXAMPLE 2.52 The lattice of Example 2.48 is not distributive, and there indeed exist at least two operations of complement. For instance $\mu(\bot) = \top$, $\mu(\top) = \bot$, $\mu(a) = c$, $\mu(b) = a$, $\mu(c) = b$.

CHAPTER 3

RECURSION AND INDUCTION

Inductive and recursive definitions are the construction of finite objects from other finite objects, according to some given rules. Inductive definitions also provide us with a way of grasping infinite objects defined by recursive definitions: indeed, since only finite objects can be handled by computer science, such infinite objects are studied via sequences of finite approximations; usually, the finite approximations are also defined by an inductive definition.

Inductive proofs enable one to reason about inductively defined objects. Because computer science makes extensive use of such objects, this chapter is essential. For instance, recursive definitions constantly occur in data structures and in the conception of recursive programs (in functional languages such as LISP, but also in logic programming and PROLOG). The proofs of such recursive programs are then inductive proofs, as are the proofs of termination of iterative programs (sometimes called top–down induction).

However, the various induction principles are not stated in detail in textbooks (to our knowledge); this is why we cannot recommend any handbook for the present chapter.

In this chapter we review the two basic induction principles on the integers: the induction principle and the complete induction principle. We introduce the notion of definition of a set by induction and we show how to prove properties of sets defined by induction. As a special case, we introduce the concept of 'set of terms' which is a major tool in computer science. Finally, we present the concept of closure, which is a general way of looking at inductive definitions.

3.1 Reasoning by induction in $\mathbb{N}$

3.1.1 First induction principle

In $\mathbb{N}$, the first induction principle, also called the mathematical induction principle, is a most useful way of reasoning. We will use both terminologies 'proof by induction' and 'proof by mathematical induction' for proofs using this first induction principle.

Theorem 3.1 *Let $P(n)$ be a predicate (a property) depending on the integer n. If both the following conditions hold:*

(B) *$P(0)$ is true, and*
(I) *$\forall n \in \mathbb{N}$, the implication $(P(n) \Longrightarrow P(n+1))$ is true,*

then $\forall n \in \mathbb{N}$, $P(n)$ is true.

(B) is called the *basis step* of the induction and (I) is called the *inductive step* (or sometimes 'going from n to $n+1$'). Here we give a direct proof of this result, but it is worth while noting that it can also be justified by using Proposition 3.11 and the inductive definition of $\mathbb{N}$ given in Example 3.9.

Proof. By contradiction. We consider the set

$$X = \{k \in \mathbb{N} \:/\: P(k) \text{ is false}\}.$$

If X is non-empty, it has a least element n. By condition (B), $n \neq 0$. Thus $n-1$ is an integer and $n-1 \notin X$, namely, $P(n-1)$ is true. Using (I), we then obtain: $P(n)$ is true, which contradicts $n \in X$. Therefore, X is empty, and this proves the theorem. □

Z (I) does not assert that $P(n+1)$ or $P(n)$ hold, but only that **if** $P(n)$ is true, **then** $P(n+1)$ must be true. Only after proving (I) **and** (B) can we conclude that, for all $n \geq 0, P(n)$ is true. Usually the basis (B) is easy to prove, and the difficult part is the inductive step (I). However, one should not forget to prove the basis (B), otherwise one will obtain false results; for instance, we verify immediately that $\forall n \geq 0, (n > 10 \Longrightarrow n+1 > 10)$. It is none the less false that $\forall n \geq 0, n > 10$. (See also Exercise 3.6.)

REMARK 3.2 We can prove a slightly more general form of Theorem 3.1 similarly. Let n_0 be an integer greater than or equal to 0, if both following conditions hold:

(B_{n_0}) $P(n_0)$ is true, and
(I_{n_0}) $\forall n \geq n_0$, the implication $(P(n) \Longrightarrow P(n+1))$ is true,

then $\forall n \geq n_0$, $P(n)$ is true.

EXAMPLE 3.3 We wish to compute the sum $S_n = 1 + 2 + \cdots + n$. We note that $2S_1 = 2 = 1 \times 2$, $2S_2 = 2 + 4 = 2 \times 3$, $2S_3 = 2 + 4 + 6 = 3 \times 4$. We then conjecture that $\forall n > 0, 2S_n = n(n+1)$. We prove this by induction. Let $P(n)$ be the property '$2S_n = n(n+1)$', we verify that

(B) $2S_1 = 1 \times 2$,
(I) Let $n \geq 1$. We assume $P(n)$. We have

$$2S_{n+1} = 2S_n + 2(n+1) = n(n+1) + 2(n+1) = (n+1)(n+2),$$

hence $P(n+1)$ is true.

We can then conclude that $\forall n \geq 1, P(n)$.

EXERCISE 3.1 Adopting the convention that $\forall r \in \mathbb{R}, r^0 = 1$, prove by induction that:

1. $\forall r \in \mathbb{R}, \forall n \in \mathbb{N}, \quad S_n = \sum_{i=0}^{n} r^i = \begin{cases} n+1 & \text{if } r = 1, \\ \dfrac{r^{n+1} - 1}{r - 1} & \text{if } r \neq 1. \end{cases}$

2. $\forall r \in \mathbb{R}, \forall n \in \mathbb{N}, \quad T_n = \sum_{i=0}^{n} i r^i = \begin{cases} n(n+1)/2 & \text{if } r = 1, \\ \dfrac{nr^{n+2} - (n+1)r^{n+1} + r}{(r-1)^2} & \text{if } r \neq 1. \end{cases}$ ◇

EXERCISE 3.2
1. Show that $\forall n \geq 1, S_n = 1^3 + 3^3 + \cdots + (2n-1)^3 = 2n^4 - n^2$.
2. Compute $T_n = \displaystyle\sum_{k=1}^{n} \dfrac{1}{4k^2 - 1}$ for all $n \geq 1$. ◇

EXERCISE 3.3 We consider the polynomial with real-valued coefficients

$$P(x) = \frac{1}{3}x^3 + ax^2 + bx.$$

1. Find a and b such that $\forall x \in \mathbb{R}, P(x+1) - P(x) = x^2$. We assume that this property holds in the remainder of the exercise.
2. Show that $\forall n \in \mathbb{N}, P(n)$ is an integer.
3. $\forall n \geq 0$, let $S_n = \displaystyle\sum_{k=0}^{n} k^2$. Show that

$$\forall n \geq 0, \quad S_n = P(n+1) = \frac{n(n+1)(2n+1)}{6}.$$ ◇

NOTATION We will write $p \mid n$ to denote the fact that p divides n, where p and n are integers.

EXERCISE 3.4 Let $n \geq 1$ and let $A \subseteq \{1, 2, \ldots, 2n\}$ be such that $|A| \geq n+1$. Show that there exist two distinct integers a and b in A such that $a \mid b$. ◇

EXERCISE 3.5 Let $\mathcal{R}$ be a binary relation on a set E. Let

$$\mathcal{R}^0 = Id_E, \quad \mathcal{R}^{i+1} = \mathcal{R}.\mathcal{R}^i.$$

Show that $\forall i, j \geq 0$, $\mathcal{R}^{i+j} = \mathcal{R}^i . \mathcal{R}^j$.

EXERCISE 3.6 We consider the properties $P(n)$: '$9 \mid 10^n - 1$' and $Q(n)$: '$9 \mid 10^n + 1$'.
1. Show that $\forall n \in \mathbb{N}$, $P(n) \Longrightarrow P(n+1)$ and $Q(n) \Longrightarrow Q(n+1)$.
2. Find the values of n for which $P(n)$ (resp. $Q(n)$) is true.

EXERCISE 3.7 Find the error in the following proof by induction. Let $P(n)$ be the property 'in any group consisting of n individuals, all the people are of the same age'.

(B) $P(1)$ is clearly true.
(I) Let n be such that $P(n)$ is true. Let G be a group of $n+1$ individuals numbered from 1 to $n+1$. Let G_1 (resp. G_2) be the group consisting of the n first (resp. last) individuals in G. Since $P(n)$ is true, all the people of G_1 (resp. G_2) are of the same age. Moreover, individual number n is a member of both G_1 and G_2. Thus all the people of G are of the same age as individual number n, and this proves $P(n+1)$.

We hence deduce that $\forall n \geq 1$, $P(n)$.

3.1.2 Second induction principle

In the first induction principle (see Theorem 3.1), the truth of $P(n+1)$ depends only upon that of $P(n)$, i.e. if proposition P is true at step n it is also true at step $(n+1)$. More complex cases may occur, where in order to establish that P is true at step $(n+1)$ we have to explicitly use the fact that P is true at steps $0, 1, \ldots, n-1, n$. In such a case, it is more convenient to use the second induction principle, which is stated as follows.

Theorem 3.4 Let $P(n)$ be a property depending on the integer n. If the following proposition is verified:

$$(\text{I}') \quad \forall n \in \mathbb{N}, \quad \Big((\forall k < n, P(k)) \quad \Longrightarrow \quad P(n) \Big),$$

then $\forall n \in \mathbb{N}, P(n)$ is true.

This second induction principle is a consequence of Theorem 2.31 because the usual ordering on $\mathbb{N}$ is a well ordering (see Section 2.4).

REMARK 3.5
1. The fact that the second induction principle has no basis step may seem suspicious; in fact, the basis step is 'hidden' in (I'). Indeed, verifying (I') implies proving that for $n = 0$ $(\forall k < 0, P(k)) \Longrightarrow P(0)$. But $(\forall k < 0, P(k))$ is true because there is no negative integer $k < 0$, hence we must prove that $P(0)$ is true. Here we see a typical instance of reasoning with the empty set: $(\forall k < 0, P(k))$ can be rewritten as $(k < 0 \Longrightarrow P(k))$, or, since there is no negative integer $k < 0$ in $\mathbb{N}$, $(k \in \emptyset \Longrightarrow P(k))$, which is true because $k \in \emptyset$ is always false; more generally, any 'empty' statement of the form $(\forall x \in \emptyset, P(x))$ always holds.

2. As for the first induction principle, we may start from any integer n_0. We must then check that:

$$(I'_{n_0}) \quad \forall n \geq n_0, \quad \Big((\forall k \in \{n_0, \ldots, n-1\}, P(k)) \implies P(n)\Big)$$

and deduce $\forall n \geq n_0$, $P(n)$.

3. On $\mathbb{N}$, the two induction principles are equivalent (i.e. each can be shown to be valid from the other), but (see Section 2.4) only the second induction principle can be generalized to more general ordered sets.

EXERCISE 3.8 Verify that the two induction principles entail the same properties on $\mathbb{N}$, i.e. that, if $P(n)$ is a property depending on the integer n, P verifies (I') if and only if P verifies (B) and (I). The two induction principles thus have the same power on $\mathbb{N}$; we will say that they are equivalent on $\mathbb{N}$. ◇

EXAMPLE 3.6 The second principle is simpler to use when the property of the elements at step n involves simultaneously the property of the elements at steps $n-1, n-2, \ldots$, etc. For instance, we can quite easily show that any integer $n \geq 2$ can be written as a product of primes. Denote by $P(n)$ the property 'n can be written as a product of primes'; it suffices to verify (I'_2), see Remark 3.5 (2). Let $n \geq 2$. Assume $\forall k \in \{2, \ldots, n-1\}$, $P(k)$. Two cases can occur:

- n is a prime. Then n can clearly be written as a product of primes (a single prime is also considered as a product).
- n is not a prime. Then we can write $n = ab$, where a and b are two integers between 2 and $n-1$. $P(a)$ and $P(b)$ are true by hypothesis, and so we deduce that n can also be written as the product of the decompositions of a and b.

EXERCISE 3.9
1. Show that $\forall n \in \mathbb{N}$, $(n+1)^2 - (n+2)^2 - (n+3)^2 + (n+4)^2 = 4$.
2. Deduce that any integer m can be written as sums and differences of squares $1^2, 2^2, \ldots, n^2$ for an n, i.e.

$$\forall m \in \mathbb{N}, \exists n \in \mathbb{N}, \exists \varepsilon_1, \ldots, \varepsilon_n \in \{-1, 1\}, \ m = \varepsilon_1 1^2 + \varepsilon_2 2^2 + \cdots + \varepsilon_n n^2.$$

(Hint: first show the result for $m \in \{0, 1, 2, 3\}$.) ◇

EXERCISE 3.10 Let A^* be the free monoid on the alphabet A (see Definition 1.15). Show that $\forall u, v \in A^*$, $u \cdot v = v \cdot u \iff \exists w \in A^*, \exists p, q \in \mathbb{N}: u = w^p$ and $v = w^q$. ◇

EXERCISE 3.11 Let A^* be the free monoid on the alphabet A (see Definition 1.15 and Example 2.4). A *language* is a subset of A^*. If L_1 and L_2 are two languages of A^*, we define their *concatenation* by:

$$L_1 \cdot L_2 = \{u \cdot v \ / \ u \in L_1, \ v \in L_2\}.$$

Language concatenation is an associative operation with unit $\{\varepsilon\}$. We can then define the powers of language L as follows:

$$L^0 = \{\varepsilon\} \text{ and } \forall n \in \mathbb{N}, L^{n+1} = L^n \cdot L = L \cdot L^n.$$

Finally, the star of language L is the submonoid of A^* generated by L, i.e.

$$L^* = \bigcup_{n \in \mathbb{N}} L^n.$$

Let L and M be two languages on A^* such that $\varepsilon \notin L$. Show that in $\mathcal{P}(A^*)$, the equation $X = L \cdot X \cup M$ has as its unique solution the language $L^* \cdot M$. ◇

3.2 Inductive definitions and proofs by structural induction

In the present section, we introduce inductive definitions of sets and functions and proofs by induction on inductively defined structures.

3.2.1 Inductively defined sets

Quite often in computer science subsets are inductively (recursively) defined. In particular, many data structures may be so defined. Intuitively, the inductive definition of a subset X of a set explicitly gives some elements of the set X together with ways of constructing new elements of X from already known elements. Such a definition will hence have the following intuitive generic form:

(B) Some elements of the set X are explicitly given (*basis* of the recursive definition).
(I) The other elements of the set X are defined in terms of elements already in the set X (*inductive steps* of the recursive definition).

Formally, we have the following definition.

Definition 3.7 *Let E be a set. An inductive definition of a subset X of E consists of giving:*

- *a subset B of E and*
- *a set K of operations $\Phi \colon E^{a(\Phi)} \longrightarrow E$, where $a(\Phi) \in \mathbb{N}$ is the arity (or rank) of Φ.*

X is defined as the least set verifying the following assertions (B) *and* (I):

(B) $B \subseteq X$.
(I) $\forall \Phi \in K, \forall x_1, \ldots, x_{a(\Phi)} \in X, \Phi(x_1, \ldots, x_{a(\Phi)}) \in X$.

The set thus defined is

$$X = \bigcap_{Y \in \mathcal{F}} Y,$$

where $\mathcal{F} = \{Y \subseteq E \ / \ B \subseteq Y, \text{ and } Y \text{ verifies (I) with } X \text{ replaced by } Y\}$.

Henceforth, we modify assertions (B) and (I) slightly, and we denote an inductive definition by the form

(B) $x \in X \quad (\forall x \in B)$,
(I) $x_1, \ldots, x_{a(\Phi)} \in X \implies \Phi(x_1, \ldots, x_{a(\Phi)}) \in X \quad (\forall \Phi \in K)$.

REMARK 3.8 The set $\mathcal{F}$ is non-empty because it contains E; indeed, E clearly verifies (B) ($B \subseteq E$) and (I). Moreover, if subsets of a set verify a condition then their intersection also verifies that condition. Indeed, let $\mathcal{Y}$ be a set of subsets Y of E verifying (B) and (I), and let $Z = \bigcap_{Y \in \mathcal{Y}} Y$. Since B is included in any set Y of $\mathcal{Y}$, B is also included in $Z = \bigcap_{Y \in \mathcal{Y}} Y$ and hence Z verifies (B); if $x_1, \ldots, x_{a(\Phi)} \in Z$, then for any $Y \in \mathcal{Y}$, $x_1, \ldots, x_{a(\Phi)} \in Y$, whence $\Phi(x_1, \ldots, x_{a(\Phi)}) \in Y$, and hence $\Phi(x_1, \ldots, x_{a(\Phi)}) \in Z$ and Z verifies (I). Thus $\bigcap_{Y \in \mathcal{F}} Y$, where $\mathcal{F}$ is the above-defined set of subsets of E, is indeed the least subset of E verifying the conditions (B) and (I).

Note that, in general, many sets verify these conditions. Consider, for instance, the conditions:

(B) $0 \in P$,
(I) $n \in P \implies n + 2 \in P$.

There are infinitely many subsets of $\mathbb{N}$ verifying these properties : $\mathbb{N}$, $\mathbb{N} \setminus \{1\}$, $\mathbb{N} \setminus \{1, 3\}$, $\mathbb{N} \setminus \{1, 3, 5\}$, etc., are such subsets. The subset P defined by (B) and (I) is not among them because it consists of the set of even integers.

We consider now examples of inductive definitions.

EXAMPLE 3.9
1. The subset X of $\mathbb{N}$ inductively defined by
   (B) $\quad 0 \in X$,
   (I) $\quad n \in X \implies n + 1 \in X$,
is identical to $\mathbb{N}$. (B) and (I) thus constitute an inductive definition of $\mathbb{N}$.
2. The subset X of the free monoid A^* (see Definition 1.15) inductively defined by
   (B) $\quad \varepsilon \in X$,
   (I) $\quad u \in X \implies \forall a \in A,\ u \cdot a \in X$,
is identical to A^*. (B) and (I) thus constitute an inductive definition of A^*.

3. Let $A = \{ (,) \}$ be the alphabet consisting of two parentheses (left and right). The set $D \subseteq A^*$ of strings of balanced parentheses, the so-called Dyck language, is defined by
   (B) $\varepsilon \in D$,
   (I)   if x and y belong to D, then (x) and xy also belong to D.
4. Let E be the set of expressions all of whose subexpressions are included in parentheses and which are formed from identifiers in a set A and the two operators $+$ and $\times$. E is the subset of $\left(A \cup \{ +, \times, (,) \} \right)^*$ inductively defined by
   (B) $A \subseteq E$,
   (I)   if e and f are in E then $(e + f)$ and $(e \times f)$ are also in E.

We note that in computer science syntactic definitions are almost always inductive. We often use the BNF (Backus–Naur Form) notation for describing them. For instance, the set E is defined by

$$E ::= A \mid (E + E) \mid (E \times E),$$

where the symbol '$\mid$' is read 'or'.

5. The set BT of labelled binary trees on the alphabet A is the subset of $\left(A \cup \{ \emptyset, (,) , , \} \right)^*$ inductively defined by
   (B) $\emptyset \in BT$ (the empty tree),
   (I) $l, r \in BT \implies \forall a \in A, (a, l, r) \in BT$ (the tree with root a, left child l and right child r).

The set BT thus defined is a language on the alphabet $A \cup \{\emptyset\} \cup \{(\} \cup \{)\} \cup \{,\}$. In general, we use a very intuitive graphical representation of trees. To simplify, tree $(a, \emptyset, \emptyset)$ will simply be denoted by a. For instance, the trees a, (a, a, b), $(a, \emptyset, (b, c, \emptyset))$ and $(a, (a, b, c), d)$ can be drawn as in Figure 3.1

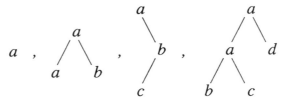

Figure 3.1

Binary trees are extensively used in algorithmics.

EXERCISE 3.12 Let A be an alphabet. We recursively define the sets $(BT_n)_{n \in \mathbb{N}}$ by
- $BT_0 = \{\emptyset\}$,
- $BT_{n+1} = BT_n \cup \{(a, l, r) \, / \, a \in A, l, r \in BT_n\}$.

Show that $X = \bigcup_{n \in \mathbb{N}} BT_n$ is the set BT of binary trees on alphabet A. ◇

Inductive definitions and proofs by structural induction 43

The preceding exercise illustrates a more general phenomenon. Indeed, in most cases, the elements of an inductively defined set can be obtained from the basis by applying finitely many inductive steps. We have the following theorem.

Theorem 3.10 *If X is defined by the conditions (B) and (I), any element of X can be obtained from the basis by applying finitely many inductive steps.*

Proof. We define the sets
- $X_0 = B$,
- $X_{n+1} = X_n \cup \{\Phi(x_1, \ldots, x_{a(\Phi)}) \,/\, x_1, \ldots, x_{a(\Phi)} \in X_n \text{ and } \Phi \in K\}$.

As in Exercise 3.12, we show by induction that $\forall n \in \mathbb{N}$, $X_n \subseteq X$, and we deduce that $X_\omega = \bigcup_{n \in \mathbb{N}} X_n \subseteq X$. The set of elements obtainable from the basis by applying finitely many inductive steps is exactly X_ω. We must now show that X_ω verifies (B) and (I). As $B = X_0 \subseteq X_\omega$, X_ω verifies (B). Let $\Phi \in K$ and let $x_1, \ldots, x_{a(\Phi)} \in X_\omega$. Each x_i belongs to a set $X_{n_i} \subseteq X_\omega$. Let $n = \sup\{n_1, \ldots, n_{a(\Phi)}\}$. Then $x_i \in X_n$, thus $\Phi(x_1, \ldots, x_{a(\Phi)}) \in X_{n+1} \subseteq X_\omega$, and X_ω verifies (I). □

3.2.2 Inductive prooofs

The induction principle is a generalization of the induction principle on the integers and is designed to prove the properties of inductively defined sets. The proof by induction exactly follows the inductive definition of the set; this is why it is also called *proof by structural induction*.

Proposition 3.11 *Let X be an inductively defined set (see Definition 3.7), and let $P(x)$ be a predicate expressing a property of the elements x of X. If the following conditions hold:*

(B″) $P(x)$ *is true for each* $x \in B$, *and*
(I″) $(P(x_1), \ldots, P(x_{a(\Phi)})) \Longrightarrow P(\Phi(x_1, \ldots, x_{a(\Phi)}))$ *for each* $\Phi \in K$,

then $P(x)$ is true for any x in X.

Verifying (B″) and (I″) constitutes a proof by induction of property P on X.

Proof. Let Y be the set of xs such that $P(x)$ is true. We have that $B \subseteq Y$ (by (B″)), and that Y verifies the inductive clauses (I) of the definition of X (by (I″)); hence $Y \supseteq X$ (see Definition 3.7). □

REMARK 3.12 If we consider that the non-negative integers are defined as in Example 3.9, the first induction principle on the integers corresponds to the above definition. All the proofs by mathematical induction seen in Section 3.1 are hence examples of proofs by induction according to the present definition.

EXAMPLE 3.13 We show by induction that any string of the Dyck language has as many left parentheses as right parentheses (see Example 3.9). For x in D, we denote by $l(x)$ (resp. $r(x)$) the number of left (resp. right) parentheses in x. (The inductive definition of these functions is left to the reader.) Finally, let $P(x)$ be the property '$r(x) = l(x)$'. We prove by induction that $P(x)$ holds for any x in D.

(B) The only element of the basis is ε, and it satisfies P because

$$r(\varepsilon) = l(\varepsilon) = 0.$$

(I) Let $x, y \in D$ be such that $r(x) = l(x)$ and $r(y) = l(y)$ and let $z = xy$. We have that $r(z) = r(x) + r(y) = l(x) + l(y) = l(z)$, and so $P(z)$ is thus verified. The case where $z = (x)$ can be verified in the same way: $r(z) = r(x) + 1 = l(x) + 1 = l(z)$.

We deduce that $\forall x \in D$, $l(x) = r(x)$.

EXERCISE 3.13 Characterization of the Dyck language. We use the notations of Examples 3.9 and 3.13. Show that $D = L$, where
$$L = \{x \in A^* \mid l(x) = r(x) \text{ and } l(y) \geq r(y) \text{ for any prefix } y \text{ of } x\}. \qquad \diamond$$

EXERCISE 3.14 Let BT be the set of binary trees and let h, n, f be the functions that give the height (see Example 3.24), the number of nodes (nodes are also called vertices) and the number of leaves of a tree respectively. Show that

1. $\forall x \in BT$, $n(x) \leq 2^{h(x)} - 1$,
2. $\forall x \in BT$, $f(x) \leq 2^{h(x)-1}$. $\qquad \diamond$

EXERCISE 3.15 A binary tree is *strict* if it is non-empty and if it has no node with a single non-empty child. For instance, the trees of the Figure 3.5 (page 51) are strict, while the tree of the Figure 3.2 is non-strict.

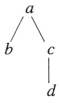

Figure 3.2

1. Give a definition of the set SBT of strict trees.
2. Show that $\forall x \in SBT$, $n(x) = 2f(x) - 1$. $\qquad \diamond$

EXERCISE 3.16 A binary tree is said to be *balanced* if for each node in the tree, the difference between the heights of its left and right subtrees is at most 1. For instance, Figure 3.3 represents balanced trees of height $3, 4$, and 5. (The labels of nodes are not represented.)

Figure 3.3

1. Give a definition of the set BBT of balanced binary trees.
2. We define $(u_n)_{n\in\mathbb{N}}$ by: $u_0 = 0, u_1 = 1$ and

$$\forall n \geq 0, \quad u_{n+2} = u_{n+1} + u_n + 1.$$

Show that $\forall x \in BBT$, $n(x) \geq u_{h(x)}$, where h and n are the functions that give the height and the number of nodes of a tree respectively. ◇

EXERCISE 3.17 Let A^* be the free monoid on alphabet A (see Definition 1.15, Example 2.4 and Exercise 3.11). The set Rat of rational languages is defined inductively by:
(B) $\emptyset \in Rat$ and $\forall a \in A$, $\{a\} \in Rat$,
(I_1) $L, M \in Rat \Longrightarrow L \cup M \in Rat$,
(I_2) $L, M \in Rat \Longrightarrow L \cdot M \in Rat$,
(I_3) $L \in Rat \Longrightarrow L^* \in Rat$.

1. The mirror image (or reverse) of language L is the set $\widetilde{L} = \{\widetilde{u} \,/\, u \in L\}$, where, if u is the string $u = a_1 a_2 \cdots a_n$, then $\widetilde{u}$ is the string $\widetilde{u} = a_n \cdots a_2 a_1$, see Exercise 3.18. Show that $L \in Rat \Longrightarrow \widetilde{L} \in Rat$.
2. We denote by $LF(L)$ the set of prefixes (left factors) of strings in the language L, i.e. $LF(L) = \{v \in A^* \,/\, \exists u \in L \text{ such that } v \text{ is a prefix of } u\}$. Show that $L \in Rat \Longrightarrow LF(L) \in Rat$. ◇

3.3 Terms

In the present section we study a particular instance of definition by structural induction that is quite useful in computer science: the definition of terms. Many structures use terms in their representation.

3.3.1 Definition

Let $F = \{f_0, \ldots, f_n, \ldots\}$ be a set of operation symbols. With each symbol f is associated a finite arity (or rank) $a(f) \in \mathbb{N}$ representing the number of arguments of f. F_n denotes the set of arity n operation symbols.

Let U be the set of all strings of symbols in $F \cup \{`(`, `)`, `,`\}$. Let F_i be the set of symbols of arity i.

Definition 3.14 The set T of terms built on F is inductively defined by:
(B) $B = F_0 \subseteq T$,
(I) $\forall f \in F_n$, $\Phi_f(t_1, \ldots, t_n) = f(t_1, \ldots, t_n)$ for $t_1, \ldots, t_n$ in T.

$\mathbb{Z}$ $\Phi_f(t_1, \ldots, t_n)$ represents the result of operation Φ applied to the n-tuple of terms $(t_1, \ldots, t_n)$, i.e. a *semantic* object, whilst $f(t_1, \ldots, t_n)$ represents a string of formal symbols constituting a term, i.e. a *syntactic* object.

A term may be represented as a tree; for instance, $f(t_1, \ldots, t_n)$ may be pictured as in Figure 3.4.

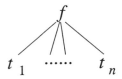

Figure 3.4

3.3.2 Interpretations of terms

Let V be an arbitrary set. With each element f of F_0 we associate an element $h(f)$ of V. With each element f of F_i with $i > 0$ we associate a mapping $h_f \colon V^i \longrightarrow V$.

Proposition 3.15 There exists a unique function h^* from T to V such that:
(B') If $t \in F_0$, $h^*(t) = h(t)$.
(I') If $t = f(t_1, \ldots, t_n)$, $h^*(t) = h_f(h^*(t_1), \ldots, h^*(t_n))$.

If t is a term, the element $h^*(t)$ of V will be called the *interpretation* of t by h^*.

Proof. By structural induction (or induction on the construction of terms). Let $P(t)$ be the property: 'there exists a unique $y = h^*(t)$ verifying (B') and (I')'.

(B) $P(t)$ is true if $t = f \in F_0$ because $y = h(t)$ by (B').
(I) If $P(t_1), \ldots, P(t_n)$ are true, and if $t = f(t_1, \ldots, t_n)$, then $P(t)$ is true because
  • on the one hand, there is a unique way of decomposing t in the form $f(t_1, \ldots, t_n)$: if $f(t_1, \ldots, t_n) = g(t'_1, \ldots, t'_p)$ then $f = g$, $n = p$, and $t_i = t'_i$, $\forall i = 1, \ldots, n$,
  • on the other hand, by (I') if $P(t_1), \ldots, P(t_n)$ are true then $P(t)$ must be true because $h^*(t)$ is entirely defined by
$$h^*(t) = h_f(h^*(t_1), \ldots, h^*(t_n)).$$

Another proof will be given in Section 3.4. □

EXAMPLE 3.16 Let $F_0 = \{a\}$, $F_1 = \{s\}$, $F = F_0 \bigcup F_1$. We have
$$T = \{a, s(a), s(s(a)), \ldots\}.$$
Let $V = \mathbb{N}$.
1. If $h_1(a) = 0$ and $h_{1s}(n) = n+1$, then
$$h_1^*(s^n(a)) = h_1^*(\underbrace{s(s\ldots(s(a))\ldots)}_{n \text{ times}}) = n.$$
2. If $h_2(a) = 1$ and $h_{2s}(n) = 2n$, then
$$h_2^*(s^n(a)) = h_2^*(\underbrace{s(s\ldots(s(a))\ldots)}_{n \text{ times}}) = 2^n.$$
3. If $h_3(a) = 1$ and $h_{3s}(n) = n+2$, then
$$h_3^*(s^n(a)) = h_3^*(\underbrace{s(s\ldots(s(a))\ldots)}_{n \text{ times}}) = 2n+1.$$

Indeed, we verify by induction that:
1. $h_1(a) = 0$ and
$h_1^*(s^{n+1}(a)) = h_1^*(s(s^n(a))) = h_{1s}(h_1^*(s^n(a))) = h_{1s}(n) = n+1$,
2. $h_2(a) = 1$ and
$h_2^*(s^{n+1}(a)) = h_2^*(s(s^n(a))) = 2 \times h_2^*(s^n(a)) = 2 \times 2 = 2^{n+1}$,
3. $h_3(a) = 1$ and
$h_3^*(s^{n+1}(a)) = h_3^*(s(s^n(a))) = h_3^*(s^n(a)) + 2 = 2n+1+2 = 2(n+1)+1$.

Let E be an arbitrary set, and let X be the subset of E inductively defined by the conditions (B) and (I). Theorem 3.10 asserts that each element of X is obtained from the basis by applying a finite number of inductive steps. We refine this result by describing by a term how the element x is obtained.

With each element b of the basis B, we associate a nullary symbol denoted by $\overline{b}$. With each function Φ of K, we associate the arity $a(\Phi)$ symbol $\overline{\Phi}$. Let T be the set of all terms constructed with these symbols.

We consider the interpretation $h^* : T \to E$ defined by
- $h(\overline{b}) = b$,
- $h_{\overline{\Phi}}(x_1, \ldots, x_{a(\Phi)}) = \Phi(x_1, \ldots, x_{a(\Phi)})$.

Proposition 3.17 $X = \{h^*(t) \,/\, t \in T\}$.

Proof. For an element x of E, let $P(x)$ be the property: 'there exists a term t such that $x = h^*(t)$'. It is easy to see that P has properties (B″) and (I″) of Proposition 3.11, and thus $X \subseteq h^*(T)$.

For a term t of T, let $Q(t)$ be the property: '$h^*(t) \in X$'. Here also Proposition 3.11 enables us to conclude that $h^*(T) \subseteq X$. □

3.3.3 Unambiguous definitions

Definition 3.18 *An inductive definition of a set X is said to be unambiguous if the mapping h^* of Proposition 3.17 is injective, i.e. for any $x \in X$ there exists a unique term t such that $x = h^*(t)$.*

More intuitively, this means that there is a unique way of building up an element x of X.

EXAMPLE 3.19   The following definition of $\mathbb{N}^2$ is ambiguous:

(B) $(0,0) \in \mathbb{N}^2$,
(I_1) $(n,m) \in \mathbb{N}^2 \implies (n+1, m) \in \mathbb{N}^2$,
(I_2) $(n,m) \in \mathbb{N}^2 \implies (n, m+1) \in \mathbb{N}^2$.

Indeed, the pair $(1,1)$ can be obtained from $(0,0)$ by using the rule (I_1) first then the rule (I_2), or by using the rule (I_2) first then the rule (I_1).

More formally, we consider the terms built up from

- the arity 0 symbol $\bar{b}$ whose interpretation $h(\bar{b})$ is $(0,0)$,
- the unary symbols $\bar{f}$ and $\bar{g}$ whose interpretations are defined by
    (i) $h_{\bar{f}}(n,m) = (n+1, m)$,
    (ii) $h_{\bar{g}}(n,m) = (n, m+1)$.

Then $(1,1) = h^*(\bar{f}(\bar{g}(\bar{b}))) = h^*(\bar{g}(\bar{f}(\bar{b})))$.

3.3.4 Inductively defined functions

In order to define a function on an inductively defined set unambiguously, it is convenient to use an inductive definition. Intuitively, we define the function on the elements of the basis directly, and then define new elements inductively, building them up from elements already defined.

Definition 3.20   *Let $X \subseteq E$ be an unambiguous inductively defined set (see Definitions 3.7 and 3.18), and let F be any set. The inductive definition of mapping ψ from X to F consists of*

(B) *specifying $\psi(x) \in F$ for each element $x \in B$,*
(I) *specifying the expression of $\psi\bigl(\Phi(x_1, \ldots, x_{a(\Phi)})\bigr)$ in terms of $x_1, \ldots, x_{a(\Phi)}$ and of $\psi(x_1), \ldots, \psi(x_{a(\Phi)})$ for each $\Phi \in K$. We will write*

$$\psi\bigl(\Phi(x_1, \ldots, x_{a(\Phi)})\bigr) = \psi_\Phi\bigl(x_1, \ldots, x_{a(\Phi)}, \psi(x_1), \ldots, \psi(x_{a(\Phi)})\bigr),$$

where ψ_Φ is a mapping from $E^{a(\Phi)} \times F^{a(\Phi)}$ to F.

The definition is illustrated by the following examples.

EXAMPLE 3.21 The factorial function from $\mathbb{N}$ to $\mathbb{N}$ is defined inductively by

(B) $\text{Fact}(0) = 1$,
(I) $\text{Fact}(n+1) = (n+1) \times \text{Fact}(n)$.

Here we use the inductive definition of $\mathbb{N}$ given in Example 3.9. First, the factorial function for the unique element of the basis (0) is defined directly, and then the factorial applied to the new element $n+1$ is expressed in terms of n and $\text{Fact}(n)$.

Henceforth, we will also write inductive definitions of functions as follows:

$$\text{Fact}(n) = \begin{cases} 1 & \text{if } n = 0, \\ n \times \text{Fact}(n-1) & \text{otherwise.} \end{cases}$$

EXERCISE 3.18 Let A^* be the free monoid on the alphabet A (see Definition 1.15). The mirror image (or reverse) of a string $u = a_1 a_2 \cdots a_n$ is the string $\widetilde{u} = a_n \cdots a_2 a_1$. Give an inductive definition of the mirror image. ◇

EXERCISE 3.19 Let the lists L of letters from the alphabet A be defined inductively by:
(B) $\varepsilon \in L$,
(I) $\forall l \in L, \forall a \in A, (al) \in L$.
We define $g(x,y)$ on $L \times L$ by, $\forall a \in A, \forall l \in L, \forall y \in L$,

$$g(\varepsilon, y) = y,$$
$$g((al), y) = g(l, (ay)).$$

1. Let $Q(x)$ be the predicate '$\forall y$, $g(x,y)$ is defined'. Prove by induction on x that $Q(x)$ holds on L.
2. Compute $g((a_1), y)$, for $a_1 \in A$, $y \in L$.
3. Prove by induction on n (for $n \geq 1$) that $g\big((a_n(a_{n-1}(\ldots(a_1)\ldots))), y\big) = g\big(\varepsilon, (a_1(\ldots(a_{n-1}(a_n y))\ldots))\big)$.
4. Let $rev(x) = g(x, \varepsilon)$. Deduce from 3 that, for $a_1, \ldots, a_n \in A$,
$$rev\big((a_n(a_{n-1}(\ldots(a_1)\ldots)))\big) = (a_1(\ldots(a_{n-1}(a_n))\ldots)).$$
◇

We now justify Definition 3.20 and explain why we have assumed the definition of the set X to be unambiguous.

Instead of defining a function ψ from X to F, we will define a function ψ' from T to F, where T is the set of terms whose interpretation is in X (see Proposition 3.17). ψ' is defined as follows:

- $\psi'(\overline{b}) = \psi(b)$,
- $\psi'(\overline{\Phi}(t_1, \ldots, t_{a(\Phi)})) = \psi_\Phi\big(h^*(t_1), \ldots, h^*(t_{a(\Phi)}), \psi'(t_1), \ldots, \psi'(t_{a(\Phi)})\big)$.

As in the proof of Proposition 3.15 we show that such a function exists and is unique.

If the inductive definition of X is unambiguous, then for each element x of X there exists a unique term t such that $h^*(t) = x$. Then let $\psi(x) = \psi'(t)$. ψ is thus indeed a mapping from X to F and it is easy to prove that ψ verifies the conditions (B) and (I) of Definition 3.20.

If the definition of X is ambiguous, then there exist several terms $t_1, \ldots, t_n$ whose interpretation is the same element x of X and, according to the chosen term, Definition 3.20 will give different values $\psi(t_1), \ldots, \psi(t_n)$ to $\psi(x)$. This is illustrated by the following example.

EXAMPLE 3.22 Let us consider the following inductive definition of ψ from $\mathbb{N}^2$ to $\mathbb{N}$, where the inductive definition of $\mathbb{N}^2$ is given in Example 3.19:

(B) $\psi(0,0) = 1$,
(I'_1) $\psi(n+1, m) = \psi(n,m)^2$,
(I'_2) $\psi(n, m+1) = 3 \times \psi(n,m)$.

The thus defined ψ is not a mapping because by using the rule (I'_1) first and then the rule (I'_2), we obtain $\psi(1,1) = \psi(0,1)^2 = (3 \times \psi(0,0))^2 = 3^2 = 9$, whilst by using the rule (I'_2) first and then the rule (I'_1), we obtain

$$\psi(1,1) = 3 \times \psi(1,0) = 3 \times \psi(0,0)^2 = 3.$$

More generally, we can consider that Definition 3.20 in fact defines a relation $\mathcal{R}$ from X to F by: $x\mathcal{R}y$ if and only if there exists a term t such that $x = h^*(t)$ and $y = \psi'(t)$. If h^* is injective then this relation is functional, as we just saw. We should, however, note that this is not the only case when $\mathcal{R}$ is functional. In fact $\mathcal{R}$ is functional if and only if

$$\forall t, t' \in T, \quad h^*(t) = h^*(t') \implies \psi'(t) = \psi'(t').$$

EXAMPLE 3.23 We consider again the ambiguous definition of $\mathbb{N}^2$ given in Example 3.19 and we consider the inductive definition

(B) $g(0,0) = 1$,
(I_1) $g(n+1, m) = 2 \times g(n,m)$,
(I_2) $g(n, m+1) = 3 \times g(n,m)$.

Using Proposition 3.11, we easily show by induction that there exists a unique mapping g verifying these conditions and that this unique mapping is defined by $\forall (n,m) \in \mathbb{N}^2$, $g(n,m) = 2^n 3^m$.

EXERCISE 3.20 Let $\mathbb{N}^* = \mathbb{N} \setminus \{0\}$. We give an inductive definition of the function 'modulo' defined on $\mathbb{N} \times \mathbb{N}^*$, that, when applied to the pair (n, m), gives the remainder of the Euclidean division of n by m

$$n \bmod m = \begin{cases} n & \text{if } n < m, \\ (n-m) \bmod m & \text{otherwise.} \end{cases}$$

Give the corresponding unambiguous inductive definition of $\mathbb{N} \times \mathbb{N}^*$. ◊

EXERCISE 3.21 Inductively define the gcd function on $X = \mathbb{N} \times \mathbb{N} \setminus \{(0,0)\}$ (i.e. the greatest common divisor). What is the corresponding unambiguous inductive definition of X? ◊

We now show some examples of inductively defined functions on sets other than $\mathbb{N}$.

EXAMPLE 3.24
1. The expressions of the set E (see Example 3.9) use an infix notation (in which the operator is placed between its arguments). We can also use a postfix notation without parentheses (in which the operator is placed after both its arguments). For instance, the postfix notation of expression

$$\Big((a \times (b+c)) + d\Big)$$

is $abc + \times d+$. The transformation from the infix notation to the postfix notation is inductively defined by

(B) $\forall a \in A,\ \text{Post}(a) = a$,
(I) $\forall e, f \in E,\ \text{Post}((e+f)) = \text{Post}(e)\,\text{Post}(f) +$ and $\text{Post}((e \times f)) = \text{Post}(e)\,\text{Post}(f) \times$.

2. The height of a binary tree is inductively defined by

(B) $h(\emptyset) = 0$,
(I) $\forall l, r \in BT, \forall a \in A,\ h((a, l, r)) = 1 + \max(h(l), h(r))$.

A more elegant definition of this function is

$$h(x) = \begin{cases} 0 & \text{if } x = \emptyset, \\ 1 + \max(h(l), h(r)) & \text{if } x = (a, l, r). \end{cases}$$

3. The inorder traversal of a tree is the list of the labels of its nodes from left to right. We can notice that several trees may have the same inorder traversal. For instance, the two trees of Figure 3.5 have the same inorder traversal $bacad$. The inductive definition of the inorder traversal is

$$\text{Inf}(x) = \begin{cases} \varepsilon & \text{if } x = \emptyset, \\ \text{Inf}(l) \cdot a \cdot \text{Inf}(r) & \text{if } x = (a, l, r). \end{cases}$$

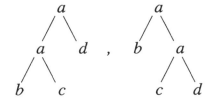

Figure 3.5

EXERCISE 3.22 Give inductive definitions of the functions n and l from BT to $\mathbb{N}$, defining respectively, the number of nodes and the number of leaves of a binary tree. For instance, if x is either tree in Figure 3.5, we have $n(x) = 5$ and $l(x) = 3$. ◇

EXERCISE 3.23 Define the preorder traversal of a binary tree. The preorder traversals of the trees of Figure 3.5 are *aabcd* and *abacd*. ◇

Note that inductive definitions are appropriate as definitions of certain algorithms: sorting algorithms, algorithms on trees such as binary search, insertion, traversal, etc.

3.4 Closure operations

In the proof of Theorem 3.10, we showed that the set X inductively defined by

(B) $B \subseteq X$, and
(I) $\forall \Phi \in K, \forall x_1, \ldots, x_{a(\Phi)} \in X, \Phi(x_1, \ldots, x_{a(\Phi)}) \in X,$

is the union of the sets X_n with $X_0 = B$ and $X_{n+1} = X_n \cup \{\Phi(x_1, \ldots, x_{a(\Phi)}) / x_1, \ldots, x_{a(\Phi)} \in X_n$ and $\Phi \in K\}$. We see that the subset (I) of the inductive definition of X is used in order to build a new set X_{n+1} from an already known set X_n. Indeed, it suffices to define X_{n+1} (or even $X_{n+1} - X_n$) from X_n and, when no new element can be added, the inductive definition is completed.

More generally, we will assume that from any given set E we can build a new set $C(E)$. We will study the properties that C should have in order to give an inductive definition which will be completed whenever C can add no new element to E. With this standpoint we will generalize the results of Section 3.2.

Let U be any set and let $C : \mathcal{P}(U) \longrightarrow \mathcal{P}(U)$ be a *monotone* mapping, i.e. a mapping verifying $\forall E, E' \subseteq U, E \subseteq E' \Longrightarrow C(E) \subseteq C(E')$.

A subset E of U is said to be C-closed if $C(E) \subseteq E$.

Proposition 3.25 *Let I be any set of indices. Let E_i be a C-closed set, for any $i \in I$. Then $\bigcap_{i \in I} E_i$ is C-closed.*

Proof. Let $E = \bigcap_{i \in I} E_i$. Since $E \subseteq E_i$ and E_i is C-closed, $C(E) \subseteq C(E_i) \subseteq E_i$ hence $C(E) \subseteq \bigcap_{i \in I} E_i = E$. □

If E is any subset of U, the intersection of all the C-closed subsets of U containing E is a C-closed subset of U containing E, denoted by $\hat{C}(E)$.

Closure operations 53

Proposition 3.26
- If E' is a C-closed subset containing E, then $\hat{C}(E) \subseteq E'$,
- $E \subseteq \hat{C}(E)$,
- $\hat{C}(\hat{C}(E)) = \hat{C}(E)$,
- $E \subseteq E' \implies \hat{C}(E) \subseteq \hat{C}(E')$.

Proof. The first two points are clear by the definition of $\hat{C}(E)$.

- On the one hand, we have that $E \subseteq \hat{C}(E) \subseteq \hat{C}(\hat{C}(E))$. And, on the other hand, since $\hat{C}(E)$ is a C-closed subset containing $\hat{C}(E)$, then $\hat{C}(\hat{C}(E)) \subseteq \hat{C}(E)$.
- If $E \subseteq E'$, then $E \subseteq \hat{C}(E')$ which is a C-closed subset containing E. We thus have that $\hat{C}(E) \subseteq \hat{C}(E')$. □

The next proposition is a generalization of the induction principles and may be called the *universal induction principle*.

Proposition 3.27 Let $P \subseteq U$ be such that $C(P) \subseteq P$. Then

$$\forall E, \quad E \subseteq P \implies \hat{C}(E) \subseteq P.$$

Proof. If $C(P) \subseteq P$, then P is C-closed. So if $E \subseteq P$, then $\hat{C}(E) \subseteq P$. □

EXAMPLE 3.28 Let X be a subset of a set U; assume that X is inductively defined by (B) and (I) (see Definition 3.7). Define $C: \mathcal{P}(U) \longrightarrow \mathcal{P}(U)$ by

$$C(Y) = \{\Phi(y_1, \ldots, y_{a(\Phi)}) \, / \, \Phi \in K, y_1, \ldots, y_{a(\Phi)} \in Y\}.$$

Then $X = \hat{C}(B)$.

EXAMPLE 3.29 Let $U = \mathbb{N}$, $C(E) = \{n + 1 \, / \, n \in E\}$. Then

$$E' = \hat{C}(E) = \{n + m \, / \, n \in E, m \in \mathbb{N}\}.$$

Indeed, $C(E') = \{n + m + 1 \, / \, n \in E', m \in \mathbb{N}\} \subseteq E'$. Assume there is a C-closed subset E'' containing E and strictly included in E'. Let k be the least integer of E' that is not in E'', i.e. $k \in E'$, $k \notin E''$ and ($k = 0$ or $k - 1 \in E''$).

- If $k = 0$ then, since $k = n + m$ with $n \in E$, $0 \in E$, and hence $0 \in E''$, a contradiction.
- Otherwise, $k - 1 \in E'' \implies k = (k-1) + 1 \in C(E'') \subseteq E''$, a contradiction.

We deduce: $\hat{C}(E) = \{m \, / \, m \geq \inf(E)\} = \hat{C}(\{\inf(E)\})$. Let P be such that $n \in P \implies n + 1 \in P$. Then $C(P) \subseteq P$, and hence $\inf(E) \in P \implies \hat{C}(E) \subseteq P$.

If $\inf(E) = 0$ then $\hat{C}(E) = \mathbb{N}$, and we again find the induction principle on the integers.

EXERCISE 3.24 Let $U = \mathbb{N}$ and $C(E) = \{n+m \,/\, n \in E, m \in E\}$. Let $k\mathbb{N} = \{kn \,/\, n \in \mathbb{N}\}$. Show that if $E \subseteq k\mathbb{N}$ then $\hat{C}(E) \subseteq k\mathbb{N}$. ◇

Let $C: \mathcal{P}(U) \longrightarrow \mathcal{P}(U)$ be such that $E \subseteq E' \Longrightarrow C(E) \subseteq C(E')$. C is finitary if it also verifies: $\forall E, \forall e \in C(E)$, there exists a finite subset F of E such that $e \in C(F)$.

Let $E \subseteq U$. Consider the monotone increasing (for inclusion) sequence

$$E_0 = E$$
$$E_1 = E_0 \cup C(E_0)$$
$$\vdots$$
$$E_{i+1} = E_i \cup C(E_i)$$
$$\vdots$$

and $\quad \hat{E} = \bigcup_{i \leq 0} E_i \, .$

Proposition 3.30 $\hat{E} \subseteq \hat{C}(E)$. If C is finitary, $\hat{E} = \hat{C}(E)$.

Proof. Let $E' = \hat{C}(E)$. We show by induction on the integers that $\forall i \geq 0$, $E_i \subseteq E'$.

- $E_0 = E \subseteq E'$.
- We assume $E_i \subseteq E'$. Then $C(E_i) \subseteq C(E') \subseteq E'$ and $E_{i+1} = E_i \cup C(E_i) \subseteq E'$. Since $\forall i \geq 0$, $E_i \subseteq E'$, we have $\hat{E} = \bigcup_{i \geq 0} E_i \subseteq E'$.

We show that if C is finitary then $\hat{E}$ is C-closed, and we will therefore deduce that $E' \subseteq \hat{E}$. Let $e \in C(\hat{E})$. Because C is finitary, there exists a finite subset $F = \{x_1, \ldots, x_p\}$ of $\hat{E}$ such that $e \in C(F)$. Since $x_j \in \bigcup_{i \geq 0} E_i$, there exists i_j such that $x_j \in E_{i_j}$; let $k = \max\{i_j \,/\, j = 1, \ldots, p\}$. We thus have that $F \subseteq E_k$ and $e \in C(F) \subseteq C(E_k) \subseteq E_{k+1} \subseteq \hat{E}$. Hence $C(\hat{E}) \subseteq \hat{E}$. □

EXAMPLE 3.31 The mapping C that inductively defines a set X (see Example 3.28) is finitary, whence Theorem 3.10.

Closure operations   55

EXAMPLE 3.32 The mapping C from $\mathcal{P}(\mathbb{R})$ to itself, which is defined by $y \in C(X)$ if and only if there exists $Y \subseteq X$ such that $y = \inf Y$, is not finitary. Indeed, let $X = \{1/n \,/\, n \in \mathbb{N}, n > 0\}$. We thus have $0 \in C(X)$. But for all finite subsets F of X, $0 \notin C(F)$ because the greatest lower bound of any finite subset of X is of the form $1/n$ for some $n > 0$.

EXERCISE 3.25 Let E be a vector space on $\mathbb{R}$. For $a, b \in E$, let
$$[a, b] = \{\lambda a + \mu b \,/\, \lambda \geq 0, \mu \geq 0, \text{ and } \lambda + \mu = 1\}$$
be the closed segment subtended by a and b. Let $C: \mathcal{P}(E) \to \mathcal{P}(E)$ be defined by
$$C(X) = \bigcup_{a, b \in X} [a, b].$$
What usual name is given to $C(X)$?
1. Is C monotone increasing and finitary?
2. Given $a \in \hat{C}(A)$ and $b \in \hat{C}(B)$, show that $[a, b] \subseteq \hat{C}(A \cup B)$.
3. Deduce that $\bigcup_{F \in \text{fin}(X)} \hat{C}(F)$ is C-closed, where $\text{fin}(X)$ is the set of finite subsets of X.
4. Is $\hat{C}$ monotone increasing and finitary?
5. Can you generalize (4) to any set transformation which is monotone increasing and finitary?   ◇

We can apply the closure operations in order to define the terms. Let U, F and F_i be as in Section 3.3. Let $C: \mathcal{P}(U) \longrightarrow \mathcal{P}(U)$ be defined by
$$C(E) = \bigcup_{i > 0} \{f(\sigma_1, \ldots, \sigma_i) \,/\, \sigma_j \in E, f \in F_i\}.$$
Then C is finitary and the set T of terms built on F is identical to $\hat{C}(F_0)$.

EXERCISE 3.26 Let $C' = C(E) \cup F_0$. Show that $\hat{C}(F_0) = \hat{C}'(\emptyset)$.   ◇

EXERCISE 3.27 Let $\hat{E} = \hat{C}(F_0)$. Show that $T = \hat{E}$.   ◇

EXERCISE 3.28 Show that there exists a unique function h^* verifying conditions (B') and (I') of Proposition 3.15.   ◇

EXERCISE 3.29 Let $U = \mathbb{N}$ and let $C: \mathcal{P}(\mathbb{N}) \longrightarrow \mathcal{P}(\mathbb{N})$ be defined by
$$C(X) = \begin{cases} \{x + 1 \,/\, x \in X\} & \text{if } X \text{ is finite,} \\ \{x + 1 \,/\, x \in X\} \cup \{0\} & \text{if } X \text{ is infinite.} \end{cases}$$
1. Show that the limit of the sequence
$$E_0 = \{1\}$$
$$E_1 = E_0 \cup C(E_0)$$
$$\vdots$$
$$E_{i+1} = E_i \cup C(E_i)$$
$$\vdots$$

is equal to $\mathbb{N} \setminus \{0\}$.
2. Show that $\hat{C}(\{1\}) = \mathbb{N}$.
3. Explain this result.

◇

CHAPTER 4

BOOLEAN ALGEBRAS

In the middle of the nineteenth century the English mathematician George Boole introduced the algebras which are now named after him. These algebras give a mathematical basis to logical reasoning (see Chapter 5); they are also the basis of electronic computer design via the physical implementation of Boolean operations.

In this chapter we introduce Boolean algebras and their operations, and we define Boolean functions which specify the operation performed by a Boolean circuit, or the truth value of a logical formula. Finally, we show how to represent a Boolean function by a polynomial expression built up using the basic operations of Boolean algebras.

We recommend the following textbooks:

Garrett Birkhoff, Thomas Bartee, *Modern Applied Algebra*, McGraw Hill, New York (1970).

Kenneth Ross, Charles Wright, *Discrete Mathematics*, Prentice Hall, London (1988).

4.1 Boolean algebras

A Boolean algebra is a distributive and complemented lattice having at least two elements.

EXAMPLE 4.1 If E is a non-empty set, $\mathcal{P}(E)$, ordered by inclusion, is a Boolean algebra. The condition $E \neq \emptyset$ ensures that $\mathcal{P}(E)$ has at least two elements.

4.1.1 Algebraic definition

A Boolean algebra can also be viewed as an algebraic structure, and this yields the following definition.

Definition 4.2 *A Boolean algebra B consists of*

- *a set E,*
- *two distinct elements of E, denoted by $\bot$ and $\top$,*
- *two binary operations on E, denoted by $\sqcap$ and $\sqcup$,*
- *a unary operation on E, denoted by $\bar{\ }$,*

satisfying the following conditions:

- *the operations $\sqcap$ and $\sqcup$ are idempotent, associative, commutative and distributive,*
- $x \sqcap (x \sqcup y) = x = (y \sqcap x) \sqcup x$,
- $x \sqcap \bot = \bot$, $x \sqcup \bot = x$, $x \sqcap \top = x$, $x \sqcup \top = \top$,
- $x \sqcap \bar{x} = \bot$, $x \sqcup \bar{x} = \top$.

Of course this list of properties is redundant: some of them are consequences of the others. This will be shown in Exercise 4.1.

If $B = (E, \sqcap, \sqcup, \bar{\ }, \bot, \top)$ is a Boolean algebra, then $\widetilde{B} = (E, \sqcup, \sqcap, \bar{\ }, \top, \bot)$ is also a Boolean algebra, called the *dual* Boolean algebra of B. This algebra is obtained by interchanging $\sqcup$ and $\sqcap$, as well as $\bot$ and $\top$. The result of this operation is a Boolean algebra because the interchanged elements play symmetric roles in the definition. Considering B as an ordered set, $\widetilde{B}$ is the set ordered by the inverse ordering.

EXERCISE 4.1 The purpose of this exercise is to prove that the following conditions ensure that $B = (E, \sqcap, \sqcup, \bar{\ }, \bot, \top)$ is a Boolean algebra:

N_0 : $x \sqcup x = x$,
N_1 : $x \sqcap (y \sqcup z) = (x \sqcap y) \sqcup (x \sqcap z)$,
N_1' : $(x \sqcup y) \sqcap z = (x \sqcap z) \sqcup (y \sqcap z)$,
N_{2a} : $x \sqcap \top = x$,
N_3 : $x \sqcup \bot = x = \bot \sqcup x$,
N_4 : $x \sqcap \bar{x} = \bot$ and $x \sqcup \bar{x} = \top$.

Note that $\sqcap$ and $\sqcup$ should not be assumed to be associative or commutative, and that $\sqcap$ should not be assumed to be idempotent; associativity, commutativity and idempotence will be shown to be consequences of the hypotheses.

1. First show the following equalities:
   N_0' : $x \sqcap x = x$,
   N_{2b} : $\top \sqcap x = x$,
   N_2' : $x \sqcup \top = \top = \top \sqcup x$.
2. Show that $y \sqcap x = \bot$ and $y \sqcup x = \top$ imply that $y = \bar{x}$. Infer that N_6 : $\bar{\bar{x}} = x$.
3. Show that N_3' : $x \sqcap \bot = \bot = \bot \sqcap x$.
4. Next show that N_5 : $x = x \sqcup (x \sqcap y) = x \sqcup (y \sqcap x) = (x \sqcap y) \sqcup x = (y \sqcap x) \sqcup x$. Infer the associativity and commutativity of $\sqcup$. Show that N_1'' : $x \sqcup (y \sqcap z) = (x \sqcup y) \sqcap (x \sqcup z)$.
5. Finally, show that N_7 : $\overline{x \sqcap y} = \bar{x} \sqcup \bar{y}$, and N_7' : $\overline{x \sqcup y} = \bar{x} \sqcap \bar{y}$. Infer the associativity and commutativity of $\sqcap$. ◇

Boolean algebras

In the rest of the present chapter we will sometimes use more algebraic and more familiar notations, denoting lattice operations by sum and product. (We have avoided their use until now, because such well-known symbols might have misled the reader into believing that the operations ⊔ and ⊓ are exactly identical to sum and product.)

We will therefore denote ⊔ by the addition symbol +, and ⊓ will be denoted as a product, by simply concatenating its arguments. We will thus write $(x+y)z$ instead of $(x \sqcup y) \sqcap z$. As is usual, we will also assume that products take precedence over sums; this will allow us to denote $(x \sqcap y) \sqcup z$ by $xy + z$, instead of $(xy) + z$. Finally, ⊥ will be denoted by 0 and ⊤ will be denoted by 1.

Sum and product operations being associative and commutative, and product being distributive over sum, this notation is quite natural. In a Boolean algebra, the usual equalities also hold: $x + 0 = x$, $x0 = 0$, and $x1 = x$. But the following equalities, which are particular to Boolean algebras, also hold:

- $x + x = x = xx$,
- $x + 1 = 1$,
- $(x + y)(x + z) = x + yz$,
- $x + xy = x$.

EXERCISE 4.2
1. Show that $x = ax + b\overline{x} \iff b \leq x \leq a$ (Poretsky's formula).
2. Show that $ax + b\overline{x} = 0 \iff b \leq x \leq \overline{a}$ (Schröder's formula).
3. Show that
    (i)   if $b \leq a$, then $\exists u : x = au + b\overline{u} \implies b \leq x \leq a$,
    (ii) if $b \leq x \leq a$, then $\forall u : \overline{b}x \leq u \leq \overline{a} + x$, $x = au + b\overline{u}$. (Hint: first show that with these hypotheses, u can be written $\overline{b}x + by + \overline{a}z$.) ◇

4.1.2 Homomorphisms

Let $B = (E, \sqcap, \sqcup, \overline{}, \bot, \top)$ and $B' = (E', \sqcap', \sqcup', \widehat{}, \bot', \top')$ be two Boolean algebras. A *homomorphism* from B to B' is a mapping h from E to E' satisfying

- $h(x \sqcup y) = h(x) \sqcup' h(y)$,
- $h(x \sqcap y) = h(x) \sqcap' h(y)$,
- $h(\overline{x}) = \widehat{h(x)}$,
- $h(\bot) = \bot'$,
- $h(\top) = \top'$.

An *antihomomorphism* from B to B' is a homomorphism from B to the dual $\widetilde{B'}$ of B'.

EXERCISE 4.3 Show that a homomorphism is a monotone mapping with respect to the order relation defining the Boolean algebra.
Show that not all monotone mappings are homomorphisms.

EXERCISE 4.4 Let $B = (E, ., +, ^-, 0, 1)$ be a Boolean algebra and e an element of E different from 0. Let $E' = \{x \in E / xe = x\}$. Show that $B' = (E', ., +, \hat{\ }, 0, e)$, where $\hat{x} = \overline{x}e$ is a Boolean algebra and that the mapping h from E to E' defined by $h(x) = xe$ is a homomorphism. ◇

Proposition 4.3 Let $B = (E, ., +, ^-, 0, 1)$ be a Boolean algebra. The mapping h from E to E defined by $h(x) = \overline{x}$ is an antihomomorphism.

Proof. $\overline{1} = 0$ and $\overline{0} = 1$ hold. By the De Morgan laws, $\overline{x+y} = \overline{x}\,\overline{y}$ and $\overline{xy} = \overline{x}+\overline{y}$. Finally, $h(\overline{x}) = \overline{\overline{x}} = x$. □

EXERCISE 4.5 Let F be a finite set. Let $\mathcal{F} = \mathcal{P}(\mathcal{P}(F))$. This is a Boolean algebra if $\mathcal{F}$ is ordered by inclusion, namely,

$$\mathcal{X} = \{X_1, \ldots, X_n\} \subseteq \mathcal{Y} = \{Y_1, \ldots, Y_m\}$$

if and only if $\forall X \in \mathcal{X}, \exists Y \in \mathcal{Y} : X = Y$ (see Exercise 1.29).

We consider the mapping i from f to $\mathcal{F}$ defined by $i(x) = \{X \subseteq F / x \in X\}$.

Let $B = (E, ., +, ^-, 0, 1)$ be a Boolean algebra, and g a mapping from F to E. Show that there exists a unique homomorphism h from $\mathcal{F}$ to B such that $\forall x \in F$, $h(i(x)) = g(x)$. ◇

4.1.3 The minimal Boolean algebra

A very special Boolean algebra, denoted by $\mathbb{B}$, is the Boolean algebra containing only the two elements 0 and 1. (It is usually referred to as *The* Boolean algebra.) The sum, product and complement operations on this two element algebra are described in the following table:

x y	$x + y$	xy	$\overline{x}$
0 0	0	0	1
0 1	1	0	1
1 0	1	0	0
1 1	1	1	0

The two values of this Boolean algebra can be given various special interpretations:

- 0 and 1, not forgetting that $1 + 1 = 1$!
- if E is a set, 0 is its empty subset and 1 its full subset. Unions, intersections and complements will be performed only on these two subsets.
- 'true' and 'false'. Sum is then logical disjunction (read 'or') and product is logical conjunction (read 'and'). This will be studied in more detail in Chapter 5.

4.2 Boolean rings

4.2.1 Exclusive 'or'

Let $(E, \cdot, +, \bar{}, 0, 1)$ be a Boolean algebra. We define a new binary operation denoted by $\oplus$ and called 'exclusive or' by

$$x \oplus y = x\bar{y} + \bar{x}y.$$

Note that $x\bar{y} + \bar{x}y$ is also equal to $(x+y)\overline{xy}$.

EXAMPLE 4.4 Considering the Boolean algebra consisting of the subsets of a set, this operation coincides exactly with the symmetrical difference (see Section 1.1.2).

The following properties arise from the definition of this operation.

Proposition 4.5
- $\oplus$ *is associative and commutative,*
- $x \oplus x = 0$,
- $x \oplus 0 = x$,
- $x \oplus 1 = \bar{x}$,
- $(x \oplus y)z = xz \oplus yz$,
- $x + y = x \oplus y \oplus xy$.

Proof. Let us merely show the associativity of $\oplus$ and the last two points (the rest is straightforward).

1. $(x \oplus y) \oplus z = (x\bar{y} + \bar{x}y) \oplus z = (x\bar{y} + \bar{x}y)\bar{z} + \overline{(x\bar{y} + \bar{x}y)}z$
$= (x\bar{y} + \bar{x}y)\bar{z} + (\bar{x} + y)(x + \bar{y})z = x\bar{y}\bar{z} + \bar{x}yz + \bar{x}\bar{y}z + xyz,$
$x \oplus (y \oplus z) = x \oplus (y\bar{z} + \bar{y}z) = x(\bar{y} + z)(y + \bar{z}) + \bar{x}(y\bar{z} + \bar{y}z)$
$= x\bar{y}\bar{z} + xyz + \bar{x}y\bar{z} + \bar{x}\bar{y}z.$

2. $xz \oplus yz = xz\overline{yz} + \overline{xz}yz = xz(\bar{y} + \bar{z}) + (\bar{x} + \bar{z})yz$
$= xz\bar{y} + \bar{x}yz = (x\bar{y} + \bar{x}y)z = (x \oplus y)z.$

3. $(x \oplus y) \oplus xy = (x\bar{y} + \bar{x}y) \oplus xy = (x\bar{y} + \bar{x}y)(\bar{x} + \bar{y}) + (\bar{x} + y)(x + \bar{y})xy$
$= \bar{x}y + x\bar{y} + xy = (x + \bar{x})y + x(y + \bar{y}) = x + y$. $\square$

4.2.2 Boolean rings

It can be seen that the 'exclusive or' has the properties of an addition operation in an additive group where each element is its own inverse, since $x \oplus x = 0$. By also taking into account the product, we obtain a ring structure. Thus the following definition holds: a *Boolean ring* is a structure $A = (E, \pm, ., 0, 1)$ satisfying the following conditions:

- sum and product are associative and commutative,
- product is distributive over sum,
- 0 is the unit (or identity element) for sum and 1 is the identity element for product,
- $x0 = 0$,

and, moreover,

- $xx = x$,
- $x \pm x = 0$.

The notation $\pm$ for addition is designed to remind us that this addition could also be thought of as a subtraction!

EXAMPLE 4.6   The ring $\mathbb{Z}/2\mathbb{Z}$ of the integers modulo 2 is a Boolean ring.

Here again some of the hypotheses are redundant. For instance, we can substitute for the condition $x \pm x = 0$ the weaker condition $x \pm y = x \implies y = 0$, and not assume commutativity of the product. Then $x \pm y = (x \pm y)(x \pm y) = x \pm y \pm xy \pm yx$, whence $\forall x, y$, $xy \pm yx = 0$. Letting $x = y$, we have $x \pm x = 0$, and adding xy to $xy \pm yx = 0$ we obtain $xy = yx$.

Proposition 4.7   *If $B = (E, ., +, ^-, 0, 1)$ is a Boolean algebra then $A = (E, \oplus, ., 0, 1)$ is a Boolean ring.*

Proposition 4.5 shows that sum and complement can be retrieved from $\oplus$. The converse of this result also holds: let $A = (E, \pm, ., 0, 1)$ be a Boolean ring. Let us define the two operations

- $x + y = x \pm y \pm xy$,
- $\overline{x} = 1 \pm x$.

Proposition 4.8 $B = (E, ., +, ^-, 0, 1)$ *is a Boolean algebra.*

Proof. Let us first show that the operation $+$ is associative, commutative, idempotent and has 0 as unit. It clearly is commutative since $\pm$ and the product are commutative. Moreover, $x + x = x \pm x \pm xx = 0 \pm xx = xx = x$, and $0 + x = 0 \pm x \pm x0 = x$. Finally,

$$(x+y) + z = (x \pm y \pm xy) + z = x \pm y \pm xy \pm z \pm (x \pm y \pm xy)z$$
$$= x \pm y \pm xy \pm xz \pm yz \pm xyz,$$

$$x + (y+z) = x + (y \pm z \pm yz) = x \pm y \pm z \pm yz \pm x(y \pm z \pm yz)$$
$$= x \pm y \pm xy \pm xz \pm yz \pm xyz.$$

We also have $1 + x = 1 \pm x \pm 1x = 1 \pm 0 = 1$.

Next, let us show distributivity:

$$(x+y)z = (x \pm y \pm xy)z = (xz \pm yz \pm xzyz) = xz + yz,$$

$$(x+y)(x+z) = (x \pm y \pm xy)(x \pm z \pm xz)$$
$$= xx \pm xz \pm xxz \pm xy \pm yz \pm xyz \pm xxy \pm xyz \pm xxyz$$
$$= x \pm yz \pm xyz = x + yz,$$

and absorption:

$$x + xy = x \pm xy \pm xxy = x \quad , \quad x(x+y) = x(x \pm y \pm xy) = x \pm xy \pm xy = x.$$

Let us now have a look at the properties of negation:

$$\overline{0} = 1 \pm 0 = 1, \quad \overline{1} = 1 \pm 1 = 0,$$

$$x\overline{x} = x(1 \pm x) = x \pm xx = x \pm x = 0,$$

$$x + \overline{x} = x \pm (1 \pm x) \pm x(1 \pm x) = x \pm (1 \pm x)(1 \pm x) = x \pm 1 \pm x = 1,$$

$$\overline{x} + \overline{y} = 1 \pm x \pm 1 \pm y \pm (1 \pm x)(1 \pm y) = x \pm y \pm (1 \pm x \pm y \pm xy) = 1 \pm xy = \overline{xy},$$

$$\overline{x}\,\overline{y} = (1 \pm x)(1 \pm y) = 1 \pm x \pm y \pm xy = 1 \pm (x+y) = \overline{x+y}. \qquad \square$$

4.3 The Boolean functions

Let $\mathbb{B}$ be the two-element Boolean algebra. A *Boolean function* (with n arguments) is a mapping from $\mathbb{B}^n$ to $\mathbb{B}$.

A Boolean function f with n arguments is completely defined by the n-tuples of $\mathbb{B}^n$ for which it takes the value 1. Since there are 2^n n-tuples in $\mathbb{B}^n$ and since a set with k elements has 2^k different subsets, there are 2^{2^n} Boolean functions with n arguments.

- If $n = 0$, then $2^n = 1$ and $2^1 = 2$; there are two Boolean functions with 0 arguments and these are the two constants 0 and 1.
- If $n = 1$, then $2^n = 2$ and $2^2 = 4$; there are four Boolean functions with 1 argument: the two constant functions, the identity function and the complement function.
- If $n = 2$, then $2^n = 4$ and $2^4 = 16$; see Exercise 4.6.

4.3.1 Polynomial form of the Boolean functions

A Boolean function is said to be *polynomial* if it can be written as a combination of its arguments via the sum, product and complement operations, or if it is the zero function.

EXAMPLE 4.9 The function $f(x,y) = \overline{x}y + \overline{y}$ is a polynomial function. Its values are given by the following table:

x y	$\overline{x}$	$\overline{x}y$	$\overline{y}$	$\overline{x}y + \overline{y}$
0 0	1	0	1	1
0 1	1	1	0	1
1 0	0	0	1	1
1 1	0	0	0	0

We next show that every Boolean function is polynomial.

Lemma 4.10 *Let f be a function with $k+1$ arguments. Then*

$$f(x_0, x_1, \ldots, x_k) = \overline{x_0} f(0, x_1, \ldots, x_k) + x_0 f(1, x_1, \ldots, x_k).$$

Proof. Let

$$g(x_0, x_1, \ldots, x_k) = \overline{x_0} f(0, x_1, \ldots, x_k) + x_0 f(1, x_1, \ldots, x_k)$$

and let $(b_0, b_1, \ldots, b_k)$ denote an arbitrary $(k+1)$-tuple in $\mathbb{B}^{k+1}$. Then

$$g(b_0, b_1, \ldots, b_k) = \overline{b_0} f(0, b_1, \ldots, b_k) + b_0 f(1, b_1, \ldots, b_k).$$

If $b_0 = 0$ then $b_0 f(1, b_1, \ldots, b_k) = 0 f(1, b_1, \ldots, b_k) = 0$ and

$$g(b_0, b_1, \ldots, b_k) = \overline{b_0} f(0, b_1, \ldots, b_k) = 1 f(0, b_1, \ldots, b_k)$$
$$= f(0, b_1, \ldots, b_k) = f(b_0, b_1, \ldots, b_k).$$

Similarly, if $b_0 = 1$ then

$$g(b_0, b_1, \ldots, b_k) = b_0 f(1, b_1, \ldots, b_k) = f(b_0, b_1, \ldots, b_k). \qquad \square$$

Theorem 4.11 *Every Boolean function is polynomial.*

Proof. By induction on the number of arguments of f. A function $f(x)$ with one argument can be written, by the lemma, $\overline{x} f(0) + x f(1)$.

If $f(0)$ and $f(1)$ are both equal to 0, the function f is the zero function. If they are both equal to 1 we obtain $x + \overline{x}$. If only one of the two is 0, we obtain $f(x) = x$ or $f(x) = \overline{x}$.

Let us now assume that every Boolean function with k arguments is polynomial. A Boolean function with $k+1$ arguments can be written $\overline{x_0} f(0, x_1, \ldots, x_k) + x_0 f(1, x_1, \ldots, x_k)$. The Boolean functions $g(x_1, \ldots, x_k) = f(0, x_1, \ldots, x_k)$ and $g'(x_1, \ldots, x_k) = f(1, x_1, \ldots, x_k)$ are functions with k arguments; hence they are polynomial, and f is polynomial, too. $\qquad \square$

EXAMPLE 4.12 Let us come back to the function f of the preceding example, as given by its table. We have

$$f(x, y) = \overline{x} f(0, y) + x f(1, y),$$
$$f(0, y) = \overline{y} f(0, 0) + y f(0, 1) = \overline{y} + y,$$
$$f(1, y) = \overline{y} f(1, 0) + y f(1, 1) = \overline{y},$$

whence

$$f(x, y) = \overline{x}(\overline{y} + y) + x\overline{y} = \overline{x}\,\overline{y} + \overline{x}y + x\overline{y}$$
$$= \overline{x}y + (\overline{x} + x)\overline{y} = \overline{x}y + \overline{y}.$$

The polynomial form of a Boolean function given by its table can be found very easily. Let f be a function with n arguments. Let $D_f = \{(b_1, \ldots, b_n) \in \mathbb{B}^n \ / \ f(b_1, \ldots, b_n) = 1\}$. If D_f is empty then f is the zero function. Otherwise, to each element $\vec{b} = (b_1, \ldots, b_n)$ of D_f we associate the Boolean function $M_{\vec{b}}(x_1, \ldots, x_n)$ whose polynomial form is

$$x'_1 \cdots x'_n \text{ with } x'_i = \begin{cases} x_i & \text{if } b_i = 1, \\ \overline{x_i} & \text{if } b_i = 0. \end{cases}$$

We then have $f(x_1,\ldots,x_n) = \sum_{\vec{b} \in D_f} M_{\vec{b}}(x_1,\ldots,x_n)$. Indeed, since a product of elements of $\mathbb{B}$ can take the value 1 only when all its factors are 1, $M_{\vec{b}}(\vec{c}) = 1$ if and only if $\vec{b} = \vec{c}$ and, since a sum of elements of $\mathbb{B}$ takes the value 1 as soon as one of its elements is 1, $(\sum_{\vec{b} \in D_f} M_{\vec{b}})(\vec{c}) = 1$ if and only if $\vec{c} \in D_f$, and therefore if and only if $f(\vec{c}) = 1$.

4.3.2 Dual functions

Let f be a Boolean function with n arguments. Its *dual*, denoted by $\tilde{f}$, is the Boolean function with n arguments defined by

$$\tilde{f}(x_1,\ldots,x_n) = \overline{f(\overline{x_1},\ldots,\overline{x_n})}.$$

EXAMPLE 4.13 Letting $f(x,y) = x + y$, its dual $\tilde{f}$ is defined by

$$\tilde{f}(x,y) = \overline{\overline{x} + \overline{y}} = xy.$$

Let $f(x) = \overline{x}$, its dual is $\overline{\overline{\overline{x}}} = \overline{x}$.

Proposition 4.14
1. $\tilde{f} = g \implies \tilde{g} = f$,
2. If $g(x_1,\ldots,x_n) = f(f_1(x_1,\ldots,x_n),\ldots,f_k(x_1,\ldots,x_n))$ then
$$\tilde{g}(x_1,\ldots,x_n) = \tilde{f}(\tilde{f_1}(x_1,\ldots,x_n),\ldots,\tilde{f_k}(x_1,\ldots,x_n)).$$

Proof.
1. If $\tilde{f} = g$ then
$$g(x_1,\ldots,x_n) = \tilde{f}(x_1,\ldots,x_n) = \overline{f(\overline{x_1},\ldots,\overline{x_n})}$$

and

$$\tilde{g}(x_1,\ldots,x_n) = \overline{\tilde{f}(\overline{x_1},\ldots,\overline{x_n})} = \overline{\overline{f(\overline{\overline{x_1}},\ldots,\overline{\overline{x_n}})}} = f(x_1,\ldots,x_n).$$

2. If $g(x_1,\ldots,x_k) = f(f_1(x_1,\ldots,x_n),\ldots,f_k(x_1,\ldots,x_n))$ then
$$\tilde{g}(x_1,\ldots,x_n) = \overline{f(f_1(\overline{x_1},\ldots,\overline{x_n}),\ldots,f_k(\overline{x_1},\ldots,\overline{x_n}))},$$

but $f_i(\overline{x_1},\ldots,\overline{x_n}) = \overline{\tilde{f_i}(x_1,\ldots,x_n)}$, whence

$$\tilde{g}(x_1,\ldots,x_n) = \overline{f(\overline{\tilde{f_1}(x_1,\ldots,x_n)},\ldots,\overline{\tilde{f_k}(x_1,\ldots,x_n)})}$$
$$= \tilde{f}(\tilde{f_1}(x_1,\ldots,x_n),\ldots,\tilde{f_k}(x_1,\ldots,x_n)). \qquad \square$$

Given a function in polynomial form, to find its dual we simply have to substitute sums for products and products for sums.

EXAMPLE 4.15 The dual of $xy + \overline{y}$ is $(x+y)\overline{y}$ which can be simplified into $x\overline{y}$.

EXERCISE 4.6 Give the sixteen Boolean functions with two arguments in polynomial form. For each one of them, give the dual function. ◇

CHAPTER 5

LOGIC

In the present chapter we introduce some notions of logic (propositional calculus and predicate calculus). Logic is the cornerstone of mathematical reasoning; it is widely used within computer science. In addition to formalizing reasoning rules, logic also highlights the distinction between formal manipulations of strings of symbols and their meanings or interpretations.

The notions of logic introduced are basic. They aid proofs of program correctness (termination, loop invariants, etc.), and also the design of programs in general and, in particular, programs written in languages such as PROLOG that are directly derived from the predicate calculus.

In this chapter, we define propositional and predicate calculus, their syntax and their semantics, and a proof system that is sound and complete for each. We prove in detail the completeness theorem for propositional logic. We illustrate predicate calculus by showing how Herbrand models characterize the satisfiability of Horn clauses; this is the basis of the semantics of languages such as PROLOG.

We recommend in the strongest possible terms the following handbook, which is delivered together with a software program (for Macintosh or PC) of exercises and computer aided learning:

Jon Barwise, John Etchemendy, *The Language of First-order Logic: Tarski's world*, 2nd edition, CSLI lecture notes n° 23, Stanford (1991).

We also recommend:

René Lalement, *Computation as Logic*, Prentice Hall, London (1993).

Anil Nerode, Richard Shore, *Logic for Applications*, Springer-Verlag, Berlin (1993).

Raymond Smullyan, *What is the Name of this Book?*, Prentice Hall, London (1978).

5.1 Remarks on mathematical reasoning

A *proposition* is an assertion which is either true or false, but not both: for instance '$2 + 2 = 5$' is a false proposition, '$p \Longrightarrow p$' is a true proposition. On the other hand, a *formula* states a property of an object or a relation between objects, and may take the value true or false after values are assigned to the objects; for instance, '$2 + 2 = x$' takes the value true if we assign value 4 to x, and takes the value false for any other assignment to x, and '$p \Longrightarrow q$' may take values true or false according to the values assigned to p and q. A formula which is always true is called a *theorem*.

Let p and q be two propositions concerning the same objects. We say that p implies q, denoted by '$p \Longrightarrow q$', if, whenever p is true, q is also true: $p \Longrightarrow q$ is a theorem whose hypothesis is p and whose conclusion is q; the converse of $p \Longrightarrow q$ is $q \Longrightarrow p$, which is usually not a theorem (see Exercise 5.2).

EXAMPLE 5.1 Verify the truth of the following theorems (whose converses are false):

- $a = a'$ and $b = b'$ $\quad \Longrightarrow \quad a + b = a' + b'$, where a, a', b, b' are integers,
- $A \cap B = C$ $\quad \Longrightarrow \quad C \subseteq A$ and $C \subseteq B$,
- $A \cup B = C$ $\quad \Longrightarrow \quad A \subseteq C$ and $B \subseteq C$.

5.1.1 Some useful facts

(a) Implication is transitive: $[(p \Longrightarrow q) \text{ and } (q \Longrightarrow r)] \Longrightarrow (p \Longrightarrow r)$. This transitivity is the basis of deductive arguments.

(b) The negation of proposition p is denoted by $\bar{p}$ or $\neg p$. An implication $p \Longrightarrow q$ and the *contrapositive implication* $\bar{q} \Longrightarrow \bar{p}$ or $(\neg q \Longrightarrow \neg p)$ are two different ways of stating the same theorem. This fact is the basis of *proofs by contradiction*, where in order to prove $p \Longrightarrow q$, we assume p and $\bar{q}$ and we deduce a contradiction.

(c) The following propositions are equivalent:
- (i) $\quad p \Longrightarrow q$ and $q \Longrightarrow p$,
- (ii) $\quad p \Longrightarrow q$ and $\bar{p} \Longrightarrow \bar{q}$,
- (iii) $\quad p \Longleftrightarrow q$,
- (iv) $\quad \bar{p} \Longleftrightarrow \bar{q}$.

For instance, in order to prove: $ab = 0 \Longleftrightarrow (a = 0 \text{ or } b = 0)$ on $\mathbb{R}$, it suffices to prove: $ab \neq 0 \Longleftrightarrow (a \neq 0 \text{ and } b \neq 0)$.

(d) *Modus ponens* rule: $[p \text{ and } (p \Longrightarrow q)] \Longrightarrow q$.

EXERCISE 5.1 Verify that the *modus ponens* rule is equivalent to the *modus tollens* rule

$$[\neg q \text{ and } (p \Longrightarrow q)] \Longrightarrow \neg p,$$

i.e. that we can prove the *modus tollens* rule from the *modus ponens* rule, and vice-versa. $\diamond$

Remarks on mathematical reasoning 69

5.1.2 Some confusions to be avoided

(a) While $p \Longrightarrow q$ and the contrapositive implication $\neg q \Longrightarrow \neg p$ are indeed two different ways of asserting the same fact, the converse implication $q \Longrightarrow p$ usually asserts a totally different fact.

(b) If $\neg q \Longrightarrow \neg p$ is true, and q is true, p is not necessarily true. We refer to Section 5.1.1 (b) for the explanation of this fact.

(c) If $p \Longrightarrow q$ is false, this usually does not imply that the converse $q \Longrightarrow p$ is true. Compare this with the fact that $A \not\supseteq B$ usually does not imply that $A \subseteq B$.

EXERCISE 5.2
1. Let:
$$p = \text{'it rains'},$$
$$q = \text{'there are clouds'}.$$

Write the implication $p \Longrightarrow q$ together with its contrapositive, its converse and the contrapositive of its converse. Which implications are true?

2. We consider the formulas
$$p = (\forall x \in A, \exists y \in B, P(x,y)) \quad \text{and}$$
$$q = (\exists y \in B, \forall x \in A, P(x,y)),$$

where:
- A is the set of men,
- B is the set of women and
- $P(x,y)$ means 'y loves x'.

What can be said of $p \Longrightarrow q$? Of its converse?   ◇

5.1.3 Propositional calculus *versus* predicate calculus

The initial motivation of logic is the modelling of mathematical reasoning. This needs a clear distinction between syntax (the language, the formulas) and semantics (the interpretation of the language, the truth values true or false of formulas). The case of *propositional calculus* is the simplest case because the variables, i.e. propositions, can take only one of the two values true or false. First, we study propositional calculus in Section 5.2. However, propositional calculus is not able to model all mathematical reasonings: for instance, we cannot express in propositional calculus the existence of an object having a given property. *Predicate calculus* can express the properties of objects or the relations between objects, and can formalize mathematical reasoning. We study predicate calculus in Section 5.3.

5.2 Propositional calculus

In the remainder of this chapter, logic and its language will be the object of our study. Logic is omnipresent in mathematics as a tool for proofs, and as such is an element of the meta-language. In the present chapter we will therefore try to distinguish between the symbols of the logical language as an object of study and the symbols of the logical language as a tool of the meta-language. For instance, we will denote by $\supset$ the implication, considered as a formal symbol of the logical language, and by $\Longrightarrow$ the implication that is just a notation for the word 'implies' with its intuitive meaning.

One of the fundamental goals of logic is to write correct proofs. In order to reach that goal, the concept of consequence is essential: when can we safely assert, with no possibility of error, that a formula is consequence of a set of premises? In Section 5.2.2 we will define a notion of semantical consequence, in Section 5.2.3 establish a notion of provability or syntactical consequence and, finally, in Section 5.2.4, show that both notions coincide.

5.2.1 Syntax: formulas

Let $P = \{p, p', q, q', \ldots\}$ be a set of propositional symbols and let $\supset$, $\neg$ and left and right parentheses be symbols.

Definition 5.2 *A propositional formula is a string of symbols from $P \cup \{\supset, \neg, (,)\}$ defined by:*

1. *every propositional symbol of P is a formula,*
2. *if F is a formula then $\neg F$ is a formula,*
3. *if F and F' are two formulas then $(F \supset F')$ is a formula and*
4. *every formula is obtained by repeating a finite number of times the applications of steps 1–3.*

Z Note that Definition 5.2 is an example of an inductive definition of a set. We have already seen such definitions in Chapter 3, and will see others later (see Chapter 7).

Z Formulas are strings of symbols. They have no meaning whatsoever for the time being. The assignment of a meaning, i.e. a value 'true' or 'false' to a formula, constitutes the semantics of the formula and will be studied in Section 5.2.2.

Let σ be a mapping, called a *substitution*, from the set of propositional symbols P to the set of formulas. The formula $\sigma(F)$ obtained by substitution from formula F is defined by:

- if $F = p \in P$ then $\sigma(F) = \sigma(p)$,

Propositional calculus

- if $F = \neg F'$ then $\sigma(F) = \neg \sigma(F')$ and
- if $F = (F_1 \supset F_2)$ then $\sigma(F) = (\sigma(F_1) \supset \sigma(F_2))$.

EXAMPLE 5.3 Let σ be the substitution defined by $\sigma(p) = q$ and $\sigma(q) = (p \supset q)$. Then $\sigma(p \supset q) = (q \supset (p \supset q))$.

Definition 5.4 *A sequent is a pair $(\mathcal{F}, F)$ where $\mathcal{F}$ is a finite set of formulas and F is a formula.*

The intuition is that a sequent formalizes the notion of logical consequence: if all premises of the sequent are true, i.e. if all formulas of $\mathcal{F}$ are true, then its conclusion, formula F, is true.

5.2.2 Semantics: interpretation of formulas

Let F be a formula and let I be a mapping from the set of propositional symbols to the Boolean algebra $\mathbb{B} = \{1, 0\}$, equipped with its operations $+$, '$\cdot$' denoting product, and $^-$ (see Definition 4.2). The Boolean constants 'true', 'false', here identified by 1 and 0, are sometimes also denoted by $t\!t, f\!f$ or T, F.

We define the *truth value* $I(F)$ of formula F in I by:

- if $F = p \in P$ then $I(F) = I(p) \in \mathbb{B}$;
- if $F = \neg F'$ then $I(F) = \overline{I(F')}$;
- if $F = (F_1 \supset F_2)$ then $I(F) = \overline{I(F_1)} + I(F_2)$.

If $I(F) = 1$, we say that F is true in I. I is called an *interpretation*, and we also say that $I(F)$ is the interpretation of formula F.

EXERCISE 5.3 Let us define $F \wedge F' \stackrel{\text{def}}{=} \neg(F \supset \neg F')$ and $F \vee F' \stackrel{\text{def}}{=} \neg F \supset F'$.
1. Write the tables giving the truth values of $\wedge, \vee, \supset$. Deduce that

$$I(F \wedge F') = I(F).I(F') \quad \text{and} \quad I(F \vee F') = I(F) + I(F').$$

2. Show that

$$I\big(F_n \supset (F_{n-1} \supset (\cdots (F_1 \supset F) \cdots))\big) = \overline{I(F_n)} + \overline{I(F_{n-1})} + \cdots + \overline{I(F_1)} + I(F)$$
$$= I\big(\neg F_n \vee (\neg F_{n-1} \vee (\cdots \vee (\neg F_1 \vee F) \cdots))\big).$$

Deduce that
$$I\big(F_n \supset (F_{n-1} \supset (\cdots (F_1 \supset F) \cdots))\big) = I\big((F_n \wedge (F_{n-1} \wedge (\cdots \wedge (F_2 \wedge F_1)) \cdots)) \supset F\big). \diamond$$

Definition 5.5 *Let F be a formula. We say that*

- *F is valid, or that F is a tautology, if for all I, $I(F) = 1$,*
- *F is satisfiable, if there exists an I such that $I(F) = 1$ and*
- *F is unsatisfiable, if for all I, $I(F) = 0$.*

EXAMPLE 5.6 $p \wedge \neg p$ is unsatisfiable; $p \vee \neg p$ is a tautology; $p \wedge (p \supset q)$ is satisfiable but is not valid.

EXERCISE 5.4 Verify that F is unsatisfiable if and only if $\neg F$ is valid. ◇

Definition 5.7 *A sequent $(\mathcal{F}, G)$ is true in I if*

$$(\forall F \in \mathcal{F}, I(F) = 1) \Longrightarrow I(G) = 1.$$

A sequent $(\mathcal{F}, G)$ is valid if it is true in I for each I, i.e. $(\forall F \in \mathcal{F}, I(F) = 1) \Longrightarrow I(G) = 1$. We write $\mathcal{F} \models G$ to denote the fact that sequent $(\mathcal{F}, G)$ is valid.

This definition formalizes the notion of *semantical consequence*.

EXERCISE 5.5
1. Show that formula G is true (resp. valid) in I if sequent $(\emptyset, G)$ is true (resp. valid) in I.
2. Are the following sequents valid?
   - $(\emptyset, (p \supset q))$,
   - $(\{p, (p \supset q)\}, q)$. ◇

Proposition 5.8 *If σ is a substitution and if $\mathcal{F} \models G$ is a valid sequent, then $\sigma(\mathcal{F}) \models \sigma(G)$ is a valid sequent.*

Proof. If I is an interpretation, we define the interpretation I_σ by $I_\sigma(p) = I(\sigma(p))$. We deduce that $I_\sigma(F) = I(\sigma(F))$.

Let I be an interpretation such that for all F' in $\sigma(\mathcal{F})$, $I(F') = 1$. We thus have for all F in $\mathcal{F}$, $I(\sigma(F)) = I_\sigma(F) = 1$. Because $\mathcal{F} \models G$, $I_\sigma(G) = I(\sigma(G)) = 1$. □

Proposition 5.9 $\{F_n, \ldots, F_1\} \models F$ *if and only if*

$$\emptyset \models F_n \supset (F_{n-1} \supset (\cdots (F_1 \supset F) \cdots)).$$

Proof. It suffices to show that $\mathcal{F} \cup \{F\} \models G$ if and only if $\mathcal{F} \models (F \supset G)$, and the result will follow by induction on n. We have $I(F \supset G) = 1$ if and only if $I(F) = 1 \Longrightarrow I(G) = 1$.

Let I be an interpretation. We have

$$(I(F') = 1 \text{ for all } F' \in \mathcal{F}) \text{ and } I(F) = 1 \Longrightarrow I(G) = 1$$

if and only if $\big(I(F') = 1 \text{ for all } F' \text{ in } \mathcal{F}\big) \Longrightarrow I(F \supset G) = 1$. □

We will often write $\mathcal{F}, F \models G$ instead of writing $\mathcal{F} \cup \{F\} \models G$.

EXERCISE 5.6 We can associate a formula $\phi((\mathcal{F}, G))$ with a sequent $S = (\mathcal{F}, G)$ in the following fashion:
- If $S = (\emptyset, G)$, then $\phi(S) = G$.
- If $S = (\{F\} \cup \mathcal{F}, G)$ and $\phi((\mathcal{F}, G)) = F'$, then $\phi(S) = (F \supset F')$.

Show that sequent $S = (\mathcal{F}, G)$ is true in I if and only if $\phi(S)$ is true in I. ◇

Propositional calculus

Proposition 5.10
1. $\mathcal{F} \models F$ if and only if $\mathcal{F} \models \neg\neg F$; $\mathcal{F}, F \models G$ if and only if $\mathcal{F}, \neg\neg F \models G$.
2. If $\mathcal{F} \models \neg(F \supset F')$ then $\mathcal{F} \models F$ and $\mathcal{F} \models \neg F'$.
3. If $\mathcal{F}, (F \supset F') \models G$ then $\mathcal{F}, F' \models G$, $\mathcal{F}, \neg G \models \neg F'$, and $\mathcal{F}, \neg G \models F$.
4. If $\mathcal{F}, \neg(F \supset F') \models G$ then $\mathcal{F}, \neg G, F \models F'$.

Proof.
1. It is straightforward that $I(F) = I(\neg\neg F)$.
2. Let I be such that for all G in $\mathcal{F}$, $I(G) = 1$. Then $I(F \supset F') = 0$ and thus $I(F) = 1$ and $I(F') = 0$. Hence, $I(\neg F') = 1$.
3. Let I be such that for all H in $\mathcal{F}$, $I(H) = 1$. If $I(F') = 1$, then $I(F \supset F') = 1$ and thus $I(G) = 1$. If $I(G) = 0$, then $I(F \supset F') = 0$, whence $I(F') = 0$ and $I(F) = 1$.
4. Let I be such that for all H in $\mathcal{F}$, $I(H) = 1$. If $I(G) = 0$ and $I(F) = 1$ then it must also be the case that $I(F') = 1$. Otherwise, $I(F \supset F') = 0$, $I(\neg(F \supset F')) = 1$, and $I(G) = 1$, a contradiction. □

5.2.3 Logical proofs

Definition 5.11 *A sequent $(\mathcal{F}, F)$ is said to be provable, denoted by $\mathcal{F} \vdash F$, if it is built from a finite number of the following rules:*
- *use of a hypothesis rule: $F \in \mathcal{F} \implies \mathcal{F} \vdash F$,*
- *augmentation of the hypotheses: if $G \notin \mathcal{F}$ and $\mathcal{F} \vdash F$ then $\mathcal{F} \cup \{G\} \vdash F$,*
- *detachment rule (or modus ponens): if $\mathcal{F} \vdash (F \supset F')$ and if $\mathcal{F} \vdash F$ then $\mathcal{F} \vdash F'$,*
- *synthesis rule (or hypothesis withdrawal): if $\mathcal{F}, F \vdash F'$ then $\mathcal{F} \vdash (F \supset F')$,*
- *double negation rule: $\mathcal{F} \vdash F$ if and only if $\mathcal{F} \vdash \neg\neg F$,*
- *proof by contradiction rule: if $\mathcal{F}, F \vdash F'$ and $\mathcal{F}, F \vdash \neg F'$, then $\mathcal{F} \vdash \neg F$.*

This definition formalizes the notion of *logical consequence* for propositional calculus.

Z Provable sequents are characterized only by manipulations of strings of symbols, in contrast with valid sequents, which are characterized by their interpretation.

A *proof* of a provable sequent $\mathcal{F} \vdash F$ is a finite sequence of provable sequents $\mathcal{F}_i \vdash F_i$, the last of which is $\mathcal{F} \vdash F$, such that any sequent in this sequence is obtained by applying the above rules to preceding sequents in the sequence. The first sequent of a proof must thus be obtained by the use of a hypothesis.

EXAMPLE 5.12   We omit brackets for explicit sequents.

1. $p \vdash p$ (use of a hypothesis)
       $\emptyset \vdash (p \supset p)$ (synthesis)

2. $p, q \vdash p$ (hypothesis)
       $p \vdash (q \supset p)$ (synthesis)
       $\emptyset \vdash (p \supset (q \supset p))$ (synthesis)

3.   (i) $(p \supset q), \neg q, p \vdash \neg q$ (hypothesis)
       (ii) $(p \supset q), \neg q, p \vdash p$ (hypothesis)
       (iii) $(p \supset q), \neg q, p \vdash p \supset q$ (hypothesis)
       (iv) $(p \supset q), \neg q, p \vdash q$ (*modus ponens* on (ii) and (iii))
       (v) $(p \supset q), \neg q \vdash \neg p$ (contradiction on (i) and (iv))
       (vi) $(p \supset q) \vdash (\neg q \supset \neg p)$ (synthesis)

4. $p, \neg p, \neg q \vdash p$ (hypothesis)
       $p, \neg p, \neg q \vdash \neg p$ (hypothesis)
       $p, \neg p \vdash \neg\neg q$ (contradiction)
       $p, \neg p \vdash q$ (double negation)
       $p \vdash (\neg p \supset q)$ (synthesis)

5. $p \vdash p$ (hypothesis)
       $\emptyset \vdash (p \supset p)$ (synthesis)
       $p \vdash (p \supset p)$ (augmentation)
       $\emptyset \vdash p \supset (p \supset p)$ (synthesis)

Proposition 5.13 $\mathcal{F}, F \vdash G$ if and only if $\mathcal{F}, \neg\neg F \vdash G$.

Proof.

1. $\mathcal{F}, F \vdash G$
       $\mathcal{F} \vdash (F \supset G)$ (synthesis)
       $\mathcal{F}, \neg\neg F \vdash (F \supset G)$ (augmentation)
       $\mathcal{F}, \neg\neg F \vdash F$ (hypothesis + double negation)
       $\mathcal{F}, \neg\neg F \vdash G$ (*modus ponens*)

2. $\mathcal{F}, \neg\neg F \vdash G$
       $\mathcal{F} \vdash (\neg\neg F \supset G)$ (synthesis)
       $\mathcal{F}, F \vdash (\neg\neg F \supset G)$ (augmentation)
       $\mathcal{F}, F \vdash \neg\neg F$ (hypothesis + double negation)
       $\mathcal{F}, F \vdash G$ (*modus ponens*)   □

Propositional calculus 75

Proposition 5.14 $\mathcal{F}, F \vdash G$ *if and only if* $\mathcal{F}, \neg G \vdash \neg F$.

Proof. We prove that if $\mathcal{F}, F \vdash G$, then $\mathcal{F}, \neg G \vdash \neg F$.

    1. $\mathcal{F}, F \vdash G$
    2. $\mathcal{F}, F, \neg G \vdash G$ (augmentation)
    3. $\mathcal{F}, F, \neg G \vdash \neg G$ (hypothesis)
    4. $\mathcal{F}, \neg G \vdash \neg F$ (contradiction)

Conversely, a similar proof shows that if $\mathcal{F}, \neg G \vdash \neg F$ then $\mathcal{F}, F \vdash G$. □

Proposition 5.15 *If σ is a substitution and $\mathcal{F} \vdash F$, then $\sigma(\mathcal{F}) \vdash \sigma(F)$.*

Proof. By induction. Let $(\mathcal{F}_i \vdash F_i)_{i=1,\ldots,n}$ be a proof of $\mathcal{F} \vdash F$. Then $(\sigma(\mathcal{F}_i) \vdash \sigma(F_i))_{i=1,\ldots,n}$ is a proof of $\sigma(\mathcal{F}) \vdash \sigma(F)$.
(B) $\sigma(\mathcal{F}_1) \vdash \sigma(F_1)$ is a provable sequent since, because $(\mathcal{F}_1, F_1)$ is the first sequent of a proof, $F_1 \in \mathcal{F}_1$ and thus $\sigma(F_1) \in \sigma(\mathcal{F}_1)$.
(I) We assume that $(\sigma(\mathcal{F}_i) \vdash \sigma(F_i))_{i=1,\ldots,k}$ is a proof. We show by structural induction over the form of the rules in Definition 5.11 that $(\sigma(\mathcal{F}_i) \vdash \sigma(F_i))_{i=1,\ldots,k+1}$ is a proof.

- If $\mathcal{F}_{k+1} \vdash F_{k+1}$ is obtained by the use of a hypothesis rule, $\sigma(\mathcal{F}_{k+1}) \vdash \sigma(F_{k+1})$ is also obtained by the use of a hypothesis.
- If $\mathcal{F}_{k+1} \vdash F_{k+1}$ is obtained by augmentation of the hypotheses from $\mathcal{F}_i \vdash F_i$ then $\mathcal{F}_{k+1} = \mathcal{F}_i \cup \{G\}$ and $F_{k+1} = F_i$, and thus $\sigma(\mathcal{F}_{k+1}) = \sigma(\mathcal{F}_i) \cup \{\sigma(G)\}$ and $\sigma(F_{k+1}) = \sigma(F_i)$.
  - If $\sigma(G) \notin \sigma(\mathcal{F}_i)$, then $\sigma(\mathcal{F}_{k+1}) \vdash \sigma(F_{k+1})$ is obtained by augmentation of the hypotheses.
  - If $\sigma(G) \in \sigma(\mathcal{F}_i)$, then we will have two identical sequents: deleting the second one will again yield a proof.
- If there exist $i, j \leq k$ such that $\mathcal{F}_i = \mathcal{F}_j = \mathcal{F}_{k+1}$, $F_j = (F_i \supset F_{k+1})$, i.e. $\mathcal{F}_{k+1} \vdash F_{k+1}$ is obtained by *modus ponens*, then

$$\sigma(\mathcal{F}_i) = \sigma(\mathcal{F}_j) = \sigma(\mathcal{F}_{k+1}) \text{ and } \sigma(F_j) = \big(\sigma(F_i) \supset \sigma(F_{k+1})\big),$$

and $\sigma(\mathcal{F}_{k+1}) \vdash \sigma(F_{k+1})$ is also obtained by *modus ponens*.
- We proceed in the same way for the other rules. □

Proposition 5.16

$$\{F_1, \ldots, F_n\} \vdash F \text{ if and only if } \emptyset \vdash (F_1 \supset (F_2 \cdots (F_n \supset F) \cdots)).$$

Proof. It suffices to show that $\mathcal{F}, F \vdash G$ if and only if $\mathcal{F} \vdash F \supset G$.

The 'only if' direction is true because of the synthesis rule. For the opposite direction, if $\mathcal{F} \vdash F \supset G$ then $\mathcal{F}, F \vdash F \supset G$ by augmentation of the hypotheses, $\mathcal{F}, F \vdash F$ by the use of a hypothesis rule and $\mathcal{F}, F \vdash G$ by *modus ponens*. □

5.2.4 Syntax and semantics

We will show that valid sequents and provable sequents coincide.

Theorem 5.17 (Soundness) *Every provable sequent is valid.*

Proof. By induction on the lengths of proofs. It suffices to show that each application of one of the rules given in Definition 5.11 generates only valid sequents from valid sequents. To this end, it suffices to verify that each rule of the form 'if $S_1, \ldots, S_n$, then S' of Definition 5.11 is valid, i.e. that if $S_1, \ldots, S_n$ are valid sequents, then S is also a valid sequent.

- If $\mathcal{F} \vdash G$ is obtained by use of a hypothesis, then $G \in \mathcal{F}$, and if for any F' in $\mathcal{F}$, if $I(F') = 1$, then $I(G) = 1$, and thus $\mathcal{F} \models G$.
- If $\mathcal{F} \cup \{F\} \vdash G$ is obtained by augmentation of the hypotheses, then $F \notin \mathcal{F}$ and $\mathcal{F} \vdash G$. By the induction $\mathcal{F} \models G$, and thus

$$(\forall F' \in \mathcal{F} \cup \{F\}, I(F') = 1) \implies (\forall F' \in \mathcal{F}, I(F') = 1) \quad \text{and}$$
$$(\forall F' \in \mathcal{F}, I(F') = 1) \implies I(G) = 1.$$

Hence $\mathcal{F}, F \models G$.
- If $\mathcal{F} \vdash G$ is obtained by *modus ponens*, then $\mathcal{F} \vdash (F \supset G)$, $\mathcal{F} \vdash F$, and by the induction hypothesis $\mathcal{F} \models (F \supset G)$, $\mathcal{F} \models F$; then, (for all $F' \in \mathcal{F}, I(F') = 1$) implies $I(F \supset G) = 1$ and $I(F) = 1$, whence $I(G) = 1$, and thus $\mathcal{F} \models G$.
- If $\mathcal{F} \vdash G$ is obtained by synthesis, then $G = (F \supset F')$, $\mathcal{F}, F \vdash F'$, and by the induction hypothesis $\mathcal{F}, F \models F'$. If for all $H \in \mathcal{F}, I(H) = 1$, then
    - if $I(F) = 1$, $I(F') = 1$ and $I(F \supset F') = 1$ and
    - if $I(F) = 0$, $I(F \supset F') = 1$.

Hence $\mathcal{F} \models (F \supset F')$.
- If $\mathcal{F} \vdash G$ is obtained by double negation, then $\mathcal{F} \models G$ if and only if $\mathcal{F} \models \neg\neg G$ (because $I(G) = I(\neg\neg G)$).
- If $\mathcal{F} \vdash G$ is obtained by a proof by contradiction, then $G = \neg F$, and by the induction hypothesis $\mathcal{F}, F \models F'$ and $\mathcal{F}, F \models \neg F'$. If for all $H \in \mathcal{F}, I(H) = 1$, then $I(F) = 1$ cannot occur (otherwise we would have $I(F') = I(\neg F') = 1$, a contradiction). Hence, we must have $I(F) = 0$ and thus $\mathcal{F} \models G$. □

Theorem 5.18 (Completeness) *Every valid sequent is provable.*

Proof. Let us define the *weight* of a sequent as the sum of the number of $\neg$ symbols and twice the number of $\supset$ symbols occurring in this sequent. We argue by induction on the weight of a sequent.

- If the weight of a sequent is zero, then this sequent can be written
$$\{p_1, \cdots, p_n\} \models p.$$

Propositional calculus

If $p \notin \{p_1, \cdots, p_n\}$, then the interpretation I defined by $I(p_i) = 1$, $I(p) = 0$ shows that this sequent is not valid. We thus have $p \in \{p_1, ..., p_n\}$, and the sequent is provable by use of a hypothesis rule.

- Let thus $\mathcal{F} \models F$ have weight $n+1$.

(a) If $F = \neg\neg F'$ then $\mathcal{F} \models F'$ and, because that sequent has weight $n-1$, then by the induction hypothesis $\mathcal{F} \vdash F'$ and thus $\mathcal{F} \vdash \neg\neg F'$ by double negation.

(b) If F is not of the form $\neg\neg F'$ and contains at least one symbol $\supset$ then $F = (F' \supset F'')$ or $F = \neg(F' \supset F'')$.

(b.1) If $F = (F' \supset F'')$ then $\mathcal{F} \models (F' \supset F'')$ implies $\mathcal{F}, F' \models F''$. The last sequent has weight $n-1$, and so we obtain $\mathcal{F} \vdash (F' \supset F'')$ (synthesis rule).

(b.2) If $F = \neg(F' \supset F'')$, then as we have seen, $\mathcal{F} \models \neg(F' \supset F'')$ implies $\mathcal{F} \models F'$ and $\mathcal{F} \models \neg F''$. These sequents have weight $\leq n$, and hence $\mathcal{F} \vdash F'$ and $\mathcal{F} \vdash \neg F''$. We thus have:

$\mathcal{F}, (F' \supset F'') \vdash F'$ (augmentation)
$\mathcal{F}, (F' \supset F'') \vdash \neg F''$ (augmentation)
$\mathcal{F}, (F' \supset F'') \vdash (F' \supset F'')$ (hypothesis)
$\mathcal{F}, (F' \supset F'') \vdash F''$ (modus ponens)
$\mathcal{F} \vdash \neg(F' \supset F'')$ (contradiction)

(c) If F is not of the form $\neg\neg F'$ and contains no $\supset$ symbol then $F = r$ or $F = \neg r$. Since $\mathcal{F}, \neg\neg F \models G$ if and only if $\mathcal{F}, F \models G$, we may assume that the elements of $\mathcal{F}$ have one of the four following forms: $p, \neg p, (F_1 \supset F_2), \neg(F_1 \supset F_2)$.

(c.1) If $\mathcal{F}$ contains a formula $\neg(F_1 \supset F_2)$ then

$$\mathcal{F}', \neg(F_1 \supset F_2) \models F$$

implies $\mathcal{F}', \neg F, F_1 \models F_2$. We can apply the induction hypothesis: $\mathcal{F}', \neg F, F_1 \vdash F_2$. Hence:

$\mathcal{F}', \neg F \vdash (F_1 \supset F_2)$ (synthesis)
$\mathcal{F}', \neg F, \neg(F_1 \supset F_2) \vdash (F_1 \supset F_2)$ (augmentation)
$\mathcal{F}', \neg F, \neg(F_1 \supset F_2) \vdash \neg(F_1 \supset F_2)$ (hypothesis)
$\mathcal{F}', \neg(F_1 \supset F_2) \vdash F$ (contradiction + double negation)

(c.2) If $\mathcal{F}$ contains a formula $(F_1 \supset F_2)$ then $\mathcal{F}', (F_1 \supset F_2) \models F$ implies $\mathcal{F}', \neg F \models F_1$ (Proposition 5.10, 3) and $\mathcal{F}', F_2 \models F$ which have weight $\leq n$.

We can apply the induction hypothesis: $\mathcal{F}', \neg F \vdash F_1$, $\mathcal{F}', F_2 \vdash F$. Hence:

1. $\mathcal{F}', \neg F \vdash F_1$
2. $\mathcal{F}', F_2 \vdash F$
3. $\mathcal{F}', (F_1 \supset F_2), \neg F \vdash \neg F$ (hypothesis)
4. $\mathcal{F}' \vdash (F_2 \supset F)$ (synthesis on 2)
5. $\mathcal{F}', (F_1 \supset F_2), \neg F \vdash (F_2 \supset F)$ (augmentation on 4)
6. $\mathcal{F}', (F_1 \supset F_2), \neg F \vdash F_1$ (augmentation on 1)
7. $\mathcal{F}', (F_1 \supset F_2), \neg F \vdash (F_1 \supset F_2)$ (hypothesis)
8. $\mathcal{F}', (F_1 \supset F_2), \neg F \vdash F_2$ (*modus ponens* 6,7)
9. $\mathcal{F}', (F_1 \supset F_2), \neg F \vdash F$ (*modus ponens* 5,8)
10. $\mathcal{F}', (F_1 \supset F_2) \vdash F$ (contradiction on 3, 9, + double negation)

(c.3) The problem is thus reduced to the case in which $\mathcal{F}$ contains only formulas of the form p or $\neg p$. Let us write $\mathcal{F}$ in the form $\mathcal{F}^+ \cup \mathcal{F}^-$ with $\mathcal{F}^+$ equal to the set of formulas of $\mathcal{F}$ of the form 'p', and $\mathcal{F}^-$ the set of formulas of the form '$\neg p$'. Let P^+ be the set of all propositional symbols occurring in $\mathcal{F}^+$ and let P^- be the set of all propositional symbols occurring in $\mathcal{F}^-$.

(c.3.1)   If $P^+ \cap P^- \neq \emptyset$ then $\mathcal{F} = \mathcal{F}', p, \neg p$; we deduce

$$\left. \begin{array}{l} \mathcal{F}, \neg F \vdash p \\ \mathcal{F}, \neg F \vdash \neg p \end{array} \right\} \quad \text{and thus} \quad \mathcal{F} \vdash F.$$

(c.3.2)   We assume that $P^+ \cap P^- = \emptyset$, and we let r be the propositional symbol occurring in F.

(c.3.2.1) If $r \notin P^+ \cup P^-$, we could find an interpretation I true on $\mathcal{F}$ and false on F which is impossible.

(c.3.2.2) If $r \in P^+$, then any interpretation true on $\mathcal{F}$ verifies $I(r) = 1$. As we then have $I(F) = 1$, then $F = r$ by necessity and $\mathcal{F} \vdash r$ by use of a hypothesis.

(c.3.2.3) If $r \in P^-$, then any interpretation true on $\mathcal{F}$ verifies $I(r) = 0$. Thus $F = \neg r$ and $\mathcal{F} \vdash \neg r$ by use of the hypothesis.   □

Grouping together Theorem 5.17 and Theorem 5.18, we deduce the following corollary.

Corollary 5.19 $\mathcal{F} \models F$ *if and only if* $\mathcal{F} \vdash F$.

Propositional calculus

5.2.5 Additional logical connectors

In propositional logic we can also use the connectors $\wedge$ (and) and $\vee$ (or).

Formulas are then defined by the additional rule: if F and F' are formulas then $(F \wedge F')$ and $(F \vee F')$ are formulas.

The interpretation of these formulas is defined by adding, see Exercise 5.3,
$I(F \wedge F') = I(F) \cdot I(F')$,
$I(F \vee F') = I(F) + I(F')$,
so that
$I(F \wedge F') = I(\neg(F \supset \neg F'))$,
$I(F \vee F') = I(\neg F \supset F')$.

Similarly, the definitions of provable sequents are extended by adding the rules:

- if $\mathcal{F} \vdash F$ and $\mathcal{F} \vdash F'$ then $\mathcal{F} \vdash (F \wedge F')$,
- if $\mathcal{F} \vdash (F \wedge F')$ then $\mathcal{F} \vdash F$,
- if $\mathcal{F} \vdash (F \wedge F')$ then $\mathcal{F} \vdash F'$,
- if $\mathcal{F}, G \vdash F$ and $\mathcal{F}, \neg G \vdash F'$ then $\mathcal{F} \vdash (F \vee F')$,
- if $\mathcal{F}, F \vdash G$ and $\mathcal{F}, F' \vdash G$ then $\mathcal{F}, (F \vee F') \vdash G$.

We deduce the following proposition.

Proposition 5.20
1. $(F \wedge F') \vdash \neg(F \supset \neg F')$,
2. $\neg(F \supset \neg F') \vdash (F \wedge F')$,
3. $(F \vee F') \vdash (\neg F \supset F')$,
4. $(\neg F \supset F') \vdash (F \vee F')$.

Proof. We prove in detail 2 and 3; 1 and 4 are simpler, and we just sketch their proofs.

1. $(F \wedge F'), (F \supset \neg F') \vdash F'$ (third rule for $\wedge$)
$(F \wedge F'), (F \supset \neg F') \vdash F$ (second rule for $\wedge$)
$(F \wedge F'), (F \supset \neg F') \vdash (F \supset \neg F')$ (hypothesis)
$(F \wedge F'), (F \supset \neg F') \vdash \neg F'$ (*modus ponens*)
$(F \wedge F') \vdash \neg(F \supset \neg F')$ (contradiction)

2. $\neg(F \supset \neg F'), \neg F' \vdash (F \supset \neg F')$ (augmentation of $p \vdash (q \supset p)$, see Example 5.12, 2)
$\neg(F \supset \neg F'), \neg F' \vdash \neg(F \supset \neg F')$ (hypothesis)
$\neg(F \supset \neg F') \vdash F'$ (contradiction)

$$\neg(F \supset \neg F'),\, \neg F \vdash \neg(F \supset \neg F') \quad \text{(hypothesis)}$$
$$\neg(F \supset \neg F'),\, \neg F,\, F,\, F' \vdash \neg F \quad \text{(hypothesis)}$$
$$\neg(F \supset \neg F'),\, \neg F,\, F,\, F' \vdash F \quad \text{(hypothesis)}$$
$$\neg(F \supset \neg F'),\, \neg F,\, F \vdash \neg F' \quad \text{(contradiction)}$$
$$\neg(F \supset \neg F'),\, \neg F \vdash (F \supset \neg F') \quad \text{(synthesis)}$$
$$\neg(F \supset \neg F') \vdash F \quad \text{(contradiction + double negation)}$$
$$\neg(F \supset \neg F') \vdash (F \wedge F') \quad \text{(first rule for } \wedge)$$

3. $\quad F' \vdash (\neg F \supset F') \quad \text{(Example 5.12, 2)}$
$$F,\, \neg F,\, \neg F' \vdash F \quad \text{(hypothesis)}$$
$$F,\, \neg F,\, \neg F' \vdash \neg F \quad \text{(hypothesis)}$$
$$F,\, \neg F \vdash \neg\neg F' \quad \text{(contradiction)}$$
$$F,\, \neg F \vdash F' \quad \text{(double negation)}$$
$$F \vdash (\neg F \supset F') \quad \text{(synthesis)}$$
$$(F \vee F') \vdash (\neg F \supset F') \quad \text{(second rule for } \vee)$$

4. $\quad (\neg F \supset F'),\, \neg F \vdash F' \quad \text{(twice hypothesis + } \textit{modus ponens}\text{)}$
$$(\neg F \supset F'),\, F \vdash F \quad \text{(hypothesis)}$$
$$(\neg F \supset F') \vdash F \vee F' \quad \text{(first rule for } \vee)\ \square$$

We now define the transformation η which suppresses the symbols $\wedge$ and $\vee$ from a formula:

$$\eta(p) = p,$$
$$\eta(\neg F) = \neg \eta(F),$$
$$\eta(F \supset F') = (\eta(F) \supset \eta(F')),$$
$$\eta(F \wedge F') = \neg(\eta(F) \supset \neg \eta(F')),$$
$$\eta(F \vee F') = (\neg \eta(F) \supset \eta(F')).$$

It is easy to see that $\mathcal{F} \models F$ if and only if $\eta(\mathcal{F}) \models \eta(F)$ and, using the preceding property, we show that if $\eta(\mathcal{F}) \vdash \eta(F)$ is a provable sequent which can be proved without the rules concerning $\wedge$ and $\vee$, then $\mathcal{F} \vdash F$ is a provable sequent that can be proved using these rules.

Similarly, we can introduce the equivalence symbol $\equiv$ whose interpretation is given by:

$$I(F \equiv F') = 1 \text{ if and only if } I(F) = I(F').$$

Propositional calculus

So that $I(F \equiv F') = I\bigl((F \supset F') \wedge (F' \supset F)\bigr)$.

EXERCISE 5.7 We consider the set of formulas as an algebra equipped with the binary operations $\wedge, \vee, \supset$ and with the unary operation $\neg$. Is the equivalence relation $\Longleftrightarrow$ defined by $F \Longleftrightarrow F'$ if and only if $I(F \equiv F') = 1$ a congruence? ◇

The proof rules associated with the equivalence symbol $\equiv$ are:

- if $\mathcal{F} \vdash (F \equiv F')$ then $\mathcal{F} \vdash (F \supset F')$ and $\mathcal{F} \vdash (F' \supset F)$,
- if $\mathcal{F} \vdash (F \supset F')$ and $\mathcal{F} \vdash (F' \supset F)$ then $\mathcal{F} \vdash (F \equiv F')$,

or in other words:

$$\mathcal{F} \vdash (F \equiv F') \quad \text{if and only if} \quad \mathcal{F} \vdash ((F \supset F') \wedge (F' \supset F)).$$

The 'meta-logical' use of symbol $\Longleftrightarrow$ can then be formalized by: $F \Longleftrightarrow F'$ if and only if $(F \equiv F')$ is a valid formula, or in other words if and only if $\vdash (F \equiv F')$.

The operations $\vee$ and $\wedge$ enjoy associativity, commutativity, and distributivity properties similar to those of Boolean algebras:

1. Distributivity of $\wedge$ over $\vee$: $F \wedge (G \vee H) \Longleftrightarrow (F \wedge G) \vee (F \wedge H)$.
2. Distributivity of $\vee$ over $\wedge$: $F \vee (G \wedge H) \Longleftrightarrow (F \vee G) \wedge (F \vee H)$.
3. Associativity of $\wedge$: $F \wedge (G \wedge H) \Longleftrightarrow (F \wedge G) \wedge H$.
4. Associativity of $\vee$: $F \vee (G \vee H) \Longleftrightarrow (F \vee G) \vee H$.
5. Commutativity of $\wedge$: $F \wedge G \Longleftrightarrow G \wedge F$.
6. Commutativity of $\vee$: $F \vee G \Longleftrightarrow G \vee F$.

The associativity properties allow us to omit parentheses.

The following equivalences are quite useful:

$$\begin{aligned} F \supset G &\Longleftrightarrow \neg F \vee G, \\ F \supset G &\Longleftrightarrow \neg G \supset \neg F, \\ \neg(F \supset G) &\Longleftrightarrow F \wedge \neg G, \\ F \equiv G &\Longleftrightarrow (F \supset G) \wedge (G \supset F), \\ F \equiv G &\Longleftrightarrow (F \wedge G) \vee (\neg G \wedge \neg F). \end{aligned}$$

EXERCISE 5.8 A logician tells his son: 'if you don't eat porridge, you won't watch television'; the son eats porridge, and is sent straight to bed. What was the error which caused him to expect watching television after dinner? ◇

EXERCISE 5.9
1. A logician, who is assumed to always tell the truth, is interviewed about his feelings, and says both the following statements:
    (a)   I love Mary or I love Anne.
    (b)   If I love Mary, then I love Anne.

What can you conclude: does he love Mary, Anne or both?

2. Assume the same logician had answered the question: 'Is it true that if you love Mary, then you love Anne?' by both of the following statements:
    (a)   If it's true, then I love Mary.
    (b)   If I love Mary, then it's true.
What would you conclude?

5.2.6 Deductive systems

In order to define provable sequents we defined manipulation rules for strings of symbols. There are other systems of rules that can obtain the same result.

First, we say that formula F is *provable* if and only if $\emptyset \vdash F$ is a provable sequent. By the preceding theorems, a formula is provable if and only if it is valid.

We will now show an example of another way of proving formulas containing only propositional symbols and the symbols $\supset$ and $\neg$. The formulas that we can prove with rules will be called 'logical theorems' (to distinguish them from provable formulas).

Let p, q, r be three arbitrary propositional symbols.

(i) The following three formulas, called *axioms*, are logical theorems:
- $(p \supset (q \supset p))$,
- $((p \supset (q \supset r)) \supset ((p \supset q) \supset (p \supset r)))$,
- $((\neg p \supset \neg q) \supset ((\neg p \supset q) \supset p))$.

(ii) If σ is a substitution and if F is a logical theorem, then $\sigma(F)$ is a logical theorem.

(iii) If F and $(F \supset F')$ are logical theorems, then F' is a logical theorem.

(ii) and (iii) are *rules of inference*; (iii) is called the *modus ponens* rule.

A *deduction* of formula F from the set of formulas $\mathcal{F}$ is a finite sequence of formulas $F_1, F_2, \ldots, F_n$ such that

- F_n is identical to F
- for each $i \leq n$,
  - either F_i is one of the axioms (i),
  - or $F_i \in \mathcal{F}$,
  - or F_i can be deduced from the preceding F_js by an application of one of the rules (ii) or (iii).

Formula F is a logical theorem if and only if there is a deduction of F from the empty set of formulas $\mathcal{F} = \emptyset$.

It is easy to see that every logical theorem is a valid formula (and thus a provable formula); this shows the soundness of the system for deducing logical theorems. Conversely, we can show the completeness of the system, i.e. that every valid formula is a logical theorem. The completeness is harder to prove.

Using other systems of rules, we might define other sets of provable formulas which may or may not coincide with valid formulas.

Deductive systems are systems of rules which enable us to define sets of formulas **included** in the set of valid formulas.

5.3 First order predicate calculus

A 'predicate' is an assertion about objects that may be true or false according to the objects to which it is applied. For instance, 'to be an even number' is true when applied to '2' and false when applied to '3'. A predicate can also be applied to several objects, for instance 'to be less than'. This assertion is true for the pair (2,3) and false for the pair (3,2).

Predicate calculus enables us to build complex statements from predicates. For instance, 'every prime number strictly greater than 2 is odd', which will be formally written as

$$\forall x \quad \bigl(\text{Prime}(x) \wedge x > 2\bigr) \Longrightarrow \text{odd}(x) \,.$$

Such complex statements may also be true or false.

5.3.1 Syntax: first order formulas

Let $\mathcal{G}$ be a set of *function symbols*. With each symbol f of $\mathcal{G}$ is associated an arity (or rank) $\rho(f) \in \mathbb{N}$. If $\rho(f) = 0$, then f is called a constant. Let $C = \{a, b, \ldots, a', b', \ldots, a_1, b_1, \ldots\} \subset \mathcal{G}$ be the set of constants.

Let $\mathcal{R}$ be a set of *relational symbols*. With each symbol R of $\mathcal{R}$ is associated an arity (or rank) $\rho(R) \in \mathbb{N}$. If $\rho(R) = 0$, then R is also called a propositional symbol.

Let $X = \{x, y, \ldots, x', y', \ldots, x_0, y_0, x_1, y_1, \ldots\}$ be a set of variables.

Define the language $\mathcal{L} = \mathcal{R} \cup \mathcal{G}$. We also consider the symbols $\supset, \neg, \wedge, \vee$ of propositional logic, and two symbols $\forall$ and $\exists$, called universal and existential quantifiers, together with both parentheses and the comma.

Recall that the set T of *terms built on* $\mathcal{G} \cup X$ is inductively defined by:
(B) $C \cup X \subseteq T$,
(I) for any n-ary f in $\mathcal{G}$, and for any $t_1, \ldots, t_n$ in T, $f(t_1, \ldots, t_n) \in T$.

A *ground term* is a variable-free term, i.e. a term built on $\mathcal{G}$.

First order formulas on $\mathcal{L}$ are inductively defined by:
- If R is an arity n relational symbol, and if $t_1, \ldots, t_n \in T$, then $R(t_1, \ldots, t_n)$ is a formula, called an *atomic formula*.

- If F and F' are formulas, then $\neg F$, $(F \supset F')$, $(F \wedge F')$, and $(F \vee F')$ are formulas.
- If F is a formula and x is a variable, then $\forall x F$ and $\exists x F$ are formulas.

EXAMPLE 5.21
1. $F = (\forall x \exists y R(x,y) \supset \exists x R'(x,y,a))$.
2. Because $\mathcal{R}$ may contain arity 0 relational symbols, propositional calculus is a 'subcalculus' of predicate calculus. Every propositional formula is thus a first order formula because, on the one hand, propositional symbols are arity 0 relational symbols and, on the other hand, all other symbols of propositional calculus are also symbols of predicate calculus.

Definition 5.22 *An occurrence of a variable x in a formula F is a pair (x,n) such that the nth symbol of F is x and the $(n-1)$th symbol is neither $\forall$ nor $\exists$.*

EXAMPLE 5.23 $(x,8)$ and $(x,17)$ are the two occurrences of x in the above formula F, $(x,7)$ and $(x,14)$ are not occurrences: $(x,7)$ because the 7th symbol of F is not an x, and $(x,14)$ because the 14th symbol of F, which indeed is an x, is quantified by $\exists$.

Let F be a formula. The set $SF(F)$ of the subformulas of F is the set of pairs (n, F') with $n \in \mathbb{N}$ and where

- F' is a consecutive sequence of symbols from F which is itself a formula,
- n is the occurrence of the first symbol of F' in F.

EXAMPLE 5.24 The subformulas of

$$(\forall x \exists y R(x,y) \supset \exists x R'(x,y,a))$$

are $(1, F)$, $(2, \forall x \exists y R(x,y))$, $(4, \exists y R(x,y))$, $(6, R(x,y))$, $(13, \exists x R'(x,y,a))$, and $(15, R'(x,y,a))$.

Formulas can be represented by trees; for instance the formula

$$(\forall x \exists y R(x,y) \supset \exists x R'(x,y,a))$$

is represented by the tree t depicted in Figure 5.1.

With each node of t labelled by a relational symbol, a quantifier, or one of the symbols $\supset, \neg, \wedge, \vee$, is associated a subtree t' of t; each subtree t' represents a subformula of F. The subformulas of $(\forall x \exists y R(x,y) \supset \exists x R'(x,y,a))$ are depicted in Figure 5.2.

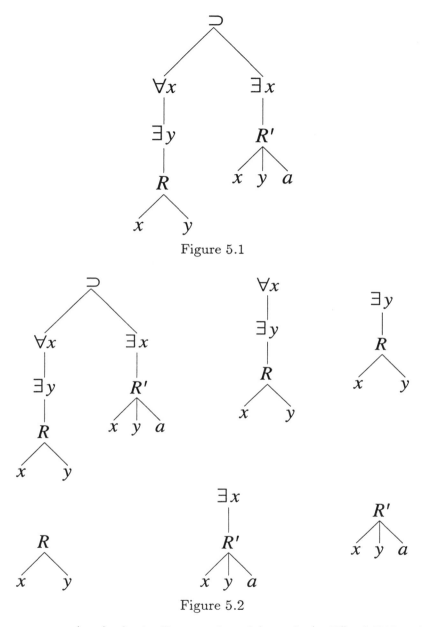

Figure 5.1

Figure 5.2

An occurrence (x, n) of x in F occurs in subformula (p, F') of F if and only if $p \leq n \leq p + |F'|$, where $|F'|$ denotes the number of symbols of F'.

EXAMPLE 5.25 $(x, 8)$ occurs in $(1, F)$, in $(2, \forall x \exists y R(x, y))$, in $(4, \exists y R(x, y))$ and in $(6, R(x, y))$.

$(y, 19)$ occurs in subformulas $(1, F)$, $(13, \exists x R'(x, y, a))$, and $(15, R'(x, y, a))$.

Definition 5.26 *An occurrence (x, n) of variable x in formula F is said to be bound if it occurs in a subformula (p, QxF'), where $Q \in \{\forall, \exists\}$. Otherwise it is said to be free.*

Variable x is said to be free in formula F if it has at least one free occurrence.

EXAMPLE 5.27 In Example 5.21, occurrences $(x, 8), (x, 17)$ and $(y, 10)$ are bound; occurrence $(y, 19)$ is free.

EXERCISE 5.10 What are the free variables and the free occurrences of variables in the following formulas:
- $\exists x \, \big(\text{logician}(x) \wedge \text{astute}(x)\big)$,
- $\big(\exists x \, \text{logician}(x)\big) \wedge \text{astute}(x)$. ◇

Let $f(F)$ be the set of free variables of F.

Proposition 5.28
- $f(R(t_1, \ldots, t_n)) = \{u_i \,/\, u_i \in X \text{ and } u_i \text{ occurs in } R(t_1, \ldots, t_n)\}$,
- $f(\neg F) = f(F)$,
- $f(F \supset F') = f(F \wedge F') = f(F \vee F') = f(F) \cup f(F')$ and
- $f(\forall x F) = f(\exists x F) = f(F) \backslash \{x\}$.

Proof. Simple: by structural induction on F (see Proposition 3.11). □

5.3.2 Semantics : Interpretation of formulas

Let $\mathcal{R}$ be a set of relational symbols and $\mathcal{G}$ a set of function symbols. Let $\mathcal{L}$ be the language $\mathcal{L} = \mathcal{R} \cup \mathcal{G}$. The language $\mathcal{L}$ will have many possible interpretations, each tailored for some domain of discourse. In order to interpret the language $\mathcal{L}$ we must specify the domain of discourse, together with the intended meanings of the predicate and function symbols; this is done by defining an $\mathcal{L}$-structure.

Definition 5.29 *An $\mathcal{L}$-structure is a triple $S = \langle E, \gamma, h \rangle$, where*

- *E is a non-empty set,*
- *γ is a mapping associating with each $R \in \mathcal{R}$ a subset $\gamma(R)$, also denoted by R_S, of $E^{\rho(R)}$ and*
- *h is a mapping associating with each $f \in \mathcal{G}$ a function $h(f) = f_S$ from $E^{\rho(R)}$ to E. (With each constant $a \in C$, h associates an element $h(a) = a_S$ of E.)*

Note that, here, E^0 has, by definition, a single element (for the same reason as $n^0 = 1$!). Hence, $\mathcal{P}(E^0)$ has only two subsets $\emptyset$ and E^0 and may be identified with the Boolean algebra with two elements.

First order predicate calculus

A valuation v is a mapping from the set of variables to E. Two valuations v and v' are congruent on a subset Y of X, which is denoted by $v \underset{Y}{=} v'$, if: for all $x \in Y$, $v(x) = v'(x)$.

Definition 5.30

(i) If t is a term and v a valuation, we define $v^*(t) \in E$ by:
- If $t = a \in C$, then $v^*(t) = a_S$.
- If $t = u \in X$, then $v^*(t) = v(u)$.
- If $t = f(t_1, \ldots, t_n)$, then $v^*(t) = f_S(v^*(t_1), \ldots, v^*(t_n))$.

(ii) If F is a formula and v a valuation, F can be assigned a unique truth value $\bar{v}(F) \in \mathbb{B}$ defined by:
- If $F = R(t_1, \ldots, t_n)$, then $\bar{v}(F) = 1$ if and only if $(v^*(t_1), \ldots, v^*(t_n)) \in R_S$. Note that if R has arity 0, then

$$\bar{v}(F) = \begin{cases} 1 & \text{if } R_S \neq \emptyset, \\ 0 & \text{if } R_S = \emptyset. \end{cases}$$

- $\bar{v}(\neg F) = \overline{\bar{v}(F)}$.
- $\bar{v}(F \supset F') = 1$ if and only if $\bar{v}(F) \leq \bar{v}(F')$.
- $\bar{v}(F \wedge F') = 1$ if and only if $\bar{v}(F) = 1$ and $\bar{v}(F') = 1$.
- $\bar{v}(F \vee F') = 1$ if and only if $\bar{v}(F) = 1$ or $\bar{v}(F') = 1$.
- $\bar{v}(\forall x F) = 1$ if and only if for all v' such that $v' \underset{X-\{x\}}{=} v$, we have $\bar{v}'(F) = 1$.
- $\bar{v}(\exists x F) = 1$ if and only if there exists v' such that $v' \underset{X-\{x\}}{=} v$ and $\bar{v}'(F) = 1$.

(iii) Two formulas F and F' are equivalent if, for any $\mathcal{L}$-structure S and for any valuation v, $\bar{v}(F) = \bar{v}(F')$. We write $F \approx F'$.

This semantics agrees with the semantics already given for propositional calculus. We have seen in Example 5.21 that any formula F of propositional calculus is a formula of predicate calculus. If S is an $\mathcal{L}$-structure, the restriction I of S to propositional symbols is a mapping from these propositional symbols to the Boolean algebra, and is thus an interpretation in the sense of propositional calculus; and we indeed have for any propositional formula F that $I(F) = \bar{v}(F)$ for any valuation v with values in S. In short, interpretations I that we have considered for propositional logic can be obtained as restrictions of $\mathcal{L}$-structures.

Proposition 5.31 $\forall x F \approx \neg \exists x \neg F$.

Proof.

$$\bar{v}(\forall x F) = 0 \iff \text{there is a } v' \text{ such that } v' \underset{X-\{x\}}{=} v \text{ and } \bar{v}'(F) = 0$$
$$\iff \text{there is a } v' \text{ such that } v' \underset{X-\{x\}}{=} v \text{ and } \bar{v}'(\neg F) = 1$$
$$\iff \bar{v}(\exists x \neg F) = 1$$
$$\iff \bar{v}(\neg \exists x \neg F) = 0 .$$
□

Proposition 5.32 *Let Y be the set of variables having a free occurrence in F. If $v \underset{Y}{=} v'$, then $\bar{v}(F) = \bar{v}'(F)$.*

Proof. For this proof we use the fact that if $Y' \subseteq Y$ then $v \underset{Y}{=} v' \implies v \underset{Y'}{=} v'$.

The proof is by induction on the structure of F.
- Basis. If $F = R(t_1, \ldots, t_n)$ then $f(F) = \{u_i \mid u_i \in X,$ and u_i occurs in some $t_j\}$, and if $v \underset{f(F)}{=} v'$ then $\bar{v}(F) = \bar{v}'(F)$.
- Inductive step.
    - If $F = (F_1 \square F_2)$ with $\square \in \{\supset, \wedge, \vee\}$, we use the induction hypothesis: because $f(F_i) \subseteq f(F)$, we have that $v \underset{f(F)}{=} v' \implies v \underset{f(F_i)}{=} v'$.
    - If $F = \exists x F'$, then $f(F) = f(F') - x$. Let $v_1 \underset{f(F)}{=} v_2$. If $\bar{v}_1(\exists x F') = 1$, there exists $v_1' \underset{X-\{x\}}{=} v_1$ such that $\bar{v}_1'(F') = 1$. Let v_2' be defined by

$$v_2'(y) = \begin{cases} v_2(y) & \text{if } y \neq x, \\ v_1'(x) & \text{otherwise.} \end{cases}$$

We have $v_2' \underset{X-\{x\}}{=} v_2$. As $f(F) \subseteq X - x$,

$$v_2' \underset{f(F)}{=} v_2 \underset{f(F)}{=} v_1 \underset{f(F)}{=} v_1',$$

and as $v_2'(x) = v_1'(x)$, $v_2' \underset{f(F) \cup \{x\}}{=} v_1'$. Because $f(F') \subseteq f(F) \cup \{x\}$, we have that $\bar{v}_2'(F') = \bar{v}_1'(F') = 1$, and hence $\bar{v}_2(\exists x F') = 1$.
    - If $F = \forall x F'$, then $f(F) = f(F') - x = f(\neg \exists x \neg F')$, and hence

$$v_1 \underset{f(F)}{=} v_2 \implies \bar{v}_1(\neg \exists x \neg F') = \bar{v}_2(\neg \exists x \neg F')$$
$$\implies \bar{v}_1(\forall x F) = \bar{v}_2(\forall x F).$$
□

First order predicate calculus

REMARK 5.33 *If a formula F contains no occurrence of a free variable (in which case it is said to be a closed or ground formula or a sentence), then its truth value in S does not depend on the valuation. Indeed, for any v, v', we have $v \underset{\emptyset}{=} v'$, and hence for any v, v', $\bar{v}(F) = \bar{v}'(F)$. This occurs if F is a propositional logic formula.*

EXERCISE 5.11 *In Aristotle's syllogisms, assertions about properties P and Q of individuals frequently occur in the following forms:*
(i)   *All Ps are Qs.*
(ii) *Some Ps are Qs.*
(iii) *No P is a Q.*
(iv) *Some Ps are not Qs.*
  Translate these assertions into predicate calculus formulas by introducing the predicates $P(x)$ and $Q(x)$. ◇

EXERCISE 5.12 *Show that if x is not free in F, then*
$$\bar{v}(\forall x F) = \bar{v}(\exists x F) = \bar{v}(F).$$ ◇

Definition 5.34 *A formula F is said to be*

- *satisfiable in S if there exists a valuation v such that $\bar{v}(F) = 1$,*
- *satisfiable if there exist a structure S and a valuation v such that $\bar{v}(F) = 1$,*
- *valid in S if for all v, $\bar{v}(F) = 1$ and*
- *universally valid if it is valid in all $\mathcal{L}$-structures.*

EXAMPLE 5.35
1. $\big((\neg \exists x P(x)) \Longleftrightarrow \forall x (\neg P(x)) \big)$ is universally valid.
   If x and y are real numbers, and S is the structure associated with $\mathbb{R}$, then $x \leq x + y$ is satisfiable in S but it is not valid in S.
   Let $F = R(x,z) \wedge Q(x,y,z)$. Consider the structure $S = (\mathbb{N}, R_\mathbb{N}, Q_\mathbb{N})$, defined by $R_\mathbb{N} = \{(n,m) \,/\, n < m\}$ and $Q_\mathbb{N} = \{(n,m,p) \,/\, n+m = p\}$. F is satisfiable in S (let, for instance, $v(x) = v(y) = 1$, $v(z) = 2$), but F is not valid in S (let for instance $v(x) = v(z) = 1$, $v(y) = 0$).
2. (Examples are given with the PROLOG terminology.) Let $\mathcal{R} = \{$male, female$\}$ be a set of two unary predicates. Then

$$A = \big((\neg \, \text{male}(x)) \Longrightarrow \text{female}(x) \big)$$

is satisfiable but not valid, whilst

$$C = \Big(\big((\neg \text{male}(x)) \Longrightarrow \text{female}(x) \big) \vee \big(\neg \text{male}(x) \wedge \neg \text{female}(x) \big) \Big)$$

is valid.

EXERCISE 5.13 Let $S = \langle E, \{R, =\}\rangle$ be a set equipped with a relation R and a predicate $=$ that we assume to be interpreted as equality. Write a formula that is valid in S if and only if R is a (total) ordering. ◇

EXERCISE 5.14
1. Is $\exists y \forall x\, r(x,y) \approx \forall x \exists y\, r(x,y)$ valid for a binary predicate r? Same question for $(\exists y p(y)) \wedge (\exists y q(y)) \approx (\exists y (p(y) \wedge q(y)))$, with p and q unary predicates. Give a proof if the answer is yes, a counterexample if the answer is no.
2. Show that
$$\exists y \forall x\, \bigl(p(x) \wedge q(y)\bigr) \approx \forall x \exists y\, \bigl(p(x) \wedge q(y)\bigr),$$
for unary predicates p and q. ◇

As for propositional logic we define sequents.

Definition 5.36 *A sequent $(\mathcal{F}, F)$ is valid in S, denoted by $\mathcal{F} \models_S F$, if*
$$\text{for any } v, \quad \Bigl(\bigl(\text{ for all } G \text{ in } \mathcal{F},\ \bar{v}(G) = 1\bigr) \implies (\bar{v}(F) = 1)\Bigr).$$
A sequent is universally valid, denoted by $\mathcal{F} \models F$, if it is valid in all Ss.

Proposition 5.37
$$\{F_1, \ldots, F_n\} \models_S F \text{ if and only if } \emptyset \models_S (F_1 \supset (F_2 \supset \cdots (F_n \supset F)\cdots)).$$

Proposition 5.38 *If $\mathcal{F} \models_S F$ and if x is not free in any formula of $\mathcal{F}$, then $\mathcal{F} \models_S \forall x F$.*

Proof. Let v be a valuation such that for any $G \in \mathcal{F}$ $\bar{v}(G) = 1$. Let v' be any valuation such that $v' =_{X-\{x\}} v$. Since x never has a free occurrence in $\mathcal{F}$, for any $G \in \mathcal{F}$ we also have $\bar{v}'(G) = 1$ and thus $\bar{v}'(F) = 1$. Since this is true for any $v' =_{X-\{x\}} v$, we have that $\bar{v}(\forall x F) = 1$. □

Let F be a formula and let x be a variable. Let t be a term. Let $F[x := t]$ be the formula where all *free* occurrences of x have been replaced by t. $F[x := t]$ is said to be an *instance* of F; if $F[x := t]$ is a formula without variables, then it is said to be a *ground instance* of F. If x has no free occurrence in F, then $F[x := t] = F$.

Let u be a term. We will say that u is *substitutable* for x in F if u is a ground term or if u is such that any occurrence of a variable in u is free in $F[x := u]$.

First order predicate calculus

EXAMPLE 5.39 Let $F = (\forall y R(x,y,z)) \vee (\forall z R'(z))$, where R, R' are relational symbols. u is substitutable for x in F if and only if y does not occur in u. For example y is not substitutable for x in F because $F[x := y] = (\forall y R(y,y,z)) \vee (\forall z R'(z))$ and occurrence $(y,5)$ becomes bound. Similarly let f be a function symbol, $f(y,z)$ is not substitutable for x in F, but $f(x,z)$ is substitutable for x in F.

From now on, when we write $F[x := u]$, we will implicitly assume that u is substitutable for x in F.

Proposition 5.40 *Let x be a variable and let u be a term substitutable for x in F. Let v be a valuation. Let v' be defined by:*

$$v'(y) = \begin{cases} v(y) & \text{if } y \neq x, \\ v^*(u) & \text{if } y = x. \end{cases}$$

Then $\bar{v}'(F) = \bar{v}(F[x := u])$.

Proof. By induction on the construction of F. □

Proposition 5.41
1. $\mathcal{F} \underset{S}{\models} \forall x F \implies \mathcal{F} \underset{S}{\models} F[x := u]$;
2. $\mathcal{F} \underset{S}{\models} F[x := u] \implies \mathcal{F} \underset{S}{\models} \exists x F$.

Proof. Let v be such that for any $G \in \mathcal{F}$, $\bar{v}(G) = 1$. Let v' be constructed as previously in Proposition 5.40; we have $v' \underset{X-\{x\}}{=} v$ and $\bar{v}'(F) = \bar{v}(F[x := u])$.

1. If $\mathcal{F} \underset{S}{\models} \forall x F$ then $\bar{v}(\forall x F) = 1$, and $\bar{v}'(F) = 1$; thus $\bar{v}(F[x := u]) = 1$ which proves that $\mathcal{F} \underset{S}{\models} F[x := u]$.
2. If $\mathcal{F} \underset{S}{\models} F[x := u]$ then $\bar{v}'(F) = 1$, and hence $\bar{v}(\exists x F) = 1$. □

5.3.3 Some particular formulas

We now give some identities of predicate calculus that are quite useful. In the present section we abbreviate $F \approx G$ by $F \iff G$ to comply with the usual notations when logic is used as meta-language and to increase readability; we will also use the notation $F \implies G$ to denote that $F \supset G$ is universally valid, i.e. that $\emptyset \models F \supset G$.

Proposition 5.42

(i) $\quad \forall x\, (p(x) \wedge q(x)) \iff \forall x\, p(x) \wedge \forall x\, q(x)$.

(ii) $\quad \exists x\, (p(x) \wedge q(x)) \implies \exists x\, p(x) \wedge \exists x\, q(x)$.

By duality between $\forall$ and $\exists$ we also have

(iii) $\quad \exists x\, (p(x) \vee q(x)) \iff \exists x\, p(x) \vee \exists x\, q(x)$.

(iv) $\quad \forall x\, p(x) \vee \forall x\, q(x) \implies \forall x\, (p(x) \vee q(x))$.

The converses of rules (ii) and (iv), namely,

$$\exists x\, p(x) \wedge \exists x\, q(x) \implies \exists x\, (p(x) \wedge q(x))$$

and

$$\forall x\, (p(x) \vee q(x)) \implies \forall x\, p(x) \vee \forall x\, q(x) ,$$

are false (see Exercise 5.14).

Lastly, the following rules, written with the same conventions as above, are useful for putting formulas in prenex form, i.e. with all quantifiers at the beginning of the formula.

Proposition 5.43 Let $* \in \{\vee, \wedge\}$, let F be a formula, let x be a variable and let G be a formula in which x has no free occurrence. We have:

(i) $\quad \neg \forall x\, F \iff \exists x\, \neg F \qquad\qquad \neg \exists x\, F \iff \forall x\, \neg F$

(ii) $\quad (\forall x\, F) * G \iff \forall x\, (F * G) \qquad (\exists x\, F) * G \iff \exists x\, (F * G)$

(iii) $\quad G * (\forall x\, F) \iff \forall x\, (G * F) \qquad G * (\exists x\, F) \iff \exists x\, (G * F)$

(iv) $\quad (\forall x\, F) \supset G \iff \exists x\, (F \supset G) \qquad (\exists x\, F) \supset G \iff \forall x\, (F \supset G)$

(v) $\quad G \supset (\forall x\, F) \iff \forall x\, (G \supset F) \qquad G \supset (\exists x\, F) \iff \exists x\, (G \supset F)$

The proofs of the two preceding propositions are straightforward.

5.3.4 Lexical variants

Let F be a formula. A lexical variant of F is a formula deduced from F by renaming some bound variables. Formally:

Definition 5.44 F' is a lexical variant of F if one of the following holds:

- $F = F'$.
- For $Q \in \{\forall, \exists\}$, $F = QxG$ and $F' = QyG'[x := y]$, where G' is a lexical variant of G and y is not free in G'.
- $F = \neg G$, $F' = \neg G'$, and G' is a lexical variant of G.
- For $\square \in \{\supset, \wedge, \vee\}$, $F = (F_1 \square F_2)$, $F' = (F_1' \square F_2')$, and F_i' is a lexical variant of F_i.

First order predicate calculus

EXAMPLE 5.45 $\forall z P(z,y)$ is a lexical variant of $\forall x P(x,y)$, but neither $\forall y P(y,y)$ nor $\forall z P(z,x)$ is.

Proposition 5.46 *If F is a lexical variant of F' then $f(F) = f(F')$, and for any $\mathcal{L}$-structure S and any valuation v in S, $\bar{v}(F) = \bar{v}(F')$.*

Proof. The first assertion is easy to prove. For the second one, it suffices to show that if for all v, $\bar{v}(F) = \bar{v}(F')$, then for all v, $\bar{v}(\exists x F) = \bar{v}(\exists y F'[x := y])$ if y is not free in F'.

Let v be a valuation, and let V be the set of valuations v' such that $v' =_{X-\{x\}} v$ and $v'(x) = v'(y)$.

Because y is not a free variable of F', y is not a free variable of F either, and: there exists v' such that $v' =_{X-\{x\}} v$ and $\bar{v}'(F) = 1$ if and only if there exists v' such that $v' \in V$ and $\bar{v}'(F) = 1$.

Similarly, x is not a free variable of $F'[x := y]$, and thus: there exists v' such that $v' =_{X-\{x\}} v$ and $\bar{v}'(F'[x := y]) = 1$ if and only if there exists v' such that $v' \in V$ and $\bar{v}'(F'[x := y])$.

Hence $\bar{v}(\exists x F) = \bar{v}(\exists y F'[x := y])$ is equivalent to: there exists $v' \in V$ such that $\bar{v}'(F) = 1$ if and only if there exists v' such that $v' \in V$ and $\bar{v}'(F'[x := y]) = 1$. Finally, we observe that this last equivalence is true. □

5.3.5 Prenex formulas

Definition 5.47 *A formula F is said to be prenex if it is in the form*

$$Q_1 x_1 Q_2 x_2 \ldots Q_n x_n F',$$

where the Q_is are quantifiers for $i = 1, 2, \ldots, n$, and where F' is a formula without quantifier.

Theorem 5.48 *Any formula F is equivalent to a prenex formula G.*

Proof. By structural induction on F we effectively build a prenex formula G equivalent to F. The inductive assumption is that there exists a prenex formula G equivalent to F.

- Basis. If F is in the form $R(t_1, \ldots, t_n)$, it is clear because F is prenex.
- Inductive step.
    - If F is in the form $\forall x F'$ (resp. $\exists x F'$), with $F' \approx F''$, where F'' is prenex, then $F \approx \forall x F''$ (resp. $F \approx \exists x F''$), which is prenex.

- If F is in the form $\neg F'$, with $F' \approx F''$, F'' prenex and

$$F'' = Q_1 x_1 Q_2 x_2 \ldots Q_n x_n G,$$

then, by Proposition 5.43 (i),

$$F' \approx Q'_1 x_1 Q'_2 x_2 \ldots Q'_n x_n \neg G,$$

with $Q'_i = \forall$ if $Q_i = \exists$ and $Q'_i = \exists$ if $Q_i = \forall$.

- If F is in the form $F_1 * F_2$, with $* \in \{\vee, \wedge\}$, we put F_1 and F_2 in prenex form, and we apply Proposition 5.43 (ii) and (iii) in order to 'pull' the quantifiers to the beginning of the formula. We must, however, proceed with care; if, for instance, $F = F_1 \wedge F_2 \approx (\forall x F'_1) \wedge F'_2$, with x free in F'_2, we must first rename variable x in F_1 and replace x by a new variable z occurring neither in F_1 nor in F'_2.

- If F is in the form $F_1 \supset F_2$, we put F_1 and F_2 in prenex form, and we apply Proposition 5.43 (iv) and (v), possibly with renamings, to 'pull' quantifiers to the beginning of the formula. □

EXERCISE 5.15 Find a prenex formula equivalent to
$$\exists x P(x) \wedge \forall x \big(\exists y Q(y) \supset R(x)\big). \qquad \diamond$$

EXERCISE 5.16 We assume that:

(a) Programmers write programs for all those who do not write programs for themselves.

(b) No programmer writes programs for someone who writes programs for him(her)self.

We then notice the paradox: if a programmer writes a program for him(her)self, he/she violates rule (b); if a programmer does not write programs for him(her)self, he/she violates rule (a) because he/she does not write programs for someone who does not write programs for him(her)self. How do you explain this paradox? (Hint: express the requirements (a) and (b) by formulas F and G of predicate calculus, and show that $F \wedge G$ implies that there is no programmer.) $\diamond$

5.4 Herbrand's theorem and consequences

5.4.1 Theories and Models

Definition 5.49 *A provable sequent is a sequent obtained by the rules for propositional logic, extended by:*

- If $\mathcal{F} \vdash \forall x F$ then: $\mathcal{F} \vdash F[x := t]$ *(instantiation rule).*
- If $\mathcal{F} \vdash F$ and if x is not free in $\mathcal{F}$, then: $\mathcal{F} \vdash \forall x F$ *(universal generalization rule).*
- $\mathcal{F} \vdash \exists x F$ *if and only if* $\mathcal{F} \vdash \neg \forall x \neg F$ *(definition of $\exists$).*

⚠ The universal generalization rule does not apply if x is free in $\mathcal{F}$. For instance, $p(x) \vdash p(x)$ is provable, but $p(x) \vdash \forall x p(x)$ is not provable. This rule is the formalization of the following reasoning: 'if a property is true for an *arbitrary* object x then it is true for any x'; x is arbitrary means that no hypothesis is made about x and that lack of knowledge about x is formally expressed by the fact that x does not occur free in $\mathcal{F}$.

EXAMPLE 5.50
- If $\mathcal{F}, F \vdash G$ and if x is not free in $\mathcal{F}$ and G then $\mathcal{F}, \exists x F \vdash G$. Indeed, we have:

1.	$\mathcal{F}, F \vdash G$	
2.	$\mathcal{F}, \neg G \vdash \neg F$	(by Proposition 5.14)
3.	$\mathcal{F}, \neg G \vdash \forall x \neg F$	(generalization)
4.	$\mathcal{F}, \neg \forall x \neg F \vdash G$	(by Proposition 5.14)
5.	$\mathcal{F}, \exists x F \vdash G$	(definition of $\exists$)

- $\exists x \forall y F \vdash \forall y \exists x F$. Indeed, we have:

1.	$\forall y F, \forall x \neg F \vdash \forall y F$	(hypothesis)
2.	$\forall y F, \forall x \neg F \vdash F[y := y]$	(instantiation)
3.	$\forall y F, \forall x \neg F \vdash \forall x \neg F$	(hypothesis)
4.	$\forall y F, \forall x \neg F \vdash \neg F[x := x]$	(instantiation)
5.	$\forall y F \vdash \neg(\forall x \neg F)$	(contradiction on 2, 4 because $F[y := y] = F[x := x] = F$)
6.	$\forall y F \vdash \exists x F$	(definition of $\exists$)
7.	$\forall y F \vdash \forall y \exists x F$	(universal generalization)
8.	$\exists x \forall y F \vdash \forall y \exists x F$	

We detail steps 7 and 8 of the proof: because y is not free in $\forall y F$, universal generalization applied to 6 gives $\forall y F \vdash \forall y \exists x F$, and because x is not free in $\forall y \exists x F$, we have, by applying to 7 the sequent obtained in the first part of the present example, $\exists x \forall y F \vdash \forall y \exists x F$.

Theorem 5.51 (Soundness) *If a sequent is provable then it is universally valid.*

EXERCISE 5.17 Prove this theorem. ◇

Theorem 5.52 (Completeness) *If a sequent is universally valid then it is provable.*

We will not give the proof of this theorem; rather, we will provide some ideas behind the proof.

Definition 5.53 *A theory is a set T of formulas such that for any finite subset $\mathcal{F}$ of T, if $\mathcal{F} \vdash F$ then $F \in T$.*

A theory T is contradictory if there exists a formula F such that $F \in T$ and $\neg F \in T$.

EXAMPLE 5.54 Let $\mathcal{F}$ be a finite set of formulas. The set

$$Th(\mathcal{F}) = \{F \ / \ \mathcal{F} \vdash F\}$$

is a theory.

Proposition 5.55 *A theory T is contradictory if and only if it contains all formulas.*

Proof. Let G be a formula.

$$\left. \begin{array}{l} F, \neg F, \neg G \vdash F \\ F, \neg F, \neg G \vdash \neg F \end{array} \right\} \quad \text{hence} \quad F, \neg F \vdash G \ .$$

Thus $G \in T$. □

Proposition 5.56 $\mathcal{F} \vdash F$ *if and only if $Th(\mathcal{F} \cup \{\neg F\})$ is contradictory.*

Proof. If $\mathcal{F} \vdash F$, then:

$$\left. \begin{array}{l} \mathcal{F}, \neg F \vdash F \\ \text{and} \quad \mathcal{F}, \neg F \vdash \neg F \end{array} \right\} \implies \text{the theory } \mathcal{F} \cup \{\neg F\} \text{ is contradictory} \ .$$

If $Th(\mathcal{F} \cup \{\neg F\})$ is contradictory, $\mathcal{F}, \neg F \vdash F$ and $\mathcal{F}, \neg F \vdash \neg F$, and hence $\mathcal{F} \vdash F$. □

An $\mathcal{L}$-structure S is a *model* of a (finite or infinite) set $\mathcal{G}$ of formulas if for any v and for any F in $\mathcal{G}$, we have $\bar{v}(F) = 1$.

We denote by $\emptyset \models_S \mathcal{G}$ the fact that S is a model of $\mathcal{G}$.

A set $\mathcal{G}$ of formulas is *satisfiable* if *there exists* an $\mathcal{L}$-structure S and *there exists* a valuation v such that for any F in $\mathcal{G}$, we have $\bar{v}(F) = 1$. A set $\mathcal{G}$ of formulas has a *model* if *there exists* an $\mathcal{L}$-structure S such that *for any* valuation v and for any F in $\mathcal{G}$, we have $\bar{v}(F) = 1$. A set $\mathcal{G}$ of formulas which does not have a model may thus be satisfiable: for example, the set $\mathcal{F}$ of formulas defined in Remark 5.58 does not have a model, but it is satisfiable.

If a theory has a model, then it is not contradictory. The converse is one of the fundamental theorems of logic. The proof of this theorem is quite long, so we will not give it; we state the theorem.

Theorem 5.57 *If $\mathcal{F}$ consists of closed formulas and $Th(\mathcal{F})$ is not contradictory, then $\mathcal{F}$ has a model.*

REMARK 5.58 Theorem 5.57 is false if non-closed formulas are allowed. Let $\mathcal{F} = \{\exists x p(x), \neg p(x)\}$; $\mathcal{F}$ is satisfiable: let $S = \langle E, \gamma \rangle$ with $E = \{0, 1\}$, and $\gamma(p) = p_S$ defined by $p_S(0) = 0$, $p_S(1) = 1$, the valuation $v(x) = 0$ is such that $\bar{v}(F) = 1$ for any F in $\mathcal{F}$. $\mathcal{F}$ does not have a model: if a structure S' is such that for any valuation v, $\bar{v}(\neg p(x)) = 1$, then, for any valuation v, $\bar{v}(\exists x p(x)) = 0$. Nevertheless $Th(\mathcal{F})$ is not contradictory: otherwise, by Proposition 5.56, we would conclude that $\exists x p(x) \vdash p(x)$, which is false because the sequent $\{\exists x p(x), p(x)\}$ is not universally valid, e.g. $\exists x p(x) \not\models_S p(x)$.

Let us deduce the completeness theorem from Theorem 5.57. We prove that if $\mathcal{F} \models F$ then $\mathcal{F} \vdash F$.

1. First, we consider closed formulas: we will assume that $\mathcal{F}$ and F consist of closed formulas, that $\mathcal{F} \models F$, and that the sequent $(\mathcal{F}, F)$ is not provable. Then $\mathcal{F} \cup \{\neg F\}$ is not contradictory and thus has a model S by Theorem 5.57. Any valuation v thus verifies for any $F_i \in \mathcal{F}$, $\bar{v}(F_i) = 1$ and $\bar{v}(\neg F) = 1$. But, because $\mathcal{F} \models F$, the sequent $(\mathcal{F}, F)$ is valid in S and thus any valuation v verifying $\bar{v}(F_i) = 1$ for any F_i in $\mathcal{F}$ also verifies $\bar{v}(F) = 1$, a contradiction.
2. Now, we will assume that $\mathcal{F} \models F$, and that $\mathcal{F} = \{F_1, \ldots, F_n\}$ and F consist of (not necessarily closed) formulas. $\mathcal{F} \models F$ and Proposition 5.37 imply that $\emptyset \models (F_1 \supset (F_2 \supset \cdots (F_n \supset F) \cdots))$. Let $x_1, \ldots, x_k$ be the free variables of $(F_1 \supset (F_2 \supset \cdots (F_n \supset F) \cdots))$; by Proposition 5.38, $\emptyset \models \forall x_1 \cdots \forall x_k (F_1 \supset (F_2 \supset \cdots (F_n \supset F) \cdots))$.

$(\emptyset, \forall x_1 \cdots \forall x_k (F_1 \supset (F_2 \supset \cdots (F_n \supset F) \cdots)))$ thus is a universally valid sequent consisting of closed formulas, hence it is provable by case 1, and $\emptyset \vdash \forall x_1 \cdots \forall x_k (F_1 \supset (F_2 \supset \cdots (F_n \supset F) \cdots))$. By the instantiation rule, $\emptyset \vdash (F_1 \supset (F_2 \supset \cdots (F_n \supset F) \cdots))$. As in Proposition 5.16, we deduce that $\{F_1, \ldots, F_n\} \vdash F$, hence the sequent $(\mathcal{F}, F)$ is provable. □

EXERCISE 5.18 Bernard and Christopher are members of the Alpine Club. Any member of the Alpine Club is either a skier, or an alpinist or both. Alpinists do not like rain, and skiers like snow. Christopher likes all that Bernard does not like, and does not like all that Bernard likes (i.e. there are things that Bernard likes and that Christopher does not like).

1. Express the requirements of the Alpine Club by a set $\mathcal{F}$ of formulas of predicate calculus.
2. Can you find a model of $\mathcal{F}$?
3. Can you prove that there is a member of the Alpine Club who is an alpinist and not a skier (or vice versa)? ◊

5.4.2 Herbrand's models

Let $\mathcal{G}$ be a set of functions whose set C of constants is non-empty. Let $\mathcal{L}$ be the language $\mathcal{L} = \mathcal{R} \cup \mathcal{G}$. The *Herbrand universe* of $\mathcal{L}$ is the set of ground (i.e. variable-free) terms built over $\mathcal{G}$. The Herbrand universe is denoted by U_H, which is inductively defined by:

(B) $C \subseteq U_H$,
(I) for any n-ary f in $\mathcal{G}$, and for any $t_1, \ldots, t_n$ in U_H, $f(t_1, \ldots, t_n) \in U_H$.

Because C is non-empty, U_H is non-empty. (Indeed U_H is empty if and only if C is empty.)

An $\mathcal{L}$-structure $S = \langle E, \gamma, h \rangle$ is a *Herbrand structure* if:

- $E = U_H$, and
- h associates with each $f \in \mathcal{G}$ the function f_S from $U_H^{\rho(f)}$ to U_H defined by $f_S(t_1, \ldots, t_n) = f(t_1, \ldots, t_n)$, see Definition 3.14. (This implies that with each constant $a \in C$, h associates the element a of U_H.)

The definition of an $\mathcal{L}$-structure (Definition 5.29) requires the domain of the structure to be non-empty. Thus U_H must be non-empty. This is why we require C to be non-empty.

The *Herbrand basis* of $\mathcal{L}$ is the set B_H of ground (i.e. variable-free) atomic formulas, i.e. formulas of the form $R(t_1, \ldots, t_n)$ with $R \in \mathcal{R}$ and $t_1, \ldots, t_n$ in U_H. For a given language $\mathcal{L}$, there is a single Herbrand universe, but on that Herbrand universe numerous Herbrand structures can be defined; a Herbrand structure H is defined by a *Herbrand interpretation* which is a subset I of the Herbrand basis B_H; I specifies the atomic formulas which are true in H. Formally, I defines $H = \langle U_H, \gamma, h \rangle$ means that for $t_1, \ldots, t_n$ in U_H and $R \in \mathcal{R}$, $(t_1, \ldots, t_n)$ is in R_H if and only if $R(t_1, \ldots, t_n) \in I$. A Herbrand structure will thus be denoted by $H = \langle U_H, I, h \rangle$, or by $H = \langle U_H, I \rangle$ or simply by I, since U_H and h are uniquely specified by the language.

EXAMPLE 5.59
1. Let $\mathcal{L}$ be the language $\mathcal{L} = \{a, p, q\}$ where a is a constant symbol and p, q are nullary relational symbols. Then the Herbrand universe is $\{a\}$, the Herbrand basis is the set $B_H = \{p, q\}$ and there are exactly four Herbrand structures specified by $I_0 = \emptyset$, $I_1 = \{p\}$, $I_2 = \{q\}$, $I_3 = \{p, q\}$.
2. Let $\mathcal{L}$ be the language $\mathcal{L} = \{a, f, p, q\}$ where p, q are unary relational symbols, a is a constant symbol and f is a unary function symbol. Then the Herbrand universe is $U_H = \{a, f(a), f^2(a), \ldots, f^n(a), \ldots\} = \{f^n(a) \,/\, n \in \mathbb{N}\}$, the Herbrand basis is the set $B_H = \{p(t), q(t) \,/\, t \in U_H\}$ and the Herbrand structures are specified by subsets I of B_H; for instance $I_0 = \emptyset$, $I_1 = B_H$, $I_2 = \{p(a), p(f(a))\}$, $I_3 = \{p(t) \,/\, t \in U_H\}$, etc., specify Herbrand structures H_0, H_1, H_2, H_3, etc.

Proposition 5.60 *For any $\mathcal{L}$-structure $A = \langle E, \gamma, h \rangle$ there is a unique Herbrand structure H and a unique mapping $h^* \colon U_H \longrightarrow E$ such that*
(i) $h^*(f(t_1, \ldots, t_n)) = f_A(h^*(t_1), \ldots, h^*(t_n))$ *and*
(ii) *for $t_1, \ldots, t_n$ in U_H, $(t_1, \ldots, t_n) \in R_H \iff (h^*(t_1), \ldots, h^*(t_n)) \in R_A$.*

Proof. The uniqueness of the mapping $h^* \colon U_H \longrightarrow E$ follows from (i) and Proposition 3.15; (ii) implies that the Herbrand structure on U_H must be defined by $I = \{R(t_1, \ldots, t_n) \in B_H \,/\, (h^*(t_1), \ldots, h^*(t_n)) \in R_A\}$. □

Definition 5.61 *Let $\mathcal{F}$ be a set of formulas of the language $\mathcal{L}$ and let H be a Herbrand structure for $\mathcal{L}$. H is said to be a Herbrand model of $\mathcal{F}$ if and only if H is a model of $\mathcal{F}$.*

EXAMPLE 5.62 Let $\mathcal{L} = \{a, f, p, q\}$, let $\mathcal{F} = \{p(a), \forall x(p(x) \supset p(f(x)))\}$. I_1 and I_3 in Example 5.59, 2, define Herbrand models of $\mathcal{F}$; I_0 and I_2 in Example 5.59, 2, do not define Herbrand models of $\mathcal{F}$; any $I = I_3 \cup \{q(f^k(a)) \,/\, k \in K \subset \mathbb{N}\}$ also defines a Herbrand model of $\mathcal{F}$.

5.4.3 Herbrand's theorem

Definition 5.63 *A prenex formula is said to be universal if and only if it has only universal quantifiers.*

Theorem 5.64 *(Herbrand's theorem) Let $\mathcal{L}$ be a language with a non-empty set C of constants, and let $\mathcal{F}$ be a set of closed universal formulas, then $\mathcal{F}$ has a model if and only if $\mathcal{F}$ has a Herbrand model.*

Proof. The 'if' direction is clear. For the 'only if' direction, assume that $\mathcal{F}$ has a model $S = \langle E, \gamma, h \rangle$ and construct a Herbrand model for $\mathcal{F}$. Let H be

the Herbrand structure defined by the following set I of atomic formulas in the Herbrand basis:
$$I = \{F \in B_H \,/\, \emptyset \models_S F\}.$$

We will prove that H is a Herbrand model of $\mathcal{F}$. Because C is non-empty, U_H is non-empty, and with any valuation $v_H\colon X \longrightarrow U_H$ we can associate a unique valuation $v = h^* \circ v_H$,
$$v\colon X \xrightarrow{v_H} U_H \xrightarrow{h^*} E,$$
where h^* is defined in Proposition 5.60. By structural induction on the formulas, it can be shown that for any quantifier-free formula G, $\bar{v}_H(G) = \bar{v}(G)$. The base case follows from Proposition 5.60 and the inductive step is straightforward. We must prove that for any $F = \forall x_1 \cdots \forall x_n G$ in $\mathcal{F}$, where G is a quantifier-free formula, $\emptyset \models_H F$ holds, i.e. for any valuation $v_H\colon \{x_1, \ldots, x_n\} \longrightarrow U_H$, $\bar{v}_H(G) = 1$. If v_H is a valuation, $v = h^* \circ v_H$ is a valuation into E, and since $\emptyset \models_S F$, $\bar{v}(G) = 1$. Hence, $\bar{v}_H(G) = 1$. $\square$

EXAMPLE 5.65 Herbrand's theorem does not hold if $\mathcal{F}$ is not a set of universal formulas. Let $\mathcal{L} = \{a, R\}$ with a a constant, R a unary relational symbol, and let $\mathcal{F} = \{R(a)\,,\, \exists x \neg R(x)\}$. $\mathcal{F}$ has a model but $\mathcal{F}$ has no Herbrand model. The structure S defined by $E = \{0, 1\}$ with $a_S = 0$ and $0 \in R_S$, $1 \notin R_S$ is a model of $\mathcal{F}$.

$\mathcal{F}$ has no Herbrand model. There are exactly two Herbrand structures on the Herbrand universe $U_H = \{a\}$, defined by, respectively, $I_0 = \emptyset$ (i.e. $R_{I_0} = \emptyset$ is always false) and $I_1 = \{R(a)\}$ (i.e. $R_{I_1} = \{a\}$ is always true), neither of which is a model of $\mathcal{F}$.

REMARK 5.66 Herbrand's theorem holds if $\mathcal{F}$ is a set of formulas without quantifiers. Indeed, if $F(x)$ is not closed, S is a model of $F(x)$ if and only if S is a model of $\forall x F(x)$.

In fact, Theorem 5.64 is a weakened form of Herbrand's theorem which asserts the following more general result.

Theorem 5.67 *Let $\mathcal{F}$ be a set of closed universal formulas, either*

- *$\mathcal{F}$ has a Herbrand model or*
- *$\mathcal{F}$ does not have a model and, moreover, there are finitely many ground instances of $\mathcal{F}$ whose conjunction is unsatisfiable.*

The proof of Theorem 5.67 will not be given here.

Herbrand's theorem and consequences

Herbrand's theorem has many useful consequences in logic programming and proof theory including:

- A satisfiable set $\mathcal{F}$ of universal formulas has a Herbrand model and so has a model which is finite or countable.
- If the set $\mathcal{F}$ of universal formulas is unsatisfiable, then Theorem 5.67 directly exhibits a finite set of unsatisfiable ground instances. Thus Theorem 5.67 gives a method for effectively producing either a Herbrand model for $\mathcal{F}$ or a particular finite counter-example to the existence of any model of $\mathcal{F}$.
- Herbrand's theorem implies the completeness of the resolution method; the resolution method is based on the following idea: the formula $F = \exists x G(x)$, where G is quantifier-free, is a consequence of the set of universal formulas $\mathcal{F}$ if and only if $\mathcal{F} \cup \{\neg F\}$ is unsatisfiable. $\mathcal{F} \cup \{\neg F\}$ is a set of universal formulas that can be proved to be unsatisfiable by exhibiting a finite set of unsatisfiable ground instances. It can be shown that exhibiting the unsatisfiable ground instances also gives valuations $v(x) = t$ such that $\mathcal{F} \vdash G[x := t]$, which are called answer substitutions.
- Herbrand's theorem can be used to prove Theorem 5.52.

EXERCISE 5.19 Find all Herbrand models of

$$\mathcal{F} = \{edge(a, b)\,,\ edge(b, c)\,,\ \forall x \forall y \big(edge(x, y) \supset path(x, y)\big)\,,$$
$$\forall x \forall y \big((edge(x, z) \land path(z, y)) \supset path(x, y)\big)\}\,,$$

where the language $\mathcal{L}$ consists of the constants a, b, c and the binary relational symbols *edge* and *path*. With the PROLOG notations, $\mathcal{F}$ would be denoted by:

$r_1:$ $\Longrightarrow edge(a, b)$
$r_2:$ $\Longrightarrow edge(b, c)$
$r_3:$ $edge(X, Y) \Longrightarrow path(X, Y)$
$r_4:$ $edge(X, Z), path(Z, Y) \Longrightarrow path(X, Y)$

where universal quantifications are omitted and the comma denotes $\land$. $\diamond$

5.4.4 Skolemization

We have seen in Example 5.65 that Herbrand's theorem does not hold for non-universal formulas. This can be remedied by constructing, for each formula F, a universal formula F' which is equisatisfiable with F: i.e. F is satisfiable if and only if F' is satisfiable. (Note: F' will not be equivalent to F, see Exercise 5.21.) Each formula

$$F = \forall x_1 \cdots \forall x_n \exists y_1 \cdots \exists y_p G(x_1, \ldots, x_n, y_1, \ldots, y_p)$$

will be replaced by

$$F' = \forall x_1 \cdots \forall x_n G(x_1, \ldots, x_n, f_1(x_1, \ldots, x_n), \ldots, f_p(x_1, \ldots, x_n)),$$

where $f_1, \ldots, f_p$ are new function symbols, called *Skolem functions*. F' is called a *Skolemization* of F.

Theorem 5.68 *Let F be a closed formula in a language $\mathcal{L}$; there exists a universal formula F' in a language $\mathcal{L}' = \mathcal{L} \cup \{f_1, \ldots, f_p\}$, where $f_1, \ldots, f_p$ are new function symbols, such that F is satisfiable if and only if F' is satisfiable.*

Proof. By Theorem 5.48 we may assume that F is in prenex form; assume

$$F = \forall x_1 ... \forall x_{n_1} \exists y_1 \forall x_{n_1+1} ... \forall x_{n_2} \exists y_2 \cdots \forall x_{n_{p-1}+1} ... \forall x_{n_p} \exists y_p \forall x_{n_p+1} ... \forall x_n G,$$

where G is a formula without quantifiers. We add p new function symbols $f_1, \ldots, f_p$ to $\mathcal{L}$; for $i = 1, \ldots, p$, each f_i is of arity n_i and depends on the x_js such that $\forall x_j$ occurs before $\exists y_i$ in F. F' is the formula

$$F' = \forall x_1 \cdots \forall x_n G[y_1 := f_1(x_1, \ldots, x_{n_1})] \cdots [y_p := f_p(x_1, \ldots, x_{n_p})].$$

F is satisfiable if and only if F' is satisfiable. By induction on p it suffices to prove Lemma 5.69. □

Lemma 5.69 *$F = \forall x_1 \cdots \forall x_n \exists y G$ is satisfiable if and only if $F' = \forall x_1 \cdots \forall x_n G[y := f(x_1, \ldots, x_n)]$ is satisfiable, where f is a new function symbol.*

EXERCISE 5.20 Prove Lemma 5.69. ◊

REMARK 5.70 F' will not be equivalent to F. See Exercise 5.21.

EXERCISE 5.21 Find Skolemizations of $F = (\forall x R(x)) \vee (\exists y R'(y))$. ◊

EXERCISE 5.22 Find Skolemizations of $F = (\forall x \exists y R(x,y)) \vee \neg(\exists x \forall y R'(x,y))$. ◊

When we are interested in the existence of a model for a formula, Skolemization enables us to suppress all existential quantifiers. By Remark 5.66, models of $F(x_1, \ldots, x_n)$ and models of $\forall x_1 \cdots \forall x_n F(x_1, \ldots, x_n)$ coincide. We can thus assume that all variables are universally quantified and omit the universal quantifiers in the denotation of the formula: this is the usual notation for PROLOG.

5.4.5 Horn clauses

Horn clauses are very useful examples of universal formulas. PROLOG and most logic programming languages are based on Horn clauses.

Definition 5.71
(i) *Literals are atomic formulas or their negations, i.e. formulas of the form $L = R(t_1,\ldots,t_n)$ (positive literal), or of the form $L = \neg R(t_1,\ldots,t_n)$ (negative literal).*
(ii) *A Horn clause is a universal formula of the form $\forall x_1 \cdots \forall x_p(L_1 \vee \cdots \vee L_n)$, where the L_is are literals, and at most one of them is positive.*
(iii) *A program clause or definite clause is a Horn clause with exactly one positive literal.*

Thus, a Horn clause can be of one of the following three forms:
(i) $\forall x_1 \cdots \forall x_p A$ (positive clause or *fact*).
(ii) $\forall x_1 \cdots \forall x_p(\neg A_1 \vee \cdots \vee \neg A_n \vee A)$, where A and the A_is are atomic formulas.
(iii) $\forall x_1 \cdots \forall x_p(\neg A_1 \vee \cdots \vee \neg A_n)$, where the A_is are atomic formulas (negative clause or *goal*).

EXAMPLE 5.72   A PROLOG program consists of Horn clauses of the form (i) or (ii), which are usually written in the form (omitting the universal quantifiers and substituting a comma for $\wedge$):

(i) $\Longrightarrow A$,
(ii) $A_1,\ldots,A_n \Longrightarrow A$.

EXAMPLE 5.73   The set $\mathcal{F}$ of formulas of Exercise 5.19 is a set of program clauses. $\mathcal{F}$ can be written as $\mathcal{F} = \{edge(a,b), edge(b,c), \forall x \forall y(\neg edge(x,y) \vee path(x,y)), \forall x \forall y(\neg edge(x,z) \vee \neg path(z,y) \vee path(x,y))\}$, With the PROLOG notations, $\mathcal{F}$ is denoted by:

$r_1:$ $\Longrightarrow edge(a,b)$,
$r_2:$ $\Longrightarrow edge(b,c)$,
$r_3:$ $edge(X,Y) \Longrightarrow path(X,Y)$,
$r_4:$ $edge(X,Z), path(Z,Y) \Longrightarrow path(X,Y)$.

Theorem 5.74 *A set P of program clauses has a least Herbrand model $M = \langle U_H, I_M \rangle$ (i.e. a Herbrand model such that I_M is contained in any other Herbrand model I_H).*

Proof. Let $P = \{C_i \mid i \in J\}$, where each C_i is either of the form (i) or (ii). We will prove that the intersection of all Herbrand models of P is itself a Herbrand model of P. Let M be defined by

$$I_M = \bigcap_{I_H} \{I_H \mid \emptyset \models_{I_H} C_i, \text{ for all } C_i \in P\},$$

i.e. I_M is the intersection of all the Herbrand models of P; then $\emptyset \models_M C_i$ for all $C_i \in P$. We verify that all the clauses C_i of P are valid in M.

(i)   If C_i is of the form $\forall x_1 \cdots \forall x_p A$, then all ground instances of A are true in all Herbrand models, hence they belong to all I_Hs, and also to I_M, and thus they are true in M.

(ii)   If C_i is of the form $\forall x_1 \cdots \forall x_p(\neg A_1 \vee \cdots \vee \neg A_n \vee A)$, let $C_i' = (\neg A_1 \vee \cdots \vee \neg A_n \vee A)$, let v be a valuation $v: \{x_1, \ldots, x_p\} \longrightarrow U_H$ and let, for $B \in \{A, A_1, \ldots, A_n\}$, $v^*(B) = B[x_1 := v(x_1)] \cdots [x_p := v(x_p)]$ be the ground atom obtained by substituting $v(x_i)$ for x_i in B; then

- either there exists an A_j such that $v^*(A_j) \notin I_M$, and then $\bar{v}(\neg A_j) = 1$ and $\bar{v}(C_i') = 1$.
- or for any A_j, $v^*(A_j) \in I_M$, and then, for any I_H defining a Herbrand model H of P: $v^*(A_j) \in I_H$, and because H is a Herbrand model of P, we also have that $v^*(A) \in I_H$, hence $v^*(A) \in I_M$, and thus $\bar{v}(A) = 1$ and $\bar{v}(C_i') = 1$.

Hence, for any valuation $v: \{x_1, \ldots, x_p\} \longrightarrow U_H$, $\bar{v}(C_i') = 1$, and thus $\emptyset \models_M C_i$. □

EXAMPLE 5.75   The least Herbrand model of the program P of Example 5.73 is defined by

$$I_M = \{edge(a,b),\ edge(b,c),\ path(a,b),\ path(b,c), path(a,c)\}.$$

REMARK 5.76
1.   Any set P of program clauses also has a greatest Herbrand model M', which is defined by the whole Herbrand basis B_H. See also Exercise 5.28 (2).
2.   A set $\mathcal{F}$ of universal formulas which are not Horn clauses may have several minimal incomparable Herbrand models and as a result no least Herbrand model. For example, let $\mathcal{F} = \{\forall x(p(x) \vee q(x))\}$, where $\mathcal{L} = \{a, p, q\}$, a a constant symbol and p, q unary relational symbols. Then $U_H = \{a\}$, $B_H = \{p(a),\ q(a)\}$, and

$\mathcal{F}$ has three Herbrand models, respectively defined by the subsets $I_1 = \{p(a)\}$, $I_2 = \{q(a)\}$ and $I_3 = B_H$; both models I_1 and I_2 are minimal (none is included in the other one) and their intersection is the Herbrand interpretation defined by $I_H = \emptyset$, which is *not* a model of $\mathcal{F}$.

EXERCISE 5.23 A set of formulas which are not universal may have minimal Herbrand models. Find the minimal Herbrand models of the set of formulas $\mathcal{F}$ of Exercise 5.18. ◇

EXERCISE 5.24 Find the least Herbrand model of the set of program clauses $P = \{\forall x \forall y (\neg edge(x, y) \lor path(x, y)), \forall x \forall y (\neg edge(x, z) \lor \neg path(z, y) \lor path(x, y))\}$. ◇

EXERCISE 5.25 Find the least Herbrand model of the set of program clauses $P = \{i(a), \forall x(i(s(x)) \lor \neg i(x))\}$, where $\mathcal{L} = \{a, s, i\}$, a a constant symbol, s a unary function symbol and i is a unary relational symbol. ◇

EXERCISE 5.26 Let P be a set of program clauses; P is a set of formulas, hence (see Definition 5.53 and Example 5.54) $Th(P) = \{F \ / \ P \vdash F\}$ is a theory.
Show that $\{A \in B_H \ / \ A \in Th(P)\}$ defines the least Herbrand model of the set of program clauses P. ◇

EXERCISE 5.27 Does any set $\mathcal{F}$ of Horn clauses have a least Herbrand model? ◇

There is a constructive proof of the existence of the least Herbrand model of a set of program clauses, which is most useful in logic programming, and which is given in the following exercise.

EXERCISE 5.28 Recall that a complete lattice is a lattice where every subset has a least upper bound and a greatest lower bound. If f is a monotone mapping from a complete lattice to itself, then we can prove as in Theorem 2.39 that f has a least fixed point defined by $e = \inf\{x \in E \ / \ f(x) \leq x\}$.

Let P be a set of program clauses. Let $\mathcal{P}(B_H)$ be the set of subsets of the Herbrand basis B_H. Then $\mathcal{P}(B_H)$ when equipped with inclusion is a complete lattice. The least element of $\mathcal{P}(B_H)$ is $\emptyset$, its greatest element is B_H, $\sup_i K_i = \cup_i K_i$, $\inf_i K_i = \cap_i K_i$.

The immediate consequence operator $T_P: \mathcal{P}(B_H) \longrightarrow \mathcal{P}(B_H)$ is defined by: $T_P(I) = \{A \in B_H | \text{ there exists } r = (B_1, \ldots, B_n \Longrightarrow B) \in P, \text{ there exists a valuation } s: X \longrightarrow U_H \text{ such that, for } i = 1, \ldots, n, \ s^*(B_i) = A_i \in I, \ s^*(B) = A\}$. ($s^*(B) = B[x_1 := s(x_1)] \cdots [x_p := s(x_p)]$ (resp. $s^*(B_i) = B_i[x_1 := s(x_1)] \cdots [x_p := s(x_p)]$) denotes the ground atom obtained by substituting the term $s(x_k)$ for x_k in B (resp. B_i), for any variable $x_k \in X$.)

In other words, $T_P(I)$ is the set of atomic formulas A, such that $A_1, \ldots, A_n \Longrightarrow A$ is a ground instance of a clause r of P and, moreover, $A_1, \ldots, A_n$ are in I.

1. Show that T_P is monotone.
2. Let I be a Herbrand interpretation; show that I is a model of P if and only if $T_P(I) \subset I$.
3. Show that the least fixpoint of T_P is the least Herbrand model of P.
4. Show that T_P is continuous (i.e. for any increasing sequence $K_1 \subset K_2 \subset \cdots \subset K_n \subset \cdots$ of $\mathcal{P}(B_H)$, $\sup_i T_P(K_i) = T_P(\sup_i K_i)$).
5. Show that the least Herbrand model of P is defined by the basis

$$I_M = \sup(\{T_P^n(\emptyset) \,/\, n \in \mathbb{N}\}).$$

6. Show that, for any $n \in \mathbb{N}$, $T_P^n(B_H)$ is a Herbrand model of P.

Let $K = \inf(\{T_P^n(B_H) \,/\, n \in \mathbb{N}\})$.

7. Is K a model of P?
8. Is K a fixpoint of P? What can you say about the greatest fixpoint of P? ◇

CHAPTER 6

COMBINATORIAL ALGEBRA

In the present chapter we are interested in tools and techniques for counting finite sets and their subsets without enumerating all their elements. Discrete probabilities (see Chapter 12) are also rooted in the study of combinatorics. We start by recalling well-known results about permutations and combinations. We then study some counting techniques for finite sets which enable us to count the number of elements in a union, in a partition and in various combinations of finite sets.

We advise the following further reading:

Ronald Graham, Donald Knuth, Oren Patashnik, *Concrete Mathematics*, Addison-Wesley, London (1989).

Donald Knuth, *The Art of Computer Programming*, Vol. 1, Addison-Wesley, London (1973).

6.1 Basics

6.1.1 Generalities

Definition 6.1 *A permutation p of a finite set E is a bijection from E to E. The number of permutations of a set with n elements will be denoted by P_n.*

Identifying E with $\{1,\ldots,n\}$, a permutation p is characterized by a bijection $\{1,\ldots,n\} \longrightarrow \{1,\ldots,n\}$ determining a total ordering on the elements of E, given by the sequence $p(1), p(2), \ldots, p(n)$. Because there are n possible choices for $p(1)$, it follows that there are $(n-1)$ possible choices for $p(2)$, etc. (the word 'etc.' hides a proof by induction), therefore, $P_n = n!$.

Definition 6.2

1. A *k-permutation*, $k \leq n$, of a finite set E with cardinality n is a totally ordered subset of E with k elements. A_n^k denotes the number of k-permutations of a set with n elements.
2. A *k-combination*, $k \leq n$, of a finite set E with cardinality n is a subset of E with k elements. $\binom{n}{k}$ denotes the number of k-combinations of a set with n elements.

EXERCISE 6.1   Show that $A_n^k = \dfrac{n!}{(n-k)!}$. $\diamond$

Combinations are unordered, whilst k-permutations are ordered; hence each k-combination yields $k!$ k-permutations, and thus $A_n^k = k!\binom{n}{k}$. We have $A_n^n = P_n = n!$, $A_n^0 = 1$, $A_n^1 = n$. We also have $A_n^k = \dfrac{n!}{(n-k)!}$ (see Exercise 6.1), and thus $\binom{n}{k} = \dfrac{n!}{(n-k)!k!} = \binom{n}{n-k}$.

EXAMPLE 6.3   Let $E = \{a, b, c\}$. Then

- $(a, b, c), (b, c, a), (c, a, b), (b, a, c), (c, b, a), (a, c, b)$ are the permutations of E,
- $(a, b), (b, a), (a, c), (c, a), (b, c), (c, b)$ are the 2-permutations of E and
- $\{a, b\}, \{a, c\}, \{c, b\}$ are the 2-combinations of E.

REMARK 6.4   A k-permutation is characterized by an *injection*

$$i\colon \{1, \ldots, k\} \longrightarrow E,$$

and a k-combination is characterized by *the image* of an injection $\{1, \ldots, k\} \longrightarrow E$. As $k!$ different injections have the same image, we return to the previously stated result: $A_n^k = k!\binom{n}{k}$.

EXERCISE 6.2   Show that $\binom{a+b}{p} = \displaystyle\sum_{k=0}^{\inf(p,a)} \binom{a}{k}\binom{b}{p-k}$, with $p \leq a + b$. $\diamond$

EXERCISE 6.3

1. Show that $\displaystyle\sum_{k=0}^{n} \binom{n}{k}^2 = \binom{2n}{n}$.
2. Show that $\displaystyle\sum_{k=0}^{n}\sum_{i=0}^{n-k} \binom{n}{k}\binom{n}{i}\binom{n}{i+k} = \binom{3n}{n}$. $\diamond$

The k-permutations (resp. the k-combinations) are also called k-permutations without repetition (resp. k-combinations without repetition). Lastly, terms of the form $\binom{n}{p}$ are also called binomial coefficients or coefficients of *Pascal's triangle*.

Basics

Proposition 6.5 (Recurrence relations on the $\binom{n}{k}$s) *The binomial coefficients verify the identities*

(i) $$\binom{n}{k} = \binom{n}{n-k} \quad \text{for } 0 \leq k \leq n,$$

(ii) $$\binom{n}{k} = \binom{n-1}{k} + \binom{n-1}{k-1} \quad \text{for } 2 \leq k, 1 \leq n. \quad (6.1)$$

Proof. We have already proved (i). To prove (ii), choose an element $e \in E$, where E has cardinality n, and divide the $\binom{n}{k}$ combinations in two disjoint sets:
- those combinations which do not contain e, of which there are $\binom{n-1}{k}$, and
- those combinations which contain e, of which there are $\binom{n-1}{k-1}$.

These two sets are disjoint, so $\binom{n}{k} = \binom{n-1}{k} + \binom{n-1}{k-1}$. □

The binomial coefficients can be represented by Pascal's triangle, using the recurrence relation (6.1) for computing the successive $\binom{n}{k}$s.

n \ k	0	1	2	3	4
0	1	0	0	0	0
1	1	1	0	0	0
2	1	2	1	0	0
3	1	3	3	1	0
4	1	4	6	4	1
5	....				

Proposition 6.6 (Binomial theorem) *Let A be a ring, $a, b \in A$ such that $ab = ba$, and $n \in \mathbb{N}$. The following identity, called the binomial identity, holds:*

$$(a+b)^n = \binom{n}{0}a^n + \binom{n}{1}a^{n-1}b + \cdots + \binom{n}{p}a^{n-p}b^p + \cdots + \binom{n}{n}b^n$$

$$= \sum_{p=0}^{n} \binom{n}{p} a^{n-p} b^p.$$

Proof. By induction on n.
(B) If $n = 0$, $(a+b)^0 = \binom{0}{0} = 1$. If $n = 1$, $a + b = \binom{1}{0}a + \binom{1}{1}b$.
(I) Assume $(a+b)^n = \sum_{p=0}^{n} \binom{n}{p} a^{n-p} b^p$. Then, taking into account that $(a+b)^{n+1} = (a+b)^n (a+b)$, we have

$$(a+b)^{n+1} = \binom{n}{0} a^{n+1} + \sum_{p=1}^{n} \left(\binom{n}{p-1} + \binom{n}{p} \right) a^{n+1-p} b^p + \binom{n}{n} b^{n+1}.$$

Noting that $\binom{n+1}{0} = \binom{n}{0} = \binom{n+1}{n+1} = \binom{n}{n} = 1$, and that $\binom{n}{p-1} + \binom{n}{p} = \binom{n+1}{p}$, we indeed have $(a+b)^{n+1} = \sum_{p=0}^{n+1} \binom{n+1}{p} a^{n+1-p} b^p$. □

EXERCISE 6.4 Compute the number N of partitions of a set with np elements in n subsets with p elements. ◇

EXERCISE 6.5 Compute $S = \sum_{q=0}^{p} (-1)^q \binom{n}{q} \binom{n-q}{p-q}$, for $p \leq n$. ◇

EXERCISE 6.6

1. Show that $\sum_{p=0}^{k} \binom{n+p}{p} = \binom{n+k+1}{k}$, for $k \geq 0$.

2. Show that $\sum_{k=p}^{n} \binom{k}{p} = \binom{p}{p} + \binom{p+1}{p} + \cdots + \binom{n}{p} = \binom{n+1}{p+1}$.

Deduce the value of $\sum_{k=1}^{n} k^p$, for $p = 1, 2, 3$. ◇

EXERCISE 6.7 Let $P(x)$ be a polynomial of degree less than or equal to n. Show that $\sum_{i=0}^{n+1} (-1)^i \binom{n+1}{i} P(x+i) = 0$. ◇

6.1.2 Applications

The notions and results of the preceding section are very basic, but they can nevertheless be applied to counting finite sets, evaluating discrete probabilities or determining complexity and feasibility of algorithms. We illustrate such applications by examples and exercises.

REMARK 6.7 Recall (see Example 1.5) that the characteristic function of a subset A of a set E is the function

$$\chi_A : E \longrightarrow \{0, 1\},$$

defined by

$$\chi_A(x) = \begin{cases} 1 & \text{if } x \in A, \\ 0 & \text{otherwise.} \end{cases}$$

Conversely, any function $\chi : E \longrightarrow \{0, 1\}$ defines the subset $A = \chi^{-1}(\{1\})$ of E.

We will consider that $\{0, 1\} \subseteq \mathbb{B}$, or $\{0, 1\} \subseteq \mathbb{N}$; the choice will be clear by the context.

Basics

EXAMPLE 6.8 We have various methods for computing
$$S_n = \binom{n}{0} + \binom{n}{1} + \cdots + \binom{n}{n}.$$
1. By Remark 6.7, there is a one-to-one correspondence between the set of subsets of E and the set of functions $E \longrightarrow \mathbb{B}$. Thus

$$S_n = |\mathcal{P}(\{1, \ldots, n\})| = 2^n,$$

since there are 2^n functions $\{1, \ldots, n\} \longrightarrow \mathbb{B}$.
2. Note that $\binom{n}{k}$ represents the number of subsets with k elements of $\{1, \ldots, n\}$. We introduce the notation $+$ for the disjoint union: if A and B are disjoint, i.e. if $A \cap B = \emptyset$, then $A \cup B$ is denoted by $A + B$, and this notation is justified by the fact that $|A + B| = |A| + |B|$. Note, finally, that if P_k denotes the set of k-element subsets of $\{1, \ldots, n\}$, $P_0, \ldots, P_n$ form a partition of the set $\mathcal{P}(\{1, \ldots, n\})$ of subsets of $\{1, \ldots, n\}$. Since $\binom{n}{k} = |P_k|$ and

$$2^n = |\mathcal{P}(\{1, \ldots, n\})| = |P_0 + P_1 + \cdots + P_n| = |P_0| + |P_1| + \cdots + |P_n|,$$

we deduce $2^n = \binom{n}{0} + \cdots + \binom{n}{n}$.
3. Check that $S_n = 2S_{n-1}$ for $n \geq 1$. We apply the recurrence relation (6.1) on the $\binom{n}{k}$s: then

$$S_n = \binom{n}{0} + \binom{n}{1} + \cdots + \binom{n}{n-1} + \binom{n}{k} + \cdots + \binom{n}{n}$$
$$= \binom{n}{0} + \binom{n-1}{1} + \binom{n-1}{2} + \cdots + \binom{n-1}{k} + \cdots + \binom{n-1}{n-1}$$
$$+ \binom{n-1}{0} + \binom{n-1}{1} + \cdots + \binom{n-1}{k-1} + \cdots + \binom{n-1}{n-2} + \binom{n}{n}$$
$$= 2S_{n-1} \quad \left(\text{since } \binom{n}{0} = \binom{n-1}{0} = 1 = \binom{n-1}{n-1} = \binom{n}{n}\right).$$

Hence, noting that $S_0 = 1$ and multiplying the equalities $S_n = 2S_{n-1}$ yields $S_n = 2^n$.
4. Finally, we can apply the binomial identity, Proposition 6.6, with $a = b = 1$, and deduce $S_n = (1 + 1)^n$.

EXERCISE 6.8 Given twenty-seven white cubes, we stack them to build a cube three times larger. The outside of the big cube is painted in red, then the big cube is pulled down and the pieces are given to a blind person who is asked to rebuild it. Compute $p = n_f/n$, where n_f is the number of ways to rebuild a red cube (number of favourable cases) and n is the total number of ways to rebuild a cube (number of possible cases)? (p is the probability that the rebuilt cube is red.) ◇

EXERCISE 6.9 How many five-card hands, chosen from a deck of thirty-two cards (four suits), are there:
1. containing a four-of-a-kind (four cards of equal face values)?
2. containing a three-of-a-kind (three cards of equal face values) and nothing else?
3. containing a pair (two cards of equal face values) and nothing else? ◇

EXERCISE 6.10 Compute the number of strings of sixteen bits containing eight bits equal to 1. ◇

EXAMPLE 6.9 This example shows how to use combinatorial algebra to prove the (in)tractability of some algorithms by evaluating their complexity *a priori*. We consider the problem of a travelling salesman who wishes to visit n pairwise connected cities $\{1, \ldots, n\}$ (i.e. forming a complete graph with n nodes). The distance between cities i and j is denoted by c_{ij}. He starts and ends his tour in city 1, visits each city exactly once, and wants to drive as few miles as possible. See also Chapter 10.

The simplest algorithm is to enumerate all the cycles starting at node 1 and to compute the length of each cycle; then choosing the shortest possible cycle will do the job. For each cycle consisting of n cities, the computation of its length needs $n-1$ additions, and since there are $(n-1)!$ cycles starting at node 1, the total cost of such an algorithm is $(n-1) \times (n-1)!$ additions. For a tour of fifty cities, we have $49 \times 49!$ (circa $3 \cdot 10^{64}$) additions (see Chapter 9 on asymptotic behaviours for the order of magnitude of $n!$). A computer performing 10^9 additions per second will need 10^{47} years to complete the computation of the optimal path. This cost is prohibitive for the sales of the travelling salesman. Practically, this algorithm will thus be excluded, and we must consider heuristic methods which will involve some cycles only. We will no longer find the shortest path but only the shortest path among the class of considered paths, the asset being that this path will be obtained after a reasonable amount of time.

6.2 Applications: counting techniques for finite sets

6.2.1 Fundamentals

Here we recall results which can be found in a slightly different form in Chapter 4. We generalize the notion of characteristic function as follows. Let $f: E \to \{0, 1, 2\}$ be defined by
$$f(x) = \begin{cases} 1 & \text{if } x \in A_1, \\ 2 & \text{if } x \in A_2 \text{ and } x \notin A_1, \\ 0 & \text{if } x \notin A_2. \end{cases}$$

EXERCISE 6.11 E is a finite set with n elements.

Applications: counting techniques for finite sets

1. What is the number N_1 of pairs (A_1, A_2) such that
$$A_1 \subseteq E, \ A_2 \subseteq E, \ \text{and} \ A_1 \subseteq A_2 \ ? \tag{6.2}$$

2. Let N_2 be the number of triples (A_1, A_2, A_3) verifying
$$A_1 \subseteq A_2 \subseteq A_3 \subseteq E. \tag{6.3}$$

Compute N_2. ◇

The operations on the subsets A of E will then correspond to operations on the corresponding characteristic functions, and to the operations on $\mathbb{B}$ through which the operations on the characteristic functions are defined (see Section 4.1.3). An operation on $\mathbb{B}$ is described by its truth table. For example, the unary operation corresponding to negation (or complement) is $\neg x = 1 - x$ with the truth table:

x	0	1
$\neg x$	1	0

The binary operations corresponding to the disjunction $x \lor y$ and the conjunction $x \land y$ are described on $\mathbb{B}$ by the tables:

$\land$	x	0	1
y			
0		0	0
1		0	1

$\lor$	x	0	1
y			
0		0	1
1		1	1

and they can be characterized on $\mathbb{N}$ by (see Example 4.6)
$$x \land y = xy \ \text{and} \ x \lor y = x + y - xy \ (= x + y \bmod 2).$$

Lemma 6.10 Let A and B be two subsets of E, and let $\alpha = \chi_A$ (resp. $\beta = \chi_B$) be the characteristic function of A (resp. B). Then $\alpha \land \beta = \alpha\beta$ (resp. $\alpha \lor \beta = \alpha + \beta - \alpha\beta$) is the characteristic function of $A \cap B$ (resp. $A \cup B$). $\neg \alpha = 1 - \alpha$ is the characteristic function of $\bar{A}$.

Lemma 6.11 $\overline{A_1 \cup \cdots \cup A_n} = \overline{A_1} \cap \cdots \cap \overline{A_n}$. That is, the complement of a union is the intersection of the complements.

Proof. See De Morgan's laws in Chapter 1 or Proposition 4.3 in Chapter 4. □

Lemma 6.12 $|A| = \sum_{e \in E} \chi_A(e)$ *for any subset A of E.*

Proof. Here, χ_A is considered to be a function with values in $\{0,1\} \subseteq \mathbb{N}$. Since $e \in A \iff \chi_A(e) = 1$,
$$\sum_{e \in E} \chi_A(e) = \sum_{e \in A} 1 = |A|. \qquad \square$$

6.2.2 Inclusion–exclusion principle and applications

We will apply the preceding techniques in order to compute the cardinality of sets (unions of subsets of E, number of surjections, injections, etc.).

Proposition 6.13 *Let $A_i \subseteq E$ be subsets of E, for $i = \{1, \ldots, m\}$. Then, we have Sylvester's identity*

$$\begin{aligned}
|A_1 \cup \cdots \cup A_m| &= |A_1| + \cdots + |A_m| - \sum_{i<j} |A_i \cap A_j| + \sum_{i<j<k} |A_i \cap A_j \cap A_k| \\
&\quad + \cdots + (-1)^{p-1} \sum_{i_1 < \cdots < i_p} |A_{i_1} \cap \cdots \cap A_{i_p}| + \cdots \\
&\quad + (-1)^{m-1} |A_1 \cap \cdots \cap A_m| \\
&= \sum_{p=1}^{m} (-1)^{p-1} \sum_{i_1 < \cdots < i_p} |A_{i_1} \cap \cdots \cap A_{i_p}|.
\end{aligned}$$

Proof. First method: We will apply the three lemmas stated at the end of the preceding section. Let $A = A_1 \cup \cdots \cup A_m$, $\chi_A = 1 - \chi_{\overline{A}}$; also let $\chi_{A_i} = \alpha_i$. Then $\chi_{\overline{A_i}} = 1 - \alpha_i$ and, since $\overline{A} = \overline{A_1} \cap \cdots \cap \overline{A_m}$, we have

$$\chi_{\overline{A}} = \prod_{i=1}^{m} \chi_{\overline{A_i}} = \prod_{i=1}^{m} (1 - \alpha_i)$$
$$= 1 - (\alpha_1 + \cdots + \alpha_m) + \sum_{i<j} \alpha_i \alpha_j - \sum_{i<j<k} \alpha_i \alpha_j \alpha_k + \cdots .$$

We will then use

$$\chi_A = 1 - \chi_{\overline{A}} = \alpha_1 + \cdots + \alpha_m - \sum_{i<j} \alpha_i \alpha_j + \sum_{i<j<k} \alpha_i \alpha_j \alpha_k - \cdots$$
$$= \sum_{p=1}^{n} (-1)^{p-1} \sum_{i_1 < i_2 < \cdots < i_p} \alpha_{i_1} \alpha_{i_2} \cdots \alpha_{i_p},$$

Applications: counting techniques for finite sets

and we will apply Lemma 6.12, which tells us that $|A| = \sum_{e \in E} \chi_A(e)$. Thus

$$|A| = \sum_{e \in E} \chi_A(e) = \sum_{e \in E} \sum_{p=1}^{m} (-1)^{p-1} \sum_{i_1 < \cdots < i_p} \alpha_{i_1}(e) \cdots \alpha_{i_p}(e)$$

$$= \sum_{p=1}^{m} (-1)^{p-1} \sum_{i_1 < \cdots < i_p} \sum_{e \in E} \alpha_{i_1}(e) \cdots \alpha_{i_p}(e)$$

$$= \sum_{p=1}^{m} (-1)^{p-1} \sum_{i_1 < \cdots < i_p} |A_{i_1} \cap \cdots \cap A_{i_p}|$$

(by noting that $\alpha_{i_1} \cdots \alpha_{i_p} = \chi_{A_{i_1} \cap \cdots \cap A_{i_p}}$).

Second method: By induction on m.
(B) Straightforward for $m = 1$ and $m = 2$.
(I) Assume

$$|A_1 \cup \cdots \cup A_m| = \sum_{p=1}^{m} (-1)^{p-1} \sum_{1 \leq i_1 < \cdots < i_p \leq m} |A_{i_1} \cap \cdots \cap A_{i_p}|$$

$$= |A_1| + \cdots + |A_m|$$

$$+ \sum_{p=2}^{m} (-1)^{p-1} \sum_{1 \leq i_1 < \cdots < i_p \leq m} |A_{i_1} \cap \cdots \cap A_{i_p}|.$$

and compute, for $m \geq 2$, $|A_1 \cup \cdots \cup A_m \cup A_{m+1}|$. It follows that

$$|A_1 \cup \cdots \cup A_m \cup A_{m+1}| = |A_1 \cup \cdots \cup A_m| + |A_{m+1}|$$
$$- |(A_1 \cup \cdots \cup A_m) \cap A_{m+1}|.$$

Moreover, letting $A'_i = A_i \cap A_{m+1}$, we have

$$(A_{i_1} \cup \cdots \cup A_{i_p}) \cap A_{m+1} = A'_{i_1} \cup \cdots \cup A'_{i_p}$$

and

$$(A_{i_1} \cap \cdots \cap A_{i_p}) \cap A_{m+1} = A'_{i_1} \cap \cdots \cap A'_{i_p};$$

hence, by the induction hypothesis

$$|(A_1 \cup \cdots \cup A_m) \cap A_{m+1}| = |A'_1 \cup \cdots \cup A'_m|$$

$$= |A'_1| + \cdots + |A'_m| + \sum_{p=2}^{m} (-1)^{p-1} \sum_{1 \leq i_1 < \cdots < i_p \leq m} |A'_{i_1} \cap \cdots \cap A'_{i_p}|$$

$$= |A'_1| + \cdots + |A'_m|$$

$$+ \sum_{p=2}^{m} (-1)^{p-1} \sum_{1 \leq i_1 < \cdots < i_p \leq m} |A_{i_1} \cap \cdots \cap A_{i_p} \cap A_{m+1}|$$

and

$$|A_1 \cup \cdots \cup A_m \cup A_{m+1}| = |A_1| + \cdots + |A_m| + |A_{m+1}| - |A_1'| - \cdots - |A_m'|$$
$$+ \sum_{p=2}^{m}(-1)^{p-1} \sum_{1 \leq i_1 < \cdots < i_p \leq m} |A_{i_1} \cap \cdots \cap A_{i_p}|$$
$$- \sum_{p=2}^{m}(-1)^{p-1} \sum_{1 \leq i_1 < \cdots < i_p \leq m} |A_{i_1} \cap \cdots \cap A_{i_p} \cap A_{m+1}|$$
$$= \sum_{p=1}^{m+1}(-1)^{p-1} \sum_{1 \leq i_1 < \cdots < i_p \leq m+1} |A_{i_1} \cap \cdots \cap A_{i_p}|. \qquad \square$$

EXERCISE 6.12 $|E| = n$, $A \cap B = \emptyset$, $|A| = n_1$, $|B| = n_2$. Compute the number N of subsets with p elements, with $p \geq 2$, and with
1. exactly one element from A and one element from B,
2. at least one element from A and one element from B. $\diamond$

EXERCISE 6.13 Let $\{a, b, c, d\}$ be a four-letter alphabet. What are:
1. the number of strings of length n over this alphabet?
2. the number of strings of length n in which each of the letters a, b, c, d occurs at least once? $\diamond$

EXERCISE 6.14 Among the permutations of $\{a, b, c, d, e, f\}$, how many contain neither 'ac' nor 'bde' ? $\diamond$

EXERCISE 6.15 What is the number u_n of binary strings with n bits containing neither 010 nor 11. $\diamond$

Proposition 6.14 *Let A and B be two sets with cardinality $|A| = m$ and $|B| = n$.*

1. *The number of mappings from A to B is n^m.*
2. *The number of injections (or one-to-one mappings) from A to B is A_n^m, if $m \leq n$.*
3. *The number of surjections (or onto mappings) from A to B is*

$$S_n^m = \begin{cases} 0 & \text{if } m < n, \\ n! & \text{if } m = n, \\ \sum_{p=0}^{n}(-1)^p \binom{n}{p}(n-p)^m & \text{if } m > n. \end{cases}$$

Proof.
1. Indeed, there are n possible choices for the image of each element $a_1, \ldots, a_m$ in A, thus n^m choices altogether (see Proposition 1.9 (iv)). As an exercise, the reader is invited to give a formal proof by induction on m.

2. See Remark 6.4.

3. It is an application of the preceding proposition. The first two cases are straightforward, and so the only case requiring a proof is the case when $m > n$. We first compute the number of mappings from A to B. We then determine, using the preceding proposition, the number of non-surjective mappings from A to B. We finally deduce by difference the number of surjections from A to B, since, clearly, the set of mappings from A to B is the disjoint union of surjections on the one hand, and of mappings that are not surjections on the other hand.

Let $N = \{f: A \longrightarrow B \;/\; f \text{ non-surjective}\}$; f is non-surjective if and only if $\exists b_i \in B$ such that $b_i \notin f(A)$. Thus let,

$$A_i = \{f: A \longrightarrow B \;/\; b_i \notin f(A)\}, \quad i = 1, \ldots, n.$$

We will have

$$N = \{f: A \longrightarrow B \;/\; f \text{ non-surjective}\}$$
$$= \{f: A \longrightarrow B \;/\; f(A) \neq B\} = A_1 \cup \cdots \cup A_n.$$

By the preceding proposition, we thus have

$$|N| = \sum_{p=1}^{n} (-1)^{p-1} \left(\sum_{i_1 < \cdots < i_p} |A_{i_1} \cap \cdots \cap A_{i_p}| \right)$$

and it suffices to compute $|A_{i_1} \cap \cdots \cap A_{i_p}|$. Note now that

$$A_{i_1} \cap \cdots \cap A_{i_p} = \{f: A \longrightarrow B \;/\; b_{i_1} \notin f(A), \ldots, b_{i_p} \notin f(A)\}$$
$$= \{f: A \longrightarrow B - \{b_{i_1}, \ldots, b_{i_p}\}\},$$

and thus $A_{i_1} \cap \cdots \cap A_{i_p}$ is the set of mappings from A, a set with m elements, to $B - \{b_{i_1}, \ldots, b_{i_p}\}$, a set with $n-p$ elements. There are $(n-p)^m$ such mappings by Proposition 6.14, 1. Since, moreover, there are $\binom{n}{p}$ possible choices of $b_{i_1}, \ldots, b_{i_p}$ in $\{b_1, \ldots, b_n\}$, we deduce $|N| = \sum_{p=1}^{n} (-1)^{p-1} \binom{n}{p} (n-p)^m$. Finally, noting that $n^m = \binom{n}{0}(n-0)^m$, we have

$$S_n^m = n^m - |N| = \sum_{p=0}^{n} (-1)^p \binom{n}{p} (n-p)^m.$$

Another way of computing S_n^m is given in Example 7.27. □

EXERCISE 6.16 A function f from $\{1, 2, \ldots, n\}$ to $\{1, 2, \ldots, m\}$ is said to be increasing if $x < y$ implies $f(x) < f(y)$.

1. What is the number of increasing functions (in terms of n and m)?
2. What is the number of increasing functions such that

$$\exists x \colon f(x) = k+1 \quad \text{for } m = 2k+1 \text{ and } k > 1 ?$$

3. What is the number of increasing functions such that, for a fixed k,

$$|\{a \,/\, f(a) < k\}| = |\{a \,/\, f(a) > k\}|\,?$$

4. What is the number of injective functions such that

$$|\{a \,/\, f(a) < k\}| = |\{a \,/\, f(a) > k\}|\,? \qquad \diamond$$

6.3 Counting sequences and partitions

We now give some other counting formulas that will be of use in probability theory.

Definition 6.15 *A k-permutation with repetition allowed of a set E with n elements is an ordered sequence with k elements from E in which each element may occur arbitrarily often.*

EXAMPLE 6.16 Let $E = \{a, b, c\}$. Then aa, ab, ba, bb are 2-permutations with repetition of E. Two k-permutations can differ by the ordering of their elements, by their elements or by both.

Proposition 6.17 *Let E be a set with cardinality n and $k \in \mathbb{N}$. (It is not assumed that $k \leq n$.) A k-permutation with repetition of E is defined by a mapping from $\{1, \ldots, k\}$ to E. There are thus n^k such k-permutations.*

Definition 6.18 *A k-combination with repetition of a set E with n elements is an unordered set with k elements of E in which each element can occur arbitrarily often.*

A set whose elements can occur arbitrarily often is called a *multiset*. The difference between a k-combination with repetition and a k-permutation with repetition is the following: a k-combination with repetition is an *unordered* multiset of elements, possibly repeated, whilst a k-permutation is an *ordered* sequence. For instance the 3-permutations aba and baa correspond to the same 3-combination: $\{a, a, b\}$. Two k-combinations with repetition can differ by their elements, by the number of repetitions or by both. A k-combination with repetition of elements of $E = \{1, \ldots, n\}$ will contain n_1 occurrences of i_1, ..., n_p occurrences of i_p, with

- $\forall j, \ 1 \leq i_j \leq n$,
- $n_1 + \cdots + n_p = k$.

Counting sequences and partitions

Proposition 6.19 *Let E be a set of cardinality n, and let $k \in \mathbb{N}$. A k-combination with repetition of elements of E is defined by a mapping f from E to $\{0, 1, \ldots, k\}$ such that $\sum_{i=1}^{n} f(e_i) = k$. An element e_i of E occurs in the k-combination j_i times if and only if $f(e_i) = j_i$.*

Equivalently, a k-combination with repetition is defined by a solution of the equation $j_1 + \cdots + j_n = k$, with $j_i \in \mathbb{N}$, $\forall i \in \{1, \ldots, n\}$.

Proposition 6.20 *The number of k-combinations with repetition of a set E with n elements is $\binom{n+k-1}{n-1}$.*

Proof. There is a one-to-one correspondence between k-combinations with repetition and the sequences $j_1, \ldots, j_n$ such that $j_1 + \cdots + j_n = k$. We can represent such a sequence by the string of length $n + k - 1$ over the alphabet $\{0, 1\}$ given by $0^{j_1} 1 0^{j_2} 1 \ldots 0^{j_{n-1}} 1 0^{j_n}$. (The rth sequence of 0s represents the number of repetitions of e_r, and the 1s act as separators.)

Such a k-combination is thus determined by a string of length $k + n - 1$ over the alphabet $\{0, 1\}$ consisting of exactly $n - 1$ occurrences of 1. It thus suffices to determine the number of such strings by characterizing them by the positions where the 1s occur. There are $\binom{n+k-1}{n-1}$ possible choices for fitting $n - 1$ 1s in a string of length $n + k - 1$. □

REMARK 6.21
1. $\binom{n+k-1}{n-1}$ is also the number of monomials of degree k on n variables.
2. $\binom{n+k-1}{n-1}$ is also the number of monotone mappings from $\{1, 2, \ldots, k\}$ to $\{1, 2, \ldots, n\}$.

Finally, we will study the partitions of a set with n elements in k disjoint sets $A_1, \ldots, A_k$ such that $|A_i| = n_i$ and $\sum_{i=1}^{k} n_i = n$, and this will lead us to the definition of multinomial coefficients.

Theorem 6.22 *The number of partitions $\binom{n}{n_1, \ldots, n_k}$ of a set E with n elements in k classes $A_1, \ldots, A_k$ each having n_i elements, with $\sum_{i=1}^{k} n_i = n$, is $\dfrac{n!}{n_1! \cdots n_k!}$.*

Proof. By induction on k.
(B) If $k = 2$, then choosing a n_1-element set A_1 defines a partition of E in two disjoint sets A_1, A_2 with $|A_2| = n_2 = n - n_1$, hence

$$\binom{n}{n_1, n_2} = \binom{n}{n_1} = \frac{n!}{n_1! n_2!}.$$

(I) Assume $\binom{n}{n_1, \ldots, n_k} = \dfrac{n!}{n_1! \cdots n_k!}$, and let $A_1, \ldots, A_{k+1}$ be a partition of E in $k + 1$ subsets. Let $B_k = A_k \cup A_{k+1}$. The number of partitions $A_1, \ldots, A_{k-1}, B_k$

of E is $\binom{n}{n_1,\ldots,m_k}$, where $m_k = n_k + n_{k+1}$, i.e. $\dfrac{n!}{n_1!\cdots n_{k-1}!(n_k+n_{k+1})!}$. Moreover, the number of partitions of B_k in $A_k \cup A_{k+1}$ is $\dfrac{(n_k+n_{k+1})!}{n_k!n_{k+1}!}$. Hence, by multiplication,

$$\binom{n}{n_1,\ldots,n_k,n_{k+1}} = \frac{n!}{n_1!\cdots n_k!n_{k+1}!}.$$
□

The $\binom{n}{n_1,\ldots,n_k}$s are also called the *multinomial coefficients*. We have the following multinomial identity.

Proposition 6.23

$$(X_1 + \cdots + X_k)^n = \sum_{n_1+\cdots+n_k=n} \binom{n}{n_1,\ldots,n_k} X_1^{n_1} \cdots X_k^{n_k}.$$

Proof. See Exercise 6.17. □

EXERCISE 6.17 We are given n letters not assumed to be pairwise distinct: q_1 letters $a_1, \ldots,$ and q_p letters a_p, with $q_1 + q_2 + \cdots + q_p = n$.
1. How many different strings of length n can be written using those n letters?
   (a) Deduce a representation of the formal polynomial $(X_1 + X_2 + \cdots + X_p)^n$.
   (b) Deduce an expression of the multinomial coefficients in terms of the binomial coefficients.
2. Deduce that $(k!)!$ is divisible by $k!^{(k-1)!}$.
3. Compute the number of strings of length 13 that can be written with $q_1 = 5$ letters a_1, and $q_2 = 8$ letters a_2. ◇

EXERCISE 6.18 Compute $\sum_{p=0}^n p^2 \binom{2n}{2p}$, for $n \geq 2$.
Hint: Let $g(x) = \sum_{p=1}^n p^2 \binom{2n}{2p} x^{2p-2}$ and $f(x) = (1+x)^{2n} + (1-x)^{2n}$, and find a relation among $g(x)$, $f'(x)$, and $f''(x)$. ◇

EXERCISE 6.19 For $n \in \mathbb{N}$ and $p \in \mathbb{N} - \{0\}$, denote by $F(n,p)$ the number of p-tuples $(x_1,\ldots,x_p) \in \mathbb{N}^p$ such that $x_1 + \cdots + x_p = n$. Compute $F(n,p)$.
Method 1
   (a) Show that $F(n, p+1) = \sum_{k=0}^n F(k,p)$.
   (b) Show that $\binom{n+p}{p} = \sum_{k=0}^n \binom{k+p-1}{p-1}$, for $n \geq 0$, and $p \geq 1$.
   (c) Compute $F(n,p)$.
Method 2
   (a) Show that $F(n, p+1) = F(n,p) + F(n-1, p+1)$.
   (b) Show that $F(n,p) = \binom{n+p-1}{n}$. ◇

CHAPTER 7

RECURRENCES

Recurrence relations define sequences of numbers. They can be obtained

- either by inductive reasoning as we have already seen in examples in Chapter 3,
- or by strategies which divide a size n problem into smaller problems (of size $\leq n-1$), each of which can be solved more easily in general; e.g. strategies of the type 'divide and conquer', solving problems by dichotomy. Several such examples will be studied below.

In the present chapter we will study some methods for explicitly finding the sequences of numbers defined by such relations.

This chapter is mainly devoted to linear recurrences for which there exist classical mathematical theories explaining how and why these recurrences can be explicitly solved. Linear algebra (vector spaces, matrices, eigenvectors and eigenvalues) is one of these theories. Its knowledge is assumed for the proof of Proposition 7.14. For fundamentals about algebra we suggest

C. Norman, *Undergraduate Algebra*, Oxford University Press, Oxford (1986).
Bartel Leenert Van der Waerden, *Algebra*, Frederick Ungar Publishing Company, New York (1970).

We recommend the following handbooks:

Gilles Brassard, Paul Bratley, *Algorithmics: Theory and Practice*, Prentice Hall, London, (1988).
Ronald Graham, Donald Knuth, Oren Patashnik, *Concrete Mathematics*, Addison-Wesley, London (1989).
Donald Knuth, *The Art of Computer Programming*, Vol. 1, Addison-Wesley, London (1973).
Chung Laung Liu, *Introduction to Combinatorial Mathematics*, Mc Graw-Hill, New York (1968).

7.1 Introduction: examples, generalities

7.1.1 Examples

The present section consists of a list of examples showing how recurrence relations are obtained, what forms they can take, and gives some ideas on how to solve them.

EXAMPLE 7.1 (Number of binary trees with n nodes) The binary trees studied here can possibly be empty, which is not the case for the trees studied in Chapter 10. Recall (see Example 3.9, 5) that binary trees B labelled by the alphabet $\{a\}$ are recursively defined by

- $\emptyset$ is a binary tree, namely, $\emptyset \in B$ (basis),
- if b_l and b_r are binary trees then (a, b_l, b_r) is also a binary tree (inductive step of the recursive definition).

We can define the number $n(b)$ of nodes of a binary tree b similarly by recurrence

- tree $\emptyset$ has no node, namely $n(\emptyset) = 0$,
- $n((a, b_l, b_r)) = 1 + n(b_l) + n(b_r)$.

We can evaluate the number b_n of binary trees with n nodes as follows. A tree with n nodes can be represented by

where f_l (the left child) is a tree with k nodes and f_r (the right child) is a tree with $n - k - 1$ nodes.

We will evaluate the number b_n of binary trees with n nodes by induction, and we will notice that

- $b_0 = 1$ (the empty tree $\emptyset$ is the only tree with 0 nodes),
- $b_n = \sum_{k=0}^{n-1} b_k \times b_{n-1-k}$: this equality follows from the fact that in order to obtain a tree with n nodes it is necessary (and sufficient) to consider all possible binary trees of the form (a, b_l, b_r) with b_l (resp. b_r) a binary tree with k (resp. $n - k - 1$) nodes; there are b_k possibilities for b_l, and b_{n-k-1} possible choices for b_r, thus $b_k \times b_{n-k-1}$ possible choices for b, and this holds for all possible ks. Hence, we have the recurrence relation

$$b_n = \sum_{k=0}^{n-1} b_k \times b_{n-1-k} \ . \tag{7.1}$$

Introduction: examples, generalities 123

We see that in this case, b_n is defined in terms of $b_0, b_1, \ldots, b_{n-1}$: this is one of the most complex cases of recurrence relations that we will encounter.

EXERCISE 7.1 Recall that in a binary tree a node is said to be internal if it has either a non-empty right child, or a non-empty left child or both. Let b_n be the number of binary trees with n internal nodes.
1. Compute b_0, b_1, b_2.
2. Find a recurrence relation giving b_n. ◇

EXERCISE 7.2 Let $\Sigma = \{a_1, a_2, \ldots, a_k\}$ be an alphabet with k elements; recall that the binary trees BIN labelled by Σ are defined inductively by
- $\emptyset$ is a binary tree,
- if $x \in \Sigma$, $b_l \in BIN$, $b_r \in BIN$, then $(x, b_l, b_r) \in BIN$.

The depth $p(b)$ of a binary tree is defined by:
- $p(\emptyset) = 0$,
- $p\big((x, b_l, b_r)\big) = 1 + \sup\{p(b_l), p(b_r)\}$.

Give recurrence relations defining
1. the number u_n of binary trees U_n of depth less than or equal to n (in terms of u_{n-1}). (Computing u_n is not required.)
2. the number v_n of binary trees AB_n of depth exactly n (in terms of v_{n-1} and of the u_is for $i \leq n$). (Computing v_n is not required.) ◇

EXAMPLE 7.2 The Fibonacci numbers are defined by the recurrence relation
$$F_{n+1} = F_n + F_{n-1}, \tag{7.2}$$
with initial conditions $F_0 = 0$, $F_1 = 1$.

EXERCISE 7.3 n lines are drawn on a plane; they intersect and thus delimit a certain number of bounded regions and of infinite regions. What is the maximum possible number of bounded regions determined? ◇

EXERCISE 7.4 n overlapping circles in the plane are assumed to intersect pairwise in two points, with neither tangential nor triple points. Show that the number of regions thus defined in the plane is determined by the recurrence relation
$$n > 1, \quad r_n = r_{n-1} + 2(n-1), \tag{7.3}$$
with $r_1 = 2$. ◇

EXAMPLE 7.3 Binary search to determine the maximum in a list of n elements: in order to find the maximum in a length n list, we divide it into two lists of length $n/2$ (assuming n of the form 2^k), we find the maximum of each one of the two lists, then we compare these two maxima. If t_n is the required time for finding the maximum of a length n list, we have
$$t_n = 2t_{n/2} + 1, \quad t_2 = 1. \tag{7.4}$$
(We assume that the unit of time complexity is the cost of a comparison, hence $t_2 = 1$.)

EXERCISE 7.5 Compute t_n defined in Example 7.3. ◇

EXAMPLE 7.4 Let $f_h(k)$ be the maximum number of leaves on a tree of height h where each node has at most k children. $f_1(k) = 1$, since a tree of height 1 is reduced to a single node being both root and leaf. We immediately check that

$$f_h(k) = k f_{h-1}(k) \, . \tag{7.5}$$

See also Chapter 10.

EXERCISE 7.6 Compute $s_p = 1 - 2 + \cdots + (2p-3) - (2p-2) + (2p-1) - 2p$, for $p \in \mathbb{N}$. ◇

EXAMPLE 7.5 The following recurrence relations are useful for studying the complexity of Quicksort (see Section 14.1):

$$p_2 = 3, \text{ and for } n > 2, \qquad p_n = n + 1 + p_{n-1} \, , \tag{7.6}$$

$a_0 = b_0 = a_1 = b_1 = c_0 = c_1 = 0$, and $\forall n \geq 2$,

$$a_n \leq c_n \leq b_n \, ,$$

$$a_n = n - 1 + 2/n \sum_{k=1}^{n-1} a_k \quad \text{and} \quad b_n = n + 1 + 2/n \sum_{k=1}^{n-1} b_k \, . \tag{7.7}$$

7.1.2 Generalities, classification

In the preceding section we saw various recurrence relations. They were all of the form

$$u_n = f(n, \{u_0, u_1, \ldots, u_{n-1})\}) , \qquad n \in J \subseteq \mathbb{N} \, ,$$

with initial conditions enabling us to start the recurrence. Several methods are available for solving them but we must first determine the type of the recurrence relation. To this end, we have three orthogonal classification criteria.

- First, the type of the function f, which can be
  - a *linear combination*, as in the case of relations (7.2), (7.3), (7.4), (7.5), (7.6), (7.7), having constant coefficients ((7.2),(7.3),(7.4),(7.5)), or coefficients depending on n ((7.7)); the recurrence relation is then said to be *linear*;
  - a *polynomial*, as in the case of relation (7.1); the recurrence relation is then said to be *polynomial*;
- Second, the set of u_ps needed to compute u_n
  - if we need $u_{n-1}, \ldots, u_{n-k}$ to compute u_n, the recurrence relation is said to be *of degree k*, and then $J \subseteq \{n \, / \, n \geq k\}$. Relation (7.2) is of degree 2, (7.3), (7.5), (7.6) are of degree 1.

Introduction: examples, generalities

– if we need $u_0, \ldots, u_{n-1}$ to compute u_n, the recurrence relation is said to be *complete*, and then $J = \{n \,/\, n \geq 1\}$. For instance, relation (7.1) is complete, and so are relations (7.7).

– if we only need $u_{n/a}$ with a constant, $a \in \mathbb{N}$, $a > 1$, to compute u_n, the recurrence relation is then said to be a *partition recurrence*; in that case $J = \{n \,/\, a \text{ divides } n\}$. For instance, relation (7.4) is of this type. Usually, partition recurrences are obtained by dividing a size n problem into one or more smaller problems, hopefully simpler to solve, e.g. using dichotomic methods, and more generally strategies of the type 'divide and conquer'.

- Third, the fact that function f does or does not have parameters and terms other than multiples of the u_js, $j < n$.

– if f depends only on the u_js, $j < n$, the recurrence relation is said to be *homogeneous*: e.g. recurrences (7.1), (7.2), (7.5).

– if f depends on terms other than the u_js, most often the recurrence relation will be of the form $u_n = f(\{u_p \,/\, p < n\}) + g(n)$, the recurrence is said to be *non-homogeneous* and $g(n)$ is called its *right-hand side*: e.g. recurrences (7.3), (7.4).

Let $v_n = f(\{v_p \,/\, p < n\})$, $n \in J \subseteq \mathbb{N}$, be a recurrence relation. Solving this recurrence consists of finding a sequence $(u_i)_{i \geq 0}$ such that $\forall n \in J$, $u_n = f(\{u_p \,/\, p < n\})$. Among the solutions, we are interested in those satisfying *initial conditions*, given by a set of values $\{a_i \,/\, i \in I\}$, where I is a subset of $\mathbb{N}$. A sequence $(u_i)_{i \geq 0}$ satisfies the initial conditions if $\forall n \in I$, $u_n = a_n$.

Many useful and systematic methods are available for solving linear recurrences of finite degree, whether homogeneous or not. Some methods are available in the case of partition recurrences, though to a lesser extent. Finally, for polynomial or complete recurrences, the solution is more complex, and will involve more elaborate tools such as generating series and differential equations. We will study some examples in Chapter 8 (generating series).

Proposition 7.6 *Let $v_n = f(\{v_p \,/\, p < n\})$ be a recurrence relation; assume that f is a mapping (i.e. its domain is the whole set $\mathbb{N}$);*

- *if the recurrence is complete, and if u_0 is given, the recurrence has at least one solution.*
- *if the recurrence is of degree k, and if we are given the initial values $u_0, \ldots, u_j$, $j < k$, the recurrence has at least one solution.*
- *if the recurrence is a partition recurrence of the form $v_n = f(v_{n/a})$, and if we are given $u_0, \ldots, u_j$, $j < a$, the recurrence has at least one solution.*

Proof. Verify, by induction on n, the property $P(n)$: $\exists u_0, u_1, u_2, \ldots, u_n$, satisfying the given recurrence; assume, for instance, that f is complete, and that u_0 is given. Then,

- u_0 exists since it is given, and u_0 trivially satisfies the recurrence;
- assume $u_0, \ldots, u_{n-1}$ satisfying the recurrence exist, and let $u_n = f(u_0, \ldots, u_{n-1})$; u_n satisfies the recurrence by construction; we have thus shown the existence of $u_0, \ldots, u_n$ satisfying the recurrence, hence the induction hypothesis and the result.

The cases when f is a linear or a partition recurrence are similar. □

REMARK 7.7 A solution may not exist if the initial conditions are too demanding, e.g. initial values $u_0, \ldots, u_{k+p}$, where $u_k, \ldots, u_{k+p}$, are incompatible with the recurrence relation in the case of a linear recurrence of degree k.

EXERCISE 7.7 Find examples of non-solvable recurrences. ◇

In spite of the preceding remark, the problem of the existence of the solutions will in general not occur for the recurrences obtained in cost or complexity evaluations. The problem of uniqueness of solution is more complex.

Proposition 7.8 *Let $v_n = f(\{v_p\ /\ p < n\})$ be a recurrence relation, then its solution u_n is uniquely defined if*

- *f is a recurrence of degree k and $u_0, \ldots, u_{k-1}$ are given,*
- *f is a complete recurrence and u_0 is given,*
- *f is a partition recurrence of the form $v_n = f(v_{n/a})$ and all the u_is such that $i < a$ or $i \geq a$ and a does not divide i are given.*

Proof. The existence of a sequence $(u_n)_{n \geq 0}$ satisfying the given conditions follows from Proposition 7.6.

Let $(u_n)_{n \geq 0}$ and $(u'_n)_{n \geq 0}$ be two solutions of $v_n = f(\{v_p\ /\ p < n\})$, $n \in J$, both satisfying the given initial conditions. Show by induction on n the property $P(n) : u_n = u'_n$ for all n. We will use the second induction principle: we must thus prove that

$$\forall n \in \mathbb{N}, \quad (\forall l < n, P(l)) \implies P(n). \tag{7.8}$$

Since $(u_n)_{n \geq 0}$ and $(u'_n)_{n \geq 0}$ satisfy $v_n = f(\{v_p\ /\ p < n\})$, $\forall n \in J$, it is clear that

$$\forall n \in J, \quad (\forall l < n, u_l = u'_l) \implies u_n = u'_n.$$

Moreover, since $(u_n)_{n \geq 0}$ and $(u'_n)_{n \geq 0}$ satisfy the same initial conditions $\{a_i\ /\ i \in I\}$, we have $\forall i \in I$, $u_i = u'_i$. Hence,

$$\forall n \in I \cup J, \quad (\forall l < n, u_l = u'_l) \implies u_n = u'_n.$$

It can easily be seen that, in all three considered cases, $I \cup J = \mathbb{N}$, hence (7.8), then $\forall n, u_n = u'_n$ and thus the solution is unique. □

REMARK 7.9 Combining the first two clauses of Proposition 7.6 with the corresponding clauses of Proposition 7.8, we obtain necessary and sufficient conditions for the existence and uniqueness of the solutions of recurrence relations of degree k and complete recurrences.

EXERCISE 7.8 Find examples of non-unique solutions for various types of recurrence.

REMARK 7.10 The condition implying uniqueness of the solution of partition recurrences is quite restrictive. Usually, we will thus try to obtain the uniqueness more simply:
- either by restricting the domain and computing the u_ns on a subset of $\mathbb{N}$.
- or by finding only estimates of the solutions, namely, by studying the asymptotic behaviour of the solutions instead of the exact solutions themselves (see Example 9.17).

EXAMPLE 7.11 Consider the partition recurrence $u_n = bu_{n/a} + d(n)$. Assuming u_1, u_n is uniquely defined on

$$S = \{n = a^k \ / \ k \in \mathbb{N}\}.$$

We will solve this recurrence by letting $v_k = u_{a^k}$; v_k is then defined by $v_0 = u_1$ and $v_{k+1} = bv_k + d(a^{k+1})$. We reduced the problem to that of solving a linear recurrence of degree 1, which will be treated in the next section.

EXAMPLE 7.12 Let the partition recurrence $u_n = 2u_{n/2}$ (corresponding, for instance, to a binary merge-sort), with $u_1 = 1$. Then u_n is uniquely determined on $S = \{n = 2^k \ / \ k \in \mathbb{N}\}$; but, for $n \notin S$ we can no longer uniquely determine u_n, i.e. for $n = 2^k v$ with v odd we will have $u_n = 2^k u_v$, namely, u_n will be determined up to the coefficient u_v.

EXERCISE 7.9 Solve the recurrence relation
$$\forall n \geq 1, \qquad u_n = 3\sum_{k=0}^{n-1} u_k + 1.$$
◇

7.2 Linear recurrences

7.2.1 Linear homogeneous recurrences with constant coefficients

A linear homogeneous recurrence relation of degree k with constant coefficients is defined by an equation of the form

$$\forall n \geq k, \qquad u_n = a_1 u_{n-1} + \cdots + a_k u_{n-k}. \tag{7.9}$$

Method of the characteristic polynomial

Definition 7.13 *Let the characteristic polynomial of the recurrence (7.9), be defined by*
$$P(r) = r^k - a_1 r^{k-1} - \cdots - a_{k-1} r - a_k$$
and the characteristic equation of the recurrence (7.9), be defined by
$$r^k = a_1 r^{k-1} + \cdots + a_{k-1} r + a_k \ . \tag{7.10}$$

Proposition 7.14
(i) *The set of solutions of equation (7.9) is a vector space of dimension k.*
(ii) 1. *If $P(r)$ has k pairwise distinct roots $r_1, \ldots, r_k$, then the k sequences $\{r_i^n \ / \ n \in \mathbb{N}\}$, $i = 1, \ldots, k$, are a basis of the vector space of the solutions of (7.9), and any solution of (7.9) is of the form $u_n = \sum_{i=1}^k \lambda_i r_i^n$, where the λ_is are determined by the initial values $u_0, \ldots, u_{k-1}$.*

2. *If the roots of $P(r)$ are r_j, such that for $j = 1, \ldots, p$, with $p < k$, each r_j is a multiple root of multiplicity m_j, then the k sequences $\{(r_j^n)_{n \in \mathbb{N}}, (n r_j^n)_{n \in \mathbb{N}}, \ldots, (n^{m_j - 1} r_j^n)_{n \in \mathbb{N}}, \ j = 1, \ldots, p\}$ are a basis of the vector space of solutions of (7.9); any solution of (7.9) will be of the form*

$$u_n = \sum_{i=1}^p P_j(n) r_j^n \ ,$$

where $P_j(n)$ is a polynomial of degree $\leq m_j - 1$.

Note that the first case is an instance of the second case, with $p = k$ and $m_j = 1$.

Proof.
(i) Clearly, if u_n and v_n are solutions of (7.9), any linear combination $w_n = \lambda u_n + \mu v_n$ is also a solution of (7.9); thus the set of solutions forms a vector space. This vector space is of dimension k at most because giving $u_1, \ldots, u_k$ suffices to uniquely determine u_n. Lastly, this vector space is of dimension k, since k linearly independent sequences can be found; for $j = 1, \ldots, k$, consider the sequences $u^1, \ldots, u^k$, where the initialization of u^j is defined by

for $i = 1, \ldots, k,$
$$u_i^j = \begin{cases} 1 & \text{if } i = j, \\ 0 & \text{otherwise.} \end{cases}$$

Then the k sequences $u^1, \ldots, u^k$ define a set of k linearly independent sequences.

Linear recurrences

(ii) Assume that r is a root of (7.10), then, clearly, $u_n = r^n$ is a solution of the recurrence relation, with the initial conditions $u_i = r^i$ for $i = 0, \ldots, k-1$.

• thus, in the case when the r_js are pairwise distinct it is simple to verify that the k sequences $u^1, \ldots, u^k$, where u^j is defined by the initial conditions $u_i^j = r_j^i$ for $i = 0, \ldots, k-1$, form a basis of the vector space of the solutions: these are indeed k linearly independent solutions since the determinant

$$\det \begin{vmatrix} 1 & 1 & \ldots & 1 \\ r_1 & r_2 & \ldots & r_k \\ \vdots & \vdots & & \vdots \\ r_1^{k-1} & r_2^{k-1} & \ldots & r_k^{k-1} \end{vmatrix} = \prod_{j<i}(r_i - r_j)$$

is non-zero (Vandermonde's determinant). Any solution of (7.9) is thus a linear combination of the solutions $u^1, \ldots, u^k$.

• Assume now that r_j is a root of multiplicity m_j of the characteristic equation $P(r) = r^k - (a_1 r^{k-1} + \cdots + a_{k-1} r + a_k) = 0$, and check that the m_j sequences $(r_j^n)_{n \in \mathbb{N}}, (n r_j^n)_{n \in \mathbb{N}}, \ldots, (n^{m_j-1} r_j^n)_{n \in \mathbb{N}}$ are all solutions of (7.9). For the sequence $(r_j^n)_{n \in \mathbb{N}}$, it is straightforward since r_j is a root of the characteristic polynomial $P(r)$. For the other sequences $(n r_j^n)_{n \in \mathbb{N}}, \ldots, (n^{m_j-1} r_j^n)_{n \in \mathbb{N}}$ we need a result from algebra (Lemma 7.15).

For any $n \geq k$ and $p \geq 0$, let $Q_{n,p}(r)$ be the polynomial

$$Q_{n,p}(r) = n^p r^n - \big(a_1 (n-1)^p r^{n-1} + a_2 (n-2)^p r^{n-2} \\ + \cdots + a_{k-1}(n-k+1)^p r^{n-k+1} + a_k (n-k)^p r^{n-k}\big).$$

We prove by induction on n that there exist polynomials $R_{n,p,i}(r)$ such that for $i = 0, \ldots, p$

$$Q_{n,p}(r) = \sum_{i=0}^{p} R_{n,p,i}(r) P^{(i)}(r), \qquad (E)$$

where $P^{(i)}$ is the ith derivative of P.

(B)   If $p = 0$, $Q_{n,0}(r) = r^{n-k} P(r)$; we thus let $R_{n,0,0}(r) = r^{n-k}$.
(I)   We assume that (E) is true for p and note that

$$Q_{n,p+1}(r) = r Q'_{n,p}(r) \\ = r \sum_{i=0}^{p} \big(R'_{n,p,i}(r) P^{(i)}(r) + R_{n,p,i}(r) P^{(i+1)}(r)\big),$$

hence
$$\begin{cases} R_{n,p+1,0}(r) = r R'_{n,p,0}(r) \\ R_{n,p+1,p+1}(r) = r R_{n,p,p}(r) \end{cases}$$

and for $0 < i \leq p$, $R_{n,p+1,i}(r) = r\big(R'_{n,p,i}(r) + R_{n,p,i-1}(r)\big).$

Hence the induction hypothesis and (E).

From (E) and Lemma 7.15 we deduce that $Q_{n,p}(r_j) = 0$ for $p < m_j$; this implies that $(n^p r_j^n)_{n \in \mathbb{N}}$ is a solution of (7.9).

The solutions thus obtained are linearly independent because the determinant

$$\begin{vmatrix} 1 & 0 & 0 & \cdots & 0 & 1 & \cdots \\ r_1 & r_1 & r_1 & \cdots & r_1 & r_2 & \cdots \\ r_1^2 & 2r_1^2 & 2^2 r_1^2 & \cdots & 2^{m_1-1} r_1^2 & r_2^2 & \cdots \\ \vdots & \vdots & \vdots & & \vdots & \vdots & \\ r_1^{k-1} & (k-1)r_1^{k-1} & (k-1)^2 r_1^{k-1} & \cdots & (k-1)^{m_1-1} r_1^{k-1} & r_2^{k-1} & \cdots \\ & & & & & & \\ 1 & 0 & 0 & \cdots & 0 & & \\ r_p & r_p & r_p & \cdots & r_p & & \\ r_p & 2r_p^2 & 2^2 r_p^2 & \cdots & 2^{m_p-1} r_p^2 & & \\ \vdots & \vdots & \vdots & & \vdots & & \\ r_p^{k-1} & (k-1)r_p^{k-1} & (k-1)^2 r_p^{k-1} & \cdots & (k-1)^{m_p-1} r_p^{k-1} & & \end{vmatrix}$$

is equal to $\left(\prod_{1 \leq j \leq p} r_j^{\binom{m_j}{2}} \right) \left(\prod_{1 \leq j < i \leq p} (r_i - r_j)^{m_i m_j} \right)$ and is non-zero (the r_is are non-zero and assume distinct values).

The general solution of (7.9) is again a linear combination of the solutions $(n^{k_j} r_j^n)_{n \in \mathbb{N}}$, $j = 1, \ldots, p$, $k_j = 0, \ldots, m_j - 1$, which form the basis of the vector space. $\square$

Lemma 7.15 *If r is a root of multiplicity m of $P(x)$, then r is also a root of $P'(x), \ldots, P^{(m-1)}(x)$, where $P^{(k)}(x)$ is the kth derivative of P.*

Proof. By induction on k, we prove that, if r is a root of multiplicity m of $P(x)$, then, for $1 \leq k \leq m-1$, r is also a root of multiplicity $m-k$ of $P^{(k)}(x)$.

(B) r is a root of multiplicity m of $P(x)$ implies that $P(x) = (x-r)^m Q(x)$, with $Q(x)$ a polynomial such that r is not a root of $Q(x)$. Since $P'(x) = m(x-r)^{m-1} Q(x) + (x-r)^m Q'(x) = (x-r)^{m-1}(mQ(x) + (x-r)Q'(x))$, r is a root of multiplicity $m-1$ of $P'(x)$.

(I) The inductive step is similar: for any $k < m-1$, we can check that if r is a root of multiplicity $m-k$ of $P^{(k)}(x)$, then r is a root of multiplicity $m-k-1$ of $P^{(k+1)}(x)$. $\square$

EXERCISE 7.10 Find the solutions of the recurrence relation

$$u_n = 5u_{n-1} - 8u_{n-2} + 4u_{n-3}, \quad n \geq 3,$$

Linear recurrences

with the initial conditions $u_0 = 0$, $u_1 = 1$, $u_2 = 2$. ◇

EXERCISE 7.11 Let $\Sigma = \{a, b\}$ and let B be the subset of Σ^* defined inductively by
(B) $a \in B$, $b \in B$,
(I) $w \in B \implies (abw \in B, baw \in B)$.
1. Write six elements of B.
2. Prove: $w \in B \implies |w|$ odd ($|w|$ denotes the length of w). Is the converse true?
3. Let $u_n = card\{w \,/\, w \in B \text{ and } |w| = n\}$. Compute u_1 and u_2. Find a recurrence relation for u_n and solve it. ◇

EXERCISE 7.12 Let $\Sigma = \{a, b, c, d\}$ and let

$$L = \{w \in \Sigma^* \,/\, ab \text{ is not a factor of } w\}$$
$$= \{w \in \Sigma^* \,/\, \not\exists w_1, w_2 \in \Sigma^* \text{ with } w = w_1 ab w_2\}.$$

$\forall n \geq 0$, let

$$L_n = L \cap \Sigma^n,$$

$$u_n = |L_n| \quad (\text{ the cardinality of } L_n).$$

Recall that Σ^n consists of the length n words on the alphabet Σ.
1. Compute $L_0, L_1, L_2, u_0, u_1, u_2$.
2. Find a necessary and sufficient condition for a word $w = xw'$, with $x \in \Sigma$, to be in L_n.
3. Express L_n in terms of L_{n-1} and L_{n-2}; deduce that the recurrence defining u_n is given by $u_n = 4u_{n-1} - u_{n-2}$.
4. Compute u_n.
5. Also compute $v_n = |L'_n|$ with $L'_n = L' \cap \Sigma^n$ and

$$L' = \{w \in \Sigma^* \,/\, w \text{ has neither } ab \text{ nor } ac \text{ as factor}\}.$$ ◇

EXERCISE 7.13 Let

$$u_n = a u_{n-1} + b u_{n-2} \tag{7.11}$$

be a linear homogeneous recurrence of degree 2 with constant coefficients whose associated polynomial $r^2 - ar - b$ has two conjugate complex roots ce^{-it} and ce^{it}. Show that the general solution of (7.11) is of the form $u_n = \lambda c^n \cos(nt) + \mu c^n \sin(nt)$. ◇

EXERCISE 7.14 Solve the recurrence relations
1. $\forall n \geq 2$, $u_n = u_{n-1} - 2u_{n-2}$.
2. $\forall n \geq 2$, $u_n = u_{n-1} - 2u_{n-2} + 4$. ◇

EXERCISE 7.15 Solve the following recurrences:
1. $\forall n \geq 2$, $2u_n = 3u_{n-1} - u_{n-2}$.
2. $\forall n \geq 2$, $u_n = 4u_{n-1} - 4u_{n-2}$. ◇

Matrix method

Another method for solving linear homogeneous recurrences with constant coefficients is available: the matrix method. Consider again the recurrence relation (7.9) (page 127)

$$\forall n \geq k, \qquad u_n = a_1 u_{n-1} + \cdots + a_k u_{n-k}. \tag{7.9}$$

It can be rewritten as

$$\begin{bmatrix} u_n \\ u_{n-1} \\ \vdots \\ u_{n-k+1} \end{bmatrix} = \begin{bmatrix} a_1 & a_2 & \cdots & a_{k-1} & a_k \\ 1 & 0 & \cdots & 0 & 0 \\ \vdots & \vdots & & \vdots & \vdots \\ 0 & 0 & \cdots & 1 & 0 \end{bmatrix} \times \begin{bmatrix} u_{n-1} \\ u_{n-2} \\ \vdots \\ u_{n-k} \end{bmatrix} = M \times \begin{bmatrix} u_{n-1} \\ u_{n-2} \\ \vdots \\ u_{n-k} \end{bmatrix}.$$

We deduce, by multiplication,

$$\begin{bmatrix} u_n \\ u_{n-1} \\ \vdots \\ u_{n-k+1} \end{bmatrix} = M^{n-k} \times \begin{bmatrix} u_k \\ u_{k-1} \\ \vdots \\ u_1 \end{bmatrix},$$

whence the following computation method: assume that M has k distinct eigenvalues, let $r_1, \ldots, r_k$ be the eigenvalues of M and let N be the associated eigenvectors matrix such that

$$M = N \times \begin{bmatrix} r_1 & 0 & \cdots & 0 \\ 0 & r_2 & \cdots & 0 \\ \vdots & \vdots & & \vdots \\ 0 & 0 & \cdots & r_k \end{bmatrix} \times N^{-1},$$

then

$$\begin{bmatrix} u_n \\ u_{n-1} \\ \vdots \\ u_{n-k+1} \end{bmatrix} = N \times \begin{bmatrix} r_1^{n-k} & 0 & \cdots & 0 \\ 0 & r_2^{n-k} & \cdots & 0 \\ \vdots & \vdots & & \vdots \\ 0 & 0 & \cdots & r_k^{n-k} \end{bmatrix} \times N^{-1} \times \begin{bmatrix} u_k \\ u_{k-1} \\ \vdots \\ u_1 \end{bmatrix},$$

namely, in order to compute the general solution of (7.9), it is enough to diagonalize M as follows:

Linear recurrences

- Find the eigenvalues of

$$M = \begin{bmatrix} a_1 & a_2 & \cdots & a_{k-1} & a_k \\ 1 & 0 & \cdots & 0 & 0 \\ 0 & 1 & \cdots & 0 & 0 \\ \vdots & \vdots & & \vdots & \vdots \\ 0 & 0 & \cdots & 1 & 0 \end{bmatrix} ;$$

this can be done by writing that the following determinant is equal to zero

$$\det \begin{vmatrix} a_1 - r & a_2 & \cdots & a_{k-1} & a_k \\ 1 & -r & \cdots & 0 & 0 \\ 0 & 1 & \cdots & 0 & 0 \\ \vdots & \vdots & & \vdots & \vdots \\ 0 & 0 & \cdots & 1 & -r \end{vmatrix},$$

which is precisely equal to $P(r)$ up to the plus or minus sign.
- Determine the eigenvector's matrix from M to its diagonal form.

The matrix method also applies for solving simultaneous linear homogeneous recurrences of degree 1 with constant coefficients; consider, for instance, the recurrence relations

$$u_n = a_1 u_{n-1} + a_2 v_{n-1} + a_3 w_{n-1},$$
$$v_n = b_1 u_{n-1} + b_2 v_{n-1} + b_3 w_{n-1},$$
$$w_n = c_1 u_{n-1} + c_2 v_{n-1} + c_3 w_{n-1}.$$

They can be written in the form

$$\begin{bmatrix} u_n \\ v_n \\ w_n \end{bmatrix} = \begin{bmatrix} a_1 & a_2 & a_3 \\ b_1 & b_2 & b_3 \\ c_1 & c_2 & c_3 \end{bmatrix} \times \begin{bmatrix} u_{n-1} \\ v_{n-1} \\ w_{n-1} \end{bmatrix} = M \times \begin{bmatrix} u_{n-1} \\ v_{n-1} \\ w_{n-1} \end{bmatrix}.$$

We deduce, by multiplication,

$$\begin{bmatrix} u_n \\ v_n \\ w_n \end{bmatrix} = M^{n-1} \times \begin{bmatrix} u_1 \\ v_1 \\ w_1 \end{bmatrix},$$

which is solved as previously.

EXERCISE 7.16 Solve the recurrences

1. $$\forall n \geq 1, \quad \begin{cases} u_n = 4u_{n-1} + 2v_{n-1}, \\ v_n = -3u_{n-1} - v_{n-1}, \end{cases}$$

with $u_0 = a$, $v_0 = b$.

2. $\quad \forall n \geq 1, \quad \begin{cases} u_n = u_{n-1}^4 v_{n-1}^2, \\ v_n = \dfrac{1}{u_{n-1}^3 v_{n-1}}, \end{cases}$

with $u_0 = a > 0$, $v_0 = b > 0$.

EXERCISE 7.17 Let $\Sigma = \{0, 1\}$. Interpret a word f in Σ^* as the binary representation of an integer. For instance, 101 represents 5, 1101 represents 13. Let $L_i = \{f \in 1 \cdot \Sigma^* \,/\, f \equiv i[3]\}$, $i = 0, 1, 2$, otherwise stated,

$$L_i = \{f \in \Sigma^* \,/\, f \text{ starts with 1 and } f \equiv i \text{ modulo } 3\}, \quad i = 0, 1, 2.$$

For all $n \geq 1$, let:
$$u_n = |L_0 \cap \Sigma^n|$$
$$v_n = |L_1 \cap \Sigma^n|$$
$$w_n = |L_2 \cap \Sigma^n|$$

1. Compute u_1, v_1, w_1 and u_2, v_2, w_2.
2. If $f \in L_i$ in which set is the word $f0$? In which set is the word $f1$? Deduce that $u_{n+1} = u_n + v_n$, $v_{n+1} = u_n + w_n$, $w_{n+1} = w_n + v_n$.
3. Compute u_n, v_n, w_n.

Method of the generating series

Finally, the third method for solving linear recurrences consists of generating series. Consider a sequence u_n defined by the recurrence relation (7.9), and define the series $u(z) = \sum_{n \geq 0} u_n z^n$, called the generating series associated with the sequence u_n. We will compute the generating series, and deduce the u_ns.

7.2.2 Non-homogeneous linear recurrences with constant coefficients

Non-homogeneous linear recurrences are also called recurrences with a right-hand side. They assume the general form

$$\forall n \geq k, \quad u_n = a_1 u_{n-1} + \cdots + a_k u_{n-k} + b(n). \tag{7.12}$$

Proposition 7.16 *The solutions of the equation (7.12) form an affine space of dimension k, whose associated vector space is the set of solutions of the associated homogeneous recurrence (7.9).*

Proof. Let v_n be a particular solution of (7.12); all other solutions v'_n are of the form $v'_n = v_n + w_n$, where w_n is a solution of the recurrence relation (7.9) (page 127)

$$w_n = a_1 w_{n-1} + \cdots + a_k w_{n-k}. \tag{7.9}$$

Linear recurrences

Indeed, if we have

$$v_n = a_1 v_{n-1} + \cdots + a_k v_{n-k} + b(n)$$

and

$$v'_n = a_1 v'_{n-1} + \cdots + a_k v'_{n-k} + b(n) .$$

We deduce by subtraction that $w_n = v'_n - v_n$ satisfies

$$w_n = a_1 w_{n-1} + \cdots + a_k w_{n-k} .$$

Conversely, it is clear that if w_n is a solution of (7.9) and v_n is a particular solution of (7.12), $u_n = w_n + v_n$ is also a solution of (7.12). □

Therefore, to obtain the general solution of (7.12), it 'suffices' to find a particular solution v_n, then to solve the associated homogeneous recurrence (7.9), and finally to add the general solution w_n of (7.9) to a particular solution v_n of (7.12). The problem is that no systematic method for finding a particular solution v_n exists except for a special type of function $b(n)$, which we will now study. First study a simple example.

EXAMPLE 7.17 Consider the recurrence relation

$$n \geq 2, \quad u_n = 2u_{n-1} + 1/n , \tag{7.13}$$

with $u_1 = 1$. We easily verify that $v_n = \sum_{i=2}^{n} 2^{n-i}/i$ is a particular solution of (7.13). Moreover, the associated homogeneous recurrence can be written $w_n = 2w_{n-1}$, and its characteristic polynomial has the root $r = 2$; thus its general solution is of the form $\lambda 2^n$, and the general solution of (7.13) is of the form $u_n = \lambda 2^n + v_n$; taking into account that $u_1 = 1$ and $v_1 = 0$, we obtain $\lambda = 1/2$, hence $u_n = 2^{n-1} + \sum_{i=2}^{n} 2^{n-i}/i$.

The generating series method also applies, see Section 8.2.

Method of the characteristic polynomial

This method can be applied when the 'right-hand side' $b(n)$ of the non-homogeneous recurrence (7.12) is of the form $\sum_{i=1}^{l} b_i^n P_i(n)$, where, for $1 \leq i \leq l$, $P_i(n)$ is a polynomial in n, $b_i \neq 0$, and the b_is are distinct numbers.

Note that when the right-hand side of a linear recurrence relation is of the form $\sum_{i=1}^{l} c_i(n)$, the solution of

$$u_n = a_1 u_{n-1} + \cdots + a_k u_{n-k} + \sum_{i=1}^{l} c_i(n)$$

is the sum of the solutions of the l recurrences
$$u_n^{(i)} = a_1 u_{n-1}^{(i)} + \cdots + a_k u_{n-k}^{(i)} + c_i(n) \ .$$

In practice, in order to solve a recurrence whose right-hand side is $\sum_{i=1}^{l} b_i^n P_i(n)$ we will apply
- either the general method given in Proposition 7.18 below,
- or solve the l recurrences with right-hand sides $b_i^n P_i(n)$ separately.

Proposition 7.18 *Let the recurrence relation be defined by*
$$u_n = a_1 u_{n-1} + \cdots + a_k u_{n-k} + \sum_{i=1}^{l} b_i^n P_i(n) \ , \tag{7.14}$$

where all the b_is are distinct, and $P_i(n)$ is a polynomial in n of degree d_i. Then u_n is the solution of a linear homogeneous recurrence relation of degree $q = k + \sum_{i=1}^{l}(1 + d_i)$,
$$u_n = c_1 u_{n-1} + \cdots + c_q u_{n-q} \ ,$$
whose characteristic polynomial $r^q - c_1 r^{q-1} - \cdots - c_q$ is
$$\left(r^k - a_1 r^{k-1} - a_2 r^{k-2} - \cdots - a_k\right) \prod_{i=1}^{l} (r - b_i)^{d_i+1} \ . \tag{7.15}$$

Any solution of (7.14) is thus of the form $u_n = \sum_{j=1}^{p} Q_j(n) r_j^n$, where r_j is a root of multiplicity m_j of (7.15) and $Q_j(n)$ is a polynomial of degree $m_j - 1$.

Preliminary remark. In order to simplify the notations in the proof, we assume here that the polynomial zero is of degree -1. Indeed, if $Z(n)$ is the polynomial zero
$$u_n = a_1 u_{n-1} + \cdots + a_k u_{n-k}$$
$$= a_1 u_{n-1} + \cdots + a_k u_{n-k} + \sum_{i=1}^{l} b_i^n Z(n)$$
and we have
$$r^k - a_1 r^{k-1} - a_2 r^{k-2} - \cdots - a_k = \left(r^k - a_1 r^{k-1} - a_2 r^{k-2} - \cdots - a_k\right) \prod_{i=1}^{l} (r - b_i)^0 \ .$$

Proof. We prove the proposition by induction on $\delta = \sum_{i=1}^{l}(1 + d_i)$.

Basis. In the preliminary remark we saw that the result holds for $\delta = 0$.

Inductive step. In order to prove that if it holds for δ then it also holds for $\delta + 1$, we note that $\delta > 0$ implies that $d_i \geq 0$ for at least one i, and we prove the following property.

Linear recurrences 137

Lemma 7.19 *Let $P(n)$ be a polynomial of degree $d \geq 0$, and let $P_i(n)$ be a polynomial of degree $d_i \geq 0$, for $i = 1, \ldots, l$. Let $b, b_1, \ldots, b_l$ be non-zero constants with $b \neq b_i$, for all $i = 1, \ldots, l$. The solution of*

$$u_n = a_1 u_{n-1} + \cdots + a_k u_{n-k} + b^n P(n) + \sum_{i=1}^{l} b_i^n P_i(n) \qquad (7.16)$$

is also the solution of

$$u_n = c_1 u_{n-1} + \cdots + c_{k+1} u_{n-k-1} + b^n Q(n) + \sum_{i=1}^{l} b_i^n Q_i(n) \, ,$$

with

- $r^{k+1} - c_1 r^k - c_2 r^{k-1} - \cdots - c_{k+1} = (r^k - a_1 r^{k-1} - a_2 r^{k-2} - \cdots - a_k)(r - b)$,
- *the degree of $Q(n)$ is $d - 1$,*
- *the degree of $Q_i(n)$ is d_i.*

Proof. For $n \geq k+1$,

$$u_n = a_1 u_{n-1} + \cdots + a_k u_{n-k} + b^n P(n) + \sum_{i=1}^{l} b_i^n P_i(n) \, ,$$

$$u_{n-1} = a_1 u_{n-2} + \cdots + a_k u_{n-k-1} + b^{n-1} P(n-1) + \sum_{i=1}^{l} b_i^{n-1} P_i(n-1) \, ;$$

multiplying the second equality by b and subtracting it from the first equality, we have

$$u_n - b u_{n-1} = a_1 (u_{n-1} - b u_{n-2}) + \cdots + a_k (u_{n-k} - b u_{n-k-1})$$

$$+ b^n Q(n) + \sum_{i=1}^{l} b_i^n Q_i(n) \, , \qquad (7.17)$$

with
$$Q(n) = P(n) - P(n-1) \, ,$$
$$Q_i(n) = P_i(n) - \frac{b}{b_i} P_i(n-1) \, .$$

The recurrence (7.17) can be rewritten as

$$u_n = c_1 u_{n-1} + \cdots + c_{k+1} u_{n-k-1} + b^n Q(n) + \sum_{i=1}^{l} b_i^n Q_i(n) \, ,$$

with

$$c_1 = a_1 + b$$
$$c_2 = a_2 - a_1 b$$
$$\vdots$$
$$c_k = a_k - a_{k-1} b$$
$$c_{k+1} = -a_k b,$$

hence $r^{k+1} - c_1 r^k - c_2 r^{k-1} - \cdots - c_k r - c_{k+1} = r^{k+1} - a_1 r^k - a_2 r^{k-1} - \cdots - a_k r - b r^k + a_1 b r^{k-1} + \cdots + a_k b = (r^k - a_1 r^{k-1} - a_2 r^{k-2} - \cdots - a_k)(r - b)$.

It remains to prove: if $P(n)$ is a polynomial of degree $d \geq 0$, $Q(n) = P(n) - cP(n-1)$ is a polynomial of degree $d-1$ if $c = 1$, of degree d otherwise.

- If $d = 0$, $P(n)$ is a constant $k \neq 0$, and $Q(n) = k(1-c)$ is 0 if and only if $c = 1$.
- Otherwise, $P(n) = kn^d + k'n^{d-1} + R(n)$ with $k \neq 0$, and $R(n)$ of degree less than or equal to $d-2$. Then,

$$Q(n) = kn^d - ck(n-1)^d + k'n^{d-1} - ck'(n-1)^{d-1} + R(n) - cR(n-1).$$

Expanding $(n-1)^d$ and $(n-1)^{d-1}$ by the binomial theorem and grouping together the terms of degree less than or equal to $d-2$ with R, we obtain

$$Q(n) = kn^d - ck(n^d - dn^{d-1}) + k'n^{d-1} - ck'n^{d-1} + R'(n),$$

which we simplify in $n^d k(1-c) + n^{d-1}\big(k'(1-c) + ckd\big) + R'(n)$.

—   If $c = 1$, $Q(n) = n^{d-1} kd + R'(n)$, and as k and d are non-zero, Q is of degree $d-1$.

—   If $c \neq 1$, Q is of degree d. □

Using Proposition 7.18 we can solve non-homogeneous recurrences of the form (7.14) in the same way as homogeneous recurrences.

EXERCISE 7.18 Let $\Sigma = \{a, b, c, d\}$. Let $L \subseteq \Sigma^*$ be the language containing no word of the form $ubvaw$, with u, v, w in Σ^* (namely, the letter a never occurs **after** the letter b).
Compute by induction on n the number u_n of length n words in the language L. ◊

Linear recurrences 139

EXAMPLE 7.20 (cf. Exercise 7.4) Let the recurrence: $u_n = 2u_{n-1}+1$, for $n > 1$, with $u_1 = 1$. This recurrence is obtained by evaluating the time complexity of the recursive algorithm for the tower of Hanoi puzzle*; u_n also represents the maximal length of a preorder traversal in a binary tree of depth $\leq n$ (see Exercise 3.23).

It is a recurrence of the form (7.14) with $l = 1$, $b_1 = 1$ and $P_1(n) = 1$. The characteristic equation is $(r-2)(r-1) = 0$ and the general solution of the recurrence is $u_n = \lambda 1^n + \mu 2^n$. We have $u_1 = \lambda + 2\mu = 1$; we need a second condition to find λ and μ; we will use the recurrence and deduce $u_2 = 3$; hence

$$\lambda + 2\mu = 1$$
$$\lambda + 4\mu = 3$$

and thus $\lambda = -1$, $\mu = 1$, i.e. $u_n = 2^n - 1$ for $n \geq 1$, which was also obtainable by a direct summation.

EXERCISE 7.19 Solve the recurrence $u_n = u_{n-1} + 2u_{n-2} + (-1)^n$, $n \geq 2$, with $u_0 = u_1 = 1$ (studied in Section 8.2.2, by the generating series method). ◊

EXERCISE 7.20 Let the recurrence $u_{n+2} = 3u_{n+1} - 2u_n + 4n$, for all $n \geq 0$ with the initial conditions $u_0 = u_1 = 0$. Compute the general term u_n. ◊

Method of undetermined coefficients

To solve the recurrence (7.14) a slightly different method can be used, consisting of

1. First finding a particular solution of the non-homogeneous recurrence, a particular solution that must be of the form $\sum_{i=1}^{l} b_i^n Q_i(n)$ with
   - $deg(Q_i) = deg(P_i)$ if b_i is not a root of the characteristic polynomial of the homogeneous recurrence associated with (7.14),
   - $deg(Q_i) = deg(P_i) + m_i$ if b_i is root of multiplicity m_i of the characteristic polynomial of the homogeneous recurrence associated with (7.14). The particular solution v_n is found by the method of *undetermined coefficients*, namely: in (7.14) substitute for v_n a term $\sum_{i=1}^{l} b_i^n Q_i(n)$, where the coefficients of the Q_is are unknown, and determine these coefficients by identifications.

---

* Recall that the tower of Hanoi puzzle consists of transferring n disks initially stacked in decreasing order on a first peg to a second peg, initially empty, moving only one disk at a time, and never moving a larger disk on top of a smaller one: on any peg the disks will at any time be stacked in decreasing order throughout the whole transfer. To this end a third peg, initially empty, is available; the idea is to transfer the $n-1$ smaller disks from peg 1 to peg 3, in time u_{n-1}, then to transfer the biggest disk from peg 1 to peg 2, where it will be at the right place, and to finally transfer the $n-1$ smaller disks from peg 3 to peg 2, in time u_{n-1}. All the disks will then be transferred, in time $u_n = 2u_{n-1} + 1$ for $n > 1$. Clearly when there is a single disk, a single manipulation will do the job, and thus $u_1 = 1$.

2. Then finding the general solution u_n in the form $u_n = v_n + w_n$, where w_n is the general solution of the associated homogeneous recurrence, and the coefficients of w_n are determined by the initial conditions.

EXAMPLE 7.21

1. Consider again the preceding example : $u_n = 2u_{n-1} + 1$, with $u_1 = 1$. First look for a particular solution $v_n = \lambda 1^n$, substituting the solution v_n in the recurrence gives $\lambda = 2\lambda + 1$ and $\lambda = -1$. Then look for the general solution in the form: $u_n = \mu 2^n - 1$, and for $n = 1$ that gives $2\mu - 1 = 1$, and thus $\mu = 1$.

2. Let the recurrence: $u_n = 2u_{n-1} + n + 2^n$, with the initial condition $u_1 = 0$. It is of the form (7.14) with $b_1 = 1$, $P_1(n) = n$, $b_2 = 2$, $P_2(n) = 1$; moreover, the characteristic polynomial of the associated homogeneous recurrence is $r = 2$. We must thus find a particular solution of the form $v_n = an + b + (cn + d)2^n$. Plugging in the recurrence relation we obtain

$$an + b + (cn + d)2^n = 2a(n-1) + 2b + (c(n-1) + d)2^n + n + 2^n,$$

i.e. simplifying
$$0 = (a+1)n + b - 2a + 2^n(1-c).$$

This equality being true for all n we deduce $a = -1$, $b = -2$, $c = 1$ and we obtain a particular solution $v_n = -2 - n + n2^n$ for the equation $u_n = 2u_{n-1} + n + 2^n$. We then try to find the general solution in the form $u_n = v_n + \lambda 2^n$, since the general solution of the homogeneous recurrence is $w_n = \lambda 2^n$; taking into account the initial conditions $u_1 = 0$ we obtain $u_1 = -2 - 1 + 1 + 2\lambda = 0$, hence $\lambda = 1$. The solution of our recurrence is thus $u_n = -2 - n + n2^n + 2^n$.

We will note that the initial conditions are of no use to find the particular solution v_n.

7.2.3 Linear recurrences with parameters

There is no general method, except for some special cases; we will study one such case, the *linear recurrences of degree 1 with parameters*. Consider the recurrence relation
$$a(n)u_n = b(n)u_{n-1} + c(n).$$

We can, without loss of generality, assume that $a(n) = 1$ (divide by $a(n)$ which is always $\neq 0$, otherwise the recurrence relation would not define u_n), and we are thus back to
$$u_n = b(n)u_{n-1} + c(n), \quad n > 0.$$

Let $f(n) = \prod_{i=1}^{n} 1/b(i)$, and let the sequence $(v_n)_{n \geq 0}$ be defined by $v_n = f(n)u_n$, for $n > 0$, and $v_0 = u_0$. We verify that, $\forall n > 0$, v_n satisfies the recurrence relation $v_n = v_{n-1} + f(n)c(n)$, which can be solved by summation (the so called *summation factors* method); hence $v_n = v_0 + \sum_{k=1}^{n} f(k)c(k)$, and thus

$$u_n = (1/f(n))v_n = \left(\prod_{i=1}^{n} b(i)\right)\left(u_0 + \sum_{k=1}^{n} f(k)c(k)\right).$$

The summation factors method can also be applied with profit to some linear recurrences with constant coefficients.

EXAMPLE 7.22 Let $u_n = u_{n-1} + 2(n-1)$, for $n > 0$, with $u_0 = 2$. We obtain $u_n = 2 + 2(0 + 1 + 2 + \cdots + (n-2) + (n-1)) = 2 + n(n-1)$.

7.3 Other recurrence relations

Non-linear recurrences are less easy to solve. However, some techniques can be applied, e.g. substitutions or image transformations.

7.3.1 Partition recurrences and substitutions

Recall that a partition recurrence is often obtained in a 'divide and conquer'-type algorithm, where a problem of size n is split into b subproblems of size n/a; if u_n represents the cost of the solution of the size n problem, and $c(n)$ the cost of the creation and the utilization of the b subproblems of size a/n, we have the recurrence relation, for $n > 1$,

$$u_n = bu_{n/a} + c(n). \tag{7.18}$$

In order to solve this recurrence exactly, the domain is restricted to integers of the form a^k, and we apply the substitution $v_k = u_{a^k}$. We have $u_{a^k} = bu_{a^{k-1}} + c(a^k)$, thus $v_k = bv_{k-1} + c(a^k)$; hence a linear recurrence of degree 1, and we obtain (see Section 7.2.3)

$$v_k = b^k \left(v_0 + \sum_{j=1}^{k} c(a^j)/b^j\right)$$

$$= v_0 b^k + \sum_{j=0}^{k-1} b^{k-j-1} c(a^{j+1})$$

$$= v_0 b^k + \sum_{j'=1}^{k} b^{j'-1} c(a^{k-j'+1}).$$

Hence, as $n = a^k$, and thus $k = \log_a n$,

$$u_n = v_0 b^{\log_a n} + \sum_{j=1}^{\log_a n} b^{j-1} c(n/a^{j-1}) . \qquad (7.19)$$

Equality (7.19) will also allow us to evaluate the order of magnitude of u_n (see Proposition 9.18).

EXERCISE 7.21 Solve the recurrence $u_n = 4u_{n/2} + n^2$, for $n = 2^k$. ◊

Note that a sequence u_n can be considered as a mapping $u : \mathbb{N} \longrightarrow \mathbb{C}$; the substitution technique that we have just seen consists of transforming the domain $\mathbb{N}$, namely, of composing u with a mapping $f : \mathbb{N} \longrightarrow \mathbb{N}$ and considering the recurrence $v = u \circ f$. In the preceding example we had $f(k) = 2^k$. It can also be fruitful to transform the image of u, namely, to compose u with $g : \mathbb{C} \longrightarrow \mathbb{C}$, and to consider the recurrence $w = g \circ u$. We will give two such examples in the next section.

7.3.2 Image transformations

EXAMPLE 7.23 Consider the recurrence $u_n = u_{n-1} - u_n u_{n-1}$, with $u_1 = 1$. Assume $u_n \neq 0 \; \forall n$, and divide by $u_n u_{n-1}$; We obtain $\dfrac{1}{u_{n-1}} = \dfrac{1}{u_n} - 1$ hence $\dfrac{1}{u_n} = \dfrac{1}{u_{n-1}} + 1$. Let $v_n = \dfrac{1}{u_n}$; this boils down to letting $v = g \circ u$ with $g(x) = \dfrac{1}{x}$. We then have $v_1 = 1$ and $v_n = v_{n-1} + 1$, hence $v_n = n$ and $u_n = 1/n$.

EXAMPLE 7.24 Let the recurrence $u_n = n(u_{n/2})^2$, with $u_1 = 6$. Assume that $n = 2^k$ and let $v_k = u_{2^k}$. We deduce $v_k = 2^k v_{k-1}^2$ and $v_0 = 6$. Now let $w_k = \log_2 v_k$; we have $w_k = k + 2w_{k-1}$ and $w_0 = \log_2 6$. The characteristic polynomial is $(r-2)(r-1)^2 = 0$, and the general solution is of the form $w_k = a2^k + b + ck$. We easily obtain, using the equality $w_k = k + 2w_{k-1}$, that $b = -2$ and $c = -1$; we obtain $a = 3 + \log_2 3$ by noting that $w_0 = 1 + \log_2 3 = a - 2$. Hence, finally, $w_k = (3 + \log_2 3)2^k - k - 2$. Then, taking into account that $v = 2^{w_k}$ and that $u_n = v_{\log_2 n}$, we obtain $u_n = 2^{3n-2} 3^n / n$.

EXERCISE 7.22 Solve the following recurrence relations:

1. $\forall n \geq 2, \quad u_n = \dfrac{2}{\dfrac{1}{u_{n-1}} + \dfrac{1}{u_{n-2}}}$, with $u_0 = a$ and $u_1 = b$.

2. $\forall n \geq 2, \quad u_n = \sqrt{u_{n-1} u_{n-2}}$, with $u_0 = 1$ and $u_1 = 2$.

3. $\forall n \geq 2, \quad u_n = \dfrac{a_n}{b_n} = \dfrac{a_{n-1} + a_{n-2}}{b_{n-1} + b_{n-2}}$, with $u_0 = \dfrac{a_0}{b_0}$ and $u_1 = \dfrac{a_1}{b_1}$. ◊

Complements and examples   143

7.3.3 Complete recurrences

They can be solved either by using generating series tools (see Chapter 8 on generating series), or by forming linear combinations of suitably chosen instances of the recurrence relation (see Section 14.1.3 for an example).

7.4 Complements and examples

7.4.1 Operations on sequences

Z   A sequence is a mapping $u: \mathbb{N} \longrightarrow \mathbb{C}$, conventionally extended into a mapping $u: \mathbb{Z} \longrightarrow \mathbb{C}$ by letting $u_n = 0$ if $n < 0$ (or $n < k_0 \in \mathbb{Z}$ if for technical reasons the sequences must start with $k_0 \neq 0$). This convention is fundamental in order for the results of the present section to hold. We can define various operations on the sequences allowing ease of manipulation (addition, difference, etc.). We give a short summary.

- E, the *predecessor* operation, is defined by
$$(Eu)_n = u_{n-1} \quad \text{and} \quad (Eu)_0 = 0 .$$

- The *product* operation is defined by
$$(uv)_n = u_n v_n \quad \text{and} \quad (uv)_0 = u_0 v_0 .$$

- Δ, the *difference* operation, is defined by:
$$(\Delta u)_n = u_n - u_{n-1} = \bigl((Id - E)u\bigr)_n \text{ and } (\Delta u)_0 = \bigl((Id - E)u\bigr)_0 = u_0 ,$$
thus $\Delta = Id - E$. Δ is linear, namely, $\Delta(\lambda u + \mu v) = \lambda \Delta u + \mu \Delta v$, $\forall \lambda, \mu \in \mathbb{C}$. Moreover, $\Delta(u \star v) = (\Delta u) \star v + (Eu) \star (\Delta v)$.

- The $\star$ *product* operation, denoted by $\star$, different from the usual products and in general non-associative, is defined by
$$(u \star v)_n = u_n v_n - u_{n-1} v_{n-1} = \bigl(\Delta(uv)\bigr)_n \text{ and } (u \star v)_0 = u_0 v_0 .$$

- Σ, the *summation* operation, is defined by
$$(\Sigma u)_n = \Sigma_{i=0}^n u_i .$$

We can verify the equalities
$$\Delta \Sigma = \Sigma \Delta = Id$$
$$\Sigma[(\Delta u) \star v] = u \star v - \Sigma[(Eu) \star (\Delta v)] \quad \text{(summation by parts)}.$$

EXAMPLE 7.25 Let k be fixed, and let $u_n = \binom{n}{k}$, then: $(\Delta u)_n = \binom{n-1}{k-1}$; similarly let $v_n = A_n^k$, i.e. $(v)_n = (A_n^k)_{n \in \mathbb{N}}$,

- for $k > 1$, we have $(\Delta v)_n = (\Delta A_n^k)_n = A_n^k - A_{n-1}^k = k A_{n-1}^{k-1}$;
- $\Delta A_n^1 = 1$, $\Delta A_n^0 = 0$;
- for $k < 0$, let $k = -l$ with $l > 0$, then letting

$$w_n = B_n^k = n(n+1) \cdots (n-k-1);$$

we have:

$$(\Delta w)_n = (\Delta B^k)_n = -k B_n^{k-1}.$$

EXERCISE 7.23 Let k be fixed, $k > 1$, compute Δu where
$$u_n = \frac{1}{n(n+1) \cdots (n+k-1)}.$$
◇

EXAMPLE 7.26 Evaluation of the order of magnitude of the sum ΣH of the harmonic numbers H_n. Recall that

$$H_n = 1 + \frac{1}{2} + \frac{1}{3} + \cdots + \frac{1}{n}, \quad H_0 = 0.$$

We have $(\Delta H)_n = \dfrac{1}{n}$, for $n > 0$.

Noting that $(\Delta u)_n = 1$ for all n, a summation by parts with $u_n = n+1$ and $v_n = H_n$ gives

$$\Sigma H = \Sigma((\Delta u) \star H) = u \star H - \Sigma(Eu) \star \Delta(H),$$

hence
$$(\Sigma H)_n = (n+1)H_n - (\Sigma \frac{n}{n})_n = (n+1)H_n - n.$$

But $H_n \sim \log n$. This comes from the following remark: the function $1/x$ being decreasing, we have the squeeze

$$\frac{1}{n} > \int_n^{n+1} \frac{dx}{x} > \frac{1}{n+1}, \quad \forall n > 0;$$

thus considering the sequence of intervals $[1, 2[, [2, 3[, \ldots, [n, n+1[$, we have

$$\int_1^{n+1} \frac{dx}{x} < H_n < \int_1^n \frac{dx}{x} + 1,$$

and thus
$$\log(n+1) < H_n < \log(n) + 1.$$

The situation is illustrated by Figure 7.1. Therefore $(\Sigma H)_n \sim n \log n$.

Complements and examples

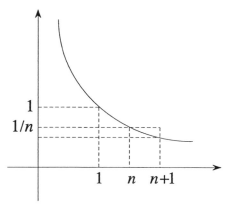

Figure 7.1

7.4.2 Applications: counting, Stirling numbers

EXAMPLE 7.27 (Number of surjections of A in B) Let A, B be such that $|A| = a$, $|B| = b$, $a \geq b$, and let S_a^b be the number of surjections $f: A \longrightarrow B$. In Proposition 6.14 we computed S_a^b by considering $E_i = \{f: A \longrightarrow B \ / \ i \notin f(A)\}$ and by noting that f is non-surjective if and only if $f \in \cup_{i \in B} E_i$, which gave

$$S_a^b = b^a - \left(\sum_{p=1}^{b} (-1)^{p+1} \binom{b}{p} (b-p)^a \right)$$

$$= b^a - \left(b(b-1)^a + \cdots + (-1)^{p+1} \binom{b}{p} (b-p)^a + \cdots + (-1)^b b \right)$$

$$= \sum_{p=0}^{b} (-1)^p \binom{b}{p} (b-p)^a.$$

We can also compute S_a^b by exhibiting a recurrence relation satisfied by S_a^b. Let $A = \{1, \ldots, a\}$ and $B = \{1, \ldots, b\}$. Let $f_{|A-\{a\}}$ be the restriction of f to $A - \{a\}$. If f is surjective, then one of the two following conditions is realized:

- Either $f' = f_{|A-\{a\}}$ is surjective, and then f' is a surjection from $\{1, \ldots, a-1\}$ onto B and there are b possibilities for choosing the image of a, thus there are bS_{a-1}^b possible choices for f in this case.
- Or $f' = f_{|A-\{a\}}$ is not surjective. Let $f(a) = j$; since f is surjective, $f'': (A-\{a\}) \longrightarrow (B - \{j\})$, also is surjective, and we thus have S_{a-1}^{b-1} possible choices for f''. Moreover there are b possible choices for the element j which we took out, hence bS_{a-1}^{b-1} possible choices for f in that case.

We finally obtain the recurrence relation

$$S_a^b = bS_{a-1}^b + bS_{a-1}^{b-1}, \quad 0 < b \leq a, \tag{7.20}$$

with the initial condition $S_1^1 = 1$.

EXAMPLE 7.28 (Number of partitions of A in b classes) Letting $|A| = a \geq b > 0$, we would like to determine the number P_a^b of partitions of A of the form $A = A_1 + \cdots + A_b$, namely, the A_is are b subsets of A, pairwise disjoint. To this end we will use S_a^b that we just computed.

Recall the following theorem first.

Theorem 7.29 *Let $f \colon A \longrightarrow B$ be a surjection, and let $\equiv_f$ be the equivalence relation associated with f which is defined by $x \equiv_f y$ if and only if $f(x) = f(y)$. f can be factored into $f = i \circ p$, where $p \colon A \longrightarrow A/\equiv_f$ is the canonical projection and $i \colon A/\equiv_f \longrightarrow B$ is an isomorphism.*

Corollary 7.30 $S_a^b = b! P_a^b$.

Proof. To each surjection $f \colon A \longrightarrow B$ corresponds a partition of A in b classes of equivalence according to the preceding theorem.

Conversely, if $A = A_1 + \cdots + A_b$ is a partition of A, $f(a) = i \iff a \in A_i$ defines a surjection $A \longrightarrow B = \{1, \ldots, b\}$. Let p be an arbitrary permutation of B, then $p \circ f \colon A \longrightarrow B$ is also a surjection corresponding to the same partition of A. Thus each partition determines $b!$ surjections. □

We can also also determine a recurrence relation satisfied by the P_a^bs. Let P be a partition of A in b classes, and let $\alpha \in A$.

- Either $\{\alpha\}$ (the singleton α) is a class of P on its own, and then the other classes of P partition $A - \{\alpha\}$ in $b-1$ classes; there are thus P_{a-1}^{b-1} possible choices for P in that case.
- Or $\{\alpha\}$ is not a class, but belongs to one of the b classes of P, and in that case P partitions $A - \{\alpha\}$ in b classes, and there are b possible choices for putting α in one of the classes, hence bP_{a-1}^b possible choices for P in that case. Finally, we obtain

$$P_a^b = bP_{a-1}^b + P_{a-1}^{b-1}, \quad 0 < b \leq a. \tag{7.21}$$

Terminology: The P_a^b are called *Stirling numbers* of the second kind. Stirling numbers of the first kind satisfy the recurrence relation

$$K_a^b = (a-1)K_{a-1}^b + K_{a-1}^{b-1}, \quad 0 < b \leq a. \tag{7.22}$$

The combinatorial interpretation of K_a^b is the following: K_a^b is the number of permutations of a letters containing b cycles. A *cycle* in a permutation p of a letters is a subset of $n < a$ letters $\{\alpha_1, \ldots, \alpha_n\}$, such that $p(\alpha_i) = \alpha_{i+1}$, for $i < n$ and $p(\alpha_n) = \alpha_1$. By a case analysis similar to the preceding ones we obtain: let $\alpha \in A$, and let p be a permutation with b cycles;

- either α is invariant by p, namely, $\{\alpha\}$ is a cycle, and there are K_{a-1}^{b-1} such possibilities
- or α is not invariant, there are then K_{a-1}^b possible choices of permutations of $a-1$ letters with b cycles, and $a-1$ possible choices for inserting α in one of the b cycles: if the b cycles have $a_1, a_2, \ldots, a_b$ letters, respectively, we have $a_1 + a_2 + \cdots + a_b = a - 1$ possibilities, hence $(a-1)K_{a-1}^b$ possibilities altogether.

Stirling numbers have numerous applications in combinatorics: two simple examples are the expansion of the binomial coefficients in powers and, conversely, the expression of powers in terms of the binomial coefficients; we have

$$x(x-1)\cdots(x-n+1) = \sum_{k=1}^{n}(-1)^{n-k}K_n^k x^k,$$

$$x^n = \sum_{k=1}^{n} P_n^k \binom{x}{k} k!.$$

CHAPTER 8

GENERATING SERIES

The basic idea for which generating series were introduced is to *represent a sequence of numbers by a function*, which on the one hand will be easier to manipulate and on the other hand will allow us to treat the sequence in its entirety. For example, if a sequence of costs is defined by a recurrence equation, the recurrence equation will correspond to a functional equation on the generating series: the latter equation may be solved by algebraic or analytic techniques.

Similarly, generating series will allow us to represent a discrete probability law globally and to handle it more easily: indeed, assuming that p_n is the probability that event n occurs, a single generating function g will represent the whole sequence p_n; moreover, the values of p_n can be recovered from the generating series g, and techniques for studying the generating series will give properties of the probability distribution (average values, means, etc.).

Generating series were introduced at the end of the eighteenth century in order to study probabilities by the French mathematicians Laplace and de Moivre (the latter took British nationality after the *édit de Nantes* was revoked).

We especially recommend the following handbooks, where many examples of computations on series and applications can be found:

Ronald Graham, Donald Knuth, Oren Patashnik, *Concrete Mathematics*, Addison-Wesley, London (1989).
William Feller, *Probability Theory*, Vol. 1, John Wiley, New York (1968).
Donald Knuth, *The Art of Computer Programming*, Vol. 1, Addison-Wesley, London (1973).

Generalities

8.1 Generalities

8.1.1 Intuitions

Let $u = (u_0, u_1, u_2, \ldots, u_p)$ be a sequence of $p+1$ numbers; they can be globally represented as a vector, or canonically associated with an element of a fixed vector space; we choose the second option and take the set of polynomials in the variable X as vector space; the sequence $u = (u_0, u_1, u_2, \ldots, u_p)$ is then represented by the polynomial

$$P(X) = u_0 + u_1 X + u_2 X^2 + \cdots + u_p X^p.$$

If $u_p \neq 0$, p is said to be the degree of the polynomial $P(X)$. The advantage of this second approach is the ability to represent polynomials of an arbitrary degree $n \in \mathbb{N}$: a polynomial is simply represented by a sequence $(u_n)_{n \in \mathbb{N}}$ such that all but a finite number of the u_ns are equal to zero. A polynomial $P(X) = u_0 + u_1 X + u_2 X^2 + \cdots + u_p X^p$, of degree p is then represented by the sequence $u = (u_0, u_1, u_2, \ldots, u_p, 0, 0, \ldots)$, where $u_n = 0$ for all $n > p$. The $+$ and $\times$ operations on sequences correspond to operations on the polynomials.

- The sum $u + v$ corresponds to the addition of polynomials. The unit for $+$ is the sequence $(0, 0, 0, \ldots)$, corresponding to the polynomial function $Z(x) = 0$ for all x. In fact, if

$$P(X) = u_0 + u_1 X + u_2 X^2 + \cdots + u_p X^p$$

and

$$Q(X) = v_0 + v_1 X + v_2 X^2 + \cdots + v_q X^q,$$

the polynomial $P(X) + Q(X)$ is given by

$$P(X) + Q(X) = (u_0 + v_0) + (u_1 + v_1)X + (u_2 + v_2)X^2 + \cdots + w_n X^n,$$

where

$$w_n X^n = \begin{cases} u_p X^p & \text{if } p > q, \\ (u_p + v_p) X^p & \text{if } p = q, \\ v_q X^q & \text{if } p < q. \end{cases}$$

- Multiplication is more complex: various problems arise with the definition $(uv)_n = u_n v_n$. Firstly, the sequence $\forall n, u_n = 1$ would be a good candidate for a unit, except for the fact that it counts an infinity of non-zero u_ns; secondly, as soon as a sequence has a term $u_n = 0$, it is not invertible; and finally, the ring of sequences is not an integral domain: $u \neq 0$, $v \neq 0$ and $uv = 0$ may occur (in case u and v are equal to zero on complementary sets of indices); for this reason the *Hadamard product* $(uv)_n = u_n v_n$ will no longer be considered in the sequel. Instead, we will use the *Cauchy product* (or *convolution product*) defined by $(uv)_n = \sum_{p+q=n} u_p v_q$, which has the additional advantage of corresponding naturally to the product of polynomials: if the sequence u is represented by $P(X) = \sum_0^N u_p X^p$ and the sequence v is represented by $Q(X) = \sum_0^M v_q X_q$, then uv is represented by $P(X)Q(X)$, since the coefficient of X^n in $P(X)Q(X)$ is $(uv)_n = \sum_{p+q=n} u_p v_q$.

EXAMPLE 8.1 Let $P(X) = u_0 + u_1 X + u_2 X^2$ and $Q(X) = v_0 + v_1 X$.

$$
\begin{aligned}
& u_0 + u_1 X + u_2 X^2 \\
\times \quad & v_0 + v_1 X \\
\hline
& u_0 v_0 + u_1 v_0 X + u_2 v_0 X^2 \\
& \quad + u_0 v_1 X + u_1 v_1 X^2 + u_2 v_1 X^3 \\
\hline
& u_0 v_0 + (u_1 v_0 + u_0 v_1) X + (u_2 v_0 + u_1 v_1) X^2 + u_2 v_1 X^3
\end{aligned}
$$

If we assume that $Q(X) = v_0 + v_1 X + v_2 X^2$ with $v_2 = 0$, we obtain $u_2 v_0 + u_1 v_1 = u_2 v_0 + u_1 v_1 + u_0 v_2$. Similarly, $u_2 v_1 = u_3 v_0 + u_2 v_1 + u_1 v_2 + u_0 v_3$ with $u_3 = v_3 = v_2 = 0$.

EXERCISE 8.1 Find the unit of the convolution product. ◇

We will generalize this approach to sequences $(u_n)_{n \in \mathbb{N}}$ possibly having an infinite number of non-zero u_ns, by associating a formal power series $\mathbf{u} = \sum_{n=0}^{\infty} u_n X^n$ to the sequence $u = \{u_n \mid n \in \mathbb{N}\}$.

Note, first, that it is a *purely formal* extension of the notion of polynomials and that we will not worry about the convergence radius of our series which can be equal to zero. A philosophical remark is called for here (for readers already familiar with series): the generating series can be considerered from two different viewpoints:

- firstly, they can be considered as analytic functions of the complex variable X, and they then inherit all the good (or bad) properties of those functions (convergence, absolute convergence, approximations, etc.) which can be found in all complex analysis handbooks,
- secondly, they can be considered as formal algebraic expressions, generalizing polynomials, where X is simply a position indicator whose values are irrelevant. We then speak of *formal power series*.

As our goal is understanding the sequence of numbers $(u_n \mid n \in \mathbb{N})$, we will be interested in the second viewpoint. In this viewpoint, all the operations we

Generalities

perform on the series can be defined and justified in a purely algebraic and formal framework, independently of their convergence.

When, moreover, the series happens to converge for some values of x and, in addition, we know how to compute the corresponding analytic function and we want to study the asymptotic behaviour of the u_ns, then the usual methods of complex analysis will allow us to find the values of the coefficients of the series, or at least their asymptotic behaviour. In that case, the first approach will be usable.

Note that, conversely, every function which admits a power series expansion having a non-zero convergence radius uniquely determines (because of the uniqueness of the power series expansion) a sequence of numbers consisting of its coefficients.

EXAMPLE 8.2 Consider the example of the Fibonacci numbers denoted by F_n. We want to determine F_n given that

$$F_n = F_{n-1} + F_{n-2} \quad \text{and} \quad F_0 = 0 \quad F_1 = 1 \, .$$

Consider the series $F(X) = \sum_{n \geq 0} F_n X^n$. From $F_n = F_{n-1} + F_{n-2}$ we deduce $\forall n \geq 2$, $F_n X^n = F_{n-1} X^n + F_{n-2} X^n = X F_{n-1} X^{n-1} + X^2 F_{n-2} X^{n-2}$. Summing the equalities $F_n X^n = X F_{n-1} X^{n-1} + X^2 F_{n-2} X^{n-2}$ for $n \geq 2$, and recalling that $F_0 = 0$ and $F_1 = 1$, yields the equation:

$$F(X) = XF(X) + X^2 F(X) + X \, .$$

Hence: $(1 - X - X^2)F(X) = X$. We can thus consider that $F(X)$ is the power series expansion of the rational function

$$F(X) = \frac{X}{1 - X - X^2} = \frac{X}{(1 - r_1 X)(1 - r_2 X)} = \frac{1}{\sqrt{5}}(\frac{1}{1 - r_1 X} - \frac{1}{1 - r_2 X}) \, ,$$

where $r_1 = \dfrac{1 + \sqrt{5}}{2}$ and $r_2 = \dfrac{1 - \sqrt{5}}{2}$ are the two roots of $1 - X - X^2 = 0$. But

$$\frac{1}{1 - Z} = 1 + Z + Z^2 + \cdots + Z^n + \cdots,$$

and hence $F(X) = 1/\sqrt{5}(\sum_{n \geq 0} (r_1^n - r_2^n) X^n)$ and $F_n = 1/\sqrt{5}(r_1^n - r_2^n)$.

For the present the formal power series will be denoted by $\sum_{n=0}^{\infty} u_n X^n$ and the analytic series by $\sum_{n=0}^{\infty} u_n x^n$. We will revert to the notation $\sum_{n=0}^{\infty} u_n x^n$ for all series from Section 8.1.5 onwards. Similarly, for the present, the series will be denoted by boldface letters, in the form $\mathbf{u} = \sum_{n=0}^{\infty} u_n X^n$ in order to distinguish them from the corresponding sequences $(u_n)_{n \in \mathbb{N}}$.

8.1.2 Definitions

Definition 8.3 *The power series* $\mathbf{u} = \sum_{n\geq 0} u_n X^n$ *is said to be a generating series of coefficients* u_n. *The set of power series with coefficients in the set* $\mathbb{C}$ *of complex numbers is denoted by* $\mathbb{C}[\![X]\!]$. *The power series* $\mathbf{u}$ *is also denoted by* $\mathbf{u}(X)$.

REMARK 8.4 Similarly, we can consider the set $\mathbb{N}[\![X]\!]$ (resp. $\mathbb{R}[\![X]\!]$, $\mathbb{Z}[\![X]\!]$,...) of power series with coefficients in $\mathbb{N}$ (resp. $\mathbb{R}$, $\mathbb{Z}$,...).

Proposition 8.5 *The set of power series, together with*

- *the addition + defined by*

$$\sum_{n\geq 0} u_n X^n + \sum_{n\geq 0} v_n X^n = \sum_{n\geq 0} (u_n + v_n) X^n \quad \text{and}$$

- *the convolution product* $\times$, *defined by*

$$\left(\sum_{n\geq 0} u_n X^n\right) \times \left(\sum_{n\geq 0} v_n X^n\right) = \sum_{n\geq 0} \left(\sum_{p+q=n} u_p v_q\right) X^n ,$$

is an integral ring (i.e. a ring without zero-divisors).

Proof. The unit for addition is the series $\mathbb{O} = (0)_{n\geq 0}$; the unit for the convolution product is the series $\mathbb{1}$ defined by: $\mathbb{1}_0 = 1$ and $\mathbb{1}_n = 0$ for all $n > 0$; $\mathbb{1}$ will also be denoted by 1 from Section 8.1.3 on. The axioms of the ring structure are easily verified (i.e. addition is associative and commutative, product is associative and commutative and distributes over addition).

Finally, recall that $\mathbf{u} \neq \mathbb{O}$ is a divisor of $\mathbb{O}$ if $\mathbf{v}$ exists such that $\mathbf{v} \neq \mathbb{O}$ and $\mathbf{u} \times \mathbf{v} = \mathbb{O}$. If $\mathbf{u} \times \mathbf{v} = \mathbb{O}$, and if one of the two series $\mathbf{u}$ or $\mathbf{v}$ is different from $\mathbb{O}$, then the other one is equal to $\mathbb{O}$; if for instance $\mathbf{u} \neq \mathbb{O}$, let $j = \min\{i \, / \, u_i \neq 0\}$; one computes the coefficients of $\mathbf{u} \times \mathbf{v}$ and shows by induction on n that $v_n = 0$ for all $n \geq 0$. □

EXERCISE 8.2 Let $r \geq 0$. Find a power series $\mathbf{u}(X)$ whose coefficient of X^n is $u_n = \sum_{k=0}^{n} \binom{r}{k}\binom{r}{n-2k}$. Hint: it can be noted that $u_n = 0$ if $n > 3r$. ◇

NOTATIONS
1. The convolution product $\mathbf{u} \times \mathbf{v}$ will be denoted simply by the concatenation $\mathbf{uv}$ when there is no ambiguity.
2. Let $i \colon \mathbb{C} \longrightarrow \mathbb{C}[\![X]\!]$ be the mapping defined by $i(a) = \mathbf{a} = a\mathbb{1} = (a_n)_{n\geq 0}$, with $a_0 = a$ and $a_n = 0$ for all $n \neq 0$; i is an injection and $i(0) = \mathbb{O}$, $i(1) = \mathbb{1}$; we will denote $\mathbb{O}$ by 0 and $\mathbb{1}$ by 1 from the Section 8.1.3 on.

Generalities 153

Shift operations can be defined on series:
- Let $u = (u_n)_{n \geq 0}$ and $a \in \mathbb{C}$, then, $\mathbf{u}_{a,r} = \mathbf{u}X + \mathbf{a}$ is the series verifying $(u_{a,r})_0 = a$, and for $n > 0$, $(u_{a,r})_n = u_{n-1}$. This operation consists of translating all the coefficients of $\mathbf{u}$ one position to the right, and adding an a at the beginning.
- Let $u = (u_n)_{n \geq 0}$, then $\mathbf{u}_l = \dfrac{\mathbf{u} - \mathbf{u}_0}{X}$ is the series defined by $\forall n \geq 0$, $(u_l)_n = u_{n+1}$. This operation consists of suppressing u_0 and translating the other coefficients of $\mathbf{u}$ one position to the left.

EXERCISE 8.3 Define a series $\mathbf{v}$ such that $\mathbf{u}_{a,r} = \mathbf{u} \times \mathbf{v} + \mathbf{a}$ and $\mathbf{u} - \mathbf{u}_0 = \mathbf{u}_l \times \mathbf{v}$.   ◊

We will characterize invertible series, which, in the sequel, will yield a convenient tool for computing terms of the form $1/f(x)$, where $f(x)$ is a function represented by a series.

Lemma 8.6 *Let* $\mathbf{u} \neq \mathbb{0}$, $\mathbf{v}$ *and* $\mathbf{v}'$ *be power series. Then*

$$\mathbf{uv} = \mathbf{uv}' \implies \mathbf{v} = \mathbf{v}'.$$

Proof. If $\mathbf{v} \neq \mathbf{v}'$, there exists a k such that $v_k \neq v'_k$. Let m be the least such k, and let n be the least integer such that $u_n \neq 0$. Then:

$$(\mathbf{uv})_{n+m} = \sum_{p=0}^{n+m} u_p v_{n+m-p}$$

$$= \sum_{p=n}^{n+m} u_p v_{m+n-p} \quad \text{(because } u_p = 0 \text{ if } p < n\text{)}$$

$$= \sum_{p=0}^{m} u_{n+p} v_{m-p}$$

$$= \sum_{p=0}^{m} u_{n+m-p} v_p \ .$$

Similarly,

$$(\mathbf{uv}')_{n+m} = \sum_{p=0}^{m} u_{n+m-p} v'_p \ ;$$

hence,

$$(\mathbf{uv} - \mathbf{uv}')_{n+m} = \sum_{p=0}^{m} u_{n+m-p}(v_p - v'_p)$$

$$= u_n(v_m - v'_m) \quad \text{(because } v_p = v'_p \text{ if } p < m\text{)}.$$

Because $u_n \neq 0$ and $v_m \neq v'_m$, $(\mathbf{uv} - \mathbf{uv}')_{n+m} \neq 0$, and that contradicts $\mathbf{uv} = \mathbf{uv}'$.   □

Definition 8.7
1. Let $\mathbf{u}, \mathbf{v} \neq \mathbb{O}$ be power series. Then $\mathbf{v}$ is said to divide $\mathbf{u}$ if and only if there exists a (necessarily unique) power series $\mathbf{w}$ such that $\mathbf{u} = \mathbf{vw}$. The unique $\mathbf{w}$ such that $\mathbf{u} = \mathbf{vw}$ is denoted by $\mathbf{u}/\mathbf{v}$.
2. A series $\mathbf{u}$ is said to be invertible if and only if there exists a series $\mathbf{v}$ such that $\mathbf{uv} = \mathbb{1}$. $\mathbf{v}$ is called the inverse of $\mathbf{u}$.

Lemma 8.8 Let $\mathbf{u} \neq \mathbb{O}$, $\mathbf{v} \neq \mathbb{O}$ be power series. Let n (resp. m) be the least integer such that $u_n \neq 0$ (resp. $v_m \neq 0$). Then $\mathbf{v}$ divides $\mathbf{u}$ if and only if $m \leq n$.

Proof. Assume, first, that $\mathbf{u} = \mathbf{vw}$. Then $u_q = (\mathbf{vw})_q = \sum_{p=0}^{q} v_p w_{q-p}$. If $q < m$ then $\sum_{p=0}^{q} v_p w_{q-p} = 0 = u_q$, hence $q < n$, and thus $m \leq n$.

Assume now that $m \leq n$. The sequence $(w_p)_{p \geq 0}$ is inductively defined by:

(B) $w_p = \begin{cases} 0 & \text{if } p < n - m, \\ u_n/v_m & \text{if } p = n - m. \end{cases}$

(I) $w_{n+m+i+1} = (u_{n+i+1} - \sum_{p=0}^{n-m+i} v_{n+i+1-p} w_p)/v_m$, for $i \geq 0$.

Let $\mathbf{w} = \sum w_p X^p$. Let us compute $\mathbf{vw}$: $(\mathbf{vw})_q = \sum_{p=0}^{q} w_p v_{q-p}$:

- If $q < n - m$, $(\mathbf{vw})_q = 0 = u_q$.

- If $q \geq n - m$, then $(\mathbf{vw})_q = \sum_{p=n-m}^{q} w_p v_{q-p} = \sum_{p=0}^{q-n+m} w_{p+n-m} v_{q-n+m-p}$.

    – If $n - m < q < n$, the sum $\sum_{p=0}^{q-n+m} w_{p+n-m} v_{q-n+m-p}$ is null and is equal to u_q.

    – If $q = n$, we have $(\mathbf{vw})_q = \sum_{p=0}^{m} w_{p+n-m} v_{m-p} = \sum_{p=0}^{m} w_{n-p} v_p = w_{n-m} v_m = u_n$.

    – If $q > n$, let $q = n + i + 1$, then

$$(\mathbf{vw})_q = \sum_{p=0}^{n+i+1} w_{n+i+1-p} v_p = \sum_{p=m}^{n+i+1} w_{n+i+1-p} v_p$$

$$= \sum_{p=0}^{n-m+i+1} w_{n-m+i+1-p} v_{p+m} = \sum_{p=0}^{n-m+i+1} w_p v_{n+i+1-p}$$

$$= \sum_{p=0}^{n-m+i} w_p v_{n+i+1-p} + w_{n-m+i+1} v_m$$

$$= u_{n+i+1} = u_q. \qquad \square$$

Generalities

Corollary 8.9 *A series $u = u_0 + u_1 X + \cdots + u_n X^n + \cdots$ in $\mathbb{C}[\![X]\!]$ is invertible if and only if $u_0 \neq 0$.*

EXERCISE 8.4 Prove that a series $u = a_0 + a_1 X + \cdots + a_n X^n + \cdots$ with coefficients in a ring is invertible if and only if a_0 is invertible. ◇

8.1.3 Operations on series

From the present section on, we will cease using the boldface notation when denoting series.

Operations on series naturally correspond to the operations on sequences introduced in Section 7.4.1. We recall these in the list given below: the first column contains the operation on sequences, and the second column the corresponding operation on series. For instance, the first line of the list should be read as: if $v = (Eu)_n = u_{n-1}$, then the operation corresponding to E on the series is the mapping $u \longmapsto Xu = v$. Indeed, it can be verified that if $u = \sum_{n \geq 0} u_n X^n$, then $Xu = \sum_{n \geq 1} u_{n-1} X^n$.

Lines 6 and 7 of the list given below can be used for *defining* the differentiation and integration operations on power series; it can then be proved that, when the convergence radii of the corresponding series are not equal to zero, the operations thus defined coincide with the usual differentiation and integration operations on analytic functions: for instance, if we *define the derivative* of the power series $u = (u_n)_{n \geq 0}$ as being the power series $v = \big((n+1) u_{n+1}\big)_{n \geq 0}$ (1st column of line 7), and if, moreover, u and v, considered as analytic series, have non-zero convergence radii and define the analytic functions $u(x)$ and $v(x)$, then we have: $u'(x) = v(x)$.

	sequences	series
1.	$v_n = (Eu)_n = u_{n-1}$, $v_0 = 0$	$v = Xu = uX + 0 = u_{0,r}$
2.	$v_n = (\Delta u)_n = \big((\mathbb{1} - E)u\big)_n = u_n - u_{n-1}$, $v_0 = u_0$	$v = (1 - X)u$
3.	$v_n = \left(\sum u\right)_n = u_0 + \cdots + u_n$	$v = \dfrac{u}{1 - X}$
4.	$w_n = u_n + v_n$	$w = u + v$
5.	$w_n = \sum_{p+q=n} u_p v_q$	$w = uv$
6.	$v_n = (n+1) u_{n+1}$	$v = u'$
7.	$v_n = \dfrac{u_{n-1}}{n}$, $v_0 = 0$	$v = \displaystyle\int_0^x u(t)\, dt$

EXERCISE 8.5 Verify lines 2–7 of the list given above. ◇

EXAMPLE 8.10 Recall that the derivative of $\frac{1}{1-x}$ is $\frac{1}{(1-x)^2}$; hence, by (6),

$$\frac{1}{(1-x)^2} = \sum_{n \geq 0} (n+1)x^n.$$

Similarly, taking the integral of $\frac{1}{1-x}$ and applying line 7, we obtain

$$\log \frac{1}{1-x} = \sum_{n \geq 1} \frac{x^n}{n}.$$

Multiplying this last equality by $\frac{1}{1-x}$ and using line 3, we obtain

$$\frac{1}{1-x} \log \frac{1}{1-x} = \sum_{n \geq 1} H_n x^n,$$

where

$$H_n = 1 + \frac{1}{2} + \cdots + \frac{1}{n}.$$

Taking derivatives again:

$$\frac{1}{(1-x)^2} \log \frac{1}{1-x} + \frac{1}{(1-x)^2} = \sum_{n \geq 0} (n+1)H_{n+1} x^n \ ;$$

hence

$$\frac{1}{1-x} \left[\sum_{n \geq 1} H_n x^n + \sum_{n \geq 0} x^n \right] = \sum_{n \geq 0} (n+1) H_{n+1} x^n .$$

Applying 3 once more:

$$\sum_{n \geq 0} \left[\left(\sum_{i=1}^{n} H_i \right) + \left(\sum_{i=0}^{n} 1 \right) \right] x^n = \sum_{n \geq 0} (n+1) H_{n+1} x^n,$$

and also $\sum_{i=1}^{n} H_i = (n+1)(H_{n+1} - 1)$.

Generalities

8.1.4 Exponential generating series

We have associated a power series with a sequence; this power series can be said to be polynomial; we can also associate an *exponential generating series* with each sequence as follows. To the sequence $u = u_0, u_1, \ldots, u_n, \ldots$ we will associate the series:
$$\hat{u} = u_0 + u_1 X + \cdots + u_n \frac{X^n}{n!} + \cdots.$$

The exponential terminology originates from the fact that we associate with the sequence $u_n = 1, \forall n$, the series $\sum_{n \geq 0} \frac{x^n}{n!}$, which is the expansion of the function e^x into a power series. As in the case of ordinary generating series, operations on the corresponding exponential generating series are associated with the operations on sequences. We give a short list below.

sequences	series
$w_n = u_n + v_n$	$(\hat{u}, \hat{v}) \longmapsto \hat{w} = \hat{u} + \hat{v}$
$v_n = u_{n+1}$	$\hat{u} \longmapsto \hat{v} = \hat{u}'$
$v_n = u_{n-1}$	$\hat{u} \longmapsto \hat{v} = \int_0^X \hat{u}(t)dt$
$w_n = \sum_{p=0}^n \binom{n}{p} u_p v_{n-p}$	$(\hat{u}, \hat{v}) \longmapsto \hat{w} = \hat{u}\hat{v}$

EXERCISE 8.6
1. Show that: $\int_0^\infty t^n e^{-t} dt = n!$.
2. Conclude from 1 that: $\int_0^\infty \hat{u}(xt)e^{-t} dt = u(x)$ provided that the series is convergent and that the integral exists. ◇

8.1.5 Partial fraction expansion of rational functions

We will now study a computation method which is very useful for obtaining the coefficients of a generating series in a simple way (this method is also used in probability theory).

Proposition 8.11 *Let $g(x) = \sum_{n \geq 0} u_n x^n$ be a generating series which is the power series expansion of a rational function of the form $g(x) = \dfrac{U(x)}{V(x)}$, where U and V are two polynomials having no common root and such that $deg(U) < deg(V) = m$. Let us assume, moreover, that $V(x)$ has m distinct roots $r_1, \ldots, r_m$, i.e. $V(x) = (x - r_1) \cdots (x - r_m)$. Then, constants $a_1, \ldots, a_m$ exist such that:*

$$g(x) = \frac{a_1}{x - r_1} + \cdots + \frac{a_m}{x - r_m}. \tag{8.1}$$

To find a_i, let us multiply the equation (8.1) by $x - r_i$; we have

$$g_i(x) = \frac{a_1(x - r_i)}{x - r_1} + \cdots + \frac{a_{i-1}(x - r_i)}{x - r_{i-1}} + a_i + \frac{a_{i+1}(x - r_i)}{x - r_{i+1}} + \cdots + \frac{a_m(x - r_i)}{x - r_m}.$$

Letting $x = r_i$, we obtain $g_i(r_i) = a_i$. Moreover,

$$g_i(x) = \frac{U(x)}{(x-r_1)\cdots(x-r_{i-1})(x-r_{i+1})\cdots(x-r_m)}$$

and, for $x = r_i$,

$$(r_i - r_1)\cdots(r_i - r_{i-1})(r_i - r_{i+1})\cdots(r_i - r_m) = V'(r_i),$$

hence

$$a_i = g_i(r_i) = \frac{U(r_i)}{V'(r_i)}.$$

Proposition 8.12 (Computation of u_n – asymptotic value) Let U and V satisfy the hypotheses of Proposition 8.11. Let r_1 be a simple root of $V(x)$, such that, for all other roots r_j, $|r_1| < |r_j|$; then when $n \to \infty$, $u_n \sim \dfrac{-a_1}{r_1^{n+1}}$.

Proof. The power series expansion of $\dfrac{1}{x - r_i}$ is known:

$$\frac{1}{x - r_i} = -\frac{1}{r_i}\left(\frac{1}{1 - \frac{x}{r_i}}\right) = -\frac{1}{r_i}\left(1 + \frac{x}{r_i} + \frac{x^2}{r_i^2} + \cdots + \frac{x^n}{r_i^n} + \cdots\right).$$

By (8.1), we have

$$g(x) = \sum_{n \geq 0} \left(\sum_{i=1}^{m} \frac{-a_i}{r_i^{n+1}}\right) x^n. \tag{8.2}$$

Identifying with $g(x) = \sum_{n \geq 0} u_n x^n$, we obtain $u_n = \sum_{i=1}^{m} \dfrac{-a_i}{r_i^{n+1}}$, whence $r_1^{n+1} u_n = -a_1 + \sum_{i=2}^{m}(-a_i)\left(\dfrac{r_1}{r_i}\right)^{n+1}$. If, moreover, $|r_1| < |r_j|$ for all the other roots r_j, then $\sum_{i=2}^{m}(-a_i)\left(\dfrac{r_1}{r_i}\right)^{n+1}$ tends to 0 as n tends to infinity, and thus $r_1^{n+1} u_n$ tends to $-a_1$; hence $u_n \sim \dfrac{-a_1}{r_1^{n+1}}$. □

EXERCISE 8.7 Can you eliminate some of the restrictions assumed for obtaining the partial fraction expansion? ◇

We state without proof the general form of Proposition 8.11.

Generalities 159

Proposition 8.13 *Let $g(x) = \sum_{n \geq 0} u_n x^n$ be a generating series which is the power series expansion of a rational function of the form $g(x) = \dfrac{U(x)}{V(x)}$, where U and V are two polynomials having no common root and such that $\deg(U) < \deg(V) = m$. Let us assume, moreover, that $V(x)$ has the m roots $r_1, \ldots, r_m$, and that $V(x) = c(x - r_1)^{d_1} \cdots (x - r_m)^{d_m}$. In this case, we have*

$$g(x) = \sum_{n \geq 0} \left(\sum_{i=1}^{m} \frac{P_i(n)}{r_i^n} \right) x^n, \tag{8.3}$$

where, for $i = 1, \ldots, m$, $P_i(n)$ is a polynomial in the variable n of degree $d_i - 1$, whose coefficient a_i of $n^{d_i - 1}$ is given, if $V^{(d_i)}$ denotes the d_ith derivative of V, by

$$a_i = \frac{d_i U(r_i)}{(-r_i)^{d_i} c V^{(d_i)}(r_i)}$$
$$= \frac{U(r_i)}{(-r_i)^{d_i} (d_i - 1)! c \prod_{j \neq i} (r_i - r_j)^{d_j}}.$$

The proof is by induction on $\max(d_1, \ldots, d_m)$, by showing that

$$g(x) - \sum_{i=1}^{m} \frac{a_i (d_i - 1)!}{(1 - x/r_i)^{d_i}}$$

is a rational fraction whose denominator is divisible by none among the $(x - r_i)^{d_i}$ s.

EXAMPLE 8.14   Let the sequence u_n be defined by $u_0 = 1$, and

$$u_n = q u_{n-1} + p(1 - u_{n-1}),$$

where $p + q = 1$ and $q \neq 1$. u_n represents the probability of obtaining an even number of 'tails' after n successive tosses of a coin. Let $g(x) = \sum_{n \geq 0} u_n x^n$. Let us multiply equation $u_n = q u_{n-1} + p(1 - u_{n-1})$ by x^n for each $n > 0$ in $\mathbb{N}$, and add the equalities thus obtained:

$$\vdots$$
$$u_n x^n = x q u_{n-1} x^{n-1} + x p (1 - u_{n-1}) x^{n-1}$$
$$\vdots$$
$$u_1 x = x q u_0 + x p (1 - u_0).$$

We obtain
$$g(x) - 1 = qx \sum_{n \geq 0} u_n x^n + px \sum_{n \geq 0} x^n - px \sum_{n \geq 0} u_n x^n$$
$$= qxg(x) + px\frac{1}{1-x} - pxg(x),$$
or
$$g(x)(1-(q-p)x) = 1 + px\frac{1}{1-x} = \frac{1-x(1-p)}{1-x} = \frac{1-qx}{1-x},$$
hence
$$g(x) = \frac{1-qx}{(1-x)(1-(q-p)x)} = \frac{a_1}{1-x} + \frac{a_2}{1-(q-p)x};$$
and, after computing a_1 and a_2, which yield
$$a_1 = a_2 = \frac{1-q}{1-(q-p)} = \frac{p}{2p} = \frac{1}{2},$$
$$g(x) = \frac{1}{2}\left(\frac{1}{1-x} + \frac{1}{1-(q-p)x}\right)$$
$$= \frac{1}{2}\left(\sum_{n \geq 0} x^n + \sum_{n \geq 0}(q-p)^n x^n\right) = \frac{1}{2}\sum_{n \geq 0}(1+(q-p)^n)x^n,$$
hence
$$u_n = \frac{1}{2}(1+(q-p)^n).$$
We deduce (see Chapter 12) the formula
$$\frac{1}{2}(1+(q-p)^n) = \sum_{k=0}^{n/2} \binom{n}{2k} p^{2k} q^{n-2k};$$
indeed, $\binom{n}{2k} p^{2k} q^{n-2k}$ represents the probability of obtaining $2k$ 'tails' in a sequence of n successive tosses of a coin (in technical terms in a sequence of n Bernoulli trials).

EXERCISE 8.8
1. Prove that
$$\sum_{k=0}^{\infty} \frac{x^{2^k}}{1-x^{2^{k+1}}} = \frac{x}{1-x}.$$
2. Deduce the following identity, where the F_ns are the Fibonacci numbers:
$$\sum_{k=0}^{\infty} \frac{1}{F_{2^k}} = \frac{7-\sqrt{5}}{2}.$$
This equality is called the Millin identity, and was proved by Dale Miller. (His name was misprinted in the first paper stating the result.) ◇

8.2 Applications of generating series to recurrences

8.2.1 Linear recurrences with constant coefficients

Let the sequence u_n be defined by the recurrence equation (7.9)
$$\forall n \geq k, \qquad u_n = a_1 u_{n-1} + \cdots + a_k u_{n-k} \qquad (7.9)$$
and let $u(z) = \sum_{n \geq 0} u_n z^n$ be the associated generating series. We will compute the generating series, and deduce the u_ns.

Multiplying (7.9) by z^n we obtain
$$\forall n \geq k, \qquad u_n z^n = a_1 z u_{n-1} z^{n-1} + a_2 z^2 u_{n-2} z^{n-2} + \cdots + a_k z^k u_{n-k} z^{n-k}$$
and summing up all these equalities for $n \geq k$,
$$u_k z^k = a_1 z u_{k-1} z^{k-1} + a_2 z^2 u_{k-2} z^{k-2} + \cdots + a_k z^k u_0$$
$$u_{k+1} z^{k+1} = a_1 z u_k z^k + a_2 z^2 u_{k-1} z^{k-1} + \cdots + a_k z^k u_1 z$$
$$\vdots$$
$$u_n z^n = a_1 z u_{n-1} z^{n-1} + a_2 z^2 u_{n-2} z^{n-2} + \cdots + a_k z^k u_{n-k} z^{n-k},$$
$$\vdots$$
we obtain
$$\sum_{n \geq k} u_n z^n = a_1 z \sum_{n \geq k-1} u_n z^n + a_2 z^2 \sum_{n \geq k-2} u_n z^n + \cdots + a_k z^k \sum_{n \geq 0} u_n z^n.$$
For $0 \leq i \leq k-1$, let $P_i(z) = u_0 + u_1 z + \cdots + u_i z^i$, so that
$$u(z) = P_{i-1}(z) + \sum_{n \geq i} u_n z^n;$$
hence
$$u(z) - P_{k-1}(z) = a_1 z \big(u(z) - P_{k-2}(z)\big) + \cdots$$
$$+ a_{k-1} z^{k-1} \big(u(z) - P_0(z)\big) + a_k z^k u(z).$$
And thus
$$u(z) = \frac{P_{k-1}(z) - a_1 z P_{k-2}(z) - \cdots - a_{k-1} z^{k-1} P_0(z)}{1 - a_1 z - a_2 z^2 - \cdots - a_k z^k}$$
$$= \frac{P(z)}{1 - a_1 z - a_2 z^2 - \cdots - a_k z^k},$$
where $P(z)$ is a polynomial in z of degree less than or equal to $k-1$; the rational fraction defining $u(z)$ has a power series expansion which can be obtained by standard techniques (see Section 8.1.5), and this allows one to find the coefficients of the series $u(z)$, i.e. the sequence u_n.

REMARK 8.15 Note that r_i is a multiple root of multiplicity d_i of the characteristic polynomial if and only if $\rho_i = 1/r_i$ is a multiple root of the same multiplicity d_i of $V(z) = 1 - a_1 z - a_2 z^2 - \cdots - a_k z^k$. The same phenomenon will occur for non-homogeneous linear recurrences with constant coefficients.

Summing up, the above general method can be split into three steps:

1. Multiply the recurrence equation by z^n and sum on n; to the left of the '=' sign is the generating series $u(z)$ where the k first terms have been deleted, i.e. the expression $u(z) - P_{k-1}(z)$, and to the right of the '=' sign is an expression of the form $u(z)P(z) + Q(z)$, where P and Q are polynomials in z.
2. Solve the equation in $u(z)$ thus obtained; we obtain a rational function;
3. Find the power series expansion of the rational function giving the coefficients of $u(z)$, namely the u_ns, for $n \geq k$. This last step is usually the step demanding most effort and computations.

Note, finally, that the methods using generating series are very suitable for some manipulations on sequences: for instance if the generating series $u(z)$ of the sequence u_n is known, and if we are interested in the subsequence $v_n = u_{2n}$ of even index terms, it is enough to remark that the sequence v_n is determined by

$$\frac{u(z) + u(-z)}{2} = \sum_{n \geq 0} u_{2n} z^{2n}.$$

Similarly, the subsequence $v_n = u_{2n+1}$ of odd index terms is determined by

$$\frac{u(z) - u(-z)}{2} = \sum_{n \geq 0} u_{2n+1} z^{2n+1}.$$

EXERCISE 8.9 The Fibonacci numbers are defined by $F_0 = 0$, $F_1 = 1$ and, for $n \geq 2$, $F_n = F_{n-1} + F_{n-2}$. Determine the generating series of the sequence F_{2n} of Fibonacci numbers of even index. ◇

EXERCISE 8.10 Solve the following recurrence equations by the generating series method.
1. $\forall n \geq 2$, $2u_n = 3u_{n-1} - u_{n-2}$.
2. $\forall n \geq 2$, $u_n = 4u_{n-1} - 4u_{n-2}$. ◇

8.2.2 Non-homogeneous linear recurrences with constant coefficients

The preceding method can easily be applied and is illustrated in this example. Consider the recurrence equation

$$u_n = u_{n-1} + 2u_{n-2} + (-1)^n, \qquad n \geq 2,$$

with $u_0 = u_1 = 1$, and let $u(z) = \sum_{n \geq 0} u_n z^n$ be the associated generating series.

Writing
$$u_0 = 1$$
$$u_1 z = z = z u_0$$
$$u_2 z^2 = z u_1 z + 2z^2 u_0 + (-1)^2 z^2$$
$$\vdots$$
$$u_n z^n = z u_{n-1} z^{n-1} + 2z^2 u_{n-2} z^{n-2} + (-1)^n z^n$$
$$\vdots$$

we deduce, by summing up the above equalities and adding the corrective term $(-1)^1 z$ on each side of the thus obtained equality, that

$$u(z) - z = z u(z) + 2z^2 u(z) + \frac{1}{1+z} ;$$

hence
$$u(z) = \frac{1 + z + z^2}{(1+z)(1 - z - 2z^2)} = \frac{1 + z + z^2}{(1 - 2z)(1 + z)^2} .$$

By Proposition 8.13, we have

$$u_n = a 2^n + (bn + c)(-1)^n,$$

with $a = 7/9$ and $b = 1/3$. Substituting these values and letting $n = 0$ in the preceding equation, we obtain $c = 2/9$.

EXERCISE 8.11 Solve the following recurrence equation:
$$\forall n \geq 2, \quad u_n = 4u_{n-1} - 4u_{n-2} + n - 1, \text{ with } u_0 = 1 \text{ and } u_1 = 1. \qquad \diamond$$

EXERCISE 8.12 Solve the simultaneous recurrence equations
$u_n = 2v_{n-1} + u_{n-2}$, with $u_0 = 1$ and $u_1 = 0$,
$v_n = u_{n-1} + v_{n-2}$, with $v_0 = 0$ and $v_1 = 1$. $\qquad \diamond$

EXERCISE 8.13 Solve the recurrence equation
$$\forall n \geq 2, \quad u_n = 3u_{n-1} - 2u_{n-2} + n/2^n, \quad \text{with } u_0 = 1 \text{ and } u_1 = 0. \qquad \diamond$$

8.2.3 Partitioning integers

The problem is that of finding the number of vectors of integers $v = (n_1, \ldots, n_p)$ which are solutions of $a_1 n_1 + \cdots + a_p n_p = n$, with fixed $a_1, \ldots, a_p \in \mathbb{N}$. We illustrate this type of problem with an example. How many ways are there to change a $100 bill for $1 and $5 bills? It boils down to finding the number of solutions of the equation: $p + 5q = 100$. Let the series

$$u(x) = 1 + x + x^2 + \cdots + x^p + \cdots$$
$$w(x) = 1 + x^5 + x^{10} + \cdots + x^{5q} + \cdots.$$

We have

$$(uw)(x) = \sum_{n \geq 0} \left(\sum_{p+5q=n} x^p x^{5q} \right).$$

Consequently, the number of solutions of $p + 5q = n$ is the coefficient of x^n in the series $v(x) = u(x)w(x)$. But

$$u(x) = \frac{1}{1-x}, \quad w(x) = \frac{1}{1-x^5},$$

and hence

$$v(x) = (uw)(x) = \frac{1}{(1-x)(1-x^5)}.$$

The standard method consists of computing the partial fraction expansion of uw, and deducing therefrom the coefficient of x^{100} in uw. Here this partial fraction expansion would have the form:

$$v(x) = \frac{a}{(1-x)^2} + \frac{b}{1-x} + \frac{c}{1-e^{i\alpha}x} + \frac{\bar{c}}{1-e^{-i\alpha}x} + \frac{d}{1-e^{2i\alpha}x} + \frac{\bar{d}}{1-e^{-2i\alpha}x},$$

where $e^{i\alpha}, e^{-i\alpha}, e^{2i\alpha}, e^{-2i\alpha}$ are the complex roots of $x^5 = 1$. Here we can, however, perform a simpler and more astute computation: noting that $1 - x^5 = (1-x)(1 + x + x^2 + x^3 + x^4)$, we can deduce

$$v(x) = \frac{1 + x + x^2 + x^3 + x^4}{(1-x^5)^2}$$
$$= (1 + x + x^2 + x^3 + x^4) \sum_{n \geq 0} (n+1) x^{5n}.$$

The number of ways to change n dollars for $1 and $5 bills is thus the coefficient of x^n in $v(x)$. Any arbitrary n can be uniquely written in the form $n = 5k + r$ with $0 \leq r \leq 4$, and the coefficient of x^{5k+r} in $v(x)$ is then $k+1$. For instance, there are $v_{20} = 21$ ways to change $100 for $1 and $5 bills.

Applications of generating series to recurrences

EXERCISE 8.14 Find the number of ways of bringing up a total of n with tokens of value 2 and 3. ◊

EXERCISE 8.15 Assuming that in Morse code a dot takes two time units, and a dash takes three time units, find the number of words taking n time units in the Morse code. ◊

EXERCISE 8.16 How can you find the number of ways of changing $100 for $1, $2 and $5 bills? ◊

8.2.4 Finite linear recurrence equations with non-constant coefficients

The method consists of associating with the recurrence equation defining the sequence u_n a functional or differential equation on the generating series $u(x)$ corresponding to u_n. This method can be applied in the case of linear recurrence equations with constant or non-constant coefficients.

- In the case of recurrence equations with constant coefficients, the functional equation will be of the form $u(x) = \dfrac{U(x)}{V(x)}$, where U and V are polynomials in x. More precisely, if the recurrence equation is of the form: $u_n = a_1 u_{n-1} + \cdots + a_k u_{n-k}$, with initial values $u_0, \ldots, u_{k-1}$, and if $u(x) = \sum_{n \geq 0} u_i x^i$, then (see Section 8.2.1) $u(x) = \dfrac{U(x)}{V(x)}$, with

$$V(x) = 1 - a_1 x - a_2 x^2 - \cdots - a_k x^k,$$

and $U(x)$ a polynomial of degree $\leq k - 1$. It is then sufficient to find the partial fraction expansion of the rational function $\dfrac{U}{V}$ by the methods described in Section 8.1.5, Proposition 8.11 and Proposition 8.13 in order to obtain the u_ns. See Example 8.14.

- In the case of recurrence equations with non-constant coefficients, we will in general obtain a differential equation involving the derivatives of $u(x)$ of order 1, 2, etc. The problem will hence be more complex. We will illustrate the method with a simple example.

EXAMPLE 8.16 Let the sequence u_n, for $n \geq 2$, be defined by

$$n u_n + (n-2) u_{n-1} - u_{n-2} = 0, \quad \text{with } u_0 = u_1 = 1.$$

Multiplying the recurrence equation by x^{n-1} and summing for $n \geq 2$, we obtain

$$\sum_{n \geq 2} n u_n x^{n-1} + \sum_{n \geq 2} (n-2) u_{n-1} x^{n-1} - \sum_{n \geq 2} u_{n-2} x^{n-1} = 0, \tag{8.4}$$

i.e.

$$\sum_{n\geq 2} nu_n x^{n-1} + x\sum_{n\geq 2}(n-1)u_{n-1}x^{n-2} - \sum_{n\geq 2} u_{n-1}x^{n-1} - x\sum_{n\geq 2} u_{n-2}x^{n-2} = 0 \,.$$

Hence, $u(x) = \sum_{n\geq 0} u_n x^n$; then

$$u'(x) = \sum_{n\geq 0}(n+1)u_{n+1}x^n = \sum_{n\geq 1} nu_n x^{n-1},$$

so that $u'(x) = u_1 + \sum_{n\geq 2} nu_n x^{n-1}$. Equation (8.4) can be written

$$(u'(x) - u_1) + xu'(x) - u(x) + u_0 - xu(x) = 0$$

or

$$(u'(x) - 1) + x(u'(x) - u(x)) - u(x) + 1 = 0$$

and, then,

$$(u'(x) - u(x))(1+x) = 0 \,;$$

hence $u(x) = u'(x)$, and $u(x) = \lambda e^x$, with $\lambda = 1$ since $u(0) = 1$. Finally,

$$u(x) = \sum_{n\geq 0} \frac{x^n}{n!} \quad \text{and} \quad u_n = \frac{1}{n!}.$$

EXERCISE 8.17 Solve the recurrence equation

$$\forall n \geq 1, \quad 2nu_n = u_{n-1} + \frac{1}{(n-1)!},$$

with $u_0 = 2$, and the convention $0! = 1$. ◇

8.2.5 Complete recurrence equations

The method of Section 8.2.1 can still be applied; of course the computation of the series can no longer be reduced to the power series expansion of a rational function because the functional equation defining the series may be more complex. Let us study an example.

Recall (Example 7.1) that the number of binary trees with n nodes is given by the recurrence equation $b_n = \sum_{k=0}^{n-1} b_k b_{n-k-1}$ for $n \geq 1$, and $b_0 = 1$. Let $b(x) = \sum_{n\geq 0} b_n x^n$ be the associated generating series; substituting for b_n in the generating series $b(x)$ the value of b_n, which is given by the recurrence equation, we have

$$b(x) = 1 + \sum_{n\geq 1}\left(\sum_{k=0}^{n-1} b_k b_{n-k-1}\right)x^n = 1 + x\sum_{n\geq 1}\left(\sum_{p+q=n-1} b_p b_q x^{n-1}\right),$$

that is
$$b(x) = 1 + x(b(x))^2.$$

$b(x)$ is thus the solution of the equation in b: $xb^2 - b + 1 = 0$, whose roots are $\dfrac{1 \pm \sqrt{1-4x}}{2x}$. Since $b(0) = b_0 = 1$, the numerator should be divisible by $2x$, and this allows a single possible solution $\dfrac{1 - \sqrt{1-4x}}{2x}$ (the other solution being undefined for $x = 0$). Now, let us recall the power series expansion of $(1+u)^\alpha$, valid for $\alpha \in \mathbb{R}$,

$$(1+u)^\alpha = 1 + \alpha u + \cdots + \binom{\alpha}{k} u^k + \cdots,$$

where, by convention,

$$\binom{\alpha}{k} = \frac{\alpha(\alpha-1)\cdots(\alpha-k+1)}{k!}.$$

Here we obtain

$$\sqrt{1-4x} = \sum_{k \geq 0} \binom{1/2}{k}(-4x)^k = \sum_{k \geq 0} \frac{1/2(-1/2)(-3/2)\cdots(3-2k)/2}{k!}(-4x)^k$$

$$= 1 + \sum_{k \geq 1} (-1)^{k-1} \frac{(2k-2)!(-4)^k x^k}{k! 2^k 2^{k-1}(k-1)!}$$

$$= 1 + \sum_{k \geq 1} -\frac{(2k-2)! 2}{k!(k-1)!} x^k.$$

Hence,

$$b(x) = \sum_{n \geq 1} \frac{(2(n-1))!}{n!(n-1)!} x^{n-1} = \sum_{n \geq 0} \frac{1}{n+1} \binom{2n}{n} x^n$$

and

$$\forall n \geq 0, \quad b_n = \frac{1}{n+1} \binom{2n}{n} = \frac{1}{n+1} \frac{(2n)!}{n! n!}.$$

The number $b_n = \dfrac{1}{n+1} \binom{2n}{n}$ is the nth *Catalan number*. The Catalan numbers are useful in combinatorics.

EXERCISE 8.18 Let $A = \{(,)\}$ be the alphabet consisting of two parentheses (left and right). The set of strings of balanced parentheses, also called the Dyck language, is the subset $D \subseteq A^*$ defined by
(B) $\varepsilon \in D$,
(I) if x and y are in D, then $(x)y$ is also in D.

1. Verify that this definition coincides with the one given in Example 3.9 and that all the strings of balanced parentheses are words of even length.
2. Let u_n be the number of strings of balanced parentheses of length $2n$; find a recurrence equation defining u_n in terms of the u_is, for $i < n$. Deduce u_n.
3. How can you generalize this result to the Dyck language on an alphabet $A_k = \{a_1, \ldots, a_k, \bar{a}_1, \ldots, \bar{a}_k\} = B_k \cup \bar{B}_k$ (assuming there are k different types of parentheses), defined by
(B) $\varepsilon \in D$,
(I) if x and y are in D, then $a_i x \bar{a}_i y$ is also in D, for all $i = 1, \ldots, k$. ◇

EXERCISE 8.19 Solve the recurrence equation
$$\forall n \geq 1, \quad u_n = u_{n-1} + 2u_{n-2} + \cdots + nu_0, \quad \text{with } u_0 = 1.$$
◇

EXERCISE 8.20 Solve the recurrence equation
$$\forall n > 1, \quad u_n = -2nu_{n-1} + \sum_{k=0}^{n} \binom{n}{k} u_k u_{n-k},$$
with $u_0 = 0$ and $u_1 = 1$, using exponential generating series. ◇

8.2.6 Average complexity of algorithms

Generally speaking, the generating series can be applied to the analysis of the average complexity of algorithms. Let A be an algorithm computing on data D, and let D_n be the set of data of size n, assumed to be all equally probable. Let $c(d)$ be the time complexity of algorithm A on data d. Then, the average time complexity of A on D_n is

$$m_n = \frac{1}{|D_n|} \sum_{d \in D_n} c(d),$$

where $|D_n|$ is the cardinality of D_n. Let us denote by $|d|$ the size of d and let $c(x) = \sum_{n \geq 0} c_n x^n$, where $c_n = \sum_{d \in D_n} c(d)$.

We can define the generating series of the enumeration of the Ds similarly:

$$d(x) = \sum_{n \geq 0} |D_n| x^n = \sum_{n \geq 0} d_n x^n.$$

We then have $m_n = \dfrac{c_n}{d_n}$.

Applications of generating series to recurrences 169

REMARK 8.17 We can adopt the notation $\sum_{d\in D} c(d) x^{|d|}$, directly introducing the object d in the definition of the generating series of the enumeration. This method enables us to directly find the equation satisfied by the generating series in many cases of combinatorial enumeration problems without considering the associated recurrence equation.

EXERCISE 8.21 Let $A = \{\text{blue, red, green}\}$ a set of colours. During a Master-mind game, one forms size n sequences with these three colours. Let a_i be the ith colour of the current solution.
1. Let t_n be the number of solutions of length n, such that, for i in $\{1,\ldots, n-2\}$, $a_i \neq a_{i+2}$.
    (a)   Compute t_1, t_2 and t_3.
    (b)   Find a recurrence equation for t_n.
    (c)   Deduce t_n.
2. Let s_n be the number of size n solutions, such that either $a_i \neq a_{i+2}$ or $a_i = a_{i+1} = a_{i+2}$, for $i = 1, \ldots, n-2$.
    (a)   Compute s_1, s_2, s_3 and s_4.
    (b)   Find a recurrence equation for s_n.
    (c)   Compute $\sum_{n=1}^{\infty} s_n z^n$ and deduce s_n.   ◇

EXERCISE 8.22 After giving a dictation to his pupils, a teacher redistributes the exercises for correction to the pupils. Let b_n be the number of ways of distributing the exercises in such a way that no pupil gets his or her own.
1. Compute b_1, b_2 and b_3.
2. Show that the following equation holds: $b_n = (n-1)(b_{n-1} + b_{n-2})$, for $n > 2$.
3. Show that $b_n - nb_{n-1} = (-1)^n$, for $n > 1$.
4. Let $b_0 = 1$, and define the exponential generating series by

$$b(z) = \sum_{n=0}^{\infty} \frac{b_n z_n}{n!}.$$

Prove that $b(z) = \dfrac{e^{-z}}{1-z}$.   ◇

CHAPTER 9

ASYMPTOTIC BEHAVIOUR

In Chapter 7 we saw that, in order to evaluate a complexity or a cost, estimating its order of magnitude could be useful. This can happen when exact computation is not possible (for instance, the average complexity of Quicksort, see Section 14.1), or when an approximate value of the cost or complexity is enough. For instance, comparing it to the cost of other algorithms or, *a priori*, excluding a too costly algorithm or determining the maximum size of data which can be dealt with by a given algorithm, etc.

This evaluation of an order of magnitude consists of finding an approximation of the behaviour of a function in limit conditions (n going to infinity, x going to zero, etc.); this is why such behaviour is called *asymptotic*. Most often, such evaluations will be used to study the complexity of an algorithm when the size n of the data goes to infinity. Note, however, that an algorithm which is optimal for large-size data is not always the best one for smaller-size data.

After the basic definitions, this chapter introduces methods to determine the asymptotic behaviour of functions, and to classify functions according to their asymptotic behaviour.

We recommend the following handbook:

Ronald Graham, Donald Knuth, Oren Patashnik, *Concrete Mathematics*, Addison-Wesley, London (1989).

9.1 Generalities

9.1.1 Definitions

Definition 9.1 *Let f, g be two mappings from $\mathbb{N}$ into $\mathbb{R}^+$. f is said to be dominated by g or to be a 'big-Oh' of g (resp. f is said to dominate g or to be a 'big-Omega' of g), and we note $f = O(g)$ (resp. $f = \Omega(g)$) if and only if*

$$\exists c \in \mathbb{R}^+, \ \exists n_0 \in \mathbb{N}, \ \forall n > n_0 \ : \ f(n) \leq cg(n) \quad (\text{resp. } f(n) \geq cg(n)).$$

Generalities 171

This definition also holds for functions with several arguments, for instance $f(n,p) = O(g(n,p))$ if and only if

$$\exists c, n_0, p_0, \ \forall n > n_0, \forall p > p_0 : \quad f(n,p) \leq cg(n,p).$$

The O-notation was introduced by the German mathematician Bachmann at the end of the last century (1894), and was made popular by his fellow countryman Landau, after whom it is named.

REMARK 9.2 The equality in the notation $f = O(g)$ is somewhat inappropriate but handy: for instance $n^2 + 2n = O(n^2)$, but no equality holds, since also $n^2 + 2n = O(n^5)$, but not $O(n^2) = O(n^5)$! To be quite precise, we should write $f \in O(g)$, and we would then have $O(n^2) \subsetneq O(n^5)$; but we will follow the standard notation by writing $f = O(g)$.

Definition 9.3 f and g are said to have the same order of magnitude, and this is denoted by $f = \theta(g)$, if and only if $f = O(g)$ and $g = O(f)$.

Proposition 9.4 $f = \theta(g) \implies f = O(g)$, but the converse is false.

Proof. Immediate as $f = \theta(g)$ if and only if $f = O(g)$ and $g = O(f)$. The converse is false, e.g. $n^p = O(n^{p+1})$. □

Proposition 9.5 $f = O(g) \iff g = \Omega(f)$.

Proof. Immediate since $\exists c \in \mathbb{R}^+, \exists n_0 \in \mathbb{N}, \forall n > n_0: f(n) \leq cg(n)$ if and only if $\exists c' = \dfrac{1}{c} \in \mathbb{R}^+, \exists n_0 \in \mathbb{N}, \forall n > n_0: g(n) \geq c'f(n)$. □

Definition 9.6 f is said to grow more slowly than g, and we write $f = o(g)$, or $f \prec g$, if and only if $\forall \varepsilon \in \mathbb{R}^+, \exists n_0 \in \mathbb{N}, \forall n > n_0: f(n) \leq \varepsilon g(n)$.

REMARK 9.7 The notations O, o, Ω, θ also hold for mappings from $\mathbb{R}$ into $\mathbb{R}$; but in this case, when speaking of limit conditions we must specify the limits we are talking about, because the same function f may be an $O(g)$ when $x \longrightarrow \infty$, and an $O(g')$ for a quite different g' when $x \longrightarrow 0$: for instance, let $f(x) = x^3 + 4x^2 + x$; then

- when $x \longrightarrow \infty$, $f = O(x^3)$, but $f = O(x)$ does not hold (on the contrary $f = \Omega(x)$, and even $x = o(f)$, since $x/f(x) \longrightarrow 0$),
- symmetrically, when $x \longrightarrow 0$, $f = O(x)$, but $f = O(x^3)$ does not hold (on the contrary $f = \Omega(x^3)$, and even $x^3 = o(f)$).

9.1.2 Operations on the orders of magnitude

Proposition 9.8 *For all g, g_1, g_2 :*

$$g(n) = O(g(n))$$
$$g(n) = \theta(g(n))$$
$$cO(g(n)) = O(g(n))$$
$$O(g(n)) + O(g(n)) = O(g(n))$$
$$O(g_1(n)) + O(g_2(n)) = O(\max(g_1(n), g_2(n)))$$
$$O(O(g(n))) = O(g(n))$$
$$O(g_1(n)) \cdot O(g_2(n)) = O(g_1(n))g_2(n) = g_1(n)O(g_2(n)) \ .$$

These equalities are abbreviated notations, for instance the third one stands for: $f = O(g(n)), f' = cf \implies f' = O(g(n))$, etc.

Proof. Let us check for instance that

$$O(g_1(n)) + O(g_2(n)) = O(\max(g_1(n), g_2(n)));$$

if $f(n) \in O(g_1(n)) + O(g_2(n))$, c_1 and c_2 exist such that $f(n) \leq c_1 g_1(n) + c_2 g_2(n)$ for $n \geq n_0$. Letting $c = \max(c_1, c_2)$, then

$$f(n) \leq c(g_1(n) + g_2(n)) \leq 2c \max(g_1(n), g_2(n)) \ . \qquad \square$$

EXERCISE 9.1 Find the error in the following argument: '$n = O(n)$ and $2n = O(n)$, and so on, hence $\sum_{k=1}^{n} kn = \sum_{k=1}^{n} O(n) = nO(n) = O(n^2)$'. $\diamond$

EXERCISE 9.2 If $f(n) = O(n)$, do the following hold?:
1. $(f(n))^2 = O(n^2)$,
2. $2^{f(n)} = O(2^n)$. $\diamond$

Proposition 9.9
1. $f = \theta(g) \implies \lambda f = \theta(g)$ for any constant $\lambda > 0$, and
2. $f_i = \theta(g_i)$ for $i = 1, 2, \ldots, k$, implies $\sum_{i=1}^{k} f_i = \theta(\max\{g_1, g_2, \ldots, g_n\})$.

Proof. 1 is immediate; to verify 2 note that

- $\sum_{i=1}^{k} f_i = O(\max\{g_1, g_2, \ldots, g_n\})$, by induction on k and applying the fifth equality of Proposition 9.8,
- for $i = 1, 2, \ldots, k$, $g_i = O(f_i)$ implies that $g_i = O(\sum_{i=1}^{k} f_i)$, and thus $\max\{g_1, g_2, \ldots, g_n\} = O(\sum_{i=1}^{k} f_i)$. $\qquad \square$

Generalities

EXAMPLE 9.10 $\forall k,\ \sum_{i=1}^{n} i^k = \theta(n^{k+1})$. The proof is by induction on n:
- Basis. For $k = 0$, $\sum_{i=1}^{n} i^0 = n = \theta(n)$;
- Inductive step. Assuming $\forall j < k$, $\sum_{i=1}^{n} i^j = \theta(n^{j+1})$, note that

$$i^{k+1} = ((i-1)+1)^{k+1} = \sum_{p=0}^{k+1} \binom{k+1}{p}(i-1)^p,$$

and let us apply this formula for $i = 1, 2, \ldots, n$:

$$n^{k+1} = (n-1)^{k+1} + (k+1)(n-1)^k + \cdots + 1$$
$$(n-1)^{k+1} = (n-2)^{k+1} + (k+1)(n-2)^k + \cdots + 1$$
$$\vdots$$
$$2^{k+1} = 1^{k+1} + (k+1)1^k + \cdots + 1$$
$$1^{k+1} = 0 + 0 + \cdots + 1.$$

By summing these n equalities, we obtain, after simplifications,

$$n^{k+1} = (k+1)\sum_{i=1}^{n-1} i^k + \binom{k+1}{2}\sum_{i=1}^{n-1} i^{k-1} + \cdots + n;$$

hence,

$$\sum_{i=1}^{n-1} i^k = \frac{n^{k+1}}{k+1} - \frac{1}{k+1}\left[\left(\binom{k+1}{2}\sum_{i=1}^{n-1} i^{k-1}\right) + \cdots + n\right],$$

and

$$\sum_{i=1}^{n} i^k = \frac{(n+1)^{k+1}}{k+1} - \frac{1}{k+1}\left[\sum_{j=2}^{k+1}\binom{k+1}{j}\left(\sum_{i=1}^{n} i^{k+1-j}\right)\right].$$

We then have by the induction, $\forall j = 2, \ldots, k+1$,

$$\frac{\binom{k+1}{j}}{k+1}\sum_{i=1}^{n} i^{k+1-j} = \theta(n^{k+2-j}),$$

and $\dfrac{n^{k+1}}{k+1} = \theta(n^{k+1})$. Hence, as $\max\{n^{k+1}, \ldots, n\} = n^{k+1}$, Proposition 9.9, 2, enables us to conclude

$$\sum_{i=1}^{n-1} i^k = \theta(n^{k+1}),$$

which proves the inductive step.

EXERCISE 9.3 Let g_1, g_2 be two mappings from $\mathbb{N}$ into $\mathbb{R}^+$; does the following equality hold?
$$O(g_1(n) + g_2(n)) = g_1(n) + O(g_2(n)).$$

◇

REMARK 9.11 The relation $f \mathcal{R} g$ if and only if $f = O(g)$ is a preorder relation, i.e. a reflexive and transitive relation. It is naturally associated with an order relation: let E be the set of mappings from $\mathbb{N}$ into $\mathbb{R}^+$, the relation $f \equiv g$ if and only if $f = \theta(g)$ is an equivalence relation, called the equivalence associated with $\mathcal{R}$; on the factor set $E/\equiv$, the relation $[\mathcal{R}/\equiv]$ defined by $[f] [\mathcal{R}/\equiv] [g]$ if and only if $f \mathcal{R} g$, is an order relation, which is the order relation canonically associated with $\mathcal{R}$ (see Proposition 2.9).

9.2 Criteria of asymptotic behaviour of functions

9.2.1 A sufficient condition

Given, now, two positive functions f and g, we will compare them to find out whether f is O, Ω or θ of function g or not. To this end, assuming f and g are not zero, we will form the quotients f/g or g/f.

Proposition 9.12 *Let f and g be two positive functions, then:*

(i) $\lim\limits_{n \to \infty} \dfrac{f(n)}{g(n)} = a \neq 0 \implies f = \theta(g),$

(ii) $\lim\limits_{n \to \infty} \dfrac{f(n)}{g(n)} = 0 \implies f = O(g)$ and $f \neq \theta(g),$

(iii) $\lim\limits_{n \to \infty} \dfrac{f(n)}{g(n)} = \infty \implies g = O(f)$ and $g \neq \theta(f).$

Proof. Let us prove (i) for instance. $\lim\limits_{n \to \infty} \dfrac{f(n)}{g(n)} = a \neq 0$ implies

$$\forall \varepsilon, \exists n_0, \forall n > n_0 : \left| \frac{f(n)}{g(n)} - a \right| < \varepsilon.$$

Since f and g are positive, this in turn implies

$$(a - \varepsilon)g(n) < f(n) < (a + \varepsilon)g(n),$$

and thus $f(n) < (a + \varepsilon)g(n)$, $g(n) < \dfrac{1}{a - \varepsilon} f(n)$, hence $f = O(g)$ and $g = O(f)$. □

EXERCISE 9.4 Prove cases (ii) and (iii). ◊

Definition 9.13 If $\lim\limits_{n\to\infty} \dfrac{f(n)}{g(n)} = 1$, f is said to be asymptotic to g and we write $f \sim g$; $\sim$ is an equivalence relation.

Proof. It can easily be checked that $\sim$ is an equivalence relation. Let us check for instance that $f \sim g$ and $g \sim h \implies f \sim h$: as h and g are not zero, we have
$$\lim_{n\to\infty} \frac{f(n)}{h(n)} = \lim_{n\to\infty} \frac{f(n)}{g(n)} \times \frac{g(n)}{h(n)} = 1. \qquad \square$$

Proposition 9.14 If $\lim\limits_{n\to\infty} \dfrac{f(n)}{g(n)} = 0$, then f grows more slowly than g, i.e. $f = o(g)$.

Proof. Straightforward. $\qquad \square$

REMARK 9.15 1. The preceding proposition gives a criterion which is sufficient but not necessary to classify the functions among O and Ω: for instance, let $g(n) = 2n$ and
$$f(n) = \begin{cases} n & \text{if } n \text{ is odd,} \\ 2n & \text{if } n \text{ is even,} \end{cases}$$
then $\dfrac{f(n)}{g(n)}$ has no limit, even though $f = \theta(g)$ since
$$\forall n, \quad \frac{g(n)}{2} \leq f(n) \leq g(n).$$

2. If $\lim\limits_{n\to\infty} \dfrac{f(n)}{g(n)}$ is one of the indeterminate forms $\dfrac{0}{0}$ or $\dfrac{\infty}{\infty}$, we can apply l'Hospital's rule in the form: if $\lim\limits_{n\to\infty} \dfrac{f'(n)}{g'(n)} = a$ exists, then $\lim\limits_{n\to\infty} \dfrac{f(n)}{g(n)} = a$.

EXAMPLE 9.16
$$\forall a > 0, \forall b > 0, \quad \lim_{n\to\infty} \frac{\log(n)^a}{n^b} = 0$$
and
$$\forall a > 1, \forall b > 0, \quad \lim_{n\to\infty} \frac{n^b}{a^n} = 0,$$
which is abbreviated to 'exponential wins over powers, and powers win over logarithms'. More precisely, for all ε and c such that $0 < \varepsilon < 1 < c$, the following holds:
$$1 \prec \log n \prec n^\varepsilon \prec n^c \prec n^{\log n} \prec c^n \prec n^n \prec c^{c^n}.$$

EXAMPLE 9.17   Let the partition recurrence $u_n = 2u_{n/2}$ with $u_1 = 1$. We have seen in Chapter 7 that u_n is not uniquely determined on $\mathbb{N}$ (see Example 7.12). We can, however, determine the asymptotic order of magnitude of u_n: here we will obtain $u_n = \theta(n)$. If we assume that u_n is ultimately increasing, namely, that there is an integer n_0 such that

$$\forall n \geq n_0, \forall p \geq n_0, \quad n \geq p \Longrightarrow u_n \geq u_p,$$

then u_n is of the same order of magnitude as n. Indeed, in this case, for a large enough k, $2^k \geq n_0$. Thus let n be such that $\exists k, \quad n_0 \leq 2^k \leq n \leq 2^{k+1}$; we deduce $2^k \leq u_n \leq 2^{k+1}$. We thus have

- on the one hand, $u_n \leq 2^{k+1}$ and $2^k \leq n$, hence $u_n \leq 2n$,
- on the other hand, $2^k \leq u_n$ and $n \leq 2^{k+1}$, hence $n \leq 2u_n$,

and thus, finally, $n/2 \leq u_n \leq 2n$, i.e. $u_n = \theta(n)$.

More generally, we can state the following proposition.

Proposition 9.18   Let $u_n = bu_{n/a} + cn^k$, with $a \geq 2, b, c > 0, k \in \mathbb{N}$. Assume, moreover, that u_n is monotone increasing for $n \geq n_0$, then

$$u_n = \begin{cases} \theta(n^k) & \text{if } b < a^k, \\ \theta(n^k \log n) & \text{if } b = a^k, \\ \theta(n^{\log_a b}) & \text{if } b > a^k. \end{cases}$$

The proof is not given here.

EXERCISE 9.5   Let the sequence u_n be defined by: $u_o = c > 1$, and

$$u_n = u_{n-1} + \frac{1}{u_{n-1}}, \quad \text{for } n \geq 1.$$

1. Show that
$$\frac{1}{u_{n-1}^2} \leq \frac{1}{u_{n-1}}.$$

2. Show that
$$2 \leq u_n^2 - u_{n-1}^2 \leq 2 + u_n - u_{n-1}. \tag{9.1}$$

3. Determine the asymptotic behaviour of u_n. ◇

Criteria of asymptotic behaviour of functions

9.2.2 Hierarchies

A *hierarchy* is a set of functions against which all other functions are measured in order to study their asymptotic behaviour. For instance, $\{n^p \ / \ p \in \mathbb{N}\}$ is such a hierarchy. The formal definition follows.

Definition 9.19 *A hierarchy E is a set of functions:*

$$E = \{g \colon \mathbb{N} \longrightarrow \mathbb{R}\}$$

such that:

(i) $\forall g \in E$, $\lim_{n \to \infty} g(n) = 0$, or $\lim_{n \to \infty} g(n) = \infty$, or g is the constant function 1,
(ii) $\forall g_1 \in E, \forall g_2 \in E$, if $g_1 \neq g_2$ then either $g_1 = o(g_2)$, or $g_2 = o(g_1)$,
(iii) $\forall g \in E$, $\forall n \in \mathbb{N}$, $g(n) > 0$.

The second condition of this definition asserts that two functions in the hierarchy never have the same order of magnitude.

EXAMPLE 9.20 $E = \{g \ / \ g(n) = n^a (\log n)^b, \ a, b \in \mathbb{R}\}$ is a hierarchy. Let $g_i = n^{a_i} (\log n)^{b_i}$, $i = 1, 2$; we have $g_1 = o(g_2)$, or $g_1 \prec g_2$ if and only if $a_1 b_1 < a_2 b_2$ in the lexicographic ordering.

EXERCISE 9.6 Study the hierarchy $E = \{g \ / \ g(n) = e^{an^b} \times n^c \times (\log n)^d$, with $a, b, c, d \in \mathbb{R}$, $b > 0\}$. ◇

9.2.3 Asymptotic approximations

Definition 9.21 *Let E be a hierarchy and f a function:*

- *If there exist $0 \neq a_1 \in \mathbb{R}$ and $g_1 \in E$ such that $f \sim a_1 g_1$, then we have $f - a_1 g_1 = o(g_1)$; we say that g_1 is the principal part of f with respect to hierarchy E and we write $f = a_1 g_1 + o(g_1)$.*
- *More generally, the asymptotic power series expansion of order k of f with respect to E is an expression of the form $f = a_1 g_1 + \cdots + a_k g_k + o(g_k)$, with $g_{i+1} = o(g_i)$ for $i - 1, \ldots, k - 1$.*

Lemma 9.22 *The principal part, if it exists, is unique.*

Proof. See Exercise 9.7. □

EXERCISE 9.7 Prove that, when it exists, the asymptotic power series expansion of order k of f is unique. ◇

EXAMPLE 9.23 The asymptotic power series expansion may not exist for various reasons:

- The hierarchy is not refined enough: for instance e^n has no principal part with respect to $E = \{n^p \mathbin{/} p \in \mathbb{N}\}$.
- The function has an irregular behaviour: for instance the function f defined by

$$f(n) = \begin{cases} 2n & \text{if } n \text{ odd} \\ n & \text{if } n \text{ even} \end{cases}$$

has no limit with respect to E.

The following asymptotic power series expansions, with respect to the hierarchy

$$\{x^k \mathbin{/} k \in \mathbb{N}\},$$

and for $x \to 0$, are quite usual and useful:

for $|x| \leq r < 1$,

$$\frac{1}{1-x} = 1 + x + \cdots + x^k + o(x^k),$$

for $\alpha \in \mathbb{R}$, $|x| \leq r < 1$,

$$(1+x)^\alpha = 1 + \alpha x + \cdots + \alpha(\alpha - 1)\cdots(\alpha - k + 1)\frac{x^k}{k!} + o(x^k),$$

for $|x| \leq r \in \mathbb{R}^+$,

$$e^x = 1 + x + \frac{x^2}{2!} + \cdots + \frac{x^k}{k!} + o(x^k),$$

for $|x| \leq r < 1$,

$$\log(1 + x) = x - \frac{x^2}{2} + \frac{x^3}{3} + \cdots + (-1)^{k+1}\frac{x^k}{k} + o(x^k).$$

The rules given in Proposition 9.24 complete those given in Proposition 9.8 and can be deduced from the asymptotic power series expansion given above.

Proposition 9.24

if $f(n) = o(1)$, $\qquad \log\bigl(1 + O(f(n))\bigr) = O\bigl(f(n)\bigr),$

if $f(n) = O(1)$, $\qquad e^{O(f(n))} = 1 + O\bigl(f(n)\bigr),$

if $f(n) = o(1)$ and $f(n)g(n) = O(1)$,

$$\bigl(1 + O(f(n))\bigr)^{O(g(n))} = 1 + O\bigl(f(n)g(n)\bigr).$$

Proof. Let $g = O(f)$, then $\exists c \in \mathbb{R}^+$, $\exists n_0 \in \mathbb{N}$, $\varepsilon \in \mathbb{R}^+$,

$$\forall n > n_0, \qquad |g(n)| \leq c|f(n)| \leq \varepsilon < 1.$$

Criteria of asymptotic behaviour of functions

The series

$$\log\left(1 + O(g(n))\right) = g(n)\left(1 - \frac{1}{2}g(n) + \frac{1}{3}g(n)^2 - \cdots\right)$$

thus converges $\forall n > n_0$, and the series $1 - \frac{1}{2}g(n) + \frac{1}{3}g(n)^2 - \cdots$ has the upper bound $1 + \frac{1}{2}\varepsilon + \frac{1}{3}\varepsilon^2 + \cdots$, whence the first rule. The second is proved similarly. The two first rules together imply the third rule. □

EXERCISE 9.8 Verify that
$$(1+n)^{1/n} = 1 + \frac{\log n}{n} + \frac{1}{2}\frac{(\log n)^2}{n^2} + \frac{1}{n^2} + \frac{1}{6}\frac{(\log n)^3}{n^3} + o\left(\frac{(\log n)^3}{n^3}\right). \quad \diamond$$

9.2.4 Asymptotic approximations by partial sums

The following lemma, is very simple, but nevertheless very useful in many cases.

Lemma 9.25

1. Let f be a monotone decreasing mapping, and $u_n = \sum_{i=p}^{n} f(i)$, then,

$$\int_p^{q+1} f(x)dx \leq \sum_{i=p}^{q} f(i) \leq \int_{p-1}^{q} f(x)dx.$$

2. Similarly let f be a monotone increasing mapping, then,

$$\int_{p-1}^{q} f(x)dx \leq \sum_{i=p}^{q} f(i) \leq \int_p^{q+1} f(x)dx.$$

Figure 9.1 is an aid to the understanding of case 2.
We have already seen an application of the squeeze obtained in Lemma 9.25, 1. If $f(x) = \dfrac{1}{x}$, then

$$\int_2^{n+1} \frac{dx}{x} \leq \sum_{i=2}^{n} \frac{1}{i} \leq \int_1^{n} \frac{dx}{x},$$

which enables us to bound $H_n = 1 + \dfrac{1}{2} + \cdots + \dfrac{1}{n}$:

$$\log(n+1) + 1 - \log 2 \leq H_n \leq \log(n) + 1.$$

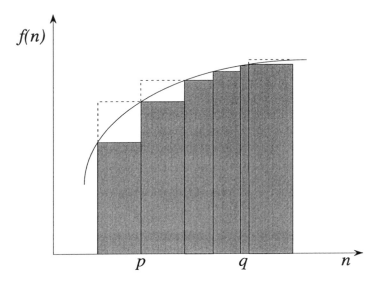

Figure 9.1 A monotone increasing mapping f.

We can prove, at the cost of more complex computations, that

$$H_n = \log n + \gamma + \frac{1}{2n} + o(n),$$

where $\gamma = 0.5772\ldots$ is Euler's constant.

We now apply Lemma 9.25, 2, to find the order of magnitude of $n!$: we reduce this problem to evaluating the order of magnitude of $\log n!$. If $f(x) = \log x$, we have $\log n! = \sum_{i=1}^{n} \log i = \sum_{i=2}^{n} \log i$ and, by the above remark,

$$\int_1^n \log x\, dx \leq \sum_{i=2}^n \log i = \sum_{i=1}^n \log i \leq \int_1^{n+1} \log x\, dx.$$

Integrating by parts $\int_1^n \log x\, dx = [x \log x]_1^n - \int_1^n dx = n \log n - (n-1)$, whence the bounds

$$n \log n - n + 1 \leq \log n! \leq (n+1)\log(n+1) - n,$$

which will enable us to find an approximation of $n!$. Forming the exponentials we have: $e^{1-n}n^n \leq n! \leq e^{-n}(n+1)^{n+1}$. The order of magnitude of $n!$ is thus $O(n^{n+1})$. A more precise asymptotic power series expansion of $n!$ can be obtained, yielding an exact equivalent of $n!$, but the proof is more complex and gives *Stirling's formula*

$$n! = \sqrt{2\pi n}\left(\frac{n}{e}\right)^n \left(1 + \frac{1}{12n} + o\left(\frac{1}{n}\right)\right),$$

whence $n! = \theta(n^{n+1/2})$.

9.2.5 'Bootstrapping'

We now introduce a technique which is both simple and powerful for finding asymptotic expansions of functions or solving recurrences asymptotically, without explicit computation but by using successive approximations. We start with a rough estimate, which we improve by plugging the estimate in the recurrence; this process gives better and better estimates, and is stopped when a reasonably good estimate is obtained (sometimes the exact solution can even be found, see Exercise 9.10). This technique is called *bootstrapping*.

We illustrate this technique with an example taken from Graham–Knuth–Patashnik. Consider the exponential generating series defined by

$$u(z) = \sum_{n \geq 0} z^n = e^{v(z)}, \quad \text{with} \quad v(z) = \sum_{k \geq 1} \frac{z^k}{k^2}, \qquad (9.2)$$

and try to find the asymptotic behaviour of the coefficients u_n of the series $u(z)$. Differentiating equation (9.2), we find

$$u'(z) = \sum_{n \geq 0} n z^{n-1} = v'(z) e^{v(z)} = \sum_{k \geq 1} \frac{z^{k-1}}{k} u(z),$$

hence the recurrence equation defining u_n, $u_0 = 1$, and for $n \geq 1$,

$$n u_n = \sum_{0 \leq k < n} \frac{u_k}{n - k}. \qquad (9.3)$$

We check by induction on n that $0 < u_n \leq 1$, since $u_0 = 1$, and $n u_n \leq \sum_{0 \leq k < n} 1 = n$. Hence,

$$u_n = O(1).$$

This fact is used to 'start up the pump' of the bootstrapping, by plugging this information in equation (9.3); this gives

$$n u_n = \sum_{0 \leq k < n} \frac{O(1)}{n - k} = H_n O(1) = O(\log n),$$

hence
$$u_n = O\Big(\frac{\log n}{n}\Big), \quad \text{for } n > 1,$$
and
$$u_n = O\Big(\frac{1+\log n}{n}\Big), \quad \text{for } n \geq 1.$$

Bootstrapping again in equation (9.3) gives, for $n > 1$,
$$nu_n = \frac{1}{n} + \sum_{0<k<n} \frac{1}{n-k} O\Big(\frac{1+\log k}{k}\Big),$$
since, for $1 \leq k < n$, $O(1+\log k) = O(\log n)$,
$$nu_n = \frac{1}{n} + \sum_{0<k<n} \frac{O(\log n)}{k(n-k)} = \frac{1}{n} + \sum_{0<k<n} \Big(\frac{1}{k} + \frac{1}{n-k}\Big)\frac{O(\log n)}{n}$$
$$= \frac{1}{n} + \frac{2}{n} H_{n-1} O(\log n) = \frac{1}{n} O((\log n)^2)$$
and, finally,
$$u_n = O\Big(\Big(\frac{\log n}{n}\Big)^2\Big), \quad \text{for } n > 1. \tag{9.4}$$

EXERCISE 9.9
1. What would come out of one more bootstrapping step using equation (9.4)?
2. Find the principal part of u_n. ◇

EXERCISE 9.10 Solve the recurrence $u_n = n + u_{n-1}$, with $u_0 = 0$, by bootstrapping. ◇

CHAPTER 10

GRAPHS AND TREES

Graphs are used in many domains. Some examples are:

- network conception and management,
- routing in VLSI circuits, or more general routing problems (e.g. finding shortest paths),
- task scheduling in parallel systems.

And, from a more theoretical standpoint, many data structures can be modelled as graphs.

This chapter gives the basic definitions for directed or undirected graphs, using as an example the proof of the celebrated Euler theorem. As a special case of graphs this chapter introduces the concept of a 'tree', probably one of the most useful concepts for a computer scientist.

Claude Berge, *Graphs and Hypergraphs*, North-Holland, Amsterdam, (1976).
Kenneth Ross, Charles Wright, *Discrete Mathematics*, Prentice Hall, London (1988).

10.1 Graphs

10.1.1 Definitions

We define two types of graph, directed graphs and undirected graphs.

Definition 10.1 *A directed graph is a quadruple (V, E, α, β) where:*
- *V is a (possibly infinite) set of vertices (or nodes),*
- *E is a set of edges, disjoint from V,*
- *α and β are two mappings from E to V associating with each edge e its initial vertex (also called origin) $\alpha(e)$ and its terminal or end vertex (also called target) $\beta(e)$. $\alpha(e)$ and $\beta(e)$ are called the endpoints of edge e.*

The fact that an edge has an origin and a target enables us to orient edges. Normal transit will go from the origin towards the target. Several edges may have the same origin and the same target; such graphs are thus sometimes called *multigraphs*.

EXAMPLE 10.2

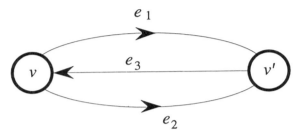

Figure 10.1

Figure 10.1 is a directed graph with two vertices, v and v', and three edges, e_1, e_2 and e_3, with

$$\alpha(e_1) = \alpha(e_2) = v,$$
$$\beta(e_1) = \beta(e_2) = v',$$
$$\alpha(e_3) = v',$$
$$\beta(e_3) = v.$$

e_1 and e_2 have the same origin and the same target.

Definition 10.3 *An undirected graph is a graph in which we cannot distinguish the origin and the target of an edge. It is a triple (V, E, δ), where δ associates two not necessarily distinct vertices with each edge.*

EXAMPLE 10.4

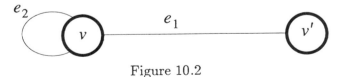

Figure 10.2

Figure 10.2 is an undirected graph with two vertices, v and v', and two edges, e_1 and e_2, with

$$\delta(e_1) = \{v, v'\},$$
$$\delta(e_2) = \{v\}.$$

Graphs

A graph (directed or undirected) is said to be *finite* if it has a finite number of vertices and edges, i.e. if both sets V and E are finite.

A directed graph can always be transformed into an undirected graph by 'forgetting' the edge orientations. If $G = (V, E, \alpha, \beta)$ is a directed graph then (V, E, δ) is an undirected graph denoted by $\gamma(G)$, where $\delta(e) = \{\alpha(e), \beta(e)\}$.

Conversely, if a graph is undirected, it may be transformed into a directed graph by assigning an arbitrary direction to each edge:

- If $\delta(e) = \{v\}$, then $\alpha(e) = v = \beta(e)$.
- If $\delta(e) = \{v, v'\}$, with $v \neq v'$, then we may let $(\alpha(e) = v$ and $\beta(e) = v')$ or $(\alpha(e) = v'$ and $\beta(e) = v)$.

Clearly, there are several ways of orienting a graph.

Definition 10.5 *If G is an undirected graph, then a directed graph G' is an orientation of G if $\gamma(G') = G$.*

10.1.2 Isomorphic graphs

Sometimes, the actual name of a vertex or an edge of a graph does not really matter and we may consider that two graphs differing only in the names of their vertices and their edges are in fact identical. The notion of *isomorphism* formally expresses this idea.

Two directed graphs $G = (V, E, \alpha, \beta)$ and $G' = (V', E', \alpha', \beta')$ are said to be *isomorphic* if there are two bijections $h_{vtx}: V \longrightarrow V'$ and $h_{edg}: E \longrightarrow E'$ such that

$$\forall e \in E, \quad \alpha'(h_{edg}(e)) = h_{vtx}(\alpha(e)) \text{ and } \beta'(h_{edg}(e)) = h_{vtx}(\beta(e)).$$

Two undirected graphs $G = (V, E, \delta)$ and $G' = (V', E', \delta')$ are said to be *isomorphic* if there are two bijections $h_{vtx}: V \longrightarrow V'$ and $h_{edg}: E \longrightarrow E'$ such that

$$\forall e \in E, \quad \delta'(h_{edg}(e)) = \begin{cases} \{h_{vtx}(v)\} & \text{if } \delta(e) = \{v\}, \\ \{h_{vtx}(v), h_{vtx}(v')\} & \text{if } \delta(e) = \{v, v'\}. \end{cases}$$

Given two (directed or undirected) graphs G and G' whose sets of vertices and of edges, respectively, are the pairwise disjoint sets V, E and V', E', the *disjoint union* of G and G' is the graph whose set of vertices is $V \cup V'$, whose set of edges is $E \cup E'$ and in which the edges and vertices are connected exactly as in graphs G and G'.

In some cases we may wish to construct the disjoint union of two graphs while their sets of vertices and edges are not disjoint; we will nevertheless be able to construct this disjoint union by substituting an isomorphic graph for one of the two graphs.

10.1.3 Simple graphs

In a graph, directed or undirected, a *loop* is an edge e, both endpoints of which are equal. This can be formally stated as follows: for a directed graph, $\alpha(e) = \beta(e)$ and, for an undirected graph, $\delta(e)$ is a singleton.

A directed or undirected graph is said to contain *multiple edges* when several edges are allowed between pairs of vertices, i.e. there may be edges e and e' with $\alpha(e) = \alpha(e')$ and $\beta(e) = \beta(e')$ for directed graphs, and $\delta(e) = \delta(e')$ for undirected graphs.

If a directed graph $G = (V, E, \alpha, \beta)$ has no multiple edges, we may identify the set E of its edges with a subset of the Cartesian product $V \times V$. Indeed, in this case, the mapping $(\alpha, \beta) : E \longrightarrow V \times V$ is injective.

A directed or undirected graph is said to be *simple* if it contains neither loops nor multiple edges. If an undirected graph $G = (V, E, \delta)$ is simple, we may identify the set of its edges with a subset of the set $\mathcal{P}_2(V)$ of two-element subsets of V. Indeed, since G has no loop, then for any edge e, $\delta(e)$ has two elements and, because G has no multiple edges, the mapping $\delta : E \longrightarrow \mathcal{P}_2(V)$ is injective.

10.1.4 Subgraphs and partial graphs

Intuitively, a partial graph is obtained from graph G by deleting some edges, and a subgraph is obtained from G by deleting some vertices, together with any edge whose origin or target is one of the deleted vertices.

Let $G = (V, E, \alpha, \beta)$ be a directed graph. The directed graph $G' = (V', E', \alpha', \beta')$ is a *partial graph* of G if

- $V' = V$,
- $E' \subseteq E$ and
- $\forall e \in E', \quad \alpha'(e) = \alpha(e)$ and $\beta'(e) = \beta(e)$.

It is a *subgraph* of G if

- $V' \subseteq V$,
- $E' = \{e \in E \;/\; \{\alpha(e), \beta(e)\} \subseteq V'\}$ and
- $\forall e \in E', \quad \alpha'(e) = \alpha(e)$ and $\beta'(e) = \beta(e)$.

It is a *subpartial graph* if it is a subgraph of a partial graph.

Let $G = (V, E, \delta)$ and $G' = (V', E', \delta')$ be undirected graphs. The undirected graph $G' = (V', E', \delta')$ is a partial graph of G if $V' = V$, $E' \subseteq E$, and $\forall e \in E'$, $\delta'(e) = \delta(e)$. It is a subgraph of G if $V' \subseteq V$, $E' = \{e \in E \;/\; \delta(e) \subseteq V'\}$, and $\forall e \in E', \delta'(e) = \delta(e)$. It is a subpartial graph if it is a subgraph of a partial graph.

EXERCISE 10.1 Show that graph G' is a subpartial graph of G if and only if it is a partial graph of a subgraph of G. Show that a subgraph and a partial graph of a subpartial graph of G are also subpartial graphs of G. ◇

10.1.5 Degree of a vertex

In the present subsection, we consider only graphs such that for any vertex the number of edges leading into that vertex or going out of that vertex is finite; all finite graphs clearly satisfy this requirement.

The *degree* $d(v)$ of a vertex v of an undirected graph $G = (V, E, \delta)$ is equal to the number of edges e such that $\delta(e) = \{v, v'\}$ with $v \neq v'$ plus twice the number of edges e such that $\delta(e) = \{v\}$. If $\delta(e) = \{v\}$, edge e will be counted twice in the degree of v! The degree $d(v)$ of a vertex v of a directed graph is equal to the number of edges e such that $v = \alpha(e)$ or $v = \beta(e)$ (if $\alpha(e) = \beta(e)$, edge e will be counted twice); in other words, the degree of a vertex is the sum of the number of ingoing edges and the number of outgoing edges. The *indegree* $d^-(v)$ of a vertex v of a directed graph G is equal to the number of edges e such that $\beta(e) = v$. The *outdegree* $d^+(v)$ is equal to the number of edges e such that $\alpha(e) = v$. We thus have, for a directed graph, $d(v) = d^+(v) + d^-(v)$.

Proposition 10.6 *The sum of the degrees of all vertices of a finite undirected graph is equal to twice the number of its edges.*

Proof. If $\delta(e) = \{v, v'\}$, with $v \neq v'$, then edge e is counted once in the degree of v and once in the degree of v'. If $\delta(e) = \{v\}$, then edge e is counted twice in the degree of v. Each edge is thus counted twice in the sum of the degrees of the vertices. □

EXERCISE 10.2 Show that in a finite directed graph, the sum of the indegrees of all vertices is equal to the sum of the outdegrees of these vertices. To what other number is this sum also equal? ◇

EXERCISE 10.3 Let G be a finite undirected graph with n vertices and m edges, where $n \geq 1$ and $m \geq 0$. For any integer $k \in \mathbb{N}$, let n_k be the number of vertices of degree k; let K be the maximum of the degrees of the vertices (i.e. $n_K > 0$ and $n_k = 0$ for $k > K$).
1. Show that
$$\sum_{k=0}^{K} k n_k = 2m \quad \text{and} \quad \sum_{k=0}^{K} n_k = n.$$
2. Show that $K \leq 2m$; give an example where equality holds.
3. Show that if G has neither loops nor multiple edges, then $K \leq n-1$; give an example where equality holds. ◇

EXERCISE 10.4 Let G be a finite simple undirected graph with n vertices ($n > 1$).
1. Show that the degree of a vertex is always strictly less than n.
2. Prove that there cannot simultaneously be a vertex of degree 0 and a vertex of degree $n-1$.
3. Deduce that there are at least two vertices having the same degree. ◇

EXERCISE 10.5 Let $G = (V, E, \delta)$ be an undirected simple finite graph. We assume that G contains no *triangles*: a triangle consists of three distinct vertices $v_1, v_2, v_3 \in V$ and three edges $e_1, e_2, e_3 \in E$ with $\delta(e_1) = \{v_2, v_3\}$, $\delta(e_2) = \{v_1, v_3\}$ and $\delta(e_3) = \{v_1, v_2\}$. Two vertices are said to be adjacent if they are connected by an edge.

1. Show that for two distinct adjacent vertices x and y, the number n_x of vertices of $V \setminus \{x, y\}$ adjacent to x and the number n_y of vertices of $V \setminus \{x, y\}$ adjacent to y satisfy the inequality
$$n_x + n_y \leq |V| - 2.$$

2. Deduce, by induction on the number $|V|$ of vertices, that the number $|E|$ of edges verifies
$$|E| \leq \frac{|V|^2}{4}. \qquad \diamond$$

10.1.6 Paths

In a directed graph G, a *path* is a sequence $c = e_1, \ldots, e_n$ of edges such that $\forall i \in \{1, \ldots, n-1\}$, $\beta(e_i) = \alpha(e_{i+1})$. Vertex $\alpha(e_1)$ will be called the *origin* of path c and vertex $\beta(e_n)$ will be called the *target* of path c. A *circuit* is a path such that $\beta(e_n) = \alpha(e_1)$. A path (or a circuit) is *simple* if it does not contain the same edge twice. It is *elementary* if it does not contain two edges with the same origin or the same target (it is hence a *fortiori* simple). In a finite directed graph, a path or circuit is said to be an *Euler* path or circuit if it is simple and contains all edges. It is said to be a *Hamiltonian* path or circuit if it is elementary and goes through all vertices (i.e. $\forall v \in V, \exists i : v = \alpha(e_i)$ or $v = \beta(e_i)$).

For undirected graphs, a *chain* is a sequence
$$c = v_0, e_1, v_1, e_2, \ldots, v_{n-1}, e_n, v_n$$
of vertices and edges such that $\delta(e_i) = \{v_{i-1}, v_i\}$. This chain is said to connect vertices v_0 and v_n. A *cycle* is a chain such that $v_0 = v_n$. If the sequence $v_0, e_1, v_1, e_2, \ldots, v_{n-1}, e_n, v_n$ is a chain, the sequence $v_n, e_n, v_{n-1}, \ldots, e_2, v_1, e_1, v_0$ is also a chain and it connects v_n to v_0. A chain (or a cycle) is *simple* if it does not contain the same edge twice and *elementary* if it does not contain the same vertex twice, with the only exception being that v_0 and v_n may be equal. In a finite undirected graph, a chain (or a cycle) is an *Euler* chain (or a cycle) if it is simple and contains all edges; it is a *Hamiltonian* chain (or a cycle) if it is elementary and contains all vertices.

EXAMPLE 10.7
1. In the graph of Figure 10.3:
    - $e_1 e_2 e_3 e_4$ is a simple circuit, and
    - $e_4 e_1 e_2 e_3 e_5$ is an Euler path but is not a Hamiltonian path.

Graphs

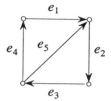

Figure 10.3

2. In the directed graph of Example 10.2:
   - $e_1 e_3$ is a simple circuit, a Hamiltonian circuit, but not an Euler circuit

   and
   - $e_1 e_3 e_2$ is an Euler path, non-elementary.
3. In the undirected graph of Example 10.4:
   - v, e_2, v is an elementary cycle,
   - v, e_2, v, e_1, v' is a simple chain, non-elementary, and an Euler chain

   and
   - $v', e_1, v, e_2, v, e_1, v'$ is a non-simple cycle.

The next result is an immediate consequence of the definitions.

Proposition 10.8 *If $e_1, \ldots, e_n$ is a simple (resp. elementary, Euler, Hamiltonian) path (circuit) of a directed graph G, then*

$$\alpha(e_1), e_1, \alpha(e_2), e_2, \ldots, \alpha(e_n), e_n, \beta(e_n)$$

is a simple (resp. elementary, Euler, Hamiltonian) chain (cycle) of $\gamma(G)$.

Conversely, if $v_0, e_1, \ldots, e_n, v_n$ is a simple chain, there is at least one orientation of G such that $e_1, \ldots, e_n$ is a path: since $\delta(e_i) = \{v_{i-1}, v_i\}$, it suffices to let $\alpha(e_i) = v_{i-1}$ and $\beta(e_i) = v_i$. The fact that the chain is simple implies that this construction is always possible because there is no index $j \neq i$ with $e_j = e_i$. Otherwise, if we had $e_i = e_j = e$, we would have $\delta(e) = \{v_{i-1}, v_i\} = \{v_{j-1}, v_j\}$; and it is easy to see that the above construction can be applied only when $\alpha(e) = v_{i-1} = v_{j-1}$ and $\beta(e) = v_i = v_j$.

Proposition 10.9 *If an undirected graph G contains two different simple chains connecting the same two distinct vertices, then it contains a simple cycle.*

Proof. Let $c = v_0, e_1, v_1, \ldots, v_{n-1}, e_n, v_n$ and $c' = v'_0, e'_1, v'_1, \ldots, v'_{n'-1}, e'_{n'}, v'_{n'}$ be two simple chains such that $v_0 = v'_0 \neq v_n = v'_{n'}$. We prove by induction on $n + n'$ that the existence of such chains implies the existence of a simple cycle:

1. Since $n \geq 1$ (because $v_0 \neq v_n$) and $n' \geq 1$ (for the same reasons), the least possible value of $n + n'$ is 2. In this case, $c = v, e, v'$, $c' = v, e', v'$, with $e \neq e'$, and v, e, v', e', v is a simple cycle.

2. (a)   If the sets $E(c) = \{e_i \,/\, 1 \leq i \leq n\}$ and $E(c') = \{e'_j \,/\, 1 \leq j \leq n'\}$ are disjoint, then $v_0, e_1, v_1, \ldots, v_{n-1}, e_n, v_n = v'_{n'}, e'_{n'}, v'_{n'-1}, \ldots, v'_1, e'_1, v'_0$ is a simple cycle.

(b)   If $e_1 = e'_1$, then $v_1 = v'_1$, and two cases must be considered:

(b.1)   If $n' = 1$, then $v_n = v'_1 = v_1$ and $v_1, \ldots, v_{n-1}, e_n, v_n$ is a simple cycle. The case $n = 1$ is similar.

(b.2)   Otherwise the two chains $v_1, e_2, \ldots, v_{n-1}, e_n, v_n$ and $v'_1, e'_2, \ldots, v'_{n'-1}, e'_{n'}, v'_{n'}$ with respective lengths $n-1$ and $n'-1$ again verify the hypotheses, and thus G contains a simple cycle.

(c)   We are thus left with the case in which $E(c) \cap E(c') \neq \emptyset$ and in which $e_1 \neq e'_1$. Hence, there exists $i > 1$ and $j > 1$ such that $e_i = e'_j$, hence $v_{i-1} = v'_{j-1}$ or $v_{i-1} = v'_j$. In the first case, the chains $v_0, e_1, v_1, \ldots, v_{i-2}, e_{i-1}, v_{i-1}$ and $v'_0, e'_1, v'_1, \ldots, v'_{j-2}, e'_{j-1}, v'_{j-1}$, with respective lengths $i-1 < n$ and $j-1 < n'$, again verify the hypotheses, and G contains a simple cycle. In the second case, the chains $v_0, e_1, v_1, \ldots, v_{i-2}, e_{i-1}, v_{i-1}$ and $v'_0, e'_1, v'_1, \ldots, v'_{j-2}, e'_{j-1}, v'_{j-1}, e'_j, v'_j$, with respective lengths $i-1 < n$ and $j \leq n$, also verify the hypotheses and G contains a simple cycle.   □

EXERCISE 10.6   Show that in an undirected graph, the shortest chain between two distinct vertices is elementary. Does this also hold for the shortest path from a vertex to another one in a directed graph?   ◇

In an undirected graph, the *distance* $d(v, v')$ between two vertices v and v' is defined as follows:

- If $v = v'$ then $d(v, v') = 0$.
- If $v \neq v'$ then $d(v, v')$ is equal to
  - the length of the shortest chain connecting these two vertices, if such a chain exists, or
  - ∞ otherwise.

The characteristic properties of a distance are indeed verified:

- $d(v, v') = 0$ if and only if $v = v'$,
- $d(v, v') = d(v', v)$,
- $d(v, v'') \leq d(v, v') + d(v', v'')$.

EXERCISE 10.7   Prove the triangular inequality $d(v, v'') \leq d(v, v') + d(v', v'')$.   ◇

The *diameter* of an undirected graph is the maximum distance between two distinct vertices, i.e. $\sup\{d(v, v') \,/\, v, v' \in V\}$.

If a graph is directed, we may also define the distance between two vertices as the length of the shortest path going from the first one to the second one. But this is no longer a distance in the mathematical sense because, while it still satisfies the triangular inequality, it is no longer symmetrical: it may well occur that $d(v, v') \neq d(v', v)$.

Graphs

EXERCISE 10.8 Let $X = \{0, 1, 2, 3, 4\}$. The *Petersen graph* is the undirected graph defined as follows: its vertices are the pairs of elements of X, and two vertices are connected by an edge if and only if they are two disjoint pairs of elements of X. For instance, if $\{0,1\}$, $\{1,2\}$ and $\{2,3\}$ are vertices, there is an edge connecting $\{0,1\}$ and $\{2,3\}$, but there is no edge connecting $\{0,1\}$ and $\{1,2\}$:
1. Determine the number of vertices, the number of edges, the degree of vertices and the diameter of this graph.
2. Draw this graph and indicate for each vertex the two elements of X constituting the pair that it contains. ◊

EXERCISE 10.9
1. Let (V, E, α, β) be a directed graph with n vertices $v_1, \ldots, v_n$. It is associated with the matrix M defined by: each entry $M_{i,j}$ of M (for $1 \le i \le n$ and $1 \le j \le n$) is the number of edges of E with origin v_i and target v_j (i.e. $\alpha(e) = v_i$ and $\beta(e) = v_j$). Prove by induction that for any integer $k > 0$, the matrix M^k has as its (i,j)th entry the number of distinct paths of length k between v_i and v_j.
2. Can you generalize this result to an undirected graph? ◊

EXERCISE 10.10 In $\mathbb{Z}^2$ we define the *4-distance* d_4 and the *8-distance* d_8 as follows:

$$d_4\big((x,y),(x',y')\big) = |x - x'| + |y - y'|,$$
$$d_8\big((x,y),(x',y')\big) = \max\big(|x - x'|, |y - y'|\big).$$

1. Define the two undirected graphs G_4 and G_8 having $\mathbb{Z}^2$ as the set of vertices, and where the distances above defined in terms of lengths of chains coincide with d_4 and d_8 respectively.
2. Draw the subgraphs of G_4 and G_8 corresponding to the set of the sixteen vertices (x, y) where $0 \le x \le 3$ et $0 \le y \le 3$. ◊

10.1.7 Connectivity

A directed graph is said to be *strongly connected* if for any pair (v, v') of distinct vertices there is a path going from v to v'.

An undirected graph is *connected* if for any pair (v, v') of distinct vertices there exists a chain connecting v and v'.

The *connected component* $CC_G(v)$ of a vertex v of an undirected graph G is equal to $\{v\}$ together with the set of vertices v' of G such that there exists a chain connecting v to v'.

Proposition 10.10 If $v' \in CC_G(v)$ then $CC_G(v) = CC_G(v')$.

Proof. We assume that $v \ne v'$. (If $v = v'$, the result is trivial.) If $v' \in CC_G(v)$ there is a chain c connecting v and v' and a chain c' connecting v' and v. Let $v'' \in CC_G(v)$:

- If $v'' = v$, then since c' connects v' to v, we have that $v'' = v \in CC(v')$.

- Otherwise, there is a chain c'' connecting v to v''; then $c'c''$ is a chain connecting v' to v'', and $v'' \in CC_G(v')$.

Thus, $CC_G(v) \subseteq CC_G(v')$. The converse inclusion $CC_G(v') \subseteq CC_G(v)$ can be proved in the same way. □

The following proposition follows immediately from the definitions.

Proposition 10.11 *An undirected graph $G = (V, E, \delta)$ is connected if and only if $\forall v \in V$, $V = CC_G(v)$.*

An *isolated* vertex is a vertex of degree 0. If a vertex v of G is isolated, then $CC_G(v) = \{v\}$. If G is a graph without loops, a vertex v of G is isolated if and only if $CC_G(v) = \{v\}$.

Proposition 10.12 *Let G be a connected undirected graph. Let G' be the graph obtained by deleting an edge e with two distinct end vertices v' and v'' (i.e. $\delta(e) = \{v', v''\}$ with $v' \neq v''$). Then $V = CC_{G'}(v') \cup CC_{G'}(v'')$.*

Proof. Let v be any vertex of G. Because G is connected, there exists a chain $v_0, e_1, v_1, \ldots, e_n, v_n$ with $v' = v_0$ and $v_n = v$. If edge e does not occur in that chain, then $v \in CC_{G'}(v')$. Otherwise, let i be the largest index such that $e_i = a$. Then $v_i \in \{v', v''\}$ and $v_i, e_{i+1}, \ldots, e_n, v_n$ is a chain of G', and hence $v = v_n \in CC_{G'}(v'') \cup CC_{G'}(v')$. □

EXERCISE 10.11 Let C_n be an undirected graph with n vertices consisting of a single cycle, i.e.
$$V = \{v_1, \ldots, v_n\},$$
$$E = \{e_1, \ldots, e_n\},$$
$$\delta(e_i) = \begin{cases} \{v_{i-1}, v_i\} & \text{if } i > 1, \\ \{v_n, v_1\} & \text{if } i = 1. \end{cases}$$

1. Show that each vertex of C_n is of degree 2.
2. Show that if G is an undirected connected graph with n vertices all of which are of degree 2, then G is isomorphic to C_n.
3. Show that if G is an undirected graph with n vertices all of which are of degree 2, then G is a disjoint union of graphs $G_{n_1}, \ldots, G_{n_k}$, where G_{n_i} is isomorphic to C_{n_i}, with $n_1 + \cdots + n_k = n$. ◊

10.1.8 An historical example: Königsberg seven bridges

The old town of Königsberg in eastern Prussia (nowadays Kaliningrad, in Russia), contains an island connected to the 'mainland' by seven bridges as shown in Figure 10.4.

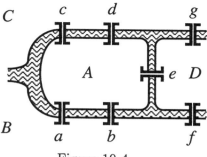

Figure 10.4

Is it possible to take a walk through Königsberg starting from some point and travelling across all the bridges exactly once?

The problem is that of finding an Euler chain in the undirected graph shown in Figure 10.5.

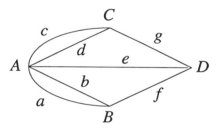

Figure 10.5

If, moreover, the walk ends at its starting point we will have an Euler cycle.

The answer to this problem can be obtained by the following theorem (proved by Euler in 1766).

Theorem 10.13 *An undirected finite graph G with no isolated vertex has an Euler chain if and only if*

(i) *it is connected*
(ii) *it has zero or two odd degree vertices.*

In the case in which there is no odd degree vertex, this Euler chain is a cycle. In the case in which there are two odd degree vertices, they are the origin and target of the chain.

Proof.
1. Assume that there is an Euler chain $v_0, e_1, v_1, e_2, \ldots, v_{n-1}, e_n, v_n$. Because there is no isolated vertex,

$$V = \cup_{e \in E} \delta(e),$$

and since $E = \{e_1, \ldots, e_n\}$,
$$v \in V \iff \exists i : v = v_i.$$

G is thus connected.

Let v be any vertex; the number $d(v)$ of edges with endpoint v is equal to

$$2 \times |\{i \,/\, v_i = v, 1 \leq i \leq n-1\}| \;+\; \begin{cases} +1 \text{ if } v = v_0, \\ +1 \text{ if } v = v_n. \end{cases}$$

Indeed, for each $i \in \{1, \ldots, n-1\}$, v_i is the endpoint of the two distinct edges e_i and e_{i+1}. Moreover, v_0 is the endpoint of edge e_1 and v_n is the endpoint of edge e_n. Note that if we have an edge e_i, both endpoints v_{i-1} and v_i of which are equal, this edge will be counted twice in the computation. We deduce the following from this characterization of the degrees of the vertices:

- If $v_0 = v_n$, i.e. if there is an Euler cycle, then all vertices are of even degree.
- If $v_0 \neq v_n$ then all vertices are of even degree except for the endpoints v_0 and v_n of the chain.

2. Consider a connected graph with zero or two vertices of odd degree. We show that it has an Euler chain.

Note first that we may add or delete on the graph any number of loops without modifying either the parity of the degrees of the vertices or the existence of Euler chains. If we add or if we delete on graph G an edge e with $\delta(e) = \{v\}$, we increase or decrease the degree of v by 2. Moreover, if we add to graph G an edge e with $\delta(e) = \{v\}$, thus obtaining a graph G', then $v_0, e_1, \ldots, e_{i-1}, v, e_{i+1}, \ldots, v_n$ is an Euler chain of G if and only if $v_0, e_1, \ldots, e_{i-1}, v, e, v, e_{i+1}, \ldots, v_n$ is an Euler chain of G'; and we have a similar result if $v = v_0$ or if $v = v_n$.

We may then reason by induction on the number of edges of a graph *without loops*.

If this number is 0, the property holds vacuously. If this number is 1, the unique edge e of this graph is such that $\delta(e) = \{v, v'\}$, with $v \neq v'$, and v, e, v' is an Euler chain.

Let G be a graph with $n + 1$ edges:

(a) If all vertices have an even degree, we consider any edge e whose endpoints are v and v'. Let G' be the partial graph obtained by deleting this edge. Both vertices v and v' now have an odd degree. G' must be connected. Indeed, let $CC_{G'}(v)$ be the connected component of v in G'. Firstly, $v' \in CC_{G'}(v)$; otherwise $CC_{G'}(v)$ would be a graph with a single vertex of odd degree, which is impossible (because the sum of the degrees of a graph is always even). There thus exists a chain c connecting v and v' in G'. We now show that $CC_{G'}(v)$ contains all vertices of G'. Let v'' be any vertex. Because G is connected, there exists in G a chain connecting v and v''. If this chain uses edge e, we substitute the chain c for edge e. We thus have a chain of G' connecting v and v''. Because G' is connected and has two odd degree vertices v and v', there exists an Euler chain c' connecting v' and v, and v, e, c' is thus an Euler cycle of G.

(b) Let v and v' be the two vertices of odd degree of G, let e be an edge such that $v \in \delta(e)$ and let v'' be the vertex such that $\delta(e) = \{v, v''\}$. Let G' be the graph obtained by deleting this edge. The degree of v in G' is even.

(b.1) If the degree of v'' in G is odd (i.e. $v'' = v'$) then the degree of v'' in G' is even.

(b.1.1) If G' is connected, there exists an Euler cycle c going from v to v, and $c, e, v' = v''$ is an Euler chain of G.

(b.1.2) Otherwise, by Proposition 10.12, $CC_{G'}(v)$ and $CC_{G'}(v')$ are two connected graphs, and all their vertices have even degrees. There exists an Euler cycle c from v to v in $CC_{G'}(v)$ and an Euler cycle c' from v' to v' in $CC_{G'}(v')$. Then c, e, c' is an Euler chain in G.

(b.2) If the degree of v'' in G is even (i.e. $v'' \neq v'$) then the degree of v'' in G' is odd and $v'' \in CC_{G'}(v')$, because v'' and v' are the only two vertices of odd degree in G'.

(b.2.1) If G' is connected, there is an Euler chain c of G' connecting v'' and v', and v, e, c is an Euler chain in G.

(b.2.2) Otherwise, all vertices of $CC_{G'}(v)$ have an even degree. There exists an Euler cycle c of $CC_{G'}(v)$ connecting v and v and an Euler chain c' of $CC_{G'}(v')$ connecting v' and v'. Then c, e, c' is an Euler chain of G connecting v and v'. □

10.1.9 Graph colouring

A *colouring* of an undirected graph $G = (V, E, \delta)$ without a loop is a mapping $\gamma: V \longrightarrow C$, where C is a finite set of 'colours', such that

$$\forall e, \delta(e) = \{v, v'\} \Longrightarrow \gamma(v) \neq \gamma(v')$$

(i.e. two different vertices connected by an edge cannot have the same colour). The *chromatic number* of a graph is the minimum number of colours needed to colour it.

EXAMPLE 10.14 Let E be a set of students and X be a set of exams. For each exam x in X, the set of students registered for that exam is $S(x)$. Each student can take at most one exam per day. What is the minimum length of the exam session?

Let G be the graph whose set of vertices is X. Two vertices x and x' are connected by an edge if and only if $S(x) \cap S(x') \neq \emptyset$. The minimal length of the session is the *chromatic number* of this graph.

10.1.10 Planar graphs

A (directed or undirected) graph is said to be *planar* if it can be drawn in the plane without any edges crossing.

REMARK 10.15 A graph, and even a planar graph, can be drawn in many ways. Figure 10.6 shows two possible drawings for the planar graph K_4.

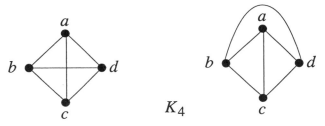

Figure 10.6

The two graphs in Figure 10.7 are not planar.

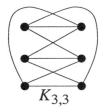

Figure 10.7

A famous problem that has been solved recently is the four-colour problem: is the chromatic number of a planar graph always less than or equal to 4? The answer is 'yes'.

EXERCISE 10.12 Find a planar graph with chromatic number 4.

10.2 Trees and rooted trees

Trees and rooted trees are particular cases of graphs. Trees are usually undirected graphs, and rooted trees are usually directed graphs. However, in computer science, the term 'tree' often means rooted tree, and sometimes very particular rooted trees.

10.2.1 Trees

Definition 10.16 *A tree is an undirected connected graph without a simple cycle.*

If an undirected graph contains at least one edge e, it contains a cycle v, e, v', e, v, where $\delta(e) = \{v, v'\}$. But this cycle is not simple. This is why the definition of trees involves simple cycles.

Proposition 10.17 *Let G be a finite graph, let n be the number of its vertices, let m be the number if its edges and let p be the number of its connected components. Then G is without a simple cycle if and only if $m - n + p = 0$.*

Proof.
1. $m - n + p$ is always non-negative. We show this by induction on the number of edges:
   - It is true for a graph with no edges because if that graph has n vertices then it has n connected components.
   - Let e be an edge of G connecting v and v' (which are thus in the same connected component) and let us delete edge e from G. The graph G' thus obtained has $n' = n$ vertices, $m' = m-1$ edges and p' connected components with $p' = p$ or $p' = p + 1$; hence $p \geq p' - 1$. By the induction hypothesis, $m' - n' + p' \geq 0$, and $m - n + p \geq m - n + p' - 1 = m' - n' + p'$.
2. Assume that G contains a simple cycle c, and let us show that $m - n + p > 0$. Let e be any edge of this cycle, with $\delta(e) = \{v, v'\}$ (v' may be equal to v). There thus exists a chain c' connecting v' and v that does not use edge e.
   Let G' be the graph obtained by deleting from G this edge e. We have for G' that $m' = m - 1$ and that $n' = n$. We also have that $p' = p$ because the number of connected components is not modified: if two vertices are connected by a chain of G using edge e, they are connected by the chain of G' obtained by substituting for each occurrence of e the chain c' which is in G'. Hence $m - n + p = m' + 1 - n' + p' > m' - n' + p' \geq 0$.
3. Conversely, we show that if $m - n + p > 0$, then G has a simple cycle. We again show this by induction on the number of edges:
   - If $m = 0$, we have shown in 1 that $m - n + p > 0$ could not occur.

- Let G' be the graph obtained by deleting from G an arbitrary edge e. We have shown in 1 that in this case $m - n + p \geq m' - n' + p'$ and that $p' = p$ or $p' = p + 1$.
  - If $m' - n' + p' > 0$, then, by the induction hypothesis, G' contains a simple cycle, and hence so does G.
  - If $m' - n' + p' = 0$, then, because $m - n + p > 0$, we have that $m - n + p > m' - n' + p'$, which implies $p' = p$. In other words, deleting edge e does not modify connected components. As the endpoints v and v' of e are in the same connected component of G, they are in the same connected component of G'. There thus exists a simple chain c connecting v and v' (see Exercise 10.6) and not using edge e. Adding edge e to this chain, we have a simple cycle. □

Theorem 10.18 *Let G be an undirected graph with n vertices $(n \geq 2)$. The following properties are equivalent:*

1. *G is connected and has no simple cycles.*
2. *G has no simple cycles and has $n - 1$ edges.*
3. *G is connected and has $n - 1$ edges.*
4. *G has no simple cycles; if we add an edge to it we form a simple cycle.*
5. *G is connected; if we delete an edge from it, it is no longer connected.*
6. *$\forall v, v' \in S$, $(v \neq v')$, there exists a unique simple chain connecting v and v'.*

Proof.
($1 \Longrightarrow 2$) Because G is connected, $p = 1$. If G has no simple cycles then $m - n + 1 = 0$ and the number of edges m of G is $n - 1$.
($2 \Longrightarrow 3$) If G has no simple cycles, $m - n + p = 0$, and if $m = n - 1$ then $p = 1$ and thus G is connected.
($3 \Longrightarrow 4$) If G is connected and has $n - 1$ edges then $m - n + p = 0$ and thus G has no simple cycles. If we add an edge it remains connected and $(m' - n' + p') = (m + 1 - n + p) > 0$. We have thus exhibited a simple cycle.
($4 \Longrightarrow 5$) If G were not connected we might add an edge to it without creating any cycles: it suffices to add an edge connecting two vertices of two distinct connected components. If deleting one edge yields another connected graph, we have $m' - n' + p' = m - 1 - n + p = 0$, and hence $m - n + p > 0$, and G would have a simple cycle.
($5 \Longrightarrow 6$) Let v and v' be two vertices of G. Because G is connected there exists a chain connecting v and v'. If there were two such chains, by Proposition 10.9 the graph would contain a simple cycle and we might thus delete one edge without destroying connectedness.

Trees and rooted trees 199

($6 \Longrightarrow 1$) G is connected. If it contained a simple cycle, we might find two distinct chains connecting two vertices. □

EXERCISE 10.13 Show that if we delete an edge from a tree then the remaining two connected components are trees. ◇

Proposition 10.19 *A tree with n vertices ($n \geq 2$) has at least two vertices with degree 1.*

Proof. In a graph, the sum of the degrees of the vertices is equal to twice the number of edges. In a tree, the number of edges is $n - 1$ and the sum of the degrees of the vertices is thus equal to $2n - 2$. Since a tree is a connected graph, there are no vertices of degree 0. Let k be the number of vertices of degree 1. There are thus $n - k$ vertices of degree at least equal to 2, and the sum of the degrees of the vertices is greater than or equal to $k + 2(n - k) = 2n - k$. As $2n - 2 \geq 2n - k$, we have $k \geq 2$. □

EXERCISE 10.14 Show that if a tree G has exactly two vertices of degree 1, then all other vertices are of degree 2. Deduce that G consists of a single elementary chain. ◇

Proposition 10.20 *A finite graph has a partial graph which is a tree if and only if it is connected.*

Proof. If a graph has a connected partial graph, it is connected. If a graph is connected and if we may delete an edge while preserving connectedness, we delete this edge. When no more edges can be deleted, we will have obtained a tree, by point 5 of Theorem 10.18. □

EXERCISE 10.15 Prove (by induction) that a tree can always be drawn in the plane in such a way that the edges form linear segments without cross-sections, except at the endpoints; in particular, it is planar. ◇

10.2.2 Rooted trees

Definition 10.21 *A rooted tree is a directed graph G such that $\gamma(G)$ is a tree and all vertices of G have indegree 1, except for a single vertex, called the root, whose indegree is 0.*

Proposition 10.22 *If G is a finite tree, then for any vertex v of G, there exists an orientation of G which is a rooted tree with root v.*

Proof. By induction on the number of vertices of G. The result is clear if the tree G has only one vertex. Otherwise, let v be any vertex of a tree G and let e be an edge with endpoints v and v'. Deleting this edge yields two connected graphs $G(v)$ and $G(v')$, which are still trees. We assign an orientation to these two trees in such a way that their respective roots are v and v', and we orient edge e from v to v'. We thus have a rooted tree with root v. In fact, for any vertex v'' different from v and v' of $G(v) \cup G(v')$, the indegree of v'' is equal to its indegree in the rooted tree constructed from $G(v)$ or from $G(v')$, i.e. 1. The indegree of v' is 1 and the indegree of v is 0. □

Proposition 10.23 *Let G be a rooted tree and let v be one of its vertices. Let $G(v)$ be the subgraph of G whose set of vertices is $\{v\}$ augmented by all the vertices that are the target of a path with origin v. Then $G(v)$ is a rooted tree with root v.*

Proof. By the construction of $G(v)$, $\gamma(G(v))$ is connected. Moreover, $\gamma(G(v))$ contains no simple cycle; otherwise, this simple cycle would also belong to $\gamma(G)$, which is impossible because $\gamma(G)$ is a tree. $\gamma(G(v))$ is thus a tree.

By the construction of $G(v)$, any vertex v' of $G(v)$ different from v is the target of an edge whose origin is in $G(v)$. This edge is the only edge of G with endpoint v'. The indegree of v' in $G(v)$ is thus 1. If the indegree of v were also equal to 1 in $G(v)$, there would exist an edge with target v and with origin in $G(v)$. There would also exist in G a simple circuit going through v, and there would thus exist in G a simple cycle, which is impossible. □

Theorem 10.24 (König's lemma) *Let G be an infinite rooted tree all of whose vertices have a finite outdegree. Then G has an infinite path originating at the root.*

Proof. Let $e_1, \ldots, e_n$ be the edges of G whose common origin is the root of G. One among the rooted trees $G(\beta(e_i))$ for $i = 1, \ldots, n$ is thus infinite. (These rooted trees are defined in Proposition 10.23.) Let i be such that $G(\beta(e_i))$ is infinite, and let $e'_1 = e_i$. Assume now that we have defined a path $e'_1 e'_2 \ldots e'_n$ such that $G(\beta(e'_n))$ is infinite and let $e''_1, \ldots, e''_k$ be the edges with origin $\beta(e'_n)$. Again, we will find a j such that $G(\beta(e''_j))$ is infinite and we will let $e'_{n+1} = e''_j$. Since we may repeat this construction indefinitely it indeed yields an infinite path in the rooted tree G. □

König's lemma is often applied in different forms that are consequences of the above theorem. Two such consequences are given in Proposition 10.25 and Proposition 10.26 below.

Proposition 10.25 *Let $G = (V, E, \alpha, \beta)$ be a directed graph such that any vertex v has a finite outdegree $d^+(v)$.*

If a vertex v of this graph is the origin of infinitely many finite paths, it must also be the origin of an infinite path.

Trees and rooted trees

Proof. Let C be the infinite set of the paths with origin v. For $n > 0$, let C_n be the set of paths with origin v and length n so that $C = \bigcup_{n>0} C_n$. Also let $C_0 = \{v\}$ and $C' = C_0 \cup C$. We define the set $R \subseteq C' \times C'$ by $(c, c') \in R$ if and only if

- either $c = v \in C_0$ and $c' \in C_1$,
- or there exists $n > 0$ and $e \in E$ such that $c \in C_n$ and $c' = ce$.

We then show that the graph $G' = (C', R, \alpha', \beta')$ with $\alpha'(c, c') = c$ and $\beta'(c, c') = c'$ is a rooted tree. The indegree of v is 0 and its outdegree is $d^+(v)$. The indegree of $c \in C$ is 1 and its outdegree is equal to $d^+(v')$, where v' is the target of the path c.

By König's lemma, this rooted tree has an infinite path $v, c_1, c_2, \ldots, c_n \ldots$ with $c_1 = e_1$, $c_{i+1} = c_i e_{i+1}$. Hence we deduce that $e_1 e_2 \cdots e_n \cdots$ is an infinite path with origin v in graph G. □

EXERCISE 10.16 Let A be a finite alphabet and let L be a subset of A^*. Show that if L is infinite, there exists at least one infinite string $a_0 a_1 a_2 \cdots a_n \cdots$ of letters of A such that $\forall n \geq 0$, $\exists w_n \in A^* : a_0 \cdots a_n w_n \in L$. ◇

Proposition 10.26 *Let E_n be a finite non-empty set, for any integer $n \geq 0$, and assume that*

(i) $n \neq m \implies E_n \cap E_m = \emptyset$.

Let R be a binary relation on $E = \cup_{n \geq 0} E_n$ such that:

(ii) *If $e\, R\, e'$ then there exists n such that $e \in E_n$ and $e' \in E_{n+1}$.*
(iii) $\forall n \geq 0, \forall e' \in E_{n+1}, \exists e \in E_n : e\, R\, e'$.

Then there exists an infinite sequence $e_0, e_1, \ldots, e_n, \ldots$ such that $\forall n \geq 0$, $e_n \in E_n$ and $e_n\, R\, e_{n+1}$.

Proof. Consider (E, R) as a directed graph. By (ii), the outdegree of an element of E_n is at most equal to the number of elements of E_{n+1}. It is thus finite. By (i), $\bigcup_{n>0} E_n$ is infinite. By (iii), each element of E_{n+1} is the target of a path whose origin is in E_0. There thus exist infinitely many paths whose origin is in E_0, and since E_0 is finite there exists an element e_0 of E_0 that is the origin of infinitely many finite paths. By Proposition 10.25, it also is the origin of an infinite path going through the sequence of vertices $e_0, e_1, \ldots, e_n, \ldots$, which is the required sequence. □

EXERCISE 10.17 Let V be a subset of $\mathbb{N} \times \mathbb{N}$ with the following properties:
(i) For any $n \geq 0$, the set $\{m \in \mathbb{N} \,/\, (n, m) \in S\}$ is finite.
(ii) For any $n \geq 0$, there exists an injection $f_n : \{0, 1, \ldots, n\} \to \mathbb{N}$ such that $\forall i \in \{0, 1, \ldots, n\}$, $(i, f_n(i)) \in V$.
Prove that there exists an injection $f : \mathbb{N} \to \mathbb{N}$ such that $\forall i \in \mathbb{N}$, $(i, f(i)) \in V$. ◇

10.2.3 Ordered rooted trees

Let G be a rooted tree and let v be one of its vertices. Vertex v' is said to be a *child* of v if there is an edge with origin v and target v'. If v is not the root of the rooted tree, v has indegree 1: there thus exists exactly one edge e with v as target. The origin vertex of this edge e will be called the *parent* of v. It is easy to show that any vertex v is the parent of its children.

A rooted tree is said to be *ordered* if, for any vertex v, the set of children of v is endowed with a total ordering. When drawing such a rooted tree in the plane (usually, counterintuitively, with the root at the top of the graph and the children below their parent), this total ordering on the children of a same parent will be materialized by writing them from left to right. This is why such rooted trees are called 'ordered'. The trees that we have studied in Chapter 3 are ordered rooted trees which may be empty, i.e. they may have empty sets of vertices and edges.

A *complete binary tree* is an ordered rooted tree in which each vertex either has two children (respectively called the *left child* and the *right child*) or none. A vertex without a child is called a *leaf*.

EXAMPLE 10.27 Two different ordered binary trees that represent the same rooted tree are shown in Figure 10.8.

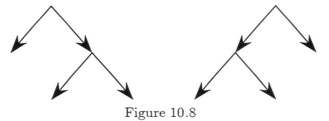

Figure 10.8

EXERCISE 10.18
1. Show that a finite complete binary tree has an odd number of vertices.
2. Show that a complete binary tree with $2n - 1$ vertices has n leaves.

CHAPTER 11

RATIONAL LANGUAGES AND FINITE AUTOMATA

Objects used in computer science are always representable as strings of symbols. We examine an algebraic structure on such strings of symbols, the free monoid. We will see how particular sets of such strings of symbols can be defined. We may think of a string of symbols as a representation of a computation. Finally, we will introduce machines, called *finite automata*, that can decide whether a given string belongs to a given set. These automata are simple examples of abstract machines. The entire study of computation theory relies mainly on generalizations and extensions of these simple machines.

This chapter defines rational subsets of a free monoid. It also defines (deterministic and non-deterministic) finite automata. It introduces some basic operations on automata: direct product, sequential product, determinization, minimization; it concludes with the proof of Kleene's theorem that rational languages are exactly those languages recognized by finite automata.

We recommend the following further reading:

Garrett Birkhoff, Thomas Bartee, *Modern Applied Algebra*, McGraw Hill, New York (1970).

Michael Harrison, *Introduction to Formal Languages*, Addison-Wesley, Reading (1978).

John Hopcroft, Jeffrey Ullman, *Introduction to Automata Theory, Languages and Computation*, Addison-Wesley, Reading (1979).

Arto Salomaa, *Formal Languages*, Academic Press, New York (1973).

11.1 The free monoid

We defined monoids and free monoids in Definition 1.12 and Definition 1.15, with slightly different notations. We recall these definitions.

Definition 11.1 *A monoid* $\mathcal{M} = (M, ., e)$ *is a set* M *equipped with an associative operation, denoted by '.', that has a unit* e.

The free monoid over A, denoted by A^*, is the set of strings over the alphabet A, equipped with concatenation (see Definition 1.15). The concatenation of two strings $u = u_1 u_2 \cdots u_n$ and $v = v_1 v_2 \cdots v_m$ is the string $u \cdot v = u_1 u_2 \cdots u_n v_1 v_2 \cdots v_m$.

EXERCISE 11.1 Are the following objects monoids?
1. The set of mappings $E \to E$ equipped with the composition operation. (The composition $g \circ f$ of f and g is denoted by $g.f$ or gf and is defined by $(gf)(x) = g(f(x))$ for $x \in E$.)
2. $\mathbb{N}$ equipped with the 'power' operation : $(x, y) \to x^y$.
3. The set of strings in A^* whose length is even.
4. The set of strings in A^* with as many occurrences of as as occurrences of bs.
5. The set of finite subsets of E, with the union operation.
6. The set of finite subsets of E, with the intersection operation. ◇

EXERCISE 11.2 Let the matrices

$$A = \begin{pmatrix} 1 & 1 \\ 0 & 1 \end{pmatrix} \text{ and } B = \begin{pmatrix} 1 & 0 \\ 1 & 1 \end{pmatrix}.$$

Consider the set $\mathcal{M}_1$ consisting of the identity matrix $I = \begin{pmatrix} 1 & 0 \\ 0 & 1 \end{pmatrix}$ and the matrices obtained by products of matrices A and B. Consider also the set $\mathcal{M}_2$ of 2×2 matrices with determinant $+1$ whose coefficients are non-negative integers.
1. Show that all the determinants of the matrices in $\mathcal{M}_1$ are equal to $+1$ and that all their coefficients are non-negative integers. Hint: try a proof by induction on the number of factors A and B in the product.
  This implies that $\mathcal{M}_1 \subseteq \mathcal{M}_2$; we now establish the reverse inclusion.
2. Show that a matrix with non-negative coefficients, and which is not the identity matrix, has determinant $+1$ *only if* all the coefficients of one of its columns are simultaneously greater than the coefficients of another column. More precisely, $\begin{pmatrix} x & y \\ z & w \end{pmatrix}$ $(x, y, z, w \in \mathbb{N})$ has determinant $+1$ only if either $x \geq y$ and $z \geq w$, or $x \leq y$ and $z \leq w$.
3. Compute the inverses of the matrices A and B. Use the result of 2 to show that any matrix $\begin{pmatrix} x & y \\ z & w \end{pmatrix}$ in $\mathcal{M}_2$ can be uniquely decomposed as a product of matrices A and B. (Use the complete induction principle and a recurrence on $x + y + z + w$.)
4. Deduce the equality $\mathcal{M}_1 = \mathcal{M}_2$. ◇

The free monoid 205

Let $\mathcal{M} = (M, ., e)$ and let M' be a subset of M. $\mathcal{M}' = (M', ., e)$ is a *submonoid* of $\mathcal{M}$ if it is a monoid, namely, if

- $e \in M'$ and
- $\forall m, m' \in M'$, $m.m' \in M'$.

If I is a set of indices, and if $\forall i \in I$, $\mathcal{M}_i = (M_i, ., e)$ is a submonoid of $\mathcal{M}$, then $(\bigcap_{i \in I} M_i, ., e)$ is a submonoid of $\mathcal{M}$. For a subset X of M, we can thus define the least submonoid of $\mathcal{M}$ containing X, called the submonoid of $\mathcal{M}$ generated by X, as the intersection of all the submonoids of $\mathcal{M}$ containing X.

EXAMPLE 11.2 Let $\mathcal{N} = (\mathbb{N}, +, 0)$. Let A be the set of even numbers and B be the set of odd numbers. $(A, +, 0)$ is the submonoid of $\mathcal{N}$ generated by $\{2\}$ while $(B \cup \{0\}, +, 0)$ is not a submonoid of $\mathcal{N}$ (e.g. $3 + 1$ is even).

If $\mathcal{M} = (M, ., e)$ and $\mathcal{M}' = (M', \times, e')$ are two monoids, a mapping h from M in M' is a *monoid homomorphism* if it satisfies

- $h(e) = e'$ and
- $\forall m, m' \in M$, $h(m.m') = h(m) \times h(m')$.

EXAMPLE 11.3 The mapping associating with each string u its length $|u|$ is a homomorphism from $(A^*, ., \varepsilon)$ to $(\mathbb{N}, +, 0)$.

The mapping associating with each $n \in \mathbb{N}$ the number 2^n is a homomorphism from $(\mathbb{N}, +, 0)$ to $(\mathbb{N}, \times, 1)$, because $2^0 = 1$ and $2^{n+m} = 2^n \times 2^m$.

EXERCISE 11.3 Let M_1 be the monoid defined in Exercise 11.2. Construct a monoid homomorphism $\{a, b\}^* \longrightarrow M_1$. Show that this homomorphism is both injective and surjective. ◇

Theorem 11.4 *Let A be an alphabet. Let $(M, ., e)$ be a monoid and let h be a mapping from A to M. There exists a unique homomorphism $h^*: A^* \longrightarrow M$ such that $\forall a \in A$, $h^*(a) = h(a)$.*

Proof. Existence: Let $h^*(\varepsilon) = e$ and $h^*(a_1 \cdots a_n) = h(a_1). \cdots .h(a_n)$. It is easy to see that h^* indeed is a homomorphism.
Uniqueness: Let g and g' be two homomorphisms from A^* in M such that $\forall a \in A$, $g(a) = g'(a)$. Then $g(\varepsilon) = g'(\varepsilon) = e$, and for any string $u = a_1 \cdots a_n$,

$$g(u) = g(a_1). \cdots .g(a_n) = g'(a_1). \cdots .g'(a_n) = g'(u). \qquad \square$$

Inspired by this theorem, we call the monoid A^* the *free monoid generated by A*.

EXERCISE 11.4 Let A be an alphabet containing at least the letters a and b.
1. Give a homomorphism from A^* to $(A \cdot A)^*$.
2. Give a homomorphism from A^* to $(A \setminus \{b\})^*$.
3. Is there a monoid isomorphism (i.e. a bijective homomorphism whose inverse is a homomorphism) from $(A \setminus \{a\})^*$ to $(A \setminus \{b\})^*$? ◇

EXERCISE 11.5 Levi's lemma: Let $u, v, x, y \in A^*$. Show that $uv = xy$ if and only if exactly one of the three following cases holds:
(i) $|u| = |x|$, $u = x$, and $v = y$.
(ii) $|u| < |x|$ and there exists a $t \in A^*$ such that $ut = x$ and $v = ty$.
(iii) $|u| > |x|$ and there exists a $t \in A^*$ such that $u = xt$ and $tv = y$. ◇

EXERCISE 11.6 Equations on strings: Let $u, v \in A^*$. Show the following equivalences:
1. $uv = vu$ if and only if there exists a $w \in A^*$ and $m, n \in \mathbb{N}$ such that $u = w^m$ and $v = w^n$.
2. $u^p = v^q$ (with $p, q \in \mathbb{N}$) if and only if there exists a $w \in A^*$ and $m, n \in \mathbb{N}$ such that $u = w^m$, $v = w^n$.
3. $uu = u$ if and only if $u = \varepsilon$.
4. $ua = au$ if and only if there exists a $p \in \mathbb{N}$ such that $u = a^p$.
5. $ua = bu$ (with $a \neq b$) if and only if $0 \neq 0$. ◇

11.2 Regular languages

In computer science we often define sets of strings of symbols, or subsets of a free monoid A^*, called languages. There are many ways of defining languages. Here we describe one such way, which also gives a simple way of determining whether a given string u is in the language or not.

We first define some operations on $\mathcal{P}(A^*)$:

- The *union*, denoted by $\cup$ or $+$: if L and L' are two subsets of A^*,
$$L + L' = L \cup L' = \{u \,/\, u \in L \text{ or } u \in L'\}.$$

- The *product*, denoted by $\cdot$: if L and L' are two subsets of A^*,
$$L \cdot L' = \{uv \,/\, u \in L, v \in L'\}.$$

- The *iteration*, denoted by $*$: if L is a subset of A^*, let
$$L^0 = \{\varepsilon\},$$
$$L^1 = L,$$
$$\vdots$$
$$L^{i+1} = L \cdot L^i,$$
$$\vdots$$
and $\quad L^* = \bigcup_{i \geq 0} L^i.$

Regular languages

We will denote by L^+ the set $\bigcup_{i>0} L^i$. We thus have
$$L^+ = L \cdot L^* = L^* \cdot L \text{ and } L^* = \{\varepsilon\} + L^+.$$

EXERCISE 11.7
1. Show that $(\mathcal{P}(A^*), \cdot, \{\varepsilon\})$ is a monoid.
2. Show that if $(L_i)_{i \in I}$ is any family of languages, then
$$\left(\bigcup_{i \in I} L_i\right) \cdot L = \bigcup_{i \in I} L_i \cdot L.$$
3. Show that $L^* = (L + \{\varepsilon\})^*$ and that $L^* = \{\varepsilon\} + L \cdot L^*$.
4. Show that $\emptyset^* = \{\varepsilon\}$. ◇

Definition 11.5 *A subset of A^* is said to be a regular set (over A) if it can be built up from the finite subsets of A^* by using the three operations $\cup, \cdot,$ and *. We denote by $Rat(A^*)$ the regular sets over A and by $Fin(A^*)$ the set of finite subsets of A^*.*

In order to specify a regular set over A it is enough to describe a procedure to construct it from finite subsets by the three operations given above. We do this by writing a *regular expression*. Since each non-empty finite subset is a finite union of singleton sets, and since any subset consisting of a single non-empty string is obtained by taking a product of sets consisting of a single one-letter string, the regular subsets can also be obtained by taking the closures under product, union and iteration of the sets consisting of a single one-letter string and of the empty set.

Definition 11.6 *We formally define the regular expressions as strings of symbols:*

- $\emptyset$ *is a regular expression.*
- *If a is a letter, a is a regular expression.*
- *If E is a regular expression, E^* is a regular expression.*
- *If E_1 and E_2 are regular expressions, $(E_1 + E_2)$ and $(E_1 E_2)$ are regular expressions.*

The set $I(E)$ specified by the regular expression E is defined by
$$I(\emptyset) = \emptyset,$$
$$I(a) = \{a\},$$
$$I(E^*) = I(E)^*,$$
$$I(E_1 E_2) = I(E_1) \cdot I(E_2) \quad \text{and}$$
$$I(E_1 + E_2) = I(E_1) + I(E_2).$$

Note that the definition of I is an inductive definition (see Definition 3.20). The rules we gave are very strict. For instance, $(a + b + c) \cdot d$ is disallowed unless it is written in the form $(((a+b)+c)\cdot d)$ or $((a+(b+c))\cdot d)$. (Both these expressions denote the same set $\{ad, bd, cd\}$.) Usually, when there is no ambiguity we allow ourselves more flexibility in denoting regular expressions, in a similar manner to the way algebraic expressions are denoted in mathematics. However, if these regular expressions are used in computer systems, we must abide strictly by the rules of the system, which might possibly differ from the rules given above. It is exactly the same situation as for arithmetic expressions, whose denotations in programming languages are also very strictly regulated.

EXERCISE 11.8 Let $X = \{b\}$ and $Y = (A \setminus \{b\}) \cdot \{b\}^*$.
1. Informally describe the elements of X^*, Y, and Y^*.
2. Show that any string of A^* starting with a letter different from b is in Y^*.
3. Show that any string u of A^* can be uniquely written in the form $u = vw$, where $v \in X^*$ and $w \in Y^*$. ◇

11.3 Finite automata

11.3.1 Labelled transition systems

Definition 11.7 *A labelled transition system over an alphabet A is a quintuple $(S, T, \alpha, \beta, \lambda)$ where*

- (S, T, α, β) *is a directed graph ($\alpha: T \longrightarrow S$ (resp. $\beta: T \longrightarrow S$) is called the source (resp. target) mapping: the elements of S are called states instead of vertices, and the elements of T are called transitions instead of edges.*
- λ *is a mapping from T to A: $\lambda(t)$ is called the label of the transition t. The triple (α, λ, β) must define an injective mapping from T to $S \times A \times S$: there can be no two different transitions with the same source, the same target and the same label. This reduces to requiring T to be a subset of $S \times A \times S$ and allows us to denote a transition system more simply by (S, T).*

A *path* is a non-empty sequence $c = t_1 \cdots t_n$ of transitions such that $\beta(t_i) = \alpha(t_{i+1})$ for $1 \leq i \leq n - 1$. The source $\alpha(c)$ of path c is $\alpha(t_1)$; the target $\beta(c)$ of path c is $\beta(t_n)$. A path is also called a *computation* of the transition system.

Consider also the empty paths: to each state s is associated an empty path ε_s, whose source and target are s. We might also have considered empty paths in graphs.

If c is a path, the *trace* of this path is the string $\lambda(c)$ of A^* defined by

- $\lambda(\varepsilon_s) = \varepsilon$, $\forall s \in S$,
- $\lambda(t_1 \cdots t_n) = \lambda(t_1) \cdots \lambda(t_n)$.

A labelled transition system is said to be *deterministic* if $\forall s \in S$, $\forall a \in A$, there exists at most one transition t with source s and label a. A labelled transition system is said to be *complete* if $\forall s \in S$, $\forall a \in A$, there exists at least one transition t with source s and label a.

Proposition 11.8 *If a transition system is deterministic, then $\forall u \in A^*$, $\forall s \in S$, there exists at most one path c with source s such that $\lambda(c) = u$. If a transition system is complete, there exists at least one such path.*

Proof. The proof proceeds by induction on the length of u.

If $u = \varepsilon$ then the unique path with source s and with trace ε is the empty path ε_s.

Let $u = av$ and let s be a given state. Consider the case when the transition system is deterministic and assume that there exist two paths c_1 and c_2 with source s. The first transition of both paths is the unique transition $t = (s, a, s')$ with source s and label a. We thus have $c_1 = tc_1'$ and $c_2 = tc_2'$. The paths c_1' and c_2' are then two distinct paths with trace v and with source s', and this is impossible by the induction hypothesis. If the transition system is complete, there exists at least one transition $t = (s, a, s')$ and at least one path c with source s' and with trace v. The path tc is then a path with source s and with trace av. □

If s and s' are two states of S, $L_{s,s'}$ is the set of traces of the paths with source s and with target s':

$$L_{s,s'} = \{\lambda(c) \ / \ \alpha(c) = s \text{ and } \beta(c) = s'\}.$$

EXERCISE 11.9 Show that $\varepsilon \in L_{s,s'}$ if and only if $s = s'$. ◇

Let Q and Q' be two subsets of S. $L_{Q,Q'} = \bigcup_{s \in Q, s' \in Q'} L_{s,s'}$.

Definition 11.9 *A finite-state automaton is an eight-tuple $(A, S, T, \alpha, \beta, \lambda, I, F)$, where*

- $(S, T, \alpha, \beta, \lambda)$ *is a labelled transition system with label alphabet A, and S is a finite set,*
- *I and F are two subsets of S whose elements are called initial states and final states respectively.*

When no ambiguity can occur, a finite-state automaton will simply be denoted by (S, T, I, F).

A finite-state automaton is said to be *complete* if (S, T) is a complete transition system. A finite-state automaton is said to be *deterministic* if (S, T) is a deterministic transition system and if there is a single initial state.

Definition 11.10 *A language $L \subseteq A^*$ is said to be recognizable if and only if there exists a finite-state automaton $\mathcal{A} = (A, S, T, \alpha, \beta, \lambda, I, F)$ such that $L = L(\mathcal{A})$, where $L(\mathcal{A})$ is, by definition, $L_{I,F}$.*

Intuitively, to check whether a string u is in $L(\mathcal{A})$, we start from an initial state of the automaton and then read the string u using each successive letter of u, from left to right. Each time it reads a letter, the automaton changes its state by taking a transition labelled by this letter. If we can thus reach a final state when all the letters of u heve been read, then the string is recognized. Note, however, that if a finite-state automaton is non-deterministic, then the same string u can yield several different evolutions of the automaton. Some of these evolutions will lead to a final state and others will not. If there is at least one evolution that leads to a final state, then u is in $L(\mathcal{A})$.

EXAMPLE 11.11 Let $\mathcal{A}$ be the finite-state automaton with two states s_1 and s_2 whose transitions are (s_1, a, s_1), (s_1, b, s_1), and (s_1, b, s_2). See Figure 11.1. The string ab is the label of two different paths: (s_1, a, s_1, b, s_1) and (s_1, a, s_1, b, s_2). If $I = \{s_1\}$ and $F = \{s_2\}$, then $ab \in L(\mathcal{A})$.

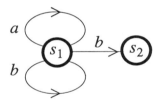

Figure 11.1

EXERCISE 11.10 Let L be a language and let $\text{Pref}(L) = \{u \in A^* \,/\, \exists v \in A^* : uv \in L\}$ be the set of prefixes of the strings in this language L. Show that if a language L is recognizable then the set $\text{Pref}(L)$ is also recognizable. ◇

EXERCISE 11.11 Show that any *finite* language is recognized by a deterministic finite-state automaton. (Explain how such an automaton can be constructed using the prefixes of the strings in the language.) ◇

EXERCISE 11.12 Let $\mathcal{A}$ be a finite-state automaton with a finite alphabet A. Show that the language recognized by $\mathcal{A}$ is finite if and only if the directed graph corresponding to the set of transitions of $\mathcal{A}$ has no circuit containing a vertex that is on a path going from an initial state to a final state. Give the least upper bound on the size of this language in terms of the number of states and of the size of the alphabet; also give this bound in the particular case when the finite-state automaton is deterministic and complete. Describe the finite-state automata for which this bound is reached. ◇

Finite automata 211

11.3.2 Completion of a finite-state automaton

In all cases, the demand that a finite-state automaton be complete is not a restriction on its language, because it is always possible to transform an incomplete finite-state automaton into a complete finite-state automaton without changing the language recognized.

Thus let $\mathcal{A} = (S, T, I, F)$ be a finite-state automaton assumed to be incomplete. We build a new finite-state automaton $\mathcal{A}' = (S', T', I, F)$ as follows:

- To obtain S', we add to S a new state, usually called the *sink state*, denoted here by p.
- To obtain T' we add to T
  - the transitions (s, a, p) for any state s and for any letter a such that there do not exist in T any transitions with source s and label a,
  - the transitions (p, a, p) for all the letters a.

$\mathcal{A}'$ is complete by construction. Moreover, paths existing in $\mathcal{A}'$ and which did not exist in $\mathcal{A}$ are paths ending in state p, which is not a final state, and thus $L(\mathcal{A}) = L(\mathcal{A}')$.

EXERCISE 11.13 Consider an incomplete finite-state automaton $\mathcal{A}$. Assume that we transform it into a complete finite-state automaton $\mathcal{A}'$ as follows: for all $a \in A$ and $s \in S$ such that there is no transition with source s and label a, we add a loop with label a in s, namely, the transition (s, a, s):
1. If $\mathcal{A}$ is deterministic, is $\mathcal{A}'$ also deterministic?
2. What is the relationship between $L(\mathcal{A})$ and $L(\mathcal{A}')$? ◇

11.3.3 Determinization of a finite-state automaton

The construction of a deterministic finite-state automaton from an arbitrary one is more complex than its completion, but it is also possible. Moreover, this construction is more interesting, because if $\mathcal{A}$ is a finite-state automaton then it is much easier to check whether any given string u is in $L(\mathcal{A})$ when $\mathcal{A}$ is deterministic than when it is not deterministic.

Let $\mathcal{A} = (S, T, I, F)$ be a finite-state automaton. Let us construct the deterministic finite-state automaton $\mathcal{A}_d$ recognizing the same language as $\mathcal{A}$. This automaton $\mathcal{A}_d = (S', T', \{i'\}, F')$ is defined by:

$$S' = \mathcal{P}(S),$$
$$T' = \{(Q, a, Q') \ / \ Q \subseteq S,\, Q' \subseteq S,\, Q' = \{q' \ / \ \exists q \in Q \colon (q, a, q') \in T\}\},$$
$$i' = I,$$
$$F' = \{Q \subseteq S \ / \ Q \cap F \neq \emptyset\}.$$

Proposition 11.12 *Let c be a path in $\mathcal{A}_d$ with source Q, target Q' and trace u. Then $Q' = \{q' \mid \exists q \in Q$: there exists in $\mathcal{A}$ a path with source q, target q' and trace $u\}$.*

Proof. By induction on the length of u:

- If $u = \varepsilon$, then $Q = Q'$ and the result trivially holds.
- If $u = au'$, then there exists in $\mathcal{A}_d$ a path tc with $t = (Q, a, Q'')$. By the definition of $\mathcal{A}_d$, $Q'' = \{q'' \mid \exists q \in Q : (q, a, q'') \in T\}$, and, by the induction hypothesis,

$$Q' = \{q' \mid \exists q'' \in Q'' \text{: there exists in } \mathcal{A} \text{ a path with source } q'',$$
$$\text{with target } q' \text{ and with trace } u'\},$$

and by combining both cases we have the result. □

Corollary 11.13 $L(\mathcal{A}) = L(\mathcal{A}_d)$.

Proof. $u \in L(\mathcal{A})$ if and only if u is the trace of a path with source $q \in I$ and with target $q' \in F$; $u \in L(\mathcal{A}_d)$ if and only if the set

$$\{q' \mid \exists q \in I : u \text{ is the trace of a path with source } q \text{ and with target } q'\}$$

contains at least an element of F. By Proposition 11.12 it is clear that these two conditions are equivalent. □

EXAMPLE 11.14 We return to the finite-state automaton of Example 11.11. It is associated with the deterministic finite-state automaton $\mathcal{A}_d$ with states $\{\emptyset, \{s_1\}, \{s_2\}, \{s_1, s_2\}\}$, whose initial state is $i = \{s_1\}$, whose set of final states is $\{\{s_2\}, \{s_1, s_2\}\}$ and whose transitions are given by the following table, where the set appearing in line Q ($Q \subseteq \{s_1, s_2\}$) and column x ($x \in \{a, b\}$) is the set $\{q' \mid \exists q \in Q : (q, x, q') \in T\}$.

s	a	b
$\emptyset$	$\emptyset$	$\emptyset$
$\{s_1\}$	$\{s_1\}$	$\{s_1, s_2\}$
$\{s_2\}$	$\emptyset$	$\emptyset$
$\{s_1, s_2\}$	$\{s_1\}$	$\{s_1, s_2\}$

The string ab is the trace of a path from the initial state $\{s_1\}$ to the final state $\{s_1, s_2\}$. In the original finite-state automaton, there are indeed two (and exactly two) different paths with trace ab and with source s_1; these lead to s_1 and s_2 respectively (Figure 11.2).

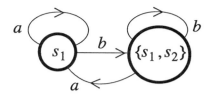

Figure 11.2

EXERCISE 11.14 Make deterministic and complete the following finite-state automaton over the alphabet $\{a, b, c\}$, where the initial state is denoted by i and the final states are denoted by f, and f'.

$$f \xleftarrow{a,c} x \xleftarrow{b} i \xleftrightarrow{b} y \xrightarrow{b,c} z \xleftrightarrow{a,c} f'.\qquad \diamond$$

EXERCISE 11.15 Let $A = \{a, b, c\}$. Give a complete deterministic finite-state automaton to recognize each of the following languages:
1. the set of strings of even length,
2. the set of strings with a number of occurrences of 'b' that is divisible by 3,
3. the set of strings ending with 'b',
4. the set of strings not ending with 'b',
5. the set of non-empty strings not ending with 'b',
6. the set of strings with at least one 'b',
7. the set of strings with at most one 'b',
8. the set of strings with exactly one 'b',
9. the set of strings with no 'b' at all,
10. the set of strings with at least one 'a' and whose first 'a' is not followed by a 'c',
11. the set of strings where at least three letters occur and whose third-from-last letter is an 'a' or a 'c'. $\diamond$

EXERCISE 11.16 (Iteration Lemma)
1. Show that if G is a directed graph with n vertices and c is a path of length $\geq n$ in G, then the subpath of c consisting of the first n edges of c contains a circuit.
2. Consider a transition system with n states labelled by the alphabet A, and let q, q' be two among these states. Show that if $z \in L_{q,q'}$ and $|z| \geq n$, then there exist $u, v, w \in A^*$ such that $z = uvw$, $v \neq \varepsilon$, $|uv| \leq n$ and v is the trace of a circuit, and for any $i \in \mathbb{N}$, $uv^i w \in L_{q,q'}$.
3. Let $\mathcal{A}$ be a finite-state automaton with n states. Then for any $z \in L(\mathcal{A})$ verifying $|z| \geq n$, there exist $u, v, w \in A^*$ such that $z = uvw$, $v \neq \varepsilon$, $|uv| \leq n$, and for any $i \in \mathbb{N}$, $uv^i w \in L(\mathcal{A})$. $\diamond$

EXERCISE 11.17 Application: Show that the following languages are not recognizable because they cannot be equal to $L(\mathcal{A})$ for any finite-state automaton $\mathcal{A}$:
1. $\{a^n b^n \mid n \in \mathbb{N}\}$.
2. $\{a^{n^2} \mid n \in \mathbb{N}\}$.
3. $\{a^p \mid p \text{ is a prime number}\}$.
4. $\{ww \mid w \in A^*\}$, when A has at least two letters.
5. The set of palindromes: the strings $w_1 \cdots w_n \in A^*$ ($n \geq 0$) such that $w_1 \cdots w_n = w_n \cdots w_1$ (i.e. $w_i = w_{n+1-i}$ for $i = 1, \ldots, n$) when A has at least two letters. $\diamond$

11.3.4 Minimal finite-state automata

Let $\mathcal{A}$ be a complete deterministic finite-state automaton whose only initial state is i. With any string u we associate the mapping $\delta_u \colon S \longrightarrow S$ defined by: $\delta_u(s)$ is the target of the unique path with source s and with trace u. Two strings u and v are said to be equivalent modulo $\mathcal{A}$ (denoted by $u \sim_{\mathcal{A}} v$) if $\delta_u(i) = \delta_v(i)$. It is easy to see that this is indeed an equivalence relation. Moreover, we have

Proposition 11.15
1. The relation $\sim_{\mathcal{A}}$ has a finite number of equivalence classes.
2. If $u \sim_{\mathcal{A}} v$, then $\forall a \in A, \quad ua \sim_{\mathcal{A}} va$.
3. If $u \sim_{\mathcal{A}} v$ and $u \in L(\mathcal{A})$, then $v \in L(\mathcal{A})$.

Proof.
1. With each equivalence class of $\sim_{\mathcal{A}}$ we can associate a state of $\mathcal{A}$; namely, the class of a string u is associated with the state $\delta_u(i)$. The number of classes is thus less than or equal to the number of states of $\mathcal{A}$.
2. If the two paths with respective traces u and v and with source i have the same target state q, the two paths with respective traces ua and va and with source i also have the same target state $q' = \delta_a(q)$.
3. If $u \in L(\mathcal{A})$, the unique path with source i and trace u has a final state as its target. If $u \sim_{\mathcal{A}} v$ then the only path with source i and trace v has the same final state as its target, and thus $v \in L(\mathcal{A})$. $\square$

An equivalence relation $\sim$ on A^* verifying
$$u \sim v \quad \Longrightarrow \quad \forall a \in A, \ ua \sim va$$
is called a *(right) semi-congruence*. If it has only a finite number of equivalence classes, it is said to have a *finite index*. Finally, if a language L is a union of equivalence classes of $\sim$, then $\sim$ is said to *saturate* L.

We have thus shown that if $\mathcal{A}$ is a complete deterministic finite-state automaton, then $\sim_{\mathcal{A}}$ is a right semi-congruence of finite index saturating $L(\mathcal{A})$. Because any language recognized by a finite-state automaton can be recognized by a complete deterministic finite-state automaton, any recognizable language is saturated by a finite index semi-congruence. We will also show the converse of this property which will give a characterization of recognizable languages.

Theorem 11.16 *A language L in A^* is recognizable if and only if it is saturated by a finite index semi-congruence.*

Proof. Let $\sim$ be a finite index semi-congruence saturating the language L. We will construct a complete deterministic finite-state automaton $\mathcal{A} = (S, T, i, F)$ recognizing L:

Finite automata 215

- S is the set of equivalence classes of A^* modulo $\sim$.
- T is the set of triples $([u]_\sim, a, [ua]_\sim)$ for all $[u]_\sim$ in S and all a in A.
- i is the class of ε.
- F is the set of classes of strings in L.

The fact that $\sim$ has a finite index implies that S is finite. The fact that it is a semi-congruence implies that the definition of T is meaningful since $[u]_\sim = [v]_\sim \implies [ua]_\sim = [va]_\sim$. It is clear that $\mathcal{A}$ is deterministic and complete.

We now show that this automaton recognizes L (i.e. $L = L(\mathcal{A})$). We easily show by induction on the length of a string $u \in A^*$ that the target of the unique path with source $i = [\varepsilon]_\sim$ and with trace u is $[u]_\sim$. It follows that if u is in L then this state is final, and hence $L \subseteq L(\mathcal{A})$. Conversely, if $[u]_\sim$ is final, then there exists a string v in L with $u \sim v$ and, since $\sim$ saturates L, u is also in L and thus $L(\mathcal{A}) \subseteq L$. □

Among the finite-index semi-congruences saturating a given recognizable language L, there is a semi-congruence that is less refined than all the others (i.e. it has fewer equivalence classes, and each equivalence class is 'fatter'). It is denoted $\sim_L$ and is defined by $u \sim_L v$ if and only if

$$\forall w \in A^*, \quad uw \in L \iff vw \in L.$$

It is clear that $\sim_L$ is an equivalence relation. It is also a semi-congruence. Indeed, if $u \sim_L v$ then $\forall w \in A^*$, $uaw \in L \iff vaw \in L$ and thus $ua \sim_L va$. It saturates L because letting $w = \varepsilon$ and $u \sim_L v$ implies that

$$u \in L \iff v \in L.$$

Finally, the fact that it has finite index follows from the next result.

Theorem 11.17 *Let L be a recognizable language and $\sim$ a finite index semi-congruence saturating L. Then $\sim_L \supseteq \sim$. That is,*

$$\forall u, v \in A^*, \quad u \sim v \implies u \sim_L v.$$

Proof. Assume that $u \sim v$ and show that

$$\forall w \in A^*, \quad uw \in L \iff vw \in L.$$

Let w be an arbitrary string in A^*. Because $\sim$ is a semi-congruence, $uw \sim vw$, and because $\sim$ saturates L, $uw \in L \iff vw \in L$. □

We saw in the proof of Theorem 11.16 that any finite-index semi-congruence $\sim$ saturating L enabled us to define a finite-state automaton recognizing L, with the equivalence classes of $\sim$ as states. Therefore, the semi-congruence $\sim_L$ enables us to define a deterministic finite-state automaton recognizing L whose number of states is minimal. We will temporarily call this automaton the canonical automaton of L.

There is a more algebraic way of characterizing this canonical automaton recognizing a language L. First we must introduce some definitions.

Definition 11.18 *A complete deterministic finite-state automaton $\mathcal{A} = (S, T, i, F)$ is said to be reachable if $\forall s \in S, \exists u \in A^*: \delta_u(i) = s$.*

It is easy to see that we can always delete from a finite-state automaton its non-reachable states (namely, those states s such that $\forall u \in A^*, \delta_u(i) \neq s$) without modifying the recognized language and also that the finite-state automaton constructed in the proof of Theorem 11.16 is reachable.

Definition 11.19 *Let $\mathcal{A} = (S, T, i, F)$ and $\mathcal{A}' = (S', T', i', F')$ be two reachable complete deterministic finite-state automata. A mapping $h: S \longrightarrow S'$ is a homomorphism of automata from $\mathcal{A}$ to $\mathcal{A}'$ if*

- *h is surjective,*
- *$\forall s \in S, \forall a \in A, h(\delta_a(s)) = \delta'_a(h(s))$,*
- *$h(i) = i'$ and*
- *$\forall s \in S, h(s) \in F' \Longleftrightarrow s \in F$.*

Proposition 11.20 *If there exists a homomorphism from $\mathcal{A}$ to $\mathcal{A}'$, then $L(\mathcal{A}) \subseteq L(\mathcal{A}')$.*

Proof. We prove, by induction on the length of $u \in A^*$, that, for any path in $\mathcal{A}$ with source s, target s' and trace u, there exists in $\mathcal{A}'$ a path with source $h(s)$, target $h(s')$ and trace u.

It follows that if there exists in $\mathcal{A}$ a path with source i, trace u and target in F, then there exists in $\mathcal{A}'$ a path with source i', trace u and target in F'. $\square$

Theorem 11.21 *For any reachable complete deterministic finite-state automaton $\mathcal{A}$, there exists a homomorphism from $\mathcal{A}$ to the canonical automaton recognizing $L(\mathcal{A})$.*

Proof. Let s be a state of $\mathcal{A}$. Since $\mathcal{A}$ is reachable, there exists a string u such that $s = \delta_u(i)$. Let $h(s) = [u]_{\sim_L}$. This mapping is well defined, because if $\delta_v(i) = s$, then $u \sim_{\mathcal{A}} v$ and, by Theorem 11.17, $u \sim_L v$. This mapping is indeed surjective ($\forall u \in A^*, [u]_{\sim_L} = h(\delta_u(i))$). It is a homomorphism because

- $h(i) = [\varepsilon]_{\sim_L}$,
- $h(\delta_a(\delta_u(i))) = h(\delta_{ua}(i)) = [ua]_{\sim_L}$ and $h(\delta_u(i)) = [u]_{\sim_L}$ and

Finite automata

- $h(\delta_u(i)) = [u]_{\sim_L} \subseteq L$ if and only if $u \in L$ if and only if $\delta_u(i) \in F$. □

Justified by this theorem, we call the canonical automaton the *minimal* automaton. It is indeed minimal, among complete deterministic finite-state automata recognizing L, for the ordering $\leq$ defined by $\mathcal{A} \leq \mathcal{A}'$ if there exists a homomorphism from $\mathcal{A}'$ to $\mathcal{A}$.

The canonical homomorphism h from $\mathcal{A}$ to the minimal automaton induces an equivalence relation on the states of $\mathcal{A}$ defined by

$$s \approx s' \iff h(s) = h(s').$$

We will show how to compute this equivalence, and this will enable us to find the minimal automaton recognizing L from any (reachable complete) deterministic finite-state automaton recognizing L.

Let $\mathcal{A} = (S, T, i, F)$ be a reachable complete deterministic finite-state automaton. Define the sequence $(\approx_i)_{i \geq 0}$ of equivalence relations on S by:

- $s \approx_0 s'$ if and only if $s \in F \iff s' \in F$ or, equivalently, $\approx_0$ has exactly two classes, F and $S \setminus F$.
- $s \approx_{i+1} s'$ if and only if $s \approx_i s'$ and $\forall a \in A, \delta_a(s) \approx_i \delta_a(s')$.

Proposition 11.22
(a) $\approx_{i+1} \subseteq \approx_i$.
(b) If $\approx_{i+1} = \approx_i$, then $\forall j \geq i, \approx_j = \approx_i$.
(c) $\exists n < |S|: \approx_{n+1} = \approx_n$.

Proof. Point (a) is straightforward by the definition of $\approx_{i+1}$.

Point (b) is proved by induction on $n = j - i$. It is straightforward if $n = 0$ and $n = 1$. Assume $\approx_j = \approx_i$. Then,

$$\begin{aligned}
s \approx_{j+1} s' &\iff s \approx_j s' \text{ and } \forall a \in A,\ \delta_a(s) \approx_j \delta_a(s') \\
&\iff s \approx_i s' \text{ and } \forall a \in A,\ \delta_a(s) \approx_i \delta_a(s') \\
&\iff s \approx_{i+1} s' \\
&\iff s \approx_i s'.
\end{aligned}$$

To show point (c), let m_i be equal to the number of classes of $\approx_i$. It is clear that $m_{i+1} \geq m_i$ and that $m_{i+1} = m_i$ implies $\approx_{i+1} = \approx_i$. Assume that $\forall n < |S|, \approx_n \neq \approx_{n+1}$. We thus have $m_0 < m_1 < \cdots < m_{|S|}$; hence $2 + |S| = m_0 + |S| \leq m_{|S|}$, and this is impossible since an equivalence relation on S can have at most $|S|$ classes. □

To minimize a finite-state automaton, we will thus construct such a sequence of equivalence relations. As soon as $\approx_{i+1} = \approx_i$, the mapping h mapping a state to its class modulo $\approx_i$ is the required homomorphism, as shown by the following proposition.

Proposition 11.23 *Let $\mathcal{A}$ be a finite-state automaton and let h be the canonical homomorphism from $\mathcal{A}$ to the minimal automaton recognizing $L(\mathcal{A})$. Let $n < |S|$ be such that $\approx_n = \approx_{n+1}$. Then*

$$\forall s, s' \in S, \qquad h(s) = h(s') \iff s \approx_n s'.$$

Proof.
1. First show by induction on i that

$$\forall i \geq 0, \forall s, s' \in S, \qquad h(s) = h(s') \implies s \approx_n s'.$$

Let u and v be two strings such that $\delta_u(i) = s$ and $\delta_v(i) = s'$. If $h(s) = h(s')$, then $u \sim_L v$, by the definition of h.
- $s \in F \iff u \in L \iff v \in L \iff s' \in F$, and thus $s \approx_0 s'$.
- Assume that $\forall s, s' \in S$, $h(s) = h(s') \iff s \approx_i s'$. If $h(s) = h(s')$ then $s \approx_i s'$; but we also have, for any a in A, $h(\delta_a(s)) = h(\delta_a(s'))$, and thus $\delta_a(s) \approx_i \delta_a(s')$. Hence, $s \approx_{i+1} s'$.
2. It is easy to show, by induction on the length $|w|$ of $w \in A^*$, that $s \approx_i s' \implies \delta_w(s) \approx_{i+|w|} \delta_w(s')$ and, in particular, since $\approx_n = \approx_{n+|w|}$, $s \approx_n s' \implies \delta_w(s) \approx_n \delta_w(s')$. Since $\approx_n \subseteq \approx_0$, we also obtain $s \approx_n s' \implies \delta_w(s) \approx_0 \delta_w(s')$. Hence,

$$\delta_u(i) \approx_n \delta_v(i) \implies \forall w \in A^*, \ \delta_{uw}(i) = \delta_w(\delta_u(i)) \approx_0 \delta_w(\delta_v(i)) = \delta_{vw}(i)$$
$$\iff \forall w \in A^*, \ \delta_{uw}(i) \in F \iff \delta_{vw}(i) \in F$$
$$\iff \forall w \in A^*, \ uw \in L \iff vw \in L$$
$$\iff u \sim_L v. \qquad \square$$

EXERCISE 11.18 Give minimal complete deterministic finite-state automata recognizing the languages over the alphabet $A = \{a, b\}$ specified by the following regular expressions:

1. $(a+b)^* b (a+b)^*$.
2. $ba^* + ab + (a+bb)ab^*$.
3. $(b + ab + aab)^* (\varepsilon + a + aaa)$.
4. $(a+b)^2 + (a+b)^3 + (a+b)^4$.
5. $\bigl((a+b)^2\bigr)^* + \bigl((a+b)^3\bigr)^*$.
6. $\varepsilon + ab^* a + (ab + ba)^*$. $\diamondsuit$

11.3.5 Operations on finite-state automata

Let $\mathcal{A} = (S, T, \{i\}, F)$ be a complete deterministic finite-state automaton. For any string u in A^*, there exists in $\mathcal{A}$ a unique path c with source i and label u. The target state of this path c is in F if and only if $u \in L(\mathcal{A})$. We immediately obtain a finite-state automaton recognizing the complement of $L(\mathcal{A})$.

Proposition 11.24 *If $\mathcal{A} = (S, T, \{i\}, F)$ is a complete deterministic finite-state automaton, the finite-state automaton $\mathcal{A}' = (S, T, \{i\}, S \setminus F)$ is again a complete deterministic finite-state automaton and $L(\mathcal{A}') = A^* \setminus L(\mathcal{A})$.*

EXERCISE 11.19   Let $\mathcal{A} = (S, T, I, F)$ be a finite-state automaton and let $\mathcal{A}' = (S, T, I, S \setminus F)$. Show, by means of examples, that $L(\mathcal{A}') = A^* \setminus L(\mathcal{A})$ does not necessarily hold if $\mathcal{A}$ is not both deterministic and complete. ◇

Let $\mathcal{A}' = (S', T', I', F')$ and $\mathcal{A}'' = (S'', T'', I'', F'')$ be two finite-state automata over the same alphabet A. The *direct product* of the transition systems (S', T') and (S'', T'') is the transition system (S, T) defined by

$$S = S' \times S'' \text{ and } T = \{((s_1', s_1''), a, (s_2', s_2'')) \,/\, (s_1', a, s_2') \in T', (s_1'', a, s_2'') \in T''\}.$$

It is easy to see that there is in (S, T) a path c with trace u, source (s_1', s_1'') and target (s_2', s_2'') if and only if there is in (S', T') a path c' with trace u, source s_1' and target s_2', and in (S'', T'') a path c'' with trace u, source s_1'' and target s_2''.

Now consider the finite-state automata

$$\mathcal{A}_1 = (S, T, I' \times I'', F' \times F'') \text{ and } \mathcal{A}_2 = (S, T, I' \times I'', F' \times S'' \cup S' \times F'').$$

Clearly, using the above remark we have

Proposition 11.25 $L(\mathcal{A}_1) = L(\mathcal{A}) \cap L(\mathcal{A}')$, $L(\mathcal{A}_2) = L(\mathcal{A}) \cup L(\mathcal{A}')$.

11.4 Equation systems

With each labelled transition system $\mathcal{S}$ we associate an *equation system* $\widehat{\mathcal{S}}$ on $\mathcal{P}(A^*)$ such that the least solution of $\widehat{\mathcal{S}}$ is the set of traces of paths of the transition system $\mathcal{S}$.

Let $\mathcal{S} = (S, T, \alpha, \beta, \lambda)$ be a labelled transition system, where S is finite. Let $\mathcal{D}$ be the set of mappings from $S \times S$ to $\mathcal{P}(A^*)$ ordered as follows: $D \leq D'$ if and only if $\forall s, s' \in S$, $D(s, s') \subseteq D'(s, s')$.

With $\mathcal{S}$ we associate the *equation system* $\widehat{\mathcal{S}}$ on $\mathcal{P}(A^*)$ whose variables are $x_{s,s'}$ for all pairs (s, s') of states of S and whose equations are

$$x_{s,s'} = \sum \{a x_{s'',s'} \,/\, \exists t \colon \alpha(t) = s, \lambda(t) = a, \beta(t) = s''\} \cup \{\varepsilon \,/\, s = s'\}.$$

Consider now this equation system as a mapping, also denoted by $\widehat{\mathcal{S}}$, from $\mathcal{D}$ to $\mathcal{D}$ defined as follows: if D is a mapping from $S \times S$ to $\mathcal{P}(A^*)$, then $\widehat{\mathcal{S}}(D)$ is the mapping D' defined by

$$D'(s,s') = \bigcup \{a \cdot D(s'',s') \mid \exists t: \alpha(t) = s, \lambda(t) = a, \beta(t) = s''\} \cup \{\varepsilon \mid s = s'\}.$$

It is easy to see that the mapping $\widehat{\mathcal{S}}$ is continuous. It thus has a least fixpoint D^μ defined by (see Theorem 2.40)

$$D^\mu(s,s') = \bigcup_{i \geq 0} D_i(s,s')$$

with

- $D_0(s,s') = \emptyset$, $\forall s, s' \in S$ and
- $D_{i+1}(s,s') = D_i(s,s') \cup \widehat{\mathcal{S}}(D_i)(s,s')$.

D^μ is also said to be the least solution of the equation system $\widehat{\mathcal{S}}$.

EXERCISE 11.20 Show that $\widehat{\mathcal{S}}$ is continuous. ◇

Theorem 11.26 $\forall s, s'$, $L_{s,s'} = D^\mu(s,s')$.

Proof. Let $D_L \in \mathcal{D}$ be defined by $D_L(s,s') = L_{s,s'}$.
We will first show that $\widehat{\mathcal{S}}(D_L) \subseteq D_L$, and hence it will follow that $D^\mu \subseteq D_L$. Since

$$\widehat{\mathcal{S}}(D_L)(s,s') = \bigcup \{a \cdot L_{s'',s'} \mid \exists t: \alpha(t) = s, \lambda(t) = a, \beta(t) = s''\} \cup \{\varepsilon \mid s = s'\},$$

we have:

1. $\varepsilon \in \widehat{\mathcal{S}}(D_L)(s,s')$ if and only if $s = s'$ and thus if and only if $\varepsilon \in L_{s,s'}$.
2. If $au \in \widehat{\mathcal{S}}(D_L)(s,s')$, there exists a path c with source s'', target s' and trace u, and there exists a transition t with source s, target s'' and label a. It follows that $t \cdot c$ is a path with source s, target s' and trace au in $L_{s,s'}$.

We now show that $L_{s,s'} \subseteq D^\mu(s,s')$. To this end, let $L^i_{s,s'}$ be the set of strings of $L_{s,s'}$ with length strictly less than i. We will show by induction that

$$L^i_{s,s'} \subseteq D_i(s,s').$$

Hence, we will immediately deduce that $L_{s,s'} \subseteq D^\mu(s,s')$.

1. If $i = 0$ then $D_i(s,s') = \emptyset$ and $L^i_{s,s'} = \{u \in L_{s,s'} \mid |u| < 0\} = \emptyset$.
2. Let u be a string in $L_{s,s'}$ with length strictly less than $i+1$.

(2.1) If its length is strictly less than i then it is in $D_i(s,s')$ and hence in $D_{i+1}(s,s')$.

(2.2) If it has length i, then the following holds:

(2.2.1) If $i = 0$ then $u = \varepsilon$ and thus $s = s'$; hence $\varepsilon \in \widehat{\mathcal{S}}(D_i)(s,s')$.

(2.2.2) If $i > 0$ then $u = au'$ with $|u'| = i$. Since u is the trace of a path with source s and with target s', there exists a path $c = t_1 \cdots t_{i+1}$ with $\alpha(c) = \alpha(t_1) = s$, $\beta(c) = \beta(t_{i+1}) = s'$ and $\lambda(t_1) = a$, $\lambda(t_2 \cdots t_{i+1}) = u'$. The transition t_1 has source s, target s'' and label a, and the path $t_2 \cdots t_{i+1}$ has source s'', target s' and trace u'. We thus have $u' \in L^i_{s'',s'} \subseteq D_i(s'',s')$ and $au' \in \widehat{\mathcal{S}}(D_i)(s,s') \subseteq D_{i+1}(s,s')$. □

We will now generalize the equation systems that we have just introduced.

Let $X = \{x_1, \ldots, x_n\}$ be variables and let $K_{i,j}$ for $i \in \{0, \ldots, n\}$ and $j \in \{1, \ldots, n\}$ be subsets of A^*.

We define the mapping $\widehat{K}$ from $\mathcal{P}(A^*)^n$ to $\mathcal{P}(A^*)^n$ by

$$\widehat{K}(D_1, \ldots, D_n) = (D'_1, \ldots, D'_n) \iff D'_j = K_{0,j} \cup \bigcup_{i=1}^n K_{i,j} D_i.$$

This mapping is continuous and hence has a least fixpoint, $(D^\mu_1, \ldots, D^\mu_n)$.

It is easy to see that the equation system $\widehat{\mathcal{S}}$ associated with a transition system $\mathcal{S}$ is of this form.

Lemma 11.27 *Let g be the mapping from $\mathcal{P}(A^*)$ to $\mathcal{P}(A^*)$ defined by $g(D) = K_1 + K_2 D$. Its least fixpoint is $K_2^* K_1$.*

Proof. (Cf. also Exercise 3.11 where we showed that if $\varepsilon \notin L$ then $L^* M$ is the only fixpoint of $X = LX + M$.)

1. $g(K_2^* K_1) = K_1 + K_2 K_2^* K_1 = K_2^* K_1$, and thus $K_2^* K_1$ is a fixpoint of g.
2. Let D^μ be the least fixpoint of g. Then $D^\mu = K_1 + K_2 D^\mu$. We thus have

$$K_1 \subseteq D^\mu, \quad K_1 + K_2 K_1 \subseteq D^\mu,$$

and, by induction,

$$K_1 + K_2 K_1 + K_2^2 K_1 + \cdots + K_2^i K_1 \subseteq D^\mu.$$

Hence, $K_2^* K_1 \subseteq D^\mu$, and thus $K_2^* K_1$ is indeed the least fixpoint of g. □

Theorem 11.28 Let $\widehat{K}$ be defined as previously, and let $(D_1^\mu, \ldots, D_n^\mu)$ be its least fixpoint. If all the $K_{i,j}$s are regular, then each D_i^μ is also regular.

Proof. The proof is by induction on n. If $n = 1$ then $\widehat{K}(D) = K_1 + K_2 D$. Its least fixpoint is $K_2^* K_1$. If K_1 and K_2 are regular, then $K_2^* K_1$ is also regular.

Deducing the result for $n+1$ from the result for n is analogous to the elimination method for solving linear systems.

The required least fixpoint is the least solution of the following equation system:

$$x_1 = K_{0,1} + \bigcup_{i=1}^{n+1} K_{i,1} x_i$$

$$\vdots$$

$$x_n = K_{0,n} + \bigcup_{i=1}^{n+1} K_{i,n} x_i$$

$$x_{n+1} = K_{0,n+1} + \bigcup_{i=1}^{n+1} K_{i,n+1} x_i \;.$$

Let us solve the last equation:

$$x_{n+1} = K_{n+1,n+1}^* (K_{0,n+1} + \bigcup_{i=1}^{n} K_{i,n+1} x_i).$$

Substituting the value of x_{n+1} in the other equations:

$$x_1 = K_{0,1} + K_{n+1,n+1}^* K_{0,n+1} + \bigcup_{i=1}^{n} (K_{i,1} + K_{n+1,n+1}^* K_{i,n+1}) x_i$$

$$\vdots$$

$$x_n = K_{0,n} + K_{n+1,n+1}^* K_{0,n+1} + \bigcup_{i=1}^{n} (K_{i,n} + K_{n+1,n+1}^* K_{i,n+1}) x_i \;.$$

By the induction hypothesis, all the components of the least solution $(D_1, \ldots, D_n)$ of this system are regular.

Substituting each D_i in the definition of x_{n+1}, we have

$$x_{n+1} = K_{n+1,n+1}^* \left(K_{0,n+1} + \bigcup_{i=1}^{n} K_{i,n+1} D_i \right),$$

which is also regular.

We claim that this elimination method indeed computes the least solutions of the considered equations. □

Corollary 11.29 Let $(S, T, \alpha, \beta, \lambda)$ be a transition system. For any $Q, Q' \subseteq S$, $L_{Q,Q'}$ is a regular language.

EXAMPLE 11.30 Consider the finite-state automaton with transitions $(1, b, 1)$, $(1, a, 2)$, $(2, b, 2)$, $(2, a, 1)$ shown in Figure 11.3.

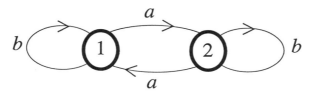

Figure 11.3

Consider the variables $x_{1,1}, x_{1,2}, x_{2,1}, x_{2,2}$. The equation system associated with this transition system is as follows:

$$x_{1,1} = \varepsilon + bx_{1,1} + ax_{2,1},$$
$$x_{2,1} = bx_{2,1} + ax_{1,1},$$
$$x_{1,2} = bx_{1,2} + ax_{2,2},$$
$$x_{2,2} = \varepsilon + bx_{2,2} + ax_{1,2}.$$

Compute $x_{2,2}$:
$$x_{2,2} = b^*(\varepsilon + ax_{1,2}) = b^* + b^*ax_{1,2}.$$

Substitute $x_{2,2}$ in the other equations:

$$x_{1,1} = \varepsilon + bx_{1,1} + ax_{2,1},$$
$$x_{2,1} = bx_{2,1} + ax_{1,1},$$
$$x_{1,2} = bx_{1,2} + ab^* + ab^*ax_{1,2} = (ab^*a + b)x_{1,2} + ab^*.$$

Compute $x_{1,2}$:
$$x_{1,2} = (ab^*a + b)^*ab^*.$$

Substitute $x_{1,2}$ in $x_{2,2}$:
$$x_{2,2} = b^* + b^*a(ab^*a + b)^*ab^* = (ab^*a + b)^*.$$

EXERCISE 11.21 Let L be the set of strings over the alphabet $\{a, b\}$ such that the number of bs is divisible by 3. There is a finite-state automaton $\mathcal{A}$ with three states such that $L = L(\mathcal{A})$. Give the equation system associated with the transitions, and solve this system for $L_{I,F}$ in order to give a regular expression denoting L. ◊

EXERCISE 11.22 Let L be the language over the alphabet $A = \{a, b\}$ consisting of all the strings without three consecutive as. Give the minimal complete deterministic finite-state automaton recognizing L and solve the corresponding equation system; deduce a regular expression for L. ◊

We will now prove the converse of Corollary 11.29.

Theorem 11.31 *If L is a regular language then there exists a finite-state automaton $\mathcal{A}$ such that $L = L(\mathcal{A})$.*

Proof.
1. If L consists of a single string, the theorem holds as follows. If $u = a_1 \cdots a_n$, consider the incomplete deterministic finite-state automaton whose states are $s_0, s_1, \ldots, s_n$, the unique initial state is s_0, the unique final state is s_n and the transitions are (s_{i-1}, a_i, s_i) for $1 \leq i \leq n$. If u is the empty string, we will let $n = 0$ and the automaton will have no transition.
2. If $L = L_1 + L_2$ then by the induction hypothesis there exist $\mathcal{A}_1$ and $\mathcal{A}_2$, with $\mathcal{A}_i = (S_i, T_i, I_i, F_i)$ such that $L_i = L(\mathcal{A}_i)$. We can assume that the states of $\mathcal{A}_1$ and of $\mathcal{A}_2$ are two disjoint sets. The finite-state automaton $\mathcal{A} = (S, T, I, F)$, with $S = S_1 \cup S_2$, $T = T_1 \cup T_2$, $I = I_1 \cup I_2$, and $F = F_1 \cup F_2$, recognizes the language $L(\mathcal{A}) = L(\mathcal{A}_1) \cup L(\mathcal{A}_2)$.
3. If $L = L_1 L_2$, there exist $\mathcal{A}_1$ and $\mathcal{A}_2$ such that $L_i = L(\mathcal{A}_i)$. Assume again that their states are two disjoint sets and let

$$S = S_1 \cup S_2,$$
$$T = T_1 \cup T_2 \cup \{(s, a, s') \,/\, s' \in I_2, \exists s'' \in F_1 : (s, a, s'') \in T_1\},$$
$$I = \begin{cases} I_1 & \text{if } I_1 \cap F_1 = \emptyset, \\ I_1 \cup I_2 & \text{otherwise,} \end{cases}$$
$$F = F_2.$$

Then $L(\mathcal{A}) = L(\mathcal{A}_1)L(\mathcal{A}_2)$. We show this as follows:

(a) Let $v \in L(\mathcal{A}_1)$ and $w \in L(\mathcal{A}_2)$. Since $w \in L(\mathcal{A}_2)$, there exists in $\mathcal{A}_2$ a path c' with trace w whose source s'_1 is in I_2 and whose target s'_2 is in $F_2 = F$.

(a.1) If $v = \varepsilon$, then $I_2 \cap F_1 \neq \emptyset$, and thus $I_2 \subseteq I$. The path c' is also a path of $\mathcal{A}$, and since $s'_1 \in I_2 \subseteq I$ and $s'_2 \in F_2 = F$, $vw = w \in L(\mathcal{A})$.

(a.2) If $v = v'a$, then there exists in $\mathcal{A}_1$ a path c with trace v', source $s_1 \in I_1$ and target s_2, and a transition $t = (s_2, a, s_3)$ with $s_3 \in F_1$. By the definition of T, there also exists a transition $t' = (s_2, a, s'_1)$. The path $ct'c'$ of $\mathcal{A}$ has trace $v'aw = vw$, source $s_1 \in I_1 \subseteq I$ and target $s'_2 \in F_2 = F$. Hence $vw \in L(\mathcal{A})$.

(b) Let $u \in L(\mathcal{A})$. There thus exists in $\mathcal{A}$ a path c with trace u, source $s_1 \in I$ and target $s'_2 \in F = F_2$.

(b.1) If $s_1 \in I_2$, then,
- on the one hand, c is a path of $\mathcal{A}_2$ and $u \in L(\mathcal{A}_2)$, and
- on the other hand, since $I_2 \cap I \neq \emptyset$, then, by the construction of I, $I_1 \cap F_1 \neq \emptyset$ and thus $\varepsilon \in L(\mathcal{A}_1)$.

We thus have $u = \varepsilon u \in L(\mathcal{A}_1)L(\mathcal{A}_2)$.

(b.2) If $s_1 \notin I_2$, then $s_1 \in I_1$, and the path c must be of the form $c_1(s_2, a, s_1')c_2$ where c_i ($i = 1, 2$) is a path of $\mathcal{A}_i$. By the construction of T, (s_2, a, s_1') is a transition of T if $s_1' \in I_2$ and $\exists s_3 \in F_1 : (s_2, a, s_3) \in T_1$. It follows that $u = vaw$, where v is the trace of c_1 and w the trace of c_2, and that $va \in L(\mathcal{A}_1)$ and $w \in L(\mathcal{A}_2)$.

4. Lastly, let $\mathcal{A} = (S, T, I, F)$ be a finite-state automaton. Since $L(\mathcal{A})^* = \{\varepsilon\} \cup L(\mathcal{A})^+$, in order to construct a finite-state automaton recognizing $L(\mathcal{A})^*$, it suffices to construct a finite-state automaton $\mathcal{A}'$ recognizing $L(\mathcal{A})^+$. To this end, it is enough to add to T the set of transitions

$$T'' = \{(s, a, s') \, / \, s' \in I \text{ and } \exists s'' \in F : (s, a, s'') \in T\}.$$

The proof is similar to the proof of point 2, in one direction by induction on the least integer i such that $u \in L(\mathcal{A})^i$, and in the other direction by induction on the number of transitions of T'' occurring in a path of $\mathcal{A}'$. □

CHAPTER 12

DISCRETE PROBABILITIES

Probability theory is a fundamental tool in many domains and, in particular computer science, where its main use is of course the average case analysis of algorithms. Combinatorics allow us to count the number of elements of a set, to count the number of operations of an algorithm on given data. However, when the set or the data are subject to change, the number of elements or operations can also change, and probabilities come into the picture to study average values, the deviation with respect to these average values, etc. Probabilities are also used in other areas in computer science, e.g.

- probabilistic algorithms: a random choice can be on average, or even almost always, more efficient than computing the exact choice,
- modelling and simulation,
- queueing theory: for instance, study the average waiting time for accessing the nerve-centres of a network,
- signal processing,
- probabilistic arguments sometimes allow us to prove properties of algorithms which are not provable otherwise.

This chapter defines the notions of discrete probability distribution, conditional probability distribution and independance, and random variables. The chapter reviews Bayes's rule and some applications, the weak law of large numbers, how to use generating series to study random variables and the main properties of the most common probability distributions.

We strongly recommend the following classic handbook:

William Feller, *Probability Theory*, Vol. 1, John Wiley, New York (1968).

12.1 Generalities

12.1.1 Terminology

Probability theory studies randomness; e.g.:

- the outcome of a coin-tossing game,
- the number of daily calls through a telephone switchboard,
- the waiting time for accessing a network,
- the length of the life span (without breakdowns) of an operating system.

Probability theory describes a mathematical model for such random experiments, where 'random' means 'depending on chance'.

12.1.2 Sample spaces

The first basic notion is the notion of *trial, or sample point or observation*: a trial is the outcome of a random experiment; such an outcome is denoted by ω and the set of all possible trials is traditionally denoted by Ω and called the *sample space* i.e.: each outcome obtained in a possible experiment will correspond to a unique element ω in Ω. The fulfillment of the random experiment is thus reduced to the random choice of ω in Ω. Of course, different experiments will lead to different sets Ω.

REMARK 12.1 Note that the same 'experiments' in the intuitive sense may correspond to different random experiments according to what we are interested in. For instance, assuming a coin-tossing game with a dime and a quarter, and assuming that the coins never fall on their edges:

- If we are interested in the outcome Heads or Tails, the sample space will be the set of possible outcomes, i.e. $\Omega = \{HH, HT, TH, TT\}$.
- If we are interested in the number of Heads, the sample space will be $\Omega = \{0, 1, 2\}$.
- If we are interested in the concordance of the outcomes, i.e. in the value of the predicate 'both coins fell on the same side', we will have $\Omega = \{$Agreement, Disagreement$\}$.

EXAMPLE 12.2
1. Consider the random experiment consisting of rolling a pair of dice (we will often use this example); the possible outcomes are pairs of integers $\omega = (x, y)$, with $1 \leq x, y \leq 6$; the sample space is thus the set $\Omega = \{1, 2, \ldots, 6\}^2$.
2. In the case of the random experiment consisting of counting the number of daily calls through a telephone switchboard, the possible outcomes are: $\{1, 2, \ldots, n, \ldots\}$. The sample space will thus be $\Omega = \mathbb{N}$.

12.1.3 Events

The second basic notion is the notion of *event*, i.e. an event whose fulfillment depends upon the outcome of a random experiment. We will thus represent an event A by the set of outcomes of the experiments fulfilling it, and we will identify

$$A = \{\omega \;/\; A \text{ occurs if } \omega \text{ is the outcome of the experiment}\}.$$

EXAMPLE 12.3
1. When throwing two dice, event A: 'the total score is ≥ 10' is represented in $\Omega = \{1,\ldots,6\}^2$ by the set of pairs (x,y) such that $x + y \geq 10$, i.e.

$$A = \{(4,6),(5,6),(6,6),(5,5),(6,5),(6,4)\}.$$

2. When counting the number of phone calls, event A: 'the switchboard can put through at most 5000 daily calls' is represented by $A = \{n \in \mathbb{N} / n \leq 5000\}$.

12.1.4 Relations among events

We briefly describe the logical operations which can be performed on events; events being represented by subsets of the sample space Ω, these logical operations will correspond to the usual set-theoretical operations, up to the terminology.

Each event A is associated with its *complementary event*, denoted by A^c (or $\neg A$, or $\overline{A}$): event A^c occurs if and only if A does not occur, and is represented in Ω by the complement of the subset representing A.

EXAMPLE 12.4 In throwing two dice, the event 'the total score is less than 10' is represented in $\Omega = \{1,\ldots,6\}^2$ by the 30 ($= 36 - 6$) pairs (x,y) such that $x+y < 10$, which is the complement of the set A representing the event 'the total score is ≥ 10'.

For each pair of events A_1 and A_2, event 'A_1 and A_2' (resp. 'A_1 or A_2') occurs if and only if both A_1 and A_2 occur (resp. either A_1 or A_2 or both occur). In the space Ω, A_1 and A_2 (resp. A_1 or A_2) is represented by the intersection (resp. the union) of A_1 and A_2. Subsequently, we will denote A_1 and A_2 (resp. A_1 or A_2) by $A_1 \cap A_2$ (resp. $A_1 \cup A_2$). *The impossible event*, denoted by $\emptyset$, is represented by the empty set in Ω. Two events A_1 and A_2 are said to be *mutually exclusive* if and only if $A_1 \cap A_2 = \emptyset$; if A_1 and A_2 are mutually exclusive, then A_1 and A_2 is impossible, and A_1 or A_2 is denoted by $A_1 + A_2$ (disjoint union). Events $\{A_n \;/\; n \in \mathbb{N}\}$ are said to be *mutually exclusive* if they are pairwise mutually exclusive. The *sure event*, which always occurs whatever the outcome of the experiment, is represented by Ω. Finally, event A is said to *imply* event B if, whenever A occurs, B also occurs, i.e. the subset of Ω representing A is contained in the subset of Ω representing B.

Probability spaces

EXAMPLE 12.5 Consider again the tossing of two dice. Define the two events A_1 'the total is ≥ 10', and A_2 'the total is < 10'. These two events are mutually exclusive and event 'A_1 or A_2' is sure. (In general, if two events A_1 and A_2 are mutually exclusive, the event 'A_1 or A_2' is not sure.)

A set $\{A_n \ / \ n \in \mathbb{N}\}$ of mutually exclusive events such that $\Omega = \cup_{n \in \mathbb{N}} A_n$ is said to be a *partition* of Ω. Summing up, the correspondence between probabilistic and set-theoretic terminologies is shown in the following table:

Probabilistic terminology	Set-theoretic terminology	Notation
Sure event	Whole space	Ω
Impossible event	Empty set	$\emptyset$
Complementary event	Complementary subset	A^c ($\overline{A}$, $\neg A$)
And	Intersection	$\cap$
Or	Union	$\cup$
Mutually exclusive events	Disjoint subsets	$A_1 \cap A_2 = \emptyset$
Partition	Partition	$\sum A_i = \Omega$, with $A_i \cap A_j = \emptyset$
Implication	Inclusion	$A \subseteq B$

12.2 Probability spaces

12.2.1 Probability space

We have seen that a random experiment can be described by a set Ω (the sample space), and a class $\mathcal{T}$ of subsets of Ω (the subsets representing events). When Ω is not denumerable, $\mathcal{T}$ cannot be the set $\mathcal{P}(\Omega)$ of all subsets of Ω: $\mathcal{P}(\Omega)$ will indeed contain too many pathological subsets which cannot represent natural events. In order to properly represent the notion of event a class $\mathcal{T}$ of subsets should, 1. contain the sure event Ω, 2. be stable by complement, 3. be stable by denumerable unions and intersections, which gives:

Definition 12.6 *A class $\mathcal{T}$ of subsets of Ω is said to be a tribe if it satisfies*

(i) $\Omega \in \mathcal{T}$,
(ii) $A \in \mathcal{T} \implies A^c \in \mathcal{T}$,
(iii) $\forall n \in \mathbb{N}, \ A_n \in \mathcal{T} \implies \bigcup_{n \in \mathbb{N}} A_n \in \mathcal{T}$.

Condition (iii) *is called the σ-additivity principle, or the additivity principle if* A_n, $n = 1, \ldots, p$ *is finite. A tribe is also called a σ-algebra, or a σ-additive class or a countably additive class.*

REMARK 12.7 $\mathcal{T} = \mathcal{P}(\Omega)$ always satisfies conditions (i)–(iii); often, for a finite or denumerable Ω, we will have $\mathcal{T} = \mathcal{P}(\Omega)$.

REMARK 12.8 Since $\mathcal{T}$ is stable by complementation and denumerable union, it also is stable by denumerable intersection, indeed

$$\bigcap_{n \in \mathbb{N}} A_n = \left(\bigcup_{n \in \mathbb{N}} A_n^c \right)^c.$$

A pair $(\Omega, \mathcal{T})$ formed by a set and a tribe on this set gives a model describing random experiments and the corresponding events. However, the goal of probability theory is not simply to find a descriptive model, but mainly to supply a tool for computing quantitative evaluations. That is what we will now do.

The notion of *probability* corresponds intuitively to the notion of 'likelihood or frequency of realization' of an event A during a sequence of repetitions of a random experiment in which A may (or may not) occur. If N is the number of repetitions of a random experiment and N_A the number of realizations of A during a sequence of N experiments, the ratio N_A/N tends to become stable when N grows. The probability of A is the limit of the ratio N_A/N when $N \to \infty$. (This notion will appear again in the law of large numbers.) Some properties are satisfied by the mapping $A \mapsto N_A/N$:

(i) $0 \leq N_A/N \leq 1$,
(ii) if $A_1, \ldots, A_N$ are mutually exclusive,

$$N_{A_1 \cup \cdots \cup A_n}/N = N_{A_1}/N + \cdots + N_{A_n}/N.$$

Hence the following definition.

Definition 12.9
1. Let $\mathcal{T}$ be a tribe in Ω. A probability distribution P on $(\Omega, \mathcal{T})$ is a mapping $P: \mathcal{T} \longrightarrow [0,1]$ such that
(i) $\forall A \in \mathcal{T}$, $0 \leq P(A) \leq 1$,
(ii) $P(\Omega) = 1$,
(iii) if $(A_n)_{n \in \mathbb{N}}$ is a family such that: for all i, j in $\mathbb{N}$, $A_i, A_j \in \mathcal{T}$ and $A_i \cap A_j = \emptyset$, then

$$P\left(\bigcup_{n \in \mathbb{N}} A_n \right) = \sum_{n \in \mathbb{N}} P(A_n).$$

2. When Ω is denumerable and $\mathcal{T} = \mathcal{P}(\Omega)$, one can substitute for the above conditions $\forall \omega \in \Omega$, $0 \leq P(\omega) \leq 1$, $P(A) = \sum_{\omega \in A} P(\omega)$ and $\sum_{\omega \in \Omega} P(\omega) = 1$, where $P(\omega)$ is an abbreviation for $P(\{\omega\})$.

EXERCISE 12.1 Verify that in case 2 of the preceding definition, the mapping $P \colon \mathcal{P}(\Omega) \longrightarrow [0,1]$ defined by $P(A) = \sum_{\omega \in A} P(\omega)$ is a probability distribution. ◇

If Ω is a finite space and $\mathcal{T} = \mathcal{P}(\Omega)$, the *uniform distribution* on Ω is often considered: it gives each elementary event ω the same probability, $p = \dfrac{1}{|\Omega|}$, since $\sum_{\omega \in \Omega} P(\omega) = \sum_{\omega \in \Omega} p = |\Omega| p = 1$ must hold. Each sample point ω is said to be equally probable, and the uniform distribution corresponds to the intuitive notion of 'random choice'. We then have, for all events A, $P(A) = \dfrac{|A|}{|\Omega|}$ (the ratio of the number of favourable cases to the total number of cases).

EXAMPLE 12.10 Consider again the tossing of two dice, and endow the sample space Ω with the uniform distribution. Each possible outcome $\omega = (x,y)$ has probability $1/36$, and event $A =$ 'the total is ≥ 10' has probability $P(A) = 6/36 = 1/6$.

If Ω is an infinite space there can no longer be a uniform distribution.

In all cases, Ω finite or infinite, an event A is said to be

- *almost sure* if it occurs *almost surely*, i.e. $P(A) = 1$,
- *almost impossible* if it almost surely will not occur, i.e. $P(A) = 0$.

EXAMPLE 12.11 Consider the random experiment consisting of tossing a coin till Tails turns up, then stopping. The space Ω is described by the sequence of outcomes, where Tails (resp. Heads) is abbreviated in T (resp. H).

$$\Omega = \{T, HT, HHT, \ldots, H^n T, \ldots, HHH \cdots\} = \Omega' \cup \{HHH \cdots\}.$$

Assume that, throwing the coin once, we have $P(T) = p$ and $P(H) = q = 1 - p$, with $0 < p < 1$. Then $P(H^n T) = q^n p$ (this will be formally proved later on), and

$$\sum_{\omega \in \Omega'} P(\omega) = \sum_{n \in \mathbb{N}} q^n p = p \sum_{n \in \mathbb{N}} q^n = \frac{p}{1-q} = 1.$$

The event $A = HHH \cdots$, i.e. an infinite sequence of Heads as outcome, is a logically possible event, corresponding to a non-empty subset of $\mathcal{P}(\Omega)$, but it is almost impossible since $P(A) = 0$.

Let us conclude with some useful formulas

Proposition 12.12
(i) For $A_1, \ldots, A_n$ mutually exclusive events,

$$P(A_1 \cup \cdots \cup A_n) = \sum_{i=1}^{n} P(A_i).$$

(ii) $\forall A, B, \quad P(A \cup B) = P(A) + P(B) - P(A \cap B)$.
(iii) $\forall A, \quad P(A^c) = 1 - P(A)$.
(iv) $\forall A, B, \quad A \subseteq B \implies P(A) \leq P(B)$.
(v) $\forall A_1, \ldots, A_n, \quad P(\cup_{i=1}^{n} A_i) \leq \sum_{i=1}^{n} P(A_i)$.
(vi) Let $(A_n)_{n \in \mathbb{N}}$ be a monotone increasing sequence, i.e. $\forall n$, $A_n \subseteq A_{n+1}$, then $P(\cup_{n \in \mathbb{N}} A_n) = \lim_{n \to \infty} P(A_n)$.
(vii) Let $(A_n)_{n \in \mathbb{N}}$ be a monotone decreasing sequence, i.e. $\forall n$, $A_{n+1} \subseteq A_n$, then $P(\cap_{n \in \mathbb{N}} A_n) = \lim_{n \to \infty} P(A_n)$.

The proof is straightforward.

Proposition 12.13 Let $A_1, \ldots, A_n$ be events, we have

$$P(\cup_{i=1}^{n} A_i) = \sum_{k=1}^{n} (-1)^{k-1} \sum_{i_1 < \cdots < i_k} P(A_{i_1} \cap \cdots \cap A_{i_k}).$$

Proof. This result is proved in the same way as Proposition 6.13 (counting techniques). □

EXERCISE 12.2 Two (not so good) snipers A and B take aim at a target. A hits the target with probability 1/4 and B hits the target with probability 2/5. What is the probability that the target is hit? ◊

12.3 Conditional probabilities and independent events

12.3.1 Intuition

Let us start with an introductory example. Consider a space Ω consisting of a population of $N = |\Omega|$ people, with uniform distribution; the corresponding experiment is the 'random' choice of an individual. Among the N people, let H be the subset of men and D be the subset of colourblind people. Let D_e (resp. H_e) be the event: 'the chosen person is colourblind (resp. male)'; we have

$$P(D_e) = \frac{|D|}{|\Omega|} \quad \text{and} \quad P(H_e) = \frac{|H|}{|\Omega|}.$$

Conditional probabilities and independent events   233

Assuming now that we are only interested in the subpopulation of males, and that we randomly choose one man, the probability that he is colourblind is $\frac{|H \cap D|}{|H|}$, i.e. the probability of being colourblind for men. It is denoted by $P(D_e/H_e)$, and we say: 'probability of D_e assuming (or given, or on the hypothesis) H_e'. In the present example, we have $P(D_e/H_e) = \frac{P(D_e \cap H_e)}{P(H_e)}$ because $\frac{|H \cap D|}{|H|} = \left(\frac{|H \cap D|}{|\Omega|}\right) / \left(\frac{|H|}{|\Omega|}\right)$.

We will see that this situation can be generalized to any subpopulation defined by an arbitrary event A. We can always consider a probability distribution restricted to the set of sample points satisfying event A; we then say the '*conditional probability given A*'.

12.3.2 Definitions

Definition 12.14 *Let $(\Omega, \mathcal{T}, P)$ be a probability space and A an event having a positive probability; the conditional probability of event B given A is defined by*

$$P(B/A) = \frac{P(A \cap B)}{P(A)}. \tag{12.1}$$

Proposition 12.15 *The mapping $B \mapsto P(B/A)$ is a probability distribution on the space $(\Omega, \mathcal{T})$, for each choice of the conditioning event A.*

Proof. We check that

(i) $0 \leq P(B/A) \leq 1$, since $P(A \cap B) \leq P(A)$,

(ii) $P(\Omega/A) = \frac{P(\Omega \cap A)}{P(A)} = \frac{P(A)}{P(A)} = 1$,

(iii) for each family $(B_i)_{i \in \mathbb{N}}$ of mutually disjoint subsets, the family $(B_i \cap A)_{i \in \mathbb{N}}$ is also formed of mutually disjoint subsets, thus:

$$P((\cup_{i \in \mathbb{N}} B_i)/A) = \frac{P(\cup_{i \in \mathbb{N}} (B_i \cap A))}{P(A)} = \sum_{i \in \mathbb{N}} \frac{P(B_i \cap A)}{P(A)}$$

$$= \sum_{i \in \mathbb{N}} P(B_i/A). \qquad \square$$

REMARK 12.16
(1) Equation (12.1) is often used in the form $P(A \cap B) = P(B/A)P(A)$.
(2) Let $(\Omega, \mathcal{P}(\Omega))$ be a finite space together with the uniform distribution, and $A \subseteq \Omega$ be such that $|A| \neq 0$, then $P(B/A) = \frac{|B \cap A|}{|A|}$, i.e. the uniform distribution restricted to A.

EXAMPLE 12.17 We again return to Example 12.11: toss a coin till the first Tails occurs. Let B be the event : 'the first Tails occurs after at least three tosses', and A: 'Tails does not occur at the first toss'; note that $B \subseteq A$. Recall that $p = P(T)$ and $q = P(H)$, where Tails (resp. Heads) is abbreviated to T (resp. H). We have

$$P(B) = \sum_{n \geq 2} q^n p = q^2 p \frac{1}{1-q} \quad \text{and} \quad P(A) = \sum_{n \geq 1} q^n p = qp \frac{1}{1-q},$$

hence

$$P(B/A) = \frac{P(A \cap B)}{P(A)} = \frac{P(B)}{P(A)} = q.$$

EXAMPLE 12.18 Consider families with two children. Letting F (resp. M) stand for girl (resp. boy), and denoting the children by decreasing age, the sample space is $\Omega = \{MM, MF, FM, FF\}$; Ω is endowed with the uniform distribution. Let A be the event 'there is at least one boy' and B the event 'there are two boys'. Assuming there is at least one boy, we compute the probability that there are two boys: $A = \{MM, MF, FM\}$ and $B = \{MM\} = A \cap B$, hence $P(B/A) = 1/3$.

Equation (12.1) can be generalized as follows

Proposition 12.19 (generalization) Let $A_1, \ldots, A_n$ be arbitrary events in $(\Omega, \mathcal{T})$, then

$$P(A_1 \cap \cdots \cap A_n) = P(A_1) P(A_2/A_1) P(A_3/A_1 \cap A_2) \cdots P(A_n/A_1 \cap \cdots \cap A_{n-1}),$$

provided that we define the right-hand side to be 0 as soon as one of its factors is 0.

Proof. By induction on n;

- Basis: for $n = 2$ we come across the definition of conditional probabilities when $P(A_1) \neq 0$, and the result is clear when $P(A_1) = 0$.
- Inductive step: assuming the result true for n, let us prove it for $n + 1$:
  (i) $P(A_1 \cap \cdots \cap A_{n+1}) > 0$, then we also have $P(A_1 \cap \cdots \cap A_n) > 0$, and thus by the definition of conditional probabilities

$$P(A_1 \cap \cdots \cap A_{n+1}) = P(A_1 \cap \cdots \cap A_n) P(A_{n+1}/A_1 \cap \cdots \cap A_n)$$

by the induction hypothesis we can replace $P(A_1 \cap \cdots \cap A_n)$ by $P(A_1)$ $P(A_2/A_1) P(A_3/A_1 \cap A_2) \cdots P(A_n/A_1 \cap \cdots \cap A_{n-1})$, hence the result.

Conditional probabilities and independent events   235

(ii) $P(A_1 \cap \cdots \cap A_{n+1}) = 0$. Note that the sequence

$$p_i = P(A_1 \cap \cdots \cap A_i)$$

is monotone decreasing; hence

(ii.1) either $P(A_1) \neq 0$, and $\exists i \in \{2,\ldots,n+1\}$, $P(A_1 \cap \cdots \cap A_{i-1}) > 0$ and $P(A_1 \cap \cdots \cap A_i) = 0$; but then $P(A_i/A_1 \cap \cdots \cap A_{i-1}) = 0$ by the definition of conditional probabilities, and the right-hand side is thus 0.

(ii.2) or $P(A_1) = 0$, and both sides are 0. □

EXERCISE 12.3 An urn contains b black balls and r red balls; k random drawings are performed, $k < r$, with the following rule: if a black ball is drawn it is replaced, if a red ball is drawn c black balls are added. What is the probability of drawing k red balls? ◇

EXERCISE 12.4 Message passing. Consider passing a message 'yes' or 'no' in a population. Each person passes the message he/she received with probability p and the opposite message with probability $q = 1 - p$. Let X_n be the message received by the nth individual I_n. Assume that I_0 passes 'yes' to I_1. What is the probability that I_n receives 'yes'? ◇

EXAMPLE 12.20 (Samplings without replacement) An urn contains $N > 2$ balls, b black balls and r red balls, $b + r = N$. Two balls are drawn and not replaced. What is the probability of drawing two red balls? Let C be the event: 'two red balls were drawn', A_1 be the event: 'the first drawn ball is red' and A_2: 'the second ball is red'. We clearly have $P(A_1) = \dfrac{r}{N}$, $P(A_2/A_1) = \dfrac{r-1}{N-1}$ and

$$P(C) = P(A_1 \cap A_2) = P(A_1)P(A_2/A_1) = \frac{r \times (r-1)}{N \times (N-1)}.$$

This should be likened to the notion of arrangement without repetitions of Chapter 6.

12.3.3 Bayes's rule

It is frequently easier to evaluate conditional probabilities for an event B than to compute its probability $P(B)$. This happens if e.g. the experiment upon which event B depends can be split into partial successive experiments, and if the probability of each partial experiment depends on the outcomes of preceding experiments; see Exercise 12.4 and Exercise 12.6 (probability of causes). There are tools enabling us to deduce $P(B)$ and $P(A_n/B)$ from the $P(B/A_n)$s, and that is what we will now show.

Lemma 12.21 (Total probabilities) *Let $(\Omega, \mathcal{T}, \mathcal{P})$ be a probability space, and let $(A_k)_{k \in \mathbb{N}}$ be a partition; for any event B, we have*

$$P(B) = \sum_{k \in \mathbb{N}} P(B/A_k) P(A_k).$$

Proof. Since $(A_k)_{k \in \mathbb{N}}$ forms a partition, we have $\cup_{k \in \mathbb{N}} A_k = \Omega$; then

$$P(B) = P(B \cap \Omega) = P(B \cap (\cup_{k \in \mathbb{N}} A_k))$$
$$= \sum_{k \in \mathbb{N}} P(B \cap A_k) \quad \text{(since the } B \cap A_k\text{s are pairwise disjoint)}$$
$$= \sum_{k \in \mathbb{N}} P(B/A_k) P(A_k). \qquad \square$$

This rule can be extended to the case when some $P(A_k)$s are 0, by attributing an arbitrary value to $P(B/A_k)$ and letting $P(B/A_k)P(A_k) = 0$.

EXERCISE 12.5 (Polya's urn model) An urn contains b black balls and r red balls. A ball is randomly drawn. It is replaced, and, moreover, c balls of the same colour are added. A new random drawing is performed and the whole procedure is iterated. Let $X^n \in \{B, R\} = \Omega$ be the colour of the ball drawn at the nth drawing. Prove by induction on n that $P(X^n = B) = b/(b+r)$ for all n. $\diamond$

The next theorem is known as Bayes's rule for the probability of causes.

Theorem 12.22 *Letting $(\Omega, \mathcal{T}, P)$ be a probability space, $(A_n)_{n \in \mathbb{N}}$ a partition and B an event with positive probability, we have*

$$P(A_n/B) = \frac{P(A_n) P(B/A_n)}{\sum_{k \in \mathbb{N}} P(A_k) P(B/A_k)}.$$

Proof. By the definition of conditional probabilities

$$P(A_n/B) = \frac{P(A_n \cap B)}{P(B)} = \frac{P(A_n) P(B/A_n)}{P(B)}$$

and, by the preceding lemma, $P(B) = \sum_{k \in \mathbb{N}} P(A_k) P(B/A_k)$, hence

$$P(A_n/B) = \frac{P(A_n) P(B/A_n)}{\sum_{k \in \mathbb{N}} P(A_k) P(B/A_k)}. \qquad \square$$

EXERCISE 12.6 (Probability of causes) Let U_i, for $i = 1, 2$, be two urns containing r_i red balls and b_i black balls. One urn is randomly chosen, and a ball is drawn from it. Assuming a red ball was drawn, what is the probability that it comes from urn 1? $\diamond$

Conditional probabilities and independent events

EXERCISE 12.7 Let A and B be two machines producing, respectively, 100 and 200 objects. A (resp. B) produces 5% (resp. 6%) of flawed objects. Given a flawed object, what is the probability that it was manufactured by A? ◇

EXERCISE 12.8 A city is divided into three areas. During a poll, the repartition of votes for candidate C was as shown in the following table.

Area number	Weight of the area in voters (%)	Score of C in the area (%)
1	30	40
2	50	48
3	20	60

1. What is the probability that a randomly chosen voter voted for C?
2. Given that e voted for C, what is the probability that e is from area 3? ◇

12.3.4 Independent events

Definition 12.23 *Two events A and B in $(\Omega, \mathcal{T}, P)$ are independent if and only if they satisfy the following equivalent conditions:*

(i) $P(B/A) = P(B)$,
(ii) $P(A/B) = P(A)$,
(iii) $P(A \cap B) = P(A)P(B)$.

That these three conditions are equivalent follows immediately from the equality $P(A \cap B) = P(B/A)P(A) = P(A/B)P(B)$.

Proposition 12.24
1. *If A and B are independent, the pairs of events (A, B^c), (A^c, B) and (A^c, B^c) are also independent.*
2. *If A and B are independent, we have:*
$$P(A \cup B) = P(A) + P(B) - P(A)P(B).$$

Proof. Check, for instance, that (A^c, B) and (A^c, B^c) are independent. We have

$$P(A^c \cap B) + P(A \cap B) = P(B)$$
$$P(A^c \cap B) = P(B) - P(A \cap B) = P(B) - P(A)P(B)$$
$$= P(B)(1 - P(A)) = P(B)P(A^c);$$
$$P(A^c \cap B^c) = 1 - P(A \cup B) = 1 - P(A) - P(B) + P(A)P(B)$$
$$= P(A^c) - P(B)(1 - P(A)) = P(A^c)(1 - P(B))$$
$$= P(A^c)P(B^c). \qquad \square$$

EXAMPLE 12.25 Return to Example 12.2 and the tossing of two dice.

(i) Events A = 'the first die is a one', and B = 'the second die is even', are independent since $P(A) = 1/6$, $P(B) = 1/2$ and $P(A \cap B) = 1/12 = P(A)P(B)$. Note that A and B are not disjoint, since $A \cap B = \{(1,2),(1,4),(1,6)\} \neq \emptyset$.
(ii) Events A = 'the first die is a one', and B = 'the first die is even' are disjoint, but not independent since $P(A) = 1/6$, $P(B) = 1/2$ and $P(A \cap B) = 0 \neq P(A)P(B)$. In general, disjoint events are never independent, unless one of them has probability 0.

Z Remember from this example that *the notions of disjoint events and of independent events are orthogonal.*

EXERCISE 12.9 Consider events A = 'there is at most one girl in a family', and B = 'there are both boys and girls'; check that events A and B are independent if the underlying space Ω_3 is the set of families with three children, but A and B are not independent in the space Ω_2 of families with two children; Ω_2 and Ω_3 are endowed with the corresponding uniform probabilities. ◇

EXERCISE 12.10 A poll among the students on a course gave the following outcomes:
(a) 2/3 of the students say they like maths, and among them:
  • 70% would rather take an exam without notes,
  • 20% are in favour of prohibiting smoking.
(b) 1/3 of the students say they do not like maths, and among them:
  • 20% would rather take an exam without notes,
  • 90% are in favour of prohibiting smoking.

Let M, D, F be the events:
  • M = liking maths,
  • D = in favour of taking an exam without notes,
  • F = in favour of prohibiting smoking,
and $\overline{M}, \overline{D}, \overline{F}$ the complementary events.

Given that a student likes maths, events D and F are independent. Similarly, given that a student does not like maths, events D and F are independent.
1. Compute $P(M)$, $P(D/M)$, $P(F/M)$, $P(\overline{M})$, $P(\overline{D}/M)$, $P(\overline{F}/M)$, $P(\overline{D}/\overline{M})$, $P(D/\overline{M})$, $P(\overline{F}/\overline{M})$, $P(F/\overline{M})$.
2. Given that student e would rather take an exam without notes and is in favour of prohibiting smoking, what is the probability that this student likes maths? ◇

One can generalize the notion of independent events to sets of events.

Definition 12.26 Let $A_1, \ldots, A_n$ be events in $(\Omega, \mathcal{T}, P)$. $A_1, \ldots, A_n$ are *mutually independent* if and only if they satisfy the following equivalent conditions

(i) For all combinations $1 \leq i_1 < \cdots < i_k \leq n$ we have

$$P(A_{i_1} \cap \cdots \cap A_{i_k}) = P(A_{i_1}) \cdots P(A_{i_k}).$$

Conditional probabilities and independent events 239

(ii) *For $1 \leq k \leq n-1$, and for all combinations $(i, i_1, \ldots, i_k)$ of $k+1$ pairwise distinct integers among $\{1, \ldots, n\}$ we have*

$$P(A_i/A_{i_1} \cap \cdots \cap A_{i_k}) = P(A_i).$$

REMARK 12.27 It is not enough to check that the A_i are pairwise independent, nor that $P(A_1 \cap \cdots \cap A_n) = P(A_1) \cdots P(A_n)$. Consider, for instance, the experiment consisting of throwing a die and looking at the outcome; $\Omega = \{1, 2, \ldots, 6\}$ with the uniform probability. Let A be 'the outcome is ≤ 3', B be 'the outcome is even' and C be 'the outcome is not divisible by 3'. $P(A) = P(B) = 1/2$, $P(C) = 2/3$. $A \cap B \cap C = \{2\}$, hence $P(A \cap B \cap C) = 1/6 = P(A)P(B)P(C)$; but A, B, C are not independent since $P(A \cap B) = 1/6 \neq P(A)P(B)$.

EXERCISE 12.11 Find an example of three events A, B, C, pairwise independent, such that $P(A \cap B \cap C) \neq P(A)P(B)P(C)$. ◇

EXERCISE 12.12 Let the set Ω consist of the eight vertices of a cube constructed on the three unit vectors of origin 0 in a Cartesian space. Each vertex is identified by its three coordinates $i, j, k \in \{0, 1\}$. The probability distribution is defined by

$$P(\omega) = \begin{cases} 0 & \text{if } i+j+k \text{ is even,} \\ \dfrac{1}{4} & \text{otherwise.} \end{cases}$$

1. Prove that this indeed defines a probability distribution.
2. Show that the three events: $A = \{i = 0\}$, $B = \{j = 0\}$, $C = \{k = 0\}$ are pairwise independent.
3. Are events $A \cap B$ and C independent? ◇

12.3.5 Product spaces

Product probability distributions correspond to the notion of repeated random experiments or of a combination of several random experiments. Consider a sequence $(\omega_1, \ldots, \omega_n)$ of random experiments, $\omega_i \in \Omega_i$, for $i = 1, \ldots, n$, and study the global experiment $\omega = (\omega_1, \ldots, \omega_n)$.

⚡ ω ranges over $\Omega = \Omega_1 \times \cdots \times \Omega_n$. It is natural to endow Ω with a tribe $\mathcal{T}$ of subsets such that all events A consisting of events $A_1, \ldots, A_n$ are in $\mathcal{T}$; however, this seemingly simplest choice for $\mathcal{T}$ which is the set-theoretical product of the $\mathcal{T}_i$ is not suitable, because it is *not* a tribe. We have to choose the least possible tribe $\mathcal{T}$ which is *generated* by this product and which we will denote by $\mathcal{T} = \prod_{i=1}^{n} \mathcal{T}_i$. Lastly, we will define a probability distribution P on $(\Omega, \mathcal{T})$.

Definition 12.28 Let $(\Omega_i, \mathcal{T}_i, P_i)$ be denumerable probability spaces with $\mathcal{T}_i = \mathcal{P}(\Omega_i)$ for $i = 1, \ldots, n$. The product space $(\Omega, \mathcal{T}, P)$ is defined by

- $\Omega = \prod_{i=n}^{n} \Omega_i$,
- $\mathcal{T} = \prod_{i=1}^{n} \mathcal{T}_i$, and thus $\mathcal{T}$ is the least tribe containing $A_1 \times \cdots \times A_n$ for all $A_i \subseteq \Omega_i$, $i = 1, \ldots, n$,
- $P(\omega_1, \ldots, \omega_n) = P_1(\omega_1) \times \cdots \times P_n(\omega_n)$, $\forall (\omega_1, \ldots, \omega_n) \in \Omega$.

EXERCISE 12.13 Prove that the above-defined P is indeed a probability distribution. ◇

The same definition can be given for the product of an infinite number of sample spaces.

This definition implicitly assumes that the sequence of represented trials $(\omega_1, \ldots, \omega_n)$ are *independent*, and that is by far the most frequent case in life (repetitions of experiments, sampling populations, etc...). Without the independence hypothesis, the only condition that can be demanded of P is that $\forall A_i \in \mathcal{T}_i$,

$$P(\Omega_1 \times \cdots \times \Omega_{i-1} \times A_i \times \Omega_{i+1} \times \cdots \times \Omega_n) = P_i(A_i) \quad (12.2)$$

and this condition is not enough to determine P, as shown by the next example.

EXAMPLE 12.29 Let $(\Omega_i, \mathcal{T}_i, P_i)$, $i = 1, 2$ be defined by $\Omega_1 = \Omega_2 = \{0, 1\}$, $\mathcal{T}_1 = \mathcal{T}_2 = \mathcal{P}(\Omega_1)$, and let $P_1 = P_2$ be defined by $P_1(0) = P_2(0) = P_1(1) = P_2(1) = 1/2$. We can define two probability distributions P and Q on $(\Omega_1 \times \Omega_2, \mathcal{T}_1 \times \mathcal{T}_2)$

$$P((0,1)) = P((1,0)) = P((0,0)) = P((1,1)) = 1/4,$$

$$Q((1,0)) = Q((0,1)) = 1/2, \quad Q((0,0)) = Q((1,1)) = 0,$$

P and Q both satisfy (12.2); P is a product distribution corresponding to two independent trials.

EXAMPLE 12.30 (Repeated independent trials. Bernoulli trials) Let $(\Omega, \mathcal{T}, P)$ be the sample space associated with an experiment. Repeat this experiment n times, independently, i.e. the outcome of the ith experiment does not depend upon experiment number j, $j \neq i$. Then, the probability distribution associated with the n successive experiments will be described by the product $(\Omega^n, \mathcal{T}^n, P^n)$. For instance, consider n successive tosses of a coin. We represent Tails by 0 and Heads by 1, and identify {Tails, Heads} by $\mathbb{B}$; with these conventions, the sample space corresponding to the coin-tossing game is $\mathbb{B} = \{0, 1\}$, $\mathcal{T} = \mathcal{P}(\mathbb{B})$, and $P(0) = P(\text{Tails}) = p$, $P(1) = P(\text{Heads}) = q$ with $p + q = 1$. The space corresponding to n successive tosses is $(\mathbb{B}^n, \mathcal{P}(\mathbb{B}^n), P)$ where P is defined for

$\omega \in \mathbb{B}^n$ by $P(\omega) = p^k q^{n-k}$ if $|\omega|_0 = k$ and $|\omega|_1 = n - k$, and where $|\omega|_i$ denotes the number of occurrences of i in ω . A sequence of n trials as given above is called a sequence of *Bernoulli trials*.

EXERCISE 12.14 Construct the sample space corresponding to n repeated independent trials, where each trial has possible outcomes r_i, $i = 1, \ldots, k$, with $P(r_i) = p_i$ and $p_1 + \cdots + p_k = 1$. ◇

EXAMPLE 12.31 (Samplings with replacement) An urn contains r red balls and b black balls; a ball is drawn and replaced. The experiment is repeated n times: then the probability of drawing a red ball k times and a black ball $(n - k)$ times is

$$\left(\frac{r}{r+b}\right)^k \left(\frac{b}{r+b}\right)^{n-k}.$$

This can be likened to the notion of permutations with repetitions of Chapter 6.

The product spaces are useful in modelling genetics (Mendel's laws), in reliability studies (life span before breakdowns, etc.) for systems.

12.4 Random variables

12.4.1 Definitions

First note that the same experiment, when repeated, does not give the same result each time. This variability is described with random variables. A *random variable* is a function whose value depends upon the outcome of a random experiment, i.e. a function defined on a random space.

EXAMPLE 12.32 Return to the example of the tossing of two dice: the sum S of the scores of the tossing of the dice is a random variable. Indeed, if $\omega = (m, n)$ is the pair of scores of the toss of the dice, we can write $S(\omega) = m + n$ for $\omega = (m, n)$, $1 \leq m, n \leq 6$.

Definition 12.33 *Let $(\Omega, \mathcal{T}, P)$ be a sample space. A discrete random variable (abbreviated to r.v.) on $(\Omega, \mathcal{T}, P)$ is a mapping $X: \Omega \longrightarrow D$, D denumerable, $D \subseteq \mathbb{R}$, such that $\forall d \in D$, $X^{-1}(d) \in \mathcal{T}$.*

The condition we demand is that the inverse image of an element of D be in $\mathcal{T}$: this condition is required in order to be able to speak of the probability that X assumes the value d. This probability will be denoted by $P(X^{-1}(d))$, or $P(X = d)$.

EXAMPLE 12.34 A mapping which is not a random variable. Let the experiment consist of throwing a die; let $\Omega = \{1,2,3,4,5,6\}$, let the tribe $\mathcal{T} = \{\Omega, \emptyset, \text{Even}, \text{Odd}\}$, where Even $= \{2,4,6\}$, Odd $= \{1,3,5\}$ and consider the sample space $(\Omega, \mathcal{T}, P)$, where $P(\Omega) = 1$, $P(\emptyset) = 0$, $P(\text{Even}) = P(\text{Odd}) = 1/2$. Define $X: \Omega \longrightarrow \mathbb{R}$ by

$$X(1) = X(3) = 4, \quad X(2) = X(4) = 6, \quad X(5) = X(6) = 11,$$

then X is not a random variable because $X^{-1}(4) = \{1,3\}$ is not in the tribe $\mathcal{T}$.

Proposition 12.35 Let $X: \Omega \longrightarrow D$ be an r.v. on $(\Omega, \mathcal{T}, P)$; X defines a probability distribution P_X on $(D, \mathcal{P}(D))$ by

$$P_X(d) = P(X = d) = P(X^{-1}(d))$$

for $d \in D$. P_X is called the probability distribution of X.

We also define the distribution function of X

$$F_X(d) = P(X < d) = P(X^{-1}]-\infty, d[) = \sum_{d' < d} P_X(d').$$

REMARK 12.36 X defines the probability distribution P_X, but P_X, conversely, does not determine X, as shown by the following example. Let X and Y be two r.v.'s $\mathbb{B} \longrightarrow \mathbb{B} \subseteq \mathbb{R}$, having the same distribution

$$P(X = 0) = P(X = 1) = P(Y = 0) = P(Y = 1) = 1/2.$$

Different pairs X, Y will satisfy this requirement, for instance:
- $P_1(X = x, Y = y) = 1/4$, $\forall x, y \in \mathbb{B}$.
- $P_2(X = x, Y = y) = 1/2$, for $x \neq y$, and $P_2(X = x, Y = x) = 0$ for $x \in \mathbb{B}$.

In the first case (P_1), X and Y are independent, in the second case (P_2), X and Y are dependent, but in both cases X and Y have the same distribution. Giving the distribution of X is thus not enough to fully describe all characteristics of X.

EXERCISE 12.15 We state some properties of distribution functions that can be verified as an exercise:
1. F is monotone.
2. $\lim_{x \to -\infty} F(x) = 0$, $\lim_{x \to \infty} F(x) = 1$. ◊

Definition 12.37 Let X and Y be two r.v.'s $X: \Omega \longrightarrow D$ and $Y: \Omega \longrightarrow D'$. The function $P_{(X,Y)}(d, d') = P(X = d, Y = d')$ is a probability distribution on $(D \times D', \mathcal{P}(D \times D'))$, called the joint distribution of the r.v.'s X and Y.

Random variables 243

REMARK 12.38 $(X = d, Y = d')$ is an abbreviated notation for the intersection $X^{-1}(d) \cap Y^{-1}(d')$ which is an event of the tribe $\mathcal{T}$ of Ω.

Given the joint distribution of (X, Y) we can determine the distributions of X and Y (called *marginal distributions*); by noting that event $X = d$ can be split into a disjoint union: $(X = d) = \sum_{d' \in D'} (X = d, Y = d')$, thus $P_X(d) = \sum_{d' \in D'} P_{(X,Y)}(d, d')$. Conversely, the marginal distributions of X and Y do not determine the joint distribution of (X, Y), except when the r.v.'s X and Y are independent (see below). The joint distributions $P_{(X,Y)}$ and $Q_{(X,Y)}$ described by the tables given below (see Example 12.29) are different but they correspond to the same marginal distribution $P_X = P_Y = Q_X = Q_Y$.

$P_{(X,Y)}$	x	0	1
y			
0		1/4	1/4
1		1/4	1/4

$Q_{(X,Y)}$	x	0	1
y			
0		0	1/2
1		1/2	0

EXERCISE 12.16 Check that for the r.v.'s X, Y the joint distribution $P_{(X,Y)}$ and $Q_{(X,Y)}$ of the above example (see Example 12.29) are indeed probability distributions, and similarly for the marginal distributions P_X and Q_X. ◇

EXAMPLE 12.39 Let X, Y be two r.v.'s assuming values in $\mathbb{B} = \{0, 1\}$. Assume that the joint distribution of (X, Y) is given by

$$P_{(X,Y)}(0,0) = \frac{1}{8}, \quad P_{(X,Y)}(1,0) = \frac{3}{8},$$
$$P_{(X,Y)}(0,1) = \frac{2}{8}, \quad P_{(X,Y)}(1,1) = \frac{2}{8}.$$

The marginal distributions P_X and P_Y of X and Y are given by

$$P_X(0) = \sum_{y \in \mathbb{B}} P_{(X,Y)}(0, y)$$
$$= P(X = 0, Y = 0) + P(X = 0, Y = 1) = \frac{3}{8},$$

and similarly,

$$P_X(1) = \frac{5}{8}, \quad P_Y(0) = P_Y(1) = \frac{1}{2}.$$

Definition 12.40 (Independence of r.v.'s) *Let X and Y be two r.v.'s defined on the same sample space and assuming values in D and D'. X and Y are said to be independent if and only if they verify the following equivalent conditions:*

(i) $\forall A \subseteq D$, $\forall A' \subseteq D'$, the events $(X \in A)$ and $(Y \in A')$ are independent, where $X \in A$ is an abbreviation for $X^{-1}(A)$.
(ii) *The joint distribution of (X,Y) is the product of the marginal distributions of X and Y.*
(iii) $\forall d \in D$, $\forall d' \in D'$, $P(X = d, Y = d') = P(X = d)P(Y = d')$.
(iv) $\forall d \in D$, $\forall d' \in D'$, $P(X = d / Y = d') = P(X = d)$.

The implications and equivalences (i) $\Longrightarrow$ (ii) $\Longrightarrow$ (iii) $\Longleftrightarrow$ (iv) are straightforward; to show (iv) $\Longrightarrow$ (i) it is enough to compute $P(X \in A, Y \in A')$.

EXAMPLE 12.41
(i) The r.v.'s X and Y of Example 12.39 are not independent, because, e.g. $P_X(0) = 3/8, P_Y(0) = 4/8$, but $P_{(X,Y)}(0,0) = 1/8 \neq (3/8) \times (4/8)$.
(ii) Consider the tossing of two dice, and let X be the score of the first die, and Y the score of the second die. X and Y are two r.v.'s defined on $(\Omega = (\{1, \ldots, 6\})^2, \mathcal{P}(\Omega))$ together with the uniform distribution into $\{1, \ldots, 6\}$. X and Y are easily seen to be independent, since $\forall i, j$, $1 \leq i, j \leq 6$, we have

$$P(X = i, Y = j) = 1/36 = (1/6) \times (1/6) = P(X = i) \times P(Y = j).$$

EXERCISE 12.17 Let X and Y be two independent r.v.'s, $X: \Omega \longrightarrow D \subseteq \mathbb{R}$ and $Y: \Omega \longrightarrow D' \subseteq \mathbb{R}$, and f and g two arbitrary functions from $\mathbb{R}$ into $\mathbb{R}$; we can define by composition two new r.v.'s $f(X) = f \circ X$ and $g(Y) = g \circ Y$; are the r.v.'s $f(X)$ and $g(Y)$ independent? ◊

EXERCISE 12.18 Let W be an r.v. assuming values 1,2,3 with the same probability. Let X, Y and Z be three independent r.v.'s, each with the same distribution as W. Let $U = X + Y$, $V = X - Z$.

1. What are the values assumed by U and by V? What are the distributions of U and V?
2. Write the table giving the probability distribution of the pair (U, V). Are the r.v.'s U and V independent? ◊

EXERCISE 12.19 How would you define the independence of n r.v.'s? ◊

Proposition 12.42 Let $X: \Omega \longrightarrow D$ and $Y: \Omega \longrightarrow D'$ be two independent r.v.'s defined on the same sample space and assuming values in D and D'. Let $f: D \longrightarrow E$ and $g: D' \longrightarrow E'$ be two functions; then $f(X): \Omega \longrightarrow E$ and $g(Y): \Omega \longrightarrow E'$ are two independent r.v.'s.

Proof.
$$\begin{aligned} P(f(X) = e, g(Y) = e') &= P(X \in f^{-1}(e), Y \in g^{-1}(e')) \\ &= P(X \in f^{-1}(e))P(Y \in g^{-1}(e')) \\ &= P(f(X) = e)P(g(Y) = e') \,. \end{aligned}$$ □

12.4.2 Mean and variance of a random variable

We wish to simplify the representation of an r.v. We can say, loosely, that the mean of an r.v. represents the average value of this r.v., and that its variance, or its standard deviation, gives a measure of the error in approximating the r.v. by its mean. For instance, if in a population consisting of n families, we have exactly n_k families with k children, then $P_k = n_k/n$ represents the probability that a 'randomly' chosen family have exactly k children, and the average number of children per family will be $1/n(\sum_{k \geq 0} k n_k)$; if we define the r.v. X as being the number of children of a 'randomly' chosen family, then $1/n(\sum_{k \geq 0} k n_k) = \sum_{k \geq 0} k(n_k/n) = \sum_{k \geq 0} k P_k$, represents the average value of X, i.e. the mean of X. The formal definition follows.

Definition 12.43 Let X be an r.v.; the mean (also called expectation, average or expected value) of X is defined by

$$E(X) = \sum_{\omega \in \Omega} X(\omega) P(\omega) = \sum_{d \in D} d P(X = d) \,, \tag{12.3}$$

provided that this sum is defined.

EXAMPLE 12.44 Return to Example 12.39. We have $E(X) = 5/8$, and similarly $E(Y) = 1/2$. Lastly, we can define the mean of (X, Y) by $E((X, Y)) = (E(X), E(Y))$ and we obtain $E((X, Y)) = (5/8, 1/2)$.

EXERCISE 12.20 Let Ω be a sample space and $A \subseteq \Omega$; recall that the characteristic function χ_A of A is a mapping $\chi_A: \Omega \longrightarrow \mathbb{B} \subseteq \mathbb{R}$ defined by

$$\chi_A(x) = \begin{cases} 1 & \text{if } x \in A, \\ 0 & \text{otherwise.} \end{cases}$$

1. Under what condition is χ_A an r.v.?
2. Check that, in this case, $E(\chi_A) = P(A)$. ◇

EXAMPLE 12.45 A computation of the average complexity of an algorithm $\mathcal{A}$ is the computation of a mean. Let $d_1, \ldots, d_k$ be all the data of size n, $P(d_i)$ the probability of datum d_i and $X(d_i)$ the complexity of $\mathcal{A}$ on the datum d_i. The average complexity of $\mathcal{A}$ on data of size n is then given by the mean of the r.v. X i.e. $E(X) = \sum_{i=1}^{k} P(d_i) X(d_i)$.

Proposition 12.46 *Let X and Y be two r.v.'s: $X \colon \Omega \longrightarrow A \subseteq \mathbb{R}$ and $Y \colon \Omega \longrightarrow B \subseteq \mathbb{R}$; we have*

(i) $X \leq Y \implies E(X) \leq E(Y)$,
(ii) $E(aX + bY) = aE(X) + bE(Y)$,
(iii) E *is linear.*

$X + Y$ (resp aX) is the r.v. defined by $(X + Y)(\omega) = X(\omega) + Y(\omega)$ (resp. $(aX)(\omega) = aX(\omega)$).

Proof. It is straighforward to show that $E(aX) = aE(X)$; similarly (i) is immediate. To check that $E(X + Y) = E(X) + E(Y)$, a computation of multiple sums is necessary; assume $X \colon \Omega \longrightarrow A$, $Y \colon \Omega \longrightarrow B$, and $X + Y \colon \Omega \longrightarrow D$.

$$E(X+Y) = \sum_{i \in D} i P(X+Y = i) = \sum_{i \in D} \sum_{x+y=i} i P(X = x, Y = y)$$
$$= \sum_{x \in A, y \in B} (x+y) P(X = x, Y = y)$$
$$= \sum_{(x,y) \in A \times B} x P(X = x, Y = y) + \sum_{(x,y) \in A \times B} y P(X = x, Y = y)$$
$$= \sum_{x \in A} x \left(\sum_{y \in B} P(X = x, Y = y) \right) + \sum_{y \in B} y \left(\sum_{x \in A} P(X = x, Y = y) \right)$$
$$= \sum_{x \in A} x P(X = x) + \sum_{y \in B} y P(Y = y). \qquad \square$$

REMARK 12.47 The mean of an r.v. is not always defined: consider $X \colon \Omega \longrightarrow D$, where $\Omega = \mathbb{N}$, $\mathcal{T} = \mathcal{P}(\mathbb{N})$ and

$$P(X = x) = \begin{cases} 1/2^{n+1} & \text{if } x = 2^n, \\ 0 & \text{otherwise.} \end{cases}$$

X is indeed an r.v. since $\sum_{n \in \mathbb{N}} P(X = x) = 1$, but $E(X)$ is not defined since $\sum_{n \in \mathbb{N}} 2^n / 2^{n+1} \longrightarrow \infty$. In technical terms $E(X)$ exists if and only if X is integrable with respect to the measure P_X.

Proposition 12.48 Let X be an r.v. $\Omega \longrightarrow D$ and $f\colon D \longrightarrow \mathbb{R}$, then $Y = f(X)$ is an r.v., and $E(Y) = \sum_{d \in D} f(d) P(X = d)$, provided that this sum is defined.

Proof. Let $D' = f(D)$, then:

$$E(Y) = \sum_{y \in D'} y P(Y = y) = \sum_{y \in D'} y P(f(X) = y)$$

$$= \sum_{y \in D'} y (\sum_{\{x \in D / f(x) = y\}} P(X = x))$$

$$= \sum_{y \in D'} (\sum_{\{x \in D / f(x) = y\}} f(x) P(X = x))$$

$$= \sum_{x \in D} f(x) P(X = x) . \qquad \square$$

Definition 12.49 $\forall n \geq 1$, we can define the nth moment of the r.v. X by

$$m_n(X) = \sum_{x \in D} x^n P(X = x) = E(X^n),$$

provided that this sum is defined.

$E(X)$ is the first moment of X. We now define the variance and standard deviation of the r.v. X which, intuitively, measure the distance between X and its mean $E(X)$; i.e. they estimate the fluctuations of X around its mean.

Definition 12.50 Let X be an r.v. such that $E(X)$ and $E(X^2)$ exist; the variance of X is defined by

$$var(X) = E((X - E(X))^2) = E(X^2) - (E(X))^2.$$

We define the standard deviation of X by $\sigma(X) = \sqrt{var(X)}$.

It is straightforward to check that

$$E((X - E(X))^2) = E(X^2 - 2X E(X) + E(X)^2)$$
$$= E(X^2) - 2E(X)^2 + E(X)^2 = E(X^2) - E(X)^2,$$

by noting that $E(X)$ is a constant; thus $E(E(X)) = E(X)$.

Proposition 12.51 Let X be an r.v. and let a and b be constants. Then $var(aX + b) = a^2 var(X)$.

Proof. We have $var(X + b) = var(X)$, hence the result. $\qquad \square$

Definition 12.52 *Let X and Y be two r.v.'s; we define:*

(i) *the co-variance of X and Y by*

$$\Gamma(X,Y) = E[(X - E(X))(Y - E(Y))] = E(XY) - E(X)E(Y) ,$$

(ii) *the correlation coefficient of X and Y by*

$$\rho(X,Y) = \frac{\Gamma(X,Y)}{\sigma(X)\sigma(Y)} .$$

Proposition 12.53 *Let X and Y be two independent r.v.'s, then*

(i) $E(XY) = E(X)E(Y)$,
(ii) $var(X + Y) = var(X) + var(Y)$,
(iii) $\Gamma(X,Y) = \rho(X,Y) = 0$.

REMARK 12.54 All converses are *false*, i.e. none of these conditions imply the independence of the r.v.'s X and Y.

Proof. Check, for instance, that (i); (ii) and (iii) are straightforward consequences of (i). Let $X\colon \Omega \longrightarrow A$, $Y\colon \Omega \longrightarrow B$, $XY\colon \Omega \longrightarrow D$:

$$\begin{aligned}
E(XY) &= \sum_{xy \in D} xy P(X = x, Y = y) \\
&= \sum_{xy \in D} xy P(X = x) P(Y = y) \quad \text{(since } X \text{ and } Y \text{ are independent)} \\
&= \sum_{x \in A} x P(X = x)(\sum_{y \in B} y P(Y = y)) = \sum_{x \in A} x P(X = x) E(Y) \\
&= E(X)E(Y). \qquad \square
\end{aligned}$$

EXERCISE 12.21
1. What are the means of the r.v.'s U and V defined in Exercise 12.18?
2. What is the correlation coefficient of the r.v.'s U and V defined in Exercise 12.18?

Random variables

EXAMPLE 12.55 Consider the two r.v.'s X, Y on $(\mathbb{N}, \mathcal{P}(\mathbb{N}))$ assuming values in $\mathbb{R}$, and such that

$$P(1,0) = P(-1,0) = P(0,1) = P(0,-1) = \frac{1}{4}$$

defines the joint distribution of (X,Y). The marginal distributions are defined by

$$P_X(1) = P_X(-1) = 1/4 \quad P_X(0) = 1/2,$$
$$P_Y(1) = P_Y(-1) = 1/4 \quad P_Y(0) = 1/2.$$

Thus $P_X = P_Y = P'$. This can be represented by the following tables:

P	x	-1	0	1
y				
-1		0	1/4	0
0		1/4	0	1/4
1		0	1/4	0

P'	x	-1	0	1
		1/4	1/2	1/4

We thus have $E(X) = E(Y) = 0$. Similarly, $E(XY) = 0$, thus also $\rho(X,Y) = 0$. The r.v.'s X and Y thus satisfy all the conditions (i), (ii), (iii) of the preceding proposition, but they are not independent, since, e.g.:

$$P(X = 1, Y = 0) = 1/4 \text{ and } P(X = 1)P(Y = 0) = 1/8.$$

EXERCISE 12.22
1. Let Z be an r.v. having a geometric distribution (see Section 12.6.3) of ratio a $(0 < a < 1)$, i.e.
$$\forall k \in \mathbb{N}^*, \quad P(Z = k) = a^{k-1}(1-a).$$
    (a)   What is the mean of Z?
    (b)   Show that $\forall k \in \mathbb{N}^*, P(Z \geq k) = a^{k-1}$.
2. Let X and Y be two independent r.v.'s defined on $(\Omega, \mathcal{T}, P)$, and such that
    X has a geometric distribution of ratio p, $(0 < p < 1)$,
    Y has a geometric distribution of ratio q, $(0 < q < 1)$.
Define an r.v. T on $(\Omega, \mathcal{T}, P)$, by

$$\forall \omega \in \Omega, \quad T(\omega) = \inf(X(\omega), Y(\omega)).$$

    (a)   Show that $\forall k \in \mathbb{N}^*$,

$$P(T = k) = P(X \geq k)P(Y \geq k) - P(X \geq k+1)P(Y \geq k+1).$$

(Hint: $P(T = k) = P(T \geq k) - P(T \geq k+1)$.)

(b) Show that T has a geometric distribution of ratio pq.

3. Consider a sequence $(X_n)_{n\geq 1}$ of independent r.v.'s having the same geometric distribution of ratio p, defined on $(\Omega, \mathcal{T}, P)$. Recall that $(X_n)_{n\geq 1}$ is a sequence of independent r.v.'s if and only if $\forall n \geq 1$, $X_1, \ldots, X_n$ are independent.

For all $n \geq 1$, define an r.v. T_n on $(\Omega, \mathcal{T}, P)$ by

$$\forall \omega \in \Omega, \; T_n(\omega) = \inf(X_1(\omega), \ldots, X_n(\omega)).$$

(a) Show, by induction on n, that T_n has a geometric distribution of ratio p^n.
(b) Show that $\lim_{n \to \infty} P(T_n > 1) = 0$. ◇

All the notions here introduced can be generalized to the case of an n-tuple $(X_1, \ldots, X_n)$ of r.v.'s defined on the same sample space. For instance,

Proposition 12.56 *Let $(X_1, \ldots, X_n)$ be a vector of n r.v.'s then*

(i) $\quad E(X_1 + \cdots + X_n) = E(X_1) + \cdots + E(X_n)$,

(ii) $\quad var(X_1 + \cdots + X_n) = var(X_1) + \cdots + var(X_n) + 2 \sum_{1 \leq i < j \leq n} \Gamma(X_i, X_j)$.

If, moreover, the n r.v.'s $(X_1, \ldots, X_n)$ are independent (such that the distribution of $(X_1, \ldots, X_n)$ is the product of the distributions of the X_is, $i = 1, \ldots, n$, i.e. $P(X_1 = x_1, \ldots, X_n = x_n) = P(X_1 = x_1) \cdots P(X_n = x_n)$), then

(iii) $\quad E(X_1 \cdots X_n) = E(X_1) \cdots E(X_n)$,

(iv) $\quad var(X_1 + \cdots + X_n) = var(X_1) + \cdots + var(X_n)$.

Equalities (ii) and (iv) are true provided that the variances $var(X_1), \ldots, var(X_n)$ are finite.

Proof. (i) is straightforward; (iv) is a consequence of (ii); (iii) is easily proved by induction on n. Let us check (ii); we have

$$X_1 + \cdots + X_n - E(X_1 + \cdots + X_n) = X_1 + \cdots + X_n$$
$$- E(X_1) - \cdots - E(X_n)$$
$$= \sum_{i=1}^n X_i - E(X_i).$$

Hence,

$$\left(X_1 + \cdots + X_n - E(X_1 + \cdots + X_n)\right)^2$$
$$= \sum_{i=1}^n \left(X_i - E(X_i)\right)^2 + 2 \sum_{1 \leq i < j \leq n} \left(X_i - E(X_i)\right)\left(X_j - E(X_j)\right).$$

Hence the result is proved by computing the mean. □

Random variables

12.4.3 Application to approximations

We will give results enabling us to bound the error in approximating an r.v. by its mean.

Theorem 12.57 (Markov's inequality) *Let $X: \Omega \longrightarrow D \subseteq \mathbb{R}^+$ be a nonnegative r.v., having mean $E(X) \neq 0$, and let λ be a positive real number, then*

$$\forall \lambda > 0, \quad P[X \geq \lambda E(X)] \leq \frac{1}{\lambda}. \tag{12.4}$$

Proof. If $0 < \lambda \leq 1$, (12.4) is trivially true, since $\forall A, P(A) \leq 1$. Assume $\lambda > 1$; since $X \geq 0$, we have

$$E(X) = \sum_{x \geq 0} xP(X=x) \geq \sum_{x \geq \lambda E(X)} xP(X=x),$$

hence

$$E(X) \geq \lambda E(X) \sum_{x \geq \lambda E(X)} P(X=x) = \lambda E(X) P(X \geq \lambda E(X)). \qquad \square$$

Theorem 12.58 (Chebyshev's inequality) *Let X be an r.v. such that $E(X)$ and $var(X)$ are both defined; then, for $\lambda > 0$,*

$$P(|X - E(X)| \geq \lambda) \leq \frac{1}{\lambda^2} var(X). \tag{12.5}$$

Proof. Apply Markov's inequality to the r.v. $Y = (X - E(X))^2$, with $\lambda' = \dfrac{\lambda^2}{var(X)}$; as $E(Y) = var(X)$, we obtain (12.5). $\qquad \square$

EXERCISE 12.23 The average height of a population is 1.65 m and the standard deviation is 0.04 m. Find an upper bound of the probability that the height of a randomly chosen individual in this population is greater than or equal to 1.80 m. ◇

EXERCISE 12.24 Consider a coin-tossing game with an ideal coin, i.e. such that $P(\text{Tails}) = P(\text{Heads}) = 1/2$. If Tails turns up, the player wins \$1, if Heads turns up, the player loses \$1. Let S_n be the average algebraic 'winnings' after n tosses of the coin. Determine an integer n such that the average winnings S_n are larger than $-1/2$ with a probability greater than or equal to $\dfrac{99}{100}$. Note that, if X_i represents the algebraic 'winnings' at the ith toss, $X_1, \ldots, X_n$ are n independent r.v.'s with the common distribution

$$P(X_i = 1) = P(X_i = -1) = \frac{1}{2}.$$

S_n is then defined by

$$S_n = \frac{X_1 + \cdots + X_n}{n}.$$ ◇

We state below the weak law of large numbers. n successive repetitions of a trial can be translated into a sequence X_n of independent r.v.'s with a common distribution. Then, the average (in the arithmetical sense) of the values assumed by the considered variables is likely to lie near the average (in the probabilistic sense), i.e. the mean $E(X)$, as n tends to infinity. We can use the weak law of large numbers to determine the mean of X up to ε, by substituting it with the arithmetical average of the X_n for n large. This justifies a *posteriori* the definition of the mean $E(X)$ of an r.v.

Theorem 12.59 (Weak law of large numbers) *Let $(X_n)_{n \in \mathbb{N}}$ be mutually independent r.v.'s, with a common distribution of mean E and of variance σ^2, let $S_n = X_1 + \cdots + X_n$, and $\varepsilon > 0$. Then*

$$\lim_{n \to \infty} P\left(\left|\frac{S_n}{n} - E\right| \geq \varepsilon\right) = 0.$$

Proof. We have $E\left(\dfrac{S_n}{n}\right) = \dfrac{nE(X_1)}{n} = E$, and $var\left(\dfrac{S_n}{n}\right) = \sum_{i=1}^{n} var\left(\dfrac{X_i}{n}\right)$ since the r.v.'s are independent; let

$$var\left(\frac{S_n}{n}\right) = n\frac{\sigma^2}{n^2} = \frac{\sigma^2}{n}.$$

Applying Chebyshev's inequality we deduce

$$P\left(\left|\frac{S_n}{n} - E\right| \geq \varepsilon\right) \leq \frac{\sigma^2}{n\varepsilon^2}.$$ □

The above proof in fact gives a slightly more precise result. If we are able to find an upper bound of σ^2, then we will be able to determine the minimum number n of experiments needed in order to substitute the arithmetic average $\dfrac{S_n}{n}$ for the mean E with an error less than ε.

Generating functions

12.5 Generating functions

As we represented a sequence by a series (Chapter 8), we can represent an integral-valued r.v. X by a series, since such an r.v. is characterized, for instance, by the sequence $\bigl(P(X = n)\bigr)_{n \in \mathbb{N}}$. The advantage is that of giving us a global representation of the r.v. together with its probability distribution and of making the computations we may have to perform much easier.

Definition 12.60 *Let X be an integral-valued r.v. $X\colon \Omega \longrightarrow \mathbb{N}$. Letting $p_n = P(X = n)$, define the generating function g_X of X by the series*

$$g_X(z) = \sum_{n=0}^{\infty} p_n z^n = \sum_{n=0}^{\infty} P(X = n) z^n = E(z^X),$$

g_X will be denoted by g when there can be no ambiguity on X.

The last equality of this definition is a straightforward consequence of Proposition 12.48. Indeed, z^X is a new discrete r.v. obtained by composing X with the function $f\colon \mathbb{N} \longrightarrow \mathbb{R}$ defined by $f(n) = z^n$.

Generating functions thus consist of generating series with non-negative coefficients, and such that $g_X(1) = \sum_{n=0}^{\infty} p_n = 1$.

The generating function of X enables us to characterize the mean and the variance of X in a simple way.

Proposition 12.61 *Let X be an integral-valued r.v. with generating function g, then*

(i) $E(X) = g'(1)$,
(ii) $\mathrm{var}(X) = g''(1) + g'(1) - (g'(1))^2$,

where g' (resp. g'') is the derivative (resp. the second derivative of g).

Proof.
(i) We have $E(X) = \sum_{n \geq 0} n p_n$, or $g'(z) = \sum_{n \geq 0} n p_n z^{n-1}$.
(ii) Similarly, we have

$$\mathrm{var}(X) = E(X^2) - (E(X))^2 = \sum_{n \geq 1} n^2 p_n - (E(X))^2$$

$$= \sum_{n \geq 1} n^2 p_n - (g'(1))^2.$$

Note that

$$g''(1) = \sum_{n \geq 2} n(n-1) p_n \quad \text{and} \quad g'(1) = \sum_{n \geq 1} n p_n$$

and deduce
$$\sum_{n\geq 1} n^2 p_n = g''(1) + g'(1).$$ □

The preceding proposition usually gives the simplest way of computing the mean and the variance of an integral-valued r.v..

EXERCISE 12.25 The Dirichlet generating function of a probability distribution is defined by
$$d(z) = \sum_{n\geq 1} \frac{p_n}{z^n}.$$
We thus have $d(1) = 1$. Let X be an r.v. such that $P(X = n) = p_n$; compute $E(X)$, $var(X)$ and $E(\log X)$ in terms of $d(z)$ and of its derivatives. ◇

Proposition 12.62 *Let X and Y be two independent integral-valued r.v.'s, we have $g_{X+Y}(z) = g_X(z)g_Y(z)$, where the product of the generating functions is the product of convolution (or Cauchy product) defined for the generating series.*

Proof. $g_{X+Y}(z) = E(z^{X+Y}) = E(z^X z^Y)$. We check that, if X and Y are two independent r.v.'s, then z^X and z^Y are also independent, since
$$P(z^X = x, z^Y = y) = P(X = \log_z x, Y = \log_z y)$$
$$= P(X = \log_z x)P(Y = \log_z y)$$
$$= P(z^X = x)P(z^Y = y).$$
then, by Proposition 12.56, $E(z^X z^Y) = E(z^X)E(z^Y)$, hence
$$g_{X+Y}(z) = g_X(z)g_Y(z).$$ □

Proposition 12.63 *Let X be an integral-valued r.v. and $a \in \mathbb{N}$, then $g_{aX}(z) = g_X(z^a)$.*

Proof. $g_{aX}(z) = E(z^{aX}) = E((z^a)^X) = g_X(z^a)$. □

EXERCISE 12.26 Directly prove the preceding result by explicitly computing the generating functions as a series. ◇

EXERCISE 12.27 Consider a coin-tossing game with probability p for Tails and probability $q = 1 - p$ for Heads. The r.v. S_1 represents the number of tosses required before the first Tails turns up, and the r.v. S_r represents the number of tosses required before the rth Tails turns up.
1. Compute the distribution of S_1 and the generating function of S_1.
2. Compute the generating function of S_r. Note that $S_r = X_1 + \cdots + X_r$, where the X_is are mutually independent and have the distribution of S_1 as common distribution.
3. Deduce the distribution of S_r, its mean and its variance. ◇

EXERCISE 12.28 Let $X_1, \ldots, X_n$ be independent r.v.'s with a common generating function g. Let U be an integral-valued r.v. independent of $X_1, \ldots, X_n$, with generating function f and assuming values in $1, \ldots, n$. Let V be the r.v. $V = \sum_{i=1}^{U} X_i$.
1. Show that $P(V = k) = \sum_{j=1}^{n} P(U = j) \times P((\sum_{i=1}^{j} X_i) = k)$.
2. Compute the generating function of the r.v. V in terms of f and g.
3. Compute the mean and the variance of V in terms of the mean and the variance of U and X_i. ◇

12.6 Common probability distributions

In the present section we give the most usual discrete probability distributions, together with their intuitive motivation, and the main results (mean, variance) will be stated. Explicit computations will be left as exercises; the reader is also advised to have a look at the many excellent handbooks of probability theory (e.g. he/she should benefit by reading Feller, Vol. 1).

12.6.1 Bernoulli trials

These consist of the distribution of a coin-tossing game, with $p = P(\text{Tails})$, $q = 1 - p = P(\text{Heads})$. The corresponding r.v. is $X\colon (\mathbb{B}, \mathcal{P}(\mathbb{B})) \longrightarrow \mathbb{B}$. If we identify Tails with 1 and Heads with 0 then X is defined by: $P(X = 1) = p$ and $P(X = 0) = q$. Its generating function is $g(z) = pz + q$, $E(X) = p$, $var(X) = pq$.

Notation: The Bernoulli distribution is denoted $B(p)$ and p is called the parameter.

12.6.2 Binomial distribution

n independent Bernoulli trials, with the common distribution $B(p)$, are repeated. We are interested in the total number k of 'Tails' produced (we also say 1, or success, for 'Tails', and 0, or failure, for 'Heads'); k is given by the r.v. $S_n = X_1 + \cdots + X_n$, where the X_is are n Bernoulli r.v.'s with the same parameter p. We thus have $g_{S_n}(z) = g_X(z)^n = (pz + q)^n = \sum_{k=0}^{n} \binom{n}{k} p^k q^{n-k} z^k$, by the binomial theorem; hence $P(S_n = k) = \binom{n}{k} p^k q^{n-k}$, $E(S_n) = np$, $var(S_n) = npq$. The binomial distribution is denoted by $b(p, n)$.

EXERCISE 12.29 Let X and Y be two independent r.v.'s with binomial distributions $b(p, m)$ and $b(p, n)$ of respective parameters (p, m) and (p, n). Let $S = X + Y$.
1. What is the distribution of S?
2. Let s be a possible value for S; the conditional distribution of X given that $S = s$ is defined by $P(X = x \,/\, S = s)$, when x ranges over the possible values for X. Find the conditional distribution of X given S. ◇

The binomial distribution is also obtained in sampling problems. Consider a size N population, partitioned into n_1 people of type 1 and n_0 people of type 0,

with
$$\frac{n_1}{N} = p, \quad \frac{n_0}{N} = q = 1 - p.$$

We randomly choose, n successive times, a person from the entire population (*sampling with replacement*); if S is the number of people drawn of type 1, the distribution of S is a binomial distribution, i.e.

$$P(S = k) = b(p, n)(k) = \binom{n}{k} p^k q^{n-k}.$$

EXERCISE 12.30 Check by a direct combinatorial computation that
$$P(S = k) = b(p, n)(k). \qquad \diamond$$

We can generalize the case of samplings in a population of N people partitioned into

- n_1 individuals of type 1,
- n_2 individuals of type 2,
- ...
- n_r individuals of type r,

with $n_1 + \cdots + n_r = N$ and $\forall i = 1, \ldots, r$: $\frac{n_i}{N} = p_i$. Choose a sample with replacement of n individuals; let X_i for $i = 1, \ldots, n$, be the r.v. representing the type of the individual obtained at the ith drawing. We have $P(X_i = j) = p_j$. If S_i, $i = 1, \ldots, r$, is the r.v. representing the number of individuals of type i obtained in the sample, we have:

$$P(S_1 = k_1, \ldots, S_r = k_r) = \frac{n!}{k_1! \cdots k_r!} p_1^{k_1} \cdots p_r^{k_r}.$$

Indeed, each sequence $(x_1, \ldots, x_n)$ contains k_i elements of type i, with $k_1 + \cdots + k_r = n$, and it has $p_1^{k_1} \cdots p_r^{k_r}$ probability of occurring. Moreover, there are $n!$ ways of permuting $(x_1, \ldots, x_n)$, but among those $n!$ ways, the $k_i!$ permutations of the k_i elements of type i give the same result, for $i = 1, \ldots, r$.

For $r = 2$, we again find the binomial distribution. We verify that $\forall i$, S_i has a binomial distribution $b(p_i, n)$.

The distribution of $(S_1, \ldots, S_k)$ is called the *multinomial distribution* with parameters $p_1, \ldots, p_k$.

EXERCISE 12.31
1. Check that S_i has a binomial distribution. For $r = 2$, what is the co-variance $\Gamma(S_1, S_2)$?
2. Generalize the generating functions in order to be able to represent the distribution of $(S_1, \ldots, S_r)$. (Hard.) $\diamond$

12.6.3 Geometric distribution

Let $(X_k)_{k\geq 1}$ be a sequence of independent r.v.'s with a Bernoulli distribution, and X the r.v. representing the number of tosses necessary for the first 'Tails' to turn up. Then, for all $k \geq 1$

$$P(X = k) = P(X_1 = \cdots = X_{k-1} = 1, X_k = 0) = p^{k-1}q.$$

We say that X has a *geometric distribution* (or *Pascal distribution*). We have the generating function

$$g(z) = \frac{1-p}{1-zp} = \frac{q}{1-zp},$$

hence

$$E(X) = \frac{p}{q}, \quad var(X) = \frac{p}{q^2}.$$

EXERCISE 12.32 What is the sample space on which X is defined? ◊

EXERCISE 12.33 A coin is tossed till two successive identical outcomes appear.
1. What is the probability that n tosses are necessary?
2. What is the probability that the experiments stop before the sixth toss?
3. What is the probability that an even number of tosses is necessary? ◊

12.6.4 Hypergeometric distribution

This can be obtained with sampling problems (*sampling without replacement*). Consider, as in 2, a size N population, consisting of n_i individuals of type i, $i = 1, \ldots, r$, $n_1 + \cdots + n_r = N$, $\frac{n_i}{N} = p_i$.

Choose a subset of n individuals from the population (sampling without replacement) at the same time. Let S_i be the number of type i individuals among those chosen, $i = 1, \ldots, r$. We verify that

$$P(S_1 = k_1, \ldots, S_r = k_r) = \frac{\binom{n_1}{k_1} \cdots \binom{n_r}{k_r}}{\binom{N}{n}}.$$

for $r = 2$, we obtain the *hypergeometric distribution* denoted by $H(N, n_1, n)$, defined by

$$P(S_1 = k) = \frac{\binom{n_1}{k}\binom{N-n_1}{n-k}}{\binom{N}{n}},$$

with $E(S_1) = np_1$, $var(S_1) = np_1(1-p_1)\left(\frac{N-n}{N-1}\right) = np_1p_2\left(\frac{N-n}{N-1}\right)$.
Note that in the case of a sampling *with* replacement, we obtain the same mean, but a slightly larger variance, namely, np_1p_2, see Section 12.6.2.

EXERCISE 12.34
1. Compute $E(S_1)$ and $var(S_1)$ when $r = 2$. Let X_i for $i = 1, \ldots, n$, be the r.v. assuming values in $\{1, 2\}$ and representing the type of the individual obtained at the ith drawing. We have $S_1 = \chi_1 + \cdots + \chi_n$, where

$$\chi_i = \chi_{(X_i=1)} = \begin{cases} 1 & \text{if the person chosen at the } i\text{th drawing is of type 1,} \\ 0 & \text{otherwise.} \end{cases}$$

2. Now let r be arbitrary; show that $\forall j = 1, \ldots, r$, S_j has a distribution $H(N, n_j, n)$. Deduce $E(S_j)$ and $var(S_j)$. ◇

Asymptotic behaviour

(a) For a fixed n, if $N \to \infty$ and $\dfrac{n_1}{N} \to p$, then $H(N, n_1, n) \to b(p, n)$. Indeed,

$$P(S_1 = k) = \frac{\binom{n_1}{k}\binom{N-n_1}{n-k}}{\binom{N}{n}} = \binom{n}{k}\frac{\binom{N-n}{n_1-k}}{\binom{N}{n_1}}$$

$$= \binom{n}{k} n_1(n_1 - 1) \cdots (n_1 - k + 1)$$
$$\times \frac{(N - n_1)(N - n_1 - 1) \cdots (N - n_1 - n + k + 1)}{N(N - 1) \cdots (N - n + 1)}.$$

Let
$$p = \frac{n_1}{N}, \quad q = 1 - p,$$

$$P(S_1 = k) = \binom{n}{k} Np(Np - 1) \cdots (Np - k + 1)$$
$$\times \frac{Nq(Nq - 1) \cdots (Nq - n + k + 1)}{N(N-1) \cdots (N-n+1)}$$
$$\sim \binom{n}{k} p^k (1-p)^{n-k} \quad \text{when } n \to \infty.$$

The intuition is as follows: for a fixed n, if $N \to \infty$, a drawing without replacement of n individuals in a very large population is close to a drawing with replacement; thus, on these asymptotic conditions, the hypergeometric distribution is close to the binomial distribution.

(b) If n, n_1, N go to infinity and $n\dfrac{n_1}{N} \to \lambda$, then

$$\lim_{n \to \infty} P(S = k) = e^{-\lambda}\frac{\lambda^k}{k!}.$$

EXERCISE 12.35 Verify this result. ◇

Common probability distributions

12.6.5 Poisson distribution

This is the distribution we have just obtained. An r.v. $X : \Omega \longrightarrow \mathbb{N}$ has a *Poisson distribution* with mean λ, denoted by $p(\lambda)$, if $P(X = k) = e^{-\lambda}\dfrac{\lambda^k}{k!}$, $\forall k \in \mathbb{N}$, with $\lambda > 0$. The generating function of X is given by:

$$g(z) = \sum_{n \geq 0} e^{-\lambda} z^n \frac{\lambda^n}{n!} = e^{\lambda(z-1)}.$$

Hence we will deduce $E(X) = \lambda$, $var(X) = \lambda$.

Proposition 12.64 *Let X and Y be two independent r.v.'s with Poisson distributions with parameters λ and μ; then $X + Y$ has a Poisson distribution with parameter $\lambda + \mu$.*

Proof. Straightforward, since the generating function g_{X+Y} of $X+Y$ is $g_{X+Y} = g_X g_Y$. □

Proposition 12.65 (Poisson approximation) *Let S_n be an r.v. with a binomial distribution $b(p_n, n)$. Assume that n goes to infinity, with $\lim_{n \to \infty} np_n = \lambda$, $0 < \lambda < 1$, then $\lim_{n \to \infty} P(S_n = k) = e^{-\lambda}\dfrac{\lambda^k}{k!}$.*

Proof. Letting $q_n = 1 - p_n$, check the result by induction on k.

- Basis: $k = 0$; then, if $n \to \infty$,

$$P(S_n = 0) = q_n^n = (1 - p_n)^n = \left(1 - \frac{\lambda}{n} + \varepsilon\left(\frac{\lambda}{n}\right)\right)^n \longrightarrow e^{-\lambda}.$$

- Inductive step: assume that $P(S_n = k) \to e^{-\lambda}\dfrac{\lambda^k}{k!}$ for $n \to \infty$.

$$\frac{P(S_n = k+1)}{P(S_n = k)} = \frac{n-k}{k+1} \times \frac{p_n}{q_n} \to \frac{\lambda}{k+1} \quad \text{when} \quad n \to \infty.$$

(Note that $p_n \to 0$ and $q_n \to 1$ for $n \to \infty$.) Thus

$$P(S_n = k+1) \to e^{-\lambda}\frac{\lambda^{k+1}}{(k+1)!} \quad \text{when} \quad n \to \infty. \qquad \square$$

Corollary 12.66 *For n 'large' and p 'small' a Poisson distribution with parameter np approximates the binomial distribution $b(p, n)$.*

12.6.6 Uniform distribution

Recall for the sake of completeness the *uniform distribution*: on a finite subset A of $\mathbb{N}$ it is defined by the uniform probability on A:

$$P(X = n) = \begin{cases} \dfrac{1}{|A|} & \text{if } n \in A, \\ 0 & \text{otherwise.} \end{cases}$$

Its generating function is $\sum_{n \in A} \dfrac{z^n}{|A|}$;

$$E(X) = \frac{1}{|A|}\left(\sum_{n \in A} n\right) \quad , \quad var(X) = \frac{1}{|A|}\left(\sum_{n \in A} n^2\right) - (E(X))^2 .$$

EXERCISE 12.36 Let X and Y be two r.v.'s such that the joint distribution of (X, Y) is given, for all $(m, n) \in \mathbb{N}^2$, such that $m \geq n$, by

$$P(X = n, Y = m) = \frac{\lambda^m}{n!(m-n)!} e^{-2\lambda}.$$

1. Show that the joint distribution of (X, Y) is completely determined.
2. Find the probability distributions of X and of Y. Are X and Y independent?
3. Find the probability distribution of $Y - X$ and the joint distribution of $(X, Y - X)$. Show that the r.v.'s X and $Y - X$ are independent.
4. What are the co-variance $\Gamma(X, Y)$ and the correlation coefficient of $\rho(X, Y)$? ◇

EXERCISE 12.37 A telephone switchboard receives N daily calls. The r.v. N is assumed to have a Poisson distribution with mean λ. Among these N calls, there are Z wrong numbers, and each number among the N numbers has the probability p of being wrong.

1. Find the probability distribution of (Z, N).
2. Find the probability distribution of Z.
3. Find the probability distribution of N conditioned by Z.
4. Compute the correlation coefficient of $\rho(N, Z)$. ◇

CHAPTER 13

FINITE MARKOV CHAINS

The theory of Markov chains is used in

- modelling and simulation,
- queueing theory: for instance, the study of the average waiting time for accessing the nerve-centres of a telematic network,
- robotics, to model the moves of the robot depending on its environment,
- signal theory.

The theory of Markov chains is based on conditional probabilities. We mainly study finite Markov chains whose underlying sample probability space is finite.

In this chapter, we define finite Markov chains, their transition matrices and their graphs. We then show how the graph of a finite Markov chain can be used to study its properties.

We recommend in the strongest possible terms the following handbook:

William Feller, *Probability Theory*, Vol. 1, John Wiley, New York (1968).

We also recommend:

Dean Isaacson, Richard Madsen, *Markov Chains Theory and Applications*, John Wiley, New York (1976).

13.1 Introduction

Up to now we have mainly studied independent experiments. For instance, a sequence of trials e_n that independently produce as results elements of the same trial space Ω yields a sequence of independent random variables X_n; we then have, by the independence, that

$$P((X_0, \ldots, X_n) = (\omega_0, \ldots, \omega_n)) = P(X_0 = \omega_0) \cdots P(X_n = \omega_n) = p_0 \cdots p_n$$

(see the binomial distribution corresponding to a coin-tossing game).

In the present chapter we will study the simplest possible generalization of this notion in which the result of the nth trial, $X_n = \omega$, no longer has a fixed probability $p(\omega)$, independent of the trials $e_0, \ldots, e_{n-1}$, but, rather, a conditional probability, completely determined by the result of the $(n-1)$th trial e_{n-1}. We associate with each pair ω_i, ω_j the probability p_{ij}, which represents the probability that $X_n = \omega_j$ given that $X_{n-1} = \omega_i$, and we assume that this probability is the same for all possible ns. We will thus have the following, assuming for instance that at the initial moment $P(X_0 = \omega_j) = q_j$:

$$P\big((X_0, X_1) = (\omega_i, \omega_j)\big) = q_i p_{ij},$$
$$P\big((X_0, X_1, X_2) = (\omega_i, \omega_j, \omega_k)\big) = q_i p_{ij} p_{jk}.$$

The situation is often summed up by saying, inaccurately, that: 'the result of the nth trial depends only on the result of the $(n-1)$th trial'. Actually, the nth trial depends (via the $(n-1)$th trial) on the $(n-2)$th trial, which in turn depends on the $(n-3)$th one, etc.

Indeed, it is more accurate, but also more cumbersome, to say that the whole past of the sequence of trials can be coded, for conditioning its future evolution, in the knowledge of its present state. Formally, we have

$$P(X_{n+1} = \omega_{n+1} \,/\, X_n = \omega_n, \ldots, X_0 = \omega_0) = P(X_{n+1} = \omega_{n+1} \,/\, X_n = \omega_n),$$

with, moreover, for all $k > 0$,

$$P(X_{n+k+1} = \omega \,/\, X_{n+k} = \omega') = P(X_{n+1} = \omega \,/\, X_n = \omega').$$

13.2 Generalities

13.2.1 Definitions

In this chapter we will adopt the usual notations for Markov chains, even though they are slightly different from the notations used in Chapter 12.

The trial space $\Omega = \{E_1, \ldots, E_r\}$ is finite, and its elements $E_1, \ldots, E_r$ are called the *states* of the system; we will abbreviate $E_1, \ldots, E_r$ to $1, \ldots, r$ when no ambiguity can occur. A sequence $(X_n)_{n \in \mathbb{N}}$ of random variables will be represented by a sequence of states of Ω, the nth state of the sequence representing the value of the random variable (X_n). The result x_n of the nth trial (or the value x_n of the nth random variable) will be called the state of the system at time n. $(X_n)_{n \in \mathbb{N}}$ is a Markov chain if the conditional probabilities at time n,

$$p_{ij} = P(X_{n+1} = E_j \,/\, X_n = E_i),$$

do not depend on n. The formal definition follows.

Generalities

Definition 13.1 *A sequence of random variables (abbreviated r.v.'s) $(X_n)_{n \in \mathbb{N}}$ with values ranging over a finite set $\Omega = \{E_1, \ldots, E_r\}$ is a Markov chain* if and only if it satisfies the following equivalent conditions:*

(i) $P(X_{n+1} = E_{i_{n+1}} / X_0 = E_{i_0}, \ldots, X_n = E_{i_n}) = P(X_{n+1} = E_{i_{n+1}} / X_n = E_{i_n})$ and

$$\forall k > 0 \quad \forall n > 0 \quad P(X_{n+k+1} = E / X_{n+k} = E') = P(X_{n+1} = E / X_n = E').$$

(ii) $P(X_0 = E_{i_0}, X_1 = E_{i_1}, \ldots, X_n = E_{i_n}) = q_{i_0} p_{i_0 i_1} p_{i_1 i_2} p_{i_{n-1} i_n}$, where $\forall i$, $q_i = P(X_0 = E_i)$ is the initial probability distribution, and $\forall i, j, n$, $p_{ij} = P(X_{n+1} = E_j / X_n = E_i)$ is the conditional probability of obtaining the result E_j given that the preceding result is E_i.

By Definition 13.1 we will thus have

1. $\forall i, \quad q_i \geq 0$ and $\sum_{i=1}^{r} q_i = 1$,
2. $\forall i \forall j, \quad p_{ij} \geq 0$ and $\sum_{k=1}^{r} p_{ik} = 1$.

EXERCISE 13.1 Show that, conversely, the q_is and the p_{ij}s satisfying 1 and 2 above, with $1 \leq i, j \leq r$, each define a Markov chain. ◇

13.2.2 Examples

We will see an example, a counterexample, and a model equivalent to the notion of Markov chain.

EXAMPLE 13.2 Message passing. Consider passing a message 'yes' or 'no' in a population. Each individual forwards the message received with probability p and the opposite message with probability $1 - p$. Let X_n be the message forwarded by the nth individual. We have:

$P(X_{n+1} = \text{'yes'} / X_n = \text{'yes'}) = P(X_{n+1} = \text{'no'} / X_n = \text{'no'}) = p,$
$P(X_{n+1} = \text{'yes'} / X_n = \text{'no'}) = P(X_{n+1} = \text{'no'} / X_n = \text{'yes'}) = 1 - p.$

---

* Here we do *not* use the usual terminology, where a Markov chain is defined by the single property:

$$P(X_{n+1} = E_{i_{n+1}} / X_0 = E_{i_0}, \ldots, X_n = E_{i_n}) = P(X_{n+1} = E_{i_{n+1}} / X_n = E_{i_n})$$
$$= p_{i_n i_{n+1}}(n),$$

i.e. $P(X_{n+1} = E_{i_{n+1}} / X_n = E_{i_n})$ can depend on n. The chains that we will consider are in fact the special case of the general Markov chains, when $P(X_{n+1} = E_{i_{n+1}} / X_n = E_{i_n}) = p_{ij}$ does not depend on n. Such chains are usually called *homogeneous Markov chains*.

These four equalities allow us, if we are given the probability distribution of the initial message, to fully describe the message passing. We see that only the message X_n forwarded by the nth individual affects X_{n+1}, and that it is not useful to memorize the whole preceding history of the message. The $(X_n)_{n\in\mathbb{N}}$s form a Markov chain, with $\Omega = \{\text{'yes'},\text{'no'}\}$. Let $E_1 = \{\text{'yes'}\}$ and $E_2 = \{\text{'no'}\}$; then $p_{11} = p$, $p_{12} = 1 - p$, $p_{21} = 1 - p$, $p_{22} = p$.

EXERCISE 13.2 The notations are as in Example 13.2. Assume, moreover, that $q_1 = P(X_0 = \text{'yes'}) = q_2 = P(X_0 = \text{'no'}) = 1/2$.
1. Show by induction on n that all the r.v.'s X_n have the same probability distribution as X_0.
2. Are the r.v.'s X_n independent? ◊

EXAMPLE 13.3 We now exhibit an example of a non-Markovian chain: the Polya urn model. An urn contains b black balls and r red balls. A ball is drawn at random. It is replaced and, moreover, c balls of the colour drawn are added. A new random drawing is performed and the whole process is iterated. Let $X^n \in \{B, R\} = \Omega$ be the colour of the ball drawn at the nth drawing. The sequence $(X^n)_{n\geq 1}$ is not a Markov chain. We thus have:

$$P(X^3 = B \,/\, X^2 = B) = \frac{b+c}{b+r+c},$$
$$P(X^3 = B \,/\, X^2 = B, X^1 = B) = \frac{b+2c}{b+r+2c}.$$

EXERCISE 13.3 Return to the Polya urn model of Example 13.3 and recall that $P(X^n = B) = \bigl(b/(b+r)\bigr)$ for all n (see Exercise 12.5).
1. For what values of c is the sequence $(X^n)_{n\in\mathbb{N}}$ defined in Example 13.3 a Markov chain?
2. Let Y_n be the r.v. giving the number of black balls in the urn at time n. Is the sequence $(Y_n)_{n\in\mathbb{N}}$ a Markov chain? ◊

EXAMPLE 13.4 Markov chains as urn models. Any Markov chain can be represented as an urn model as follows: if $\Omega = \{E_1, \ldots, E_r\}$, r urns are available; each urn contains a fixed number of balls marked $E_1, \ldots, E_r$. In the jth urn, the probability of drawing a ball marked E_k is p_{jk}. At the initial trial, an urn is chosen according to the probability distribution q_j. From that chosen urn, a ball is drawn at random, and if it is marked E_j, the next drawing is made from the jth urn, and so on. The sequence X_n of the drawn markings is a Markov chain. We thus see that Markov chains can be modelled by urns.

Generalities

13.2.3 Transition matrices

A Markov chain is thus characterized by

- on the one hand, the conditional probabilities p_{ij}, $1 \le i, j \le r$, where p_{ij} is the probability of a state i given that the state j occurred at the preceding trial, and is called the *transition probability* from i to j and
- on the other hand, the initial probability distributions q_i, $1 \le i \le r$.

The p_{ij}s form the *transition matrix* P

$$P = \begin{pmatrix} p_{11} & p_{12} & \cdots & p_{1r} \\ p_{21} & p_{22} & \cdots & p_{2r} \\ \vdots & \vdots & \ddots & \vdots \\ p_{r1} & p_{r2} & \cdots & p_{rr} \end{pmatrix}$$

and verify

$$\forall i,j, \quad p_{ij} \ge 0 \text{ and } \forall i, \quad \sum_{k=1}^{r} p_{ik} = 1. \tag{13.1}$$

A matrix verifying (13.1) is called a *stochastic matrix*.

EXERCISE 13.4 Sentry.

Assume that a sentry watches over a square stronghold having four turrets in the following way: he starts at random by one of the turrets and after each five-minute interval tosses a coin and goes to the first turret on his left (if TAILS turns up) or to the first turret on his right (if HEADS turns up).

1. Formalize the problem.
2. Let X_n be the number of the turret chosen as the nth watchtower, $n = 0, 1, \ldots$. Show that X_n is a Markov chain. What is the transition matrix? ◇

EXERCISE 13.5 Let $(X_n)_{n \in \mathbb{N}}$ be a Markov chain with transition matrix P. Let

$$L_n = \begin{pmatrix} P(X_n = 1) \\ \vdots \\ P(X_n = i) \\ \vdots \\ P(X_n = r) \end{pmatrix}$$

be the column describing the probability distribution of X_n.

1. Express L_{n+1} in terms of P and L_n.
2. Deduce L_n in terms of P and of the probability distribution L_0 of X_0.
3. Give a necessary and sufficient condition ensuring that all the X_ns have the same probability distribution. ◇

EXERCISE 13.6 Let $(X_n)_{n\in\mathbb{N}}$ be a Markov chain.
1. Show that the following properties are equivalent:
   (i)   There exists a k such that X_k and X_{k+1} are independent.
   (ii) The columns of P are constant.
   (iii) For each k, X_k and X_{k+1} are independent.
   (iv) $\forall n$, $(X_0, \ldots, X_n)$ are independent.
2. What is the distribution of X_n when the conditions of 1 are satisfied? ◇

13.2.4 Properties

In the sequel, X_n will denote a Markov chain with values in $\Omega = \{E_1, \ldots, E_r\}$. Let $q_i^{(n)} = P(X_n = E_i)$ be the (unconditional) probability that the chain is in state i at time n, and let

$$p_{ij}^{(n)} = P(X_n = E_j \,/\, X_0 = E_i)$$

be the conditional probability of a transition from E_i to E_j in exactly n steps.

Let us first state five simple lemmata about conditional probabilities that will be quite useful when computing with Markov chains.

Lemma 13.5 $P(A \cap B \,/\, C) = P(A \,/\, B \cap C) \times P(B \,/\, C)$.

Lemma 13.6 $P(A \cap B \,/\, A) = P(B \,/\, A)$.

Lemma 13.7 $q_i^{(n)} = \sum_{j=1}^{r} q_j p_{ji}^{(n)}$, $\forall i = 1, \ldots, r$.

Lemma 13.8
(i) $\forall k \geq 0$, $\forall i, j = 1, \ldots, r$, $p_{ij}^{(n)} = P(X_{n+k} = E_j \,/\, X_k = E_i)$.
(ii) $\forall n \geq 0$, $\forall i, j = 1, \ldots, r$, the $p_{ij}^{(n)}$ can be computed recursively by

$$p_{ij}^{(1)} = p_{ij}$$
$$p_{ij}^{(n+1)} = \sum_{k=1}^{r} p_{ik}\, p_{kj}^{(n)}.$$

(iii) The $p_{ij}^{(n)}$ are the coefficients of the matrix P^n (the transition matrix P multiplied by itself n times).

REMARK 13.9 All the matrices P^n are stochastic matrices.

Generalities

Lemma 13.10 (Markov property)
$$P(X_{n+1} = E_{i_{n+1}}, \ldots, X_{n+k} = E_{i_{n+k}} \,/\, X_0 = E_{i_0}, \ldots, X_n = E_{i_n})$$
$$= p_{i_n i_{n+1}} \cdots p_{i_{n+k-1} i_{n+k}}$$
$$= P(X_1 = E_{i_{n+1}}, \ldots, X_k = E_{i_{n+k}} \,/\, X_0 = E_{i_n}).$$

Proof. By induction on k.
1. If $k = 1$, it is simply the definition of Markov chains.
2. Assume that the result holds for $k \leq k_0$. Let
$$B = (X_{n+1} = E_{i_{n+1}}, \ldots, X_{n+k_0} = E_{i_{n+k_0}})$$
and $A = X_{n+k_0+1} = E_{i_{n+k_0+1}}$. We then have, by Lemma 13.5,
$$P(X_{n+1} = E_{i_{n+1}}, \ldots, X_{n+k_0} = E_{i_{n+k_0}}, X_{n+k_0+1} = E_{i_{n+k_0+1}}$$
$$\,/\, X_0 = E_{i_0}, \ldots, X_n = E_{i_n})$$
$$= P(X_{n+k_0+1} = E_{i_{n+k_0+1}}$$
$$\,/\, X_0 = E_{i_0}, \ldots, X_n = E_{i_n}, X_{n+1} = E_{i_{n+1}}, \ldots, X_{n+k_0} = E_{i_{n+k_0}})$$
$$\times P(X_{n+1} = E_{i_{n+1}}, \ldots, X_{n+k_0} = E_{i_{n+k_0}} \,/\, X_0 = E_{i_0}, \ldots, X_n = E_{i_n}).$$

We thus obtain, applying the induction hypothesis once with $k = 1$ and once with $k = k_0$,
$$P(X_{n+1} = E_{i_{n+1}}, \ldots, X_{n+k_0} = E_{i_{n+k_0}}, X_{n+k_0+1} = E_{i_{n+k_0+1}}$$
$$\,/\, X_0 = E_{i_0}, \ldots, X_n = E_{i_n})$$
$$= p_{i_{n+k_0} i_{n+k_0+1}} \times p_{i_n i_{n+1}} \cdots p_{i_{n+k_0-1} i_{n+k_0}}$$
$$= P(X_1 = E_{i_{n+k_0+1}} \,/\, X_0 = E_{i_{n+k_0}})$$
$$\times P(X_1 = E_{i_{n+1}}, \ldots, X_k = E_{i_{n+k}} \,/\, X_0 = E_{i_n}).$$

The same computation with $n = 0$ shows that
$$P(X_1 = E_{i_{n+1}}, \ldots, X_{k_0} = E_{i_{n+k_0}}, X_{k_0+1} = E_{i_{n+k_0+1}} \,/\, X_0 = E_{i_0})$$
$$= p_{i_{n+k_0} i_{n+k_0+1}} \times p_{i_n i_{n+1}} \cdots p_{i_{n+k_0-1} i_{n+k_0}}$$
$$= P(X_1 = E_{i_{n+k_0+1}} \,/\, X_0 = E_{i_{n+k_0}})$$
$$\times P(X_1 = E_{i_{n+1}}, \ldots, X_k = E_{i_{n+k_0}} \,/\, X_0 = E_{i_n}),$$
and hence the inductive step and the result. □

Lemma 13.10 states that the probability that a Markov chain which went through the states $E_{i_0}, \ldots, E_{i_n}$ goes on through the states $E_{i_{n+1}}, \ldots, E_{i_{n+k}}$ is the same as the probability that a chain starting at time 0 from state E_{i_n} then goes through the states $E_{i_{n+1}}, \ldots, E_{i_{n+k}}$.

Lemma 13.10 has several immediate consequences for expressing properties of Markov chains.

Corollary 13.11 *Let E be a state of a Markov chain. Then*

$$P(X_{n+1} \neq E, \ldots, X_{n+k} \neq E, X_{n+k+1} = E \,/\, X_0 = E_{i_0}, \ldots, X_n = E_{i_n})$$
$$= P(X_1 \neq E, \ldots, X_k \neq E, X_{k+1} = E \,/\, X_0 = E_{i_n}).$$

EXERCISE 13.7
1. Prove Corollary 13.11.
2. Show that, similarly: $P(X_{n+1} \in A_1, \ldots, X_{n+k} \in A_k \,/\, X_0 = E_{i_0}, \ldots, X_n = E_{i_k})$ $= P(X_1 \in A_1, \ldots, X_k \in A_k \,/\, X_0 = E_{i_k})$.
3. Show, finally: $P(X_{n+1} \in A_1, \ldots, X_{n+k} \in A_k \,/\, X_0 \in A'_0, \ldots, X_{n-1} \in A'_{n-1}, X_n = E_{i_k}) = P(X_1 \in A_1, \ldots, X_k \in A_k \,/\, X_0 = E_{i_k})$. This last equality can be called the generalized Markov property. ◊

13.3 Classification of states

Many properties of Markov chains are intrinsic: this means that they depend only on the transition probabilities (i.e. on the transition matrix), and they do not depend on the starting point of the chain (i.e. on the initial probability distribution). This is true for the properties studied in the present section.

13.3.1 Irreducible chains

We shall say that state E_j can be reached from state E_i if and only if there exists an n such that $p_{ij}^{(n)} > 0$. In other words, if there is a strictly positive probability of reaching E_j from E_i.

Definition 13.12 *A non-empty set of states C is said to be closed if no state outside C can be reached from any state E_i in C, i.e. C is closed if and only if $\forall E_i \in C$ and $\forall E_j \notin C$, $p_{ij} = 0$.*

If the singleton $C = \{E_i\}$ is closed, the state E_i is said to be absorbing. A Markov chain is irreducible if there exists no closed set other than the set of all states.

Lemma 13.13 *If C is closed, then $\forall n$, $p_{ij}^{(n)} = 0$.*

Proof. Straightforward by induction on n. □

Lemma 13.14 *Let C be a closed set; let P_C (resp. P_C^n) be the matrix deduced from P (resp. P^n) by deleting from P (resp. P^n) all rows and columns corresponding to the states $E_j \notin C$. Then*
1. *for all n, $P_C^n = (P_C)^n$ and*
2. *the sequence P_C^n is a sequence of stochastic matrices.*

This lemma intuitively means that we have a Markov chain on C which can be studied *per se* via the matrix P_C^n, i.e. independently of the states outside C.

Classification of states 269

Proposition 13.15 *A chain is irreducible if and only if any state can be reached from any other state.*

Proof. The 'if' part is straightforward because if there were a closed set $C \subsetneq \Omega$, then the states of the complement $\overline{C} = \Omega - C$ of C would not be reachable from the states of C.

As for the 'only if' part, we reason by contradiction and assume that there exist states E_i and E_j such that E_j is not reachable from E_i. Then, the least closed set C containing E_i cannot contain E_j, and thus C is a closed set strictly contained in the set of all states, $C \subsetneq \Omega$, a contradiction. □

13.3.2 Classification of states

A state E_i is said to be *persistent* (or *recurrent*) if the probability that the system starting from E_i eventually returns to E_i is equal to 1. Otherwise, it is said to be *transient*.

For all i, j in $\{1, \ldots, r\}$, let $f_{ij}^{(n)} = P(A_{ij}^{(n)})$, where $A_{ij}^{(n)} =$ 'the system starting from E_i reaches E_j for the first time after exactly n steps'. Let $f_{ij} = \sum_{n=1}^{\infty} f_{ij}^{(n)}$ (assuming $f_{ij}^{(0)} = 0$). Then f_{ij} is the probability that the system starting from E_i eventually reaches E_j. We thus have

$$E_i \text{ persistent} \iff f_{ii} = 1$$

and

$$E_i \text{ transient} \iff f_{ii} < 1.$$

If $f_{ii} = 1$, then the $f_{ii}^{(n)}$s for $n \in \mathbb{N}$ form a probability distribution, and we can define

$$\mu_i = \sum_{n=1}^{\infty} n f_{ii}^{(n)},$$

which represents the average waiting time in order for a system starting from E_i to come back to E_i (i.e. the mean of the r.v. $T =$ number of the first return of the chain to state E_i).

EXERCISE 13.8 We can compute the $f_{ij}^{(n)}$s recursively by the recurrence relation

$$p_{ij}^{(n)} = \sum_{k=1}^{n} f_{ij}^{(k)} p_{jj}^{(n-k)}. \qquad \diamond$$

The definition and computation suggested in the above exercise are not very simple. Fortunately, we have, for the case of finite Markov chains that are of interest to us, a simple characterization of transient states, from which we will also deduce a characterization of persistent states by noticing that any state is either transient or persistent.

Proposition 13.16 *A state E_i of a Markov chain is absorbing if and only if it satisfies either one the following equivalent conditions:*

(i) $p_{ij} = \begin{cases} 1 & \text{if } j = i, \\ 0 & \text{if } j \neq i. \end{cases}$

(ii) $P(X_n = E_i, \forall n \mid X_0 = E_i) = 1$, *where* $(X_n = E_i, \forall n)$ *denotes the event* $X_0 = X_1 = \cdots = X_n = \cdots = E_i$.

EXERCISE 13.9 Prove this result. ◇

EXERCISE 13.10 Prove that any absorbing state is persistent. ◇

The following results are immediate consequences of Proposition 13.29; we will not prove them here.

Proposition 13.17 E_i *is transient if and only if there exists an E_j such that E_j can be reached from E_i and E_i cannot be reached from E_j.*

Two states of a chain are *of the same type* if and only if either they are both transient or they are both persistent. We have:

Theorem 13.18 *All states of an irreducible chain are of the same type.*

Theorem 13.19 *If E_i is persistent, there exists a unique closed and irreducible set $C(E_i)$ containing E_i such that for any E_j, E_k in $C(E_i)$, $f_{jk} = f_{kj} = 1$. $C(E_i)$ is called the class of the persistent state E_i.*

Thus, the system, starting from any state of $C(E_i)$, will eventually reach another state of $C(E_i)$ and will never get out of $C(E_i)$.

Corollary 13.20 *A transient state cannot be reached from a persistent state.*

From now on, we will assume a technical restriction enabling us to give a simple classification of the states of a Markov chain: we will assume that all states are aperiodic, i.e. that there exists no integer $k > 1$ such that $p_{ii}^{(n)} \neq 0$ if and only if n is a multiple of k. We can then characterize transient (resp. persistent) states by the following so-called ergodicity conditions. (This terminology stems from the fact that an aperiodic persistent state of a finite Markov chain is said to be ergodic.)

Theorem 13.21 *Let X_n be a Markov chain all of whose states are aperiodic:*

(i) E_i *is transient if and only if*

$$\sum_{n=0}^{\infty} p_{ii}^{(n)} < \infty$$

Classification of states

and in that case, $\forall j$, $\sum_{n=1}^{\infty} p_{ji}^{(n)} < \infty$.
(ii) E_i is persistent (or ergodic), if and only if

$$\sum_{n=0}^{\infty} p_{ii}^{(n)} = \infty$$

and in that case $\mu_i < \infty$; moreover, for all j,

$$\sum_{n=0}^{\infty} p_{ij}^{(n)} = \begin{cases} \infty & \text{if } j \in C(E_i), \\ 0 & \text{if } j \notin C(E_i), \end{cases}$$

and $\quad \lim_{n \to \infty} p_{ji}^{(n)} = f_{ji} \mu_i^{-1}$.

The proof of this theorem is too complex to be given here, but it has many useful consequences.

Definition 13.22 *The potential matrix of a Markov chain is the matrix $U = (u_{ij})$ defined by $u_{ij} = \sum_{n=0}^{\infty} p_{ij}^{(n)}$, where $U = (1 - P)^{-1}$ is the inverse of the matrix $1 - P$.*

Corollary 13.23 *A state i of a Markov chain is persistent if and only if $u_{ii} = \infty$.*

Proposition 13.24 *Given a Markov chain X_n, we define the random variable $N_i = \sum_{n=0}^{\infty} \delta_{ni}$, where*

$$\delta_{ni} = \begin{cases} 1 & \text{if } X_n = i, \\ 0 & \text{if } X_n \neq i. \end{cases}$$

N_i represents the number of occurrences of state i from time 0 (inclusive) on:

(i) *If i is persistent, $P(N_i = \infty \mid X_0 = i) = 1$.*
(ii) *If i is transient, $P(N_i < \infty \mid X_0 = i) = 1$; N_i then has a geometric distribution and*

$$\forall k, \quad P(N_i = k \mid X_0 = i) = (1 - f_{ii}) f_{ii}^{k-1}.$$

(iii) *If i and j are transient, and $i \neq j$, $P(N_j < \infty \mid X_0 = i) = 1$; N_j then has the distribution*

$$\forall k, \quad P(N_j = k \mid X_0 = i) = f_{ij}(1 - f_{jj}) f_{jj}^{k-1}.$$

Proof. We first show (iii).

$$P(N_j = k \mid X_0 = i) = P(N_j = k, \sum_m A_m \mid X_0 = i),$$

where $A_m = (X_m = j, X_{m-1} \neq j, \ldots, X_1 \neq j, X_0 = i)$ is the event 'the system, starting from the initial state i, visits state j for the first time at the mth step'. We have the following, noting that the A_ms form a partition for $m \in \mathbb{N}$:

$$P(N_j = k \,/\, X_0 = i) = P(N_j = k, \textstyle\sum_m A_m \,/\, X_0 = i)$$

$$= \sum_m P(N_j = k, A_m \,/\, X_0 = i)$$

$$= \sum_m P(N_j = k, A_m \,/\, A_m) \times P(A_m \,/\, X_0 = i) \quad (13.2)$$

$$= \sum_m P(A_m \,/\, X_0 = i) \times P(N_j = k \,/\, A_m) \quad (13.3)$$

$$= \sum_m f_{ij}^{(m)} \times P(N_j = k \quad\quad\quad\quad\quad (13.4)$$

$$\phantom{= \sum_m f_{ij}^{(m)} \times P(N_j = k\ } /\, X_m = j, X_{m-1} \neq j, \ldots, X_1 \neq j, X_0 = i)$$

$$= \sum_m f_{ij}^{(m)} \times P(N_j = k \,/\, X_0 = j) \quad (13.5)$$

$$= P(N_j = k \,/\, X_0 = j) \times \sum_m f_{ij}^{(m)}$$

$$= P(N_j = k \,/\, X_0 = j) \times f_{ij} \,, \quad (13.6)$$

where
(13.2) follows from Lemma 13.5 applied with $A = A_m$, $B = (N_j = k)$ and $C = (X_0 = i)$,
(13.3) follows from Lemma 13.6 applied with $A = A_m$ and $B = (N_j = k)$,
(13.4) follows from the definition of the $f_{ij}^{(m)}$s and
(13.5) follows from the generalized Markov property proved in Exercise 13.7.

A similar argument shows that

$$P(N_j = k \,/\, X_0 = j) = f_{jj} \times P(N_j = k - 1, X_0 = j \,/\, X_0 = j)$$

$$= f_{jj} \times P(N_j = k - 1 \,/\, X_0 = j)$$

$$= f_{jj}^{k-1} \times P(N_j = 1 \,/\, X_0 = j)$$

$$= f_{jj}^{k-1} \times (1 - f_{jj})$$

by noting that $P(N_j = 1, X_0 = j \,/\, X_0 = j) = 1 - f_{jj}$ is the probability that the chain starting from the initial state i never comes back to that state. The probability distributions given in (ii) and (iii) follow immediately.

We then check that

$$P(N_i < \infty \,/\, X_0 = i) = \sum_{k \geq 1} P(N_i = k \,/\, X_0 = i) = \sum_{k \geq 1} f_{ii}^{k-1} \times (1 - f_{ii}) = 1. \quad \square$$

Classification of states 273

EXERCISE 13.11 Show the following properties:
1. $u_{ij} = E(N_j / X_0 = i)$. For $i \neq j$, and i and j transient, u_{ij} thus represents the average number of occurrences of j for a chain starting from the initial state i.
2. $u_{ij} = f_{ij} u_{jj}$.
3. If j is transient, then $\forall i$, $E(N_j / X_0 = i) < \infty$. ◇

Corollary 13.25 *The states of a Markov chain can be uniquely partitioned into*

$$\Omega = T \cup C_1 \cup \cdots \cup C_k ,$$

such that T consists of all the transient states, and each C_i is an irreducible closed set of persistent states. If E_j is in C_i, then $\forall E_k \in C_i$, $f_{jk} = 1$ and $\forall E_k \notin C_i$, $f_{jk} = 0$. Moreover, any finite Markov chain has at least one persistent state, and it is impossible that $\forall E_j \in C_i$, $\mu_j < \infty$.

Proposition 13.26 *Let E_i be a state of a Markov chain. The least irreducible closed set $C(E_i)$ containing E_i is*

$$C(E_i) = \{E_j \ / \ u_{ij} > 0\}.$$

Consequently, a chain is irreducible if and only if $\forall E_i, E_j$, $u_{ij} > 0$; moreover, for finite chains, we will have $\forall E_i, E_j$, $u_{ij} = \infty$.

The asset of the preceding characterizations and classifications is that they are easily generalized to infinite Markov chains. Their liability is that they are difficult to use. For finite Markov chains, there is a much simpler characterization of transient and persistent states, via a graph associated with the Markov chain. We must first recall some basic notions about graphs (see Chapter 10 for more details).

13.3.3 Graph of a finite Markov chain

Strongly connected graphs were defined in Chapter 10, Section 10.1.7. It is easy to see that a graph is *strongly connected* if it satisfies the following equivalent conditions:

(i) Any two vertices x and y are on some circuit.
(ii) For any two vertices x and y, there exist both a path with origin x and target y and a path with origin y and target x.

On the vertices of a graph, we can define an equivalence relation $x \equiv y$ if and only if
- either $x = y$,
- or x and y are on the same circuit.

We check that ≡ is indeed an equivalence relation, where the equivalence classes modulo ≡ form a partition of the set of vertices of the graph and are called the strongly connected components of the graph. Note that the equivalence classes modulo ≡ are the maximal strongly connected subgraphs of the graph, and hence are called the strongly connected components. With each graph G we can associate a *reduced graph*, whose vertices are the strongly connected components of G and whose edges are defined as follows: there is an edge whose origin is the strongly connected component C and whose target is the strongly connected component C' if and only if there exists a vertex $x \in C$ and a vertex $x' \in C'$ such that (x, x') is an edge of G (i.e. there is at least one edge going from a vertex of C to a vertex of C'). The reduced graph is, by construction, a graph without circuits.

Proposition 13.27 *Define on a graph without circuits the following relation: $x < y$ if and only if there exists a path going from x to y. The relation $<$ is an ordering.*

We now have all the tools needed for studying the classification of Markov chains by means of graphs.

Definition 13.28 (Graph of a Markov chain) *The transition matrix of a Markov chain is represented by a directed graph $G = (S, A)$, where S is a finite set of vertices and A is the 'edge' relation on S. G is equipped with a labelling $l: A \longrightarrow]0, 1]$, and is defined as follows:*

- $S = \{E_1, \ldots, E_r\} = \{1, \ldots, r\} = \Omega$.
- $(i, j) \in A$ if and only if $p_{ij} > 0$, and in this case $l((i, j)) = p_{ij}$.

EXERCISE 13.12 We consider two urns U_1 and U_2. Initially, each urn contains five balls. There are altogether (in urns U_1 and U_2) two black balls, four red balls, and four green balls. At each step, a ball is drawn from each urn and replaced in the other urn. Let X_n be the number of black balls in U_1 after n swaps.
1. Check that the X_ns form a Markov chain.
2. Draw the graph of the chain and write its transition matrix. Is the chain irreducible?
3. Find the distribution of X_0 in order for all the X_ns to have the same distribution. (Hint: refer to Exercise 13.5.) ◇

EXERCISE 13.13 Let U_1 and U_2 be two urns. Initially, each urn contains two black balls and seven red balls. At each step, we draw a ball from each urn and replace it in the other urn. Let X_n be the number of black balls in U_1 at step n, i.e. after n swaps; the initial state corresponds to $n = 0$.
1. Show that X_n is a Markov chain. What are its states? Compute the initial probabilities $q_i = P(X_0 = i)$.
2. Compute the transition probabilities of X_n. Deduce the transition matrix of the chain.

3. Draw the graph of the chain. Is the chain irreducible?
4. Let $L(n)$ be the vector of the probabilities of the states of the chain at step n. Compute $L(0), L(1)$, and $L(2)$. ◇

We have now a simple characterization of persistent and transient states.

Proposition 13.29
1. *A Markov chain is irreducible if and only if its graph is strongly connected.*
2. *A state of a Markov chain is persistent if and only if its strongly connected component is a maximal element for the order $<$ on the reduced graph associated with the graph of the Markov chain and defined in Proposition 13.27.*
3. *A state of a Markov chain is absorbing if and only if it is persistent and its strongly connected component is reduced to that single state.*

EXAMPLE 13.30 Consider the Markov chain with transition matrix

$$P = \begin{pmatrix} 1 & 0 & 0 & 0 & 0 & 0 & 0 \\ 0 & 0 & 0 & 1 & 0 & 0 & 0 \\ 2/3 & 0 & 0 & 1/3 & 0 & 0 & 0 \\ 0 & 0 & 0 & 1/2 & 1/2 & 0 & 0 \\ 0 & 0 & 1/4 & 0 & 0 & 3/4 & 0 \\ 0 & 0 & 0 & 0 & 0 & 0 & 1 \\ 0 & 0 & 0 & 0 & 0 & 1 & 0 \end{pmatrix}$$

The associated graph is described by Figure 13.1, where the strongly connected components are circled with hyphenated lines: they are thus $C_1 = \{1\}$, $C_2 = \{2\}$, $C_3 = \{3, 4, 5\}$, and $C_4 = \{6, 7\}$

The associated reduced graph is described in Figure 13.2.

The persistent states are thus the states of C_1 and C_4; we can decompose the set of states of the chain into $S = T \cup C_1 \cup C_4$, where C_1 and C_4 are the closed irreducible sets of persistent states and $T = C_2 \cup C_3$ is the set of transient states. State 1 is absorbing.

EXERCISE 13.14 Draw the graph of the Markov chain of Exercise 13.4. Which are the absorbing, persistent, transient states? ◇

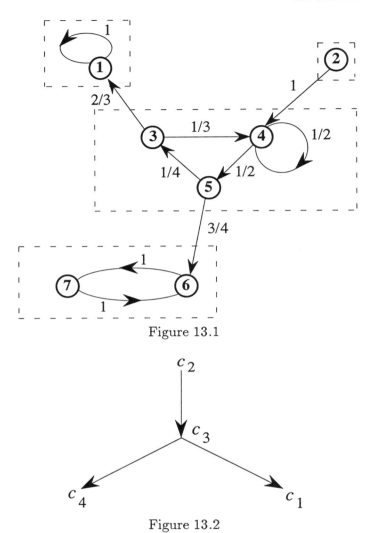

Figure 13.1

Figure 13.2

13.3.4 Probability and average waiting time for absorption

Consider a Markov chain decomposed as $S = T \cup C_1 \cup \cdots \cup C_k$. Assume that the chain starts in state E_i. If E_i is a persistent state in C_j, the chain will remain in C_j forever. If E_i is a transient state in T, then since the chain is finite, it will (almost surely) after some finite time go in one of the irreducible closed sets of persistent states C_j and will remain there forever. Then, for an arbitrary state E_i in the set $T = \{E_m \,/\, m = 1, \ldots, p\}$ of transient states, there arise the problems of determining

1. the probability that, starting from E_i, the chain ends up in C_j (called the

probability of ultimate absorption of E_i in C_j) and
2. the average waiting time after which this absorption occurs (called the *average waiting time before absorption* and corresponding to the average number of steps before absorption).

In each sequence, let C_j be a closed irreducible set of persistent states. Let λ_{in} be the probability that the chain, starting from E_i, is absorbed in C_j at the nth step exactly, i.e.

$$\lambda_{in} = P(X_n \in C_j, X_k \notin C_j \, \forall k < n \, / \, X_0 = E_i).$$

Then:

1. The probability λ_i of ultimate absorption of E_i in C_j is $\lambda_i = \sum_{n=1}^{\infty} \lambda_{in}$.
2. The average waiting time before absorption is given by $t = \sum_{n=1}^{\infty} n \lambda_{in}$.

Several methods are available for computing λ_{in} and λ_i, by which we can deduce t.

1. We note that
$$\sum_{k \in C_j} p_{ik}^{(n)} = \lambda_{i1} + \lambda_{i2} + \cdots + \lambda_{in}.$$

Hence, when n goes to infinity,

$$\lambda_i = \lim_{n \to \infty} \sum_{k \in C_j} p_{ik}^{(n)},$$

which we can obtain by studying the matrix

$$P' = \lim_{n \to \infty} P^n.$$

2. We can also write that the first passage in C_j occurs at step n if at step $n-1$ the system was still in T and if it goes from T to C_j at step n. Thus

$$\lambda_{in} = \sum_{m \in T} p_{im}^{(n-1)} \Big(\sum_{k \in C_j} p_{mk} \Big),$$

which defines a direct computation of λ_{in} in terms of the elements of the submatrix of transient elements and of its powers; the latter computation can be simpler than the former one.

3. Finally, we can write that the first passage in C_j occurs at step n if the first step took us to a transient state m and if, starting from m, we have reached C_j after $n-1$ steps exactly. This gives

$$\lambda_{in} = \sum_{m \in T} p_{im} \lambda_{m(n-1)}, \quad \forall n > 1,$$

and summing on n,

$$\lambda_i - \lambda_{i1} = \sum_{m \in T} p_{im} \lambda_m, \text{ with } \lambda_{i1} = \sum_{k \in C_j} p_{ik}. \tag{13.7}$$

We thus obtain a system of $|T| = card(T)$ linear equations in the unknowns $\lambda_1, \ldots, \lambda_{|T|}$, and we can show that this system has a unique solution.

EXERCISE 13.15 Show that when $k = 1$, i.e. when there is a single closed irreducible set C_k, it is the case that $\forall i$, $\lambda_i = 1$. ◇

4. Note, finally, that we can use the generating functions and their partial fraction expansions in order to compute the transition probabilities in n steps $p_{ij}^{(n)}$ and deduce λ_{in} and λ_i.

We introduce the generating functions $Z_i = P_{ij}(z) = \sum_{n=0}^{\infty} p_{ij}^{(n)} z^n$ and show that they are solutions of a linear equation system of the form $Z_i - z \sum_{k=1}^{r} p_{ij} Z_j = b_i$. The solutions Z_i of such a system are rational functions that are decomposed into partial fractions and then expanded into geometric power series in order to obtain the coefficients $p_{ij}^{(n)}$ of Z_i. (See Feller.)

EXERCISE 13.16 We represent the curriculum for a student on a two-year course by a Markov chain defined as follows: the success probabilities at both the first-year and second-year final exams are p; the failure probabilities at both the first-year and second-year final exams are q; the probability of dropping out of the course at the end of each year is r; we have $p + q + r = 1$. The states of the Markov chain are the two years of study denoted by 1 and 2, together with a 'drop-out' state denoted by a and a 'success' state denoted by s. The average student will doubtless be interested in the probability of success and the average waiting time for reaching success. These are computed in 4 and 7 below.

1. Draw the graph of the chain and give the transition matrix. (States may be represented by $\{1, 2, a, s\}$, where 1 and 2 represent the first and second years, and a and s represent 'drop-out' and 'success'.)
2. Is the chain irreducible? What are the transient (resp. absorbing, persistent) states?
3. Let $\lambda_{i,t}^n$ for $i = 1, 2$ and $t = a, s$ be the probabilities that the chain, given that it started from initial state i, reaches state t (with $t \in \{a, s\}$) for the first time at step n, i.e. $\lambda_{i,t}^n = P(X_0 = i, X_1 \neq t, \ldots, X_{n-1} \neq t, X_n = t / X_0 = i)$. Compute the probability that the chain, given that it started from the initial state i, reaches state a (resp. s) for the first time at time $n + 1$ in terms of the $\lambda_{i,t}^n$s.

Classification of states

4. Let $\lambda_{i,t}$ for $i = 1, 2$ and $t = a, s$ be the probabilities that the chain, given that it started from initial state i, ends up in t (with $t \in \{a, s\}$). Compute the ultimate absorption probabilities $\lambda_{i,t}$ in terms of the data p, q, r. Compute these ultimate absorption probabilities assuming $p = 0.6$, $q = 0.3$, and $r = 0.1$.

5. Let M be the matrix
$$M = \begin{pmatrix} q & p \\ 0 & q \end{pmatrix}.$$
Show by induction on n that, $\forall n \geq 0$,
$$M^n = \begin{pmatrix} q^n & nq^{n-1}p \\ 0 & q^n \end{pmatrix}$$
(with the conventions that $0^0 = 1$ and that $M^0 = \begin{pmatrix} 1 & 0 \\ 0 & 1 \end{pmatrix}$). Deduce the values of the $\lambda_{i,t}^n$s for $i = 1, 2$, $t = a, s$ and $n \in \mathbb{N}$.

6. Compute the generating function of the random variable N_i giving the number of years of study before leaving the course, assuming that the student started at year i, with $i = 1, 2$. N_i, equivalently, represents the time needed in order for the chain, given that it started from initial state i, to reach either one of the states in the set $\{s, a\}$. Deduce the average number m_i of years of study in order for a student who started at year i, with $i = 1, 2$, to leave the course.

7. Compute the average number m_i^s (resp. m_i^a) of years of study ending with final success (resp. final failure) for a student who started at year i, with $i = 1, 2$. Compute these average waiting times assuming $p = 0.6$, $q = 0.3$ and $r = 0.1$.

8. In this example, what is the intuitive meaning of the Markov chain hypothesis? ◇

CHAPTER 14

APPLICATIONS AND EXAMPLES

In the present chapter we illustrate how the various methods we have introduced in this book can be applied to

1. study complexity problems for algorithms and
2. prove correctness of programs.

There are several criteria by which to evaluate the complexity of an algorithm:

• Space complexity: for instance, the number of variables or instructions, the size of the variables, the space allowed in the memory, etc.
• Time complexity: the length of time required to execute the program; this will usually depend on the input data, and several notions of complexity may be of interest:
    − average case complexity,
    − worst-case complexity (namely, the complexity for the input data resulting in the longest possible computation),
    − best-case complexity (namely, the complexity for the input data resulting in the shortest possible computation).

One of the goals of a 'discrete mathematics' course is to build tools for evaluating these various notions of complexity. We illustrate these tools by studying the average complexity of Quicksort and the worst-case complexity of a simple algorithm: Euclid's algorithm. We will only sketch the various tools in the present chapter; for further details the reader is refferred to the chapters in which each tool is defined. This chapter owes much to the class handouts by Jean Claude Raoult on the same subject; we heartily thank him.

On the other hand, to prove the correctness of programs, most existing techniques boil down to proofs by induction and searches for loop invariants. We illustrate these techniques in the remainder of the chapter.

14.1 Quicksort

Intuitively, the complexity of an algorithm corresponds to the size $t(i_n)$ of the computations on a size n input i_n. However, $t(i_n)$ is usually not the same for all size n inputs. For i_n ranging over size n inputs:

- The worst-case complexity is the maximum value of all the $t(i_n)$s.
- The average complexity is the average value of the $t(i_n)$s.
- The best-case complexity is the least value of all the $t(i_n)$s.

We will now study a more complex example: the worst-case complexity and the average complexity of Quicksort, which is considered one of the best sorting algorithms.

14.1.1 Quicksort

Briefly recall the idea of Quicksort: for sorting a length n list, we choose a pivot p in this list and we pairwise permute the elements of the list in order to put together all the elements $\leq p$ at the beginning of the list, and all the elements $> p$ at the end of the list; the pivot is then put in its proper place (between the elements $\leq p$, and the elements $> p$), and we repeat this operation with the two sublists of elements $\leq p$, and of elements $> p$ (see Figure 14.1). Here again we have a 'divide and conquer' strategy.

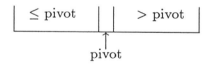

Figure 14.1

Let $T[i \ldots j]$ be the list of the elements to be sorted; the algorithm just described can be written:

```
PROGRAM Quicksort
VAR i,j,k:  integer
VAR T: integer list
BEGIN
READ i,j,T
IF i < j THEN
   pivot(T,i,j; k)
   Quicksort(T, i, k-1)
   Quicksort(T, k+1, j)
ENDIF
PRINT T
END
```

pivot is a procedure choosing $T(i)$ as pivot and permuting the elements of $T[i \ldots j]$ until $T(i)$ is put in its final place, whose index is k. All the $T(j) \leq T(i)$ are before $T(i)$ and all the $T(j) > T(i)$ are after $T(i)$. The pivot procedure uses two counters for ranging over the $T(j)$s, one of them increasing, and the other one decreasing until they meet in k, which is the final place of the pivot, then $T(i)$ and $T(k)$ are interchanged and $T(k)$ sits in its final place (Figure 14.2).

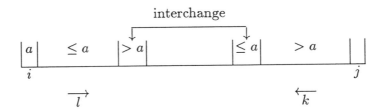

Figure 14.2

```
PROGRAM pivot
VAR i, j, k, l: integer
VAR T: integer list
BEGIN
READ i,j,T
l:= i+1
k:= j
p:= T(i)
WHILE l ≤ k DO
   WHILE T(k) > p DO k:= k-1 ENDWHILE
   WHILE T(l) ≤ p DO l:= l+1 ENDWHILE
   IF l < k THEN
      interchange (T(l),T(k))
      k:= k-1
      l:= l+1
   ENDIF
ENDWHILE
interchange (T(i),T(k))
RETURN (k)
END
```

k stops as soon as $T(k) \leq$ pivot, l stops as soon as $T(l) >$ pivot and $T(k)$ and $T(l)$ are interchanged; then, when k and l meet, namely, k becomes greater than or equal to l, we can stop and place the pivot, which will no longer move. The following example for sorting the list $100, 202, 22, 143, 53, 78, 177$, can be studied:

100	202	22	143	53	78	177
<u>53</u>	78	22	100	143	202	177
22	53	78	100	<u>143</u>	202	177
22	53	78	100	143	<u>202</u>	177
22	53	78	100	143	**177**	**202**

The pivots have been underlined and the sublists of the elements $\leq k$ and of the elements $> k$ which are left for sorting, are in bold-face.

We will study the *complexity* of Quicksort in terms of the *number of comparisons performed* on the list which we want to sort, T: if there are n elements in this list, the pivot procedure will compare the pivot to all the other elements; however, depending upon how the two counters k and l meet, we will have $n-1$, n or $n+1$ comparisons to perform.

EXERCISE 14.1 Study the various possible cases for the termination of the pivot procedure in order to check the above statement. ◇

Let $C_n(T[1,\ldots,n])$ be the complexity of Quicksort for a list $T[1,\ldots,n]$ of length n; if the pivot is put in the kth place we will have

$$C_n(T[1,\ldots,n]) = C_{k-1}(T[1,\ldots,k-1]) + C_{n-k}(T[k,\ldots,n]) \\ + \pi_n(T[1,\ldots,n]),$$

where $\pi_n(T[1,\ldots,n])$ denotes the complexity of placing the pivot for the list $(T[1,\ldots,n])$.

EXERCISE 14.2 Check that
1. $\pi_2(T[1,2]) = 2$ for $T[1,2] = (11,22)$ or $T[1,2] = (22,11)$.
2. $\pi_5(T[1,\ldots,5]) = 6$ for $T[1,\ldots,5] = (30,51,23,42,14)$. ◇

14.1.2 Worst-case complexity of Quicksort in number of comparisons

We can see that the worst-case complexity will occur when the initial list is already sorted and each call to pivot simply verifies that the first element is the least one. Let p_n be the worst-case complexity for sorting a length n list; we thus have

$$p_n = n + 1 + p_{n-1} \tag{14.1}$$

with $p_2 = 3$.

EXERCISE 14.3 Solve this recurrence relation, and evaluate p_n. ◇

14.1.3 Average complexity of Quicksort in number of comparisons

Let piv_n be the average number of comparisons performed by the procedure `pivot` on a size n list, let $c_n(k)$ be the average number of comparisons performed by the procedure `Quicksort` on a size n list assuming that the result of the `pivot` procedure was the pivot in the kth position and let c_n be the average number of comparisons performed by `Quicksort` on a size n list; we then have

$$n - 1 \leq piv_n \leq n + 1,$$

$$n - 1 + c_{k-1} + c_{n-k} \leq c_n(k) \leq n + 1 + c_{k-1} + c_{n-k}.$$

Assume a random repartition of the numbers of the list, and assume that the positions $1, \ldots, n$ are equally probable for the pivot with probability $1/n$; thus $c_n = 1/n\bigl(c_n(1) + \cdots + c_n(n)\bigr)$ and, finally, we obtain the following squeeze for c_n:

$$\forall n \geq 2, \quad n - 1 + 1/n\left(\sum_{k=1}^{n}(c_{k-1} + c_{n-k})\right) \leq c_n \leq n + 1 + 1/n\left(\sum_{k=1}^{n}(c_{k-1} + c_{n-k})\right).$$

Hence, noting that

$$\sum_{k=1}^{n} c_{k-1} = \sum_{k=1}^{n} c_{n-k} = \sum_{k=1}^{n-1} c_k \quad (\text{ since } c_0 = 0),$$

$$\forall n \geq 2, \quad n - 1 + 2/n \sum_{k=1}^{n-1} c_k \leq c_n \leq n + 1 + 2/n \sum_{k=1}^{n-1} c_k,$$

or also, letting

$$a_n = n - 1 + 2/n \sum_{k=1}^{n-1} a_k \quad \text{and} \quad b_n = n + 1 + 2/n \sum_{k=1}^{n-1} b_k, \tag{14.2}$$

$$\forall n \geq 2, \quad a_n \leq c_n \leq b_n.$$

Note, moreover, that $c_0 = c_1 = 0$, and thus we also have $a_0 = b_0 = a_1 = b_1 = 0$.

Consider the recurrence relation defining b_n. It will be evaluated by forming linear combinations of suitably chosen instances of the recurrence relation. Note, first, that

$$n(b_n - (n+1)) = 2\sum_{k=1}^{n-1} b_k,$$

hence for $n-1$: $(n-1)(b_{n-1} - n) = 2\sum_{k=1}^{n-2} b_k$ and, by subtraction,

$$nb_n - (n-1)b_{n-1} - 2n = 2b_{n-1}, \quad \text{for } n \geq 3,$$

namely, $nb_n = (n+1)b_{n-1} + 2n$, and dividing by $n(n+1)$,

$$\frac{b_n}{n+1} = \frac{b_{n-1}}{n} + \frac{2}{n+1}, \quad n \geq 3,$$

with $b_0 = b_1 = 0$ and $b_2 = 3$ (computed by the initial recurrence relation). Letting $v_n = \frac{b_n}{n+1}$, we obtain $v_n = v_{n-1} + \frac{2}{n+1}$, which immediately yields, by the summation factors method,

$$v_n = 1 + 2\sum_{k=4}^{n+1} \frac{1}{k},$$

i.e. letting $H_n = 1 + \frac{1}{2} + \frac{1}{3} + \frac{1}{4} + \cdots + \frac{1}{n}$,

$$b_n = n + 1 + 2(n+1)\left(H_{n+1} - 1 - \frac{1}{2} - \frac{1}{3}\right) = 2(n+1)\left(H_{n+1} - \frac{4}{3}\right)$$

$$= 2(n+1)\left(H_n + \frac{1}{n+1} - \frac{4}{3}\right) = 2(n+1)H_n - \frac{8n+2}{3}.$$

EXERCISE 14.4 Prove that, similarly, $a_n = 2(n+1)H_n - 4n$. ◊

Since $H_n \sim \log n$ (see Example 7.26, Definition 9.13 and Section 9.2.4), we deduce that $a_n \sim b_n \sim 2n\log n$, and thus $c_n \sim 2n\log n$.

The average complexity of Quicksort for sorting a length n list, assuming that all the possible permutations of the list are equally probable, is thus of the order of magnitude $2n\log n$. This gives us an example where c_n was in fact given by two recurrence inequations, and not by a single equation. When a recurrence c_n defining a cost is defined not by an equality but by a squeeze, we will proceed as above, and try to evaluate the order of magnitude of each part of the squeeze. If we obtain the same order of magnitude $f(n)$ for both parts, then we will have proved that c_n is also of the same order of magnitude $f(n)$.

The average complexity of Quicksort can also be computed by using generating series. Consider the recurrence relation defining a_n:

$$a_n = n - 1 + \frac{2}{n}\sum_{k=0}^{n-1} a_k \quad \text{for} \quad n \geq 2, \quad \text{and} \quad a_0 = a_1 = 0.$$

Letting $a(x) = \sum_{n\geq 0} a_n x^n$, multiplying the recurrence equation by x^n and summing for $n \geq 2$:

$$\sum_{n\geq 2} a_n x^n = \sum_{n\geq 2}(n-1)x^n + 2\sum_{n\geq 2}\left(\sum_{k=0}^{n-1} a_k\right)\frac{x^n}{n}.$$

Note that

$$\sum_{n\geq 2}(n-1)x^n = x^2 \sum_{n\geq 2}(n-1)x^{n-2} = \frac{x^2}{(1-x)^2}$$

and

$$\sum_{n\geq 2}\left(\sum_{k=1}^{n-1} a_k\right)\frac{x^n}{n} = \sum_{n\geq 1}\left(\sum_{k=0}^{n} a_k\right)\frac{x^{n+1}}{n+1} \quad \text{(since } a_0 = 0\text{)}$$

$$= \int_0^x \sum_{n\geq 0}\left(\sum_{k=0}^{n} a_k\right) s^n ds$$

$$= \int_0^x a(s) \times \frac{1}{1-s} ds.$$

Hence,

$$a(x) = \frac{x^2}{(1-x)^2} + 2\int_0^x \frac{a(s)}{1-s} ds,$$

and, taking the derivative,

$$a'(x) = \frac{2x}{(1-x)^2} + \frac{2x^2}{(1-x)^3} + \frac{2a(x)}{1-x}. \tag{14.3}$$

Equation (14.3) is a differential equation which can be solved by the usual methods:

- First solve the associated homogeneous equation

$$\frac{a'(x)}{a(x)} = \frac{2}{1-x},$$

hence

$$a(x) = \frac{\lambda}{(1-x)^2}.$$

- Then reporting the result in (14.3) and applying the method of variation of parameters, we have

$$\frac{\lambda'(x)}{(1-x)^2} + \frac{2\lambda(x)}{(1-x)^3} = \frac{2x}{(1-x)^2} + \frac{2x^2}{(1-x)^3} + \frac{2\lambda(x)}{(1-x)^3},$$

and, after simplifications,

$$\lambda'(x) = 2x + \frac{2x^2}{1-x} = 2x + \frac{2(1-x)^2 - 4(1-x) + 2}{1-x}$$
$$= -2 + \frac{2}{1-x}.$$

(We computed the partial fraction expansion of the rational function $\frac{2x^2}{1-x}$.) Integrating we obtain

$$\lambda(x) = -2x + 2\log\frac{1}{1-x} + c$$

and, since $\lambda(0) = a_0 = 0$, we have $c = 0$. Hence, finally,

$$a(x) = \frac{2}{(1-x)^2}\left(\log\frac{1}{1-x} - x\right).$$

But

$$\frac{1}{(1-x)^2}\log\frac{1}{1-x} = \sum_{n \geq 0}(n+1)H_{n+1}x^n - \frac{1}{(1-x)^2} \quad \text{(see Example 8.10)}$$
$$= \sum_{n \geq 0}(n+1)H_{n+1}x^n - \sum_{n \geq 0}(n+1)x^n$$

and

$$\frac{x}{(1-x)^2} = \sum_{n \geq 0}(n+1)x^{n+1},$$

hence,

$$a_n = 2(n+1)H_{n+1} - 2(n+1) - 2n = 2(n+1)H_n - 4n - 2.$$

We thus have

$$a_n \sim 2nH_n = \theta(n\log n)$$

since the order of magnitude of H_n is $\log n$.

EXERCISE 14.5 Verify by the same method that the upper bound b_n of the inequalities giving the average complexity of Quicksort, which is defined by the recurrence equation

$$b_n = n + 1 + \frac{2}{n}\sum_{k=1}^{n-1}b_k, \qquad n \geq 2, \qquad b_0 = b_1 = 0,$$

is equal to

$$b_n = 2(n+1)H_n - \frac{8n+2}{3}, \qquad n \geq 2. \qquad \diamond$$

14.2 Euclid's algorithm

14.2.1 Euclid's algorithm

This algorithm takes as input two integers $u, v \in \mathbb{N}$ and gives as output their *gcd*, namely, the greatest integer d such that:

$$d \mid u, \; d \mid v \quad \text{and} \quad \forall d' \; (d' \mid u \text{ and } d' \mid v \implies d' \mid d),$$

where $d \mid u$ is an abbreviation for 'd divides u'.

The idea of the algorithm is as follows: we assume $0 \leq v \leq u$. If $v = 0$ then $gcd(u, v) = d = u$. Otherwise, we use the property given below.

Lemma 14.1 *If $u = vq + r$ with $0 \leq r < v$, then $gcd(u, v) = gcd(v, r)$.*

Proof. By induction. See Exercise 14.6. □

We thus inductively substitute the computation of $gcd(v, r)$ for the computation of $gcd(u, v)$, where r is the remainder of the division of u by v. Formally, the algorithm is:

```
PROGRAM Euclid
VAR u,v,r:  integer
BEGIN
READ u,v
IF u < 0 THEN u:= -u ENDIF
IF v < 0 THEN v:= -v ENDIF
IF u = 0 and v = 0 THEN
   WRITE gcd undefined
OTHERWISE
   WHILE v ≠ 0 DO
      r:= remainder of the division of u by v
      u:= v
      v:= r
   ENDWHILE
   WRITE u
ENDIF
END
```

EXERCISE 14.6 Show that Euclid's algorithm terminates and computes the greatest common divisor of its arguments. Hint: an induction on the value of the variable v in the loop will do the job. (See Chapter 3 and Section 14.3.) ◇

The space complexity of this algorithm is quite simple: only three variables. The time complexity can be measured according to various criteria: number of operations performed, possibly multiplied by a factor measuring the cost of the operation, etc. We will choose as the complexity measure the number n of times

Euclid's algorithm

we go through the 'WHILE' loop; this implicitly assumes that the costs of the assignments and the divisions performed in a loop are integrated in the cost of executing that loop, and we count only the 'number of divisions performed' before obtaining the *gcd*.

The best-case complexity (case when $(v = 0)$) is $n = 0$. We will now study the worst-case complexity. This complexity was first studied in the eighteenth century using Fibonacci numbers.

14.2.2 Complexity of Euclid's algorithm in the worst-case

In the preceding section we studied the time complexity, and we characterized it by the number of divisions to be performed in order to obtain the *gcd*. It thus suffices to evaluate this number of divisions.

Proposition 14.2 *Let $0 < v \leq u$. Then*

(i) $\dfrac{v}{u}$ *can be represented by a continued fraction, namely, an expression of the form*

$$\frac{v}{u} = \cfrac{1}{a_1 + \cfrac{1}{a_2 + \cfrac{1}{\cdots + \cfrac{1}{a_{n-1} + \cfrac{1}{a_n}}}}}$$

where all the a_is are positive integers and $a_n \neq 1$.

(ii) *The number of divisions performed by Euclid's algorithm for obtaining the gcd of u and v is equal to the number of coefficients a_i of the continued fraction representing $\dfrac{v}{u}$.*

Before sketching the proof, note that in order to obtain the worst-case complexity it is enough to minimize u and v with respect to n, thus obtaining the least possible u and v for each n, and to this end we will see that it will suffice to minimize the a_is. We only sketch the proofs and invite the reader to fill in the details as an exercise.

Lemma 14.3 *The continued fraction*

$$e = \frac{v}{u} = \cfrac{1}{a_1 + \cfrac{1}{a_2 + \cfrac{1}{\cdots + \cfrac{1}{a_{n-1} + \cfrac{1}{a_n}}}}}$$

can be rewritten as $\dfrac{Q_{n-1}(a_2,\ldots,a_n)}{Q_n(a_1,\ldots,a_n)}$ where the Q_is are polynomials in the a_is, with positive integer coefficients, and where the a_is are the successive quotients obtained by Euclid's algorithm applied to the pair (u,v), with $u = Q_n(a_1,\ldots,a_n)$ and $v = Q_{n-1}(a_2,\ldots,a_n)$.

Proof. By induction on n.

Basis: If $n = 2$, $\dfrac{1}{a_1 + \dfrac{1}{a_2}} = \dfrac{a_2}{a_1 a_2 + 1}$ and

$$Q_1(a_2) = a_2, \quad Q_2(a_1,a_2) = a_1 a_2 + 1 \ ;$$

moreover, letting $u = a_1 v + 1$ and $v = a_2$, we have $\gcd(u,v) = 1$.

Inductive step: Assume the result holds for p, and let us prove that it holds for $n = p+1$. Let $e = \dfrac{1}{a_1 + e'}$, where

$$e' = \cfrac{1}{a_2 + \cfrac{1}{\cdots + \cfrac{1}{a_{n-1} + \cfrac{1}{a_n}}}}$$

then, by the induction hypothesis,

$$e' = \frac{v'}{u'} = \frac{Q_{p-1}(a_3,\ldots,a_n)}{Q_p(a_2,\ldots,a_n)},$$

where Q_{p-1} and Q_p are polynomials in the a_is, with positive integral coefficients, $v' < u'$, and $a_2, \ldots, a_n$ are the successive quotients obtained in the computation of $\gcd(u',v')$. Then,

$$e = \cfrac{1}{a_1 + \cfrac{v'}{u'}} = \frac{u'}{a_1 u' + v'}.$$

The integers

$$u = a_1 u' + v' = Q_n(a_1,\ldots,a_n) = a_1 Q_p(a_2,\ldots,a_n) + Q_{p-1}(a_3,\ldots,a_n),$$
$$v = u' = Q_{n-1}(a_2,\ldots,a_n) = Q_p(a_2,\ldots,a_n)$$

have the required form, and $u' < a_1 u' + v'$. The quotient of $a_1 u' + v'$ by u' is a_1, and $\gcd(a_1 u' + v', u') = \gcd(u', v')$. Hence the result. $\square$

Euclid's algorithm

REMARK 14.4
1. In the preceding proof we could have started the induction with $n = 1$ instead of $n = 2$.
2. The preceding proof, moreover, shows that the recurrence defining the polynomials Q_n is given by

$$Q_n(a_1, \ldots, a_n) = a_1 Q_{n-1}(a_2, \ldots, a_n) + Q_{n-2}(a_3, \ldots, a_n).$$

The proposition immediately follows from Lemma 14.3.

Because all the coefficients of the polynomials Q_n are positive integers, the worst-case complexity (namely, the greatest possible n for the least possible u and v) will be obtained when all the a_is with $i < n$ are equal to 1 and $a_n = 2$.

So, for a given n, the least possible u and v whose *gcd* requires n divisions are determined by the recurrence

$$u_n = Q_n(1, \ldots, 1, 2) = 1 \times Q_{n-1}(1, \ldots, 1, 2) + Q_{n-2}(1, \ldots, 1, 2),$$

and

$$v_n = Q_{n-1}(1, \ldots, 1, 2).$$

u_n and v_n are obtained by solving: $u_n = u_{n-1} + u_{n-2}$ and $v_n = u_{n-1}$. (The verification is left to the reader.)

We thus have to solve the recurrence $u_n = u_{n-1} + u_{n-2}$, with $u_0 = 1$, $u_1 = 2$.

Shortly, we sketch various methods for so doing, and refer the reader to Chapter 7 for more details about recurrence relations.

14.2.3 Fibonacci numbers

The u_ns such that $u_n = u_{n-1} + u_{n-2}$ are called the Fibonacci numbers, and, following tradition, we will denote them by F_n. We thus want to determine F_n given that

$$F_n = F_{n-1} + F_{n-2} \quad \text{and} \quad F_0 = 0 \quad F_1 = 1. \tag{14.4}$$

(We will have: $u_n = F_{n+2}$, $u_{n-1} = F_{n+1}$.)

First method: the characteristic polynomial

We look for a solution of the form $F_n = r^n$. We deduce $r^n = r^{n-1} + r^{n-2}$, and hence $r^2 = r + 1$, which has two roots $r = \dfrac{1 \pm \sqrt{5}}{2}$. Then the general solution of (14.4) is of the form: $\lambda r_1^n + \mu r_2^n$, where $r_1 = \dfrac{1 + \sqrt{5}}{2}$ and $r_2 = \dfrac{1 - \sqrt{5}}{2}$; moreover, $\lambda + \mu = F_0 = 0$ and $\lambda r_1 + \mu r_2 = 1$, and hence $\lambda = -\mu = 1/\sqrt{5}$, and $F_n = 1/\sqrt{5} \times (r_1^n - r_2^n)$.

For evaluating the complexity, we are interested in large values of n. When n tends to infinity, $\lim_{n\to\infty} r_2^n = 0$ and $\lim_{n\to\infty} r_1^n = \infty$, and thus F_n is equivalent to $r_1^n/\sqrt{5}$ when $n \to \infty$ (see Chapter 9 on asymptotic behaviours). Consequently, if u_n is the Fibonacci number F_{n+2}, and v_n is the Fibonacci number F_{n+1}, the number of divisions performed by Euclid's algorithm will be equal to n, and by the asymptotic behaviour of F_n we will have:

$$u_n = F_{n+2} \sim \frac{r_1^{n+2}}{\sqrt{5}},$$

and thus $n + 2 \sim \log_{r_1}(\sqrt{5} u_n)$ and finally $n \sim \log_{r_1}(\sqrt{5} u_n)$.

The worst-case complexity is thus at most $\log_{r_1}(\sqrt{5} u)$. This example shows us that, in order to evaluate the complexity of a very simple algorithm such as Euclid's algorithm, we already need tools as elaborate as inductive proofs, methods for solving recurrences and methods for studying asymptotic behaviours, to which chapters of this book have been devoted. Before concluding the present study, we briefly sketch another method for solving recurrences.

Second method: matrix method

$F_{n+1} = F_n + F_{n-1}$ can be written in the form of a linear system:

$$\begin{aligned} F_{n+1} &= F_n + F_{n-1} \\ F_n &= F_n \end{aligned} \quad \Longleftrightarrow \quad \begin{pmatrix} F_{n+1} \\ F_n \end{pmatrix} = \begin{pmatrix} 1 & 1 \\ 1 & 0 \end{pmatrix} \begin{pmatrix} F_n \\ F_{n-1} \end{pmatrix}.$$

Let M be the matrix

$$M = \begin{pmatrix} 1 & 1 \\ 1 & 0 \end{pmatrix}.$$

We have

$$\begin{pmatrix} F_{n+1} \\ F_n \end{pmatrix} = M \begin{pmatrix} F_n \\ F_{n-1} \end{pmatrix} = \cdots = M^n \begin{pmatrix} F_1 \\ F_0 \end{pmatrix} = M^n \begin{pmatrix} 1 \\ 0 \end{pmatrix}.$$

It thus suffices to compute the nth power of M. To this end, we compute the eigenvalues (we find r_1 and r_2) and diagonalize M. Hence,

$$N = \begin{pmatrix} r_1 & 0 \\ 0 & r_2 \end{pmatrix} = A^{-1} M A \quad \text{and} \quad M^n = A N^n A^{-1} = A \begin{pmatrix} r_1^n & 0 \\ 0 & r_2^n \end{pmatrix} A^{-1}.$$

(A is the transition matrix from M to N.)

Third method: generating series

See Example 8.2.

EXERCISE 14.7 Compute the continued fraction corresponding to v/u with $v = 8$ and $u = 29$. What is the number n of divisions needed for obtaining $gcd(8, 29)$? ◇

14.3 Proofs of program properties and termination

To conclude the present chapter, we will show how to use inductive proofs to verify properties of programs. Our goal is *not* to build a program prover but simply to show, with some examples, how to use proofs by induction in that area. Essentially, proofs by induction are useful for iterative loops as well as recursive processes (recursive procedures, functions, clauses, etc.).

Consider the following program:

```
PROGRAM square
VAR a,b,c:  integer
BEGIN
READ a
IF a < 0 THEN
    a:= -a
ENDIF
b:= a
c:= 0
WHILE b ≠ 0 DO
    c:= c + a
    b:= b - 1
ENDWHILE
WRITE c
END
```

We will show that this program computes and writes a^2. The first point to be verified is that we will eventually exit the loop. We proceed as follows:

- Before entering the loop, the value of b is a positive integer.
- The value of b is decreased at each execution of the loop. This value will thus eventually become zero.
- When the value of b is zero, we exit the loop.

Once we have proved that the loop terminates, we must check that the result is correct. For so doing, we show that the property $I(a,b,c)$: '$a^2 = c + b \times a$' holds at each execution of the loop. Let b_n and c_n be the values of the variables b and c after the nth execution of the loop. Let $P(n)$ be the property '$a^2 = c_n + b_n \times a$'. We must now verify that $\forall n \in \mathbb{N}, P(n)$. The proof is by induction:

- $b_0 = a$ and $c_0 = 0$, and thus $P(0)$ is true.
- Let $n \in \mathbb{N}$, and assume $P(n)$. We have $b_{n+1} = b_n - 1$ and $c_{n+1} = c_n + a$. We deduce $c_{n+1} + b_{n+1} \times a = c_n + a + (b_n - 1) \times a = c_n + b_n \times a = a^2$, which proves that $P(n+1)$ is true.

As a result, on exit of the loop both $b = 0$ and $I(a,b,c)$ hold. Thus $a^2 = c + 0 \times a = c$, which shows that the result of the program is a^2.

The basis for proving the correctness of a program is formalized by Hoare's *assertion method*. Let p, q be two first order formulas, and let S be a program; S is said to be *partially correct with respect to initial assertion p and final assertion q* if and only if whenever p is true for the input values of S, then q is true for the output values of S. This is denoted by: $p\{S\}q$.

S is said to be *totally correct with respect to initial assertion p and final assertion q* if and only if whenever p is true for the input values of S, then S terminates and S is partially correct with respect to p and q.

We sketch a deductive system for proving partial correctness of iterative program.

The *axioms* are:

for any assignment x:=t, and any formula p, $p(t)\{\text{x} := \text{t}\}p(x)$,

i.e. if $p(t)$ holds and we assign t to x then $p(x)$ holds.

The *inference rules* are:

composition $\qquad \dfrac{p\{S_1\}q \quad q\{S_2\}r}{p\{S_1; S_2\}r}$

conditional $\qquad \dfrac{(p \wedge b)\,\{S\}\,q \qquad ((p \wedge \neg b) \implies q)}{p\,\{\text{IF } b \text{ THEN } S \text{ ENDIF}\}\,q}$

$\qquad\qquad\quad \dfrac{(p \wedge b)\,\{S_1\}\,q \qquad (p \wedge \neg b)\,\{S_2\}\,q}{p\,\{\text{IF } b \text{ THEN } S_1 \text{ OTHERWISE } S_2 \text{ ENDIF}\}\,q}$

loop invariant $\qquad \dfrac{(p \wedge b)\,\{S\}\,p}{p\,\{\text{WHILE } b \text{ DO } S \text{ ENDWHILE}\}\,(p \wedge \neg b)}$

consequence $\qquad \dfrac{(p \implies q) \quad q\{S\}r \quad (r \implies s)}{p\{S\}s}$

The rules are to be read as: if the formula(s) above the horizontal line hold, then the formula below the horizontal line also holds. For instance, the composition rule states that: assume that if p is true and S_1 executes and terminates, then q is true and, moreover, if q is true and S_2 executes and terminates, then r is true; with those assumptions, if p is true and $S = S_1;S_2$ executes and terminates, where ';' denotes sequential composition, then r is true.

EXAMPLE 14.5

1. Consider the program segment S defined by

$$c' := c + a$$
$$b' := b - 1$$

and let p be $(a^2 = ab + c)$; then, by the consequence rule and the axioms for assignment, we have that

$$p \{c' := c + a\} (a^2 = ab + c' - a).$$

Applying the consequence rule

$$p \{c' := c + a\} (a^2 = a(b-1) + c').$$

Let q be $(a^2 = a(b-1) + c')$; then by the axioms for assignment

$$q \{b' := b-1\} (a^2 = ab' + c').$$

Applying now the composition rule, we obtain

$$(a^2 = ab + c) \{S\} (a^2 = ab' + c').$$

2. Consider the program segment S' defined by

IF $a < 0$ THEN $a := -a$ ENDIF

Applying the conditional rule and the assignment axioms, we can deduce that: $p \{S'\} q$ holds, where p is 'a is an integer' and q is $(a \geq 0)$.

3. First, we give the intuition behind the loop invariant rule. Assertion p is said to be a loop invariant if p holds before entering the loop and if p remains true at each execution of the loop. Because the loop is executed until condition b becomes false, then, when exiting the loop (if this occurs), $\neg b \wedge p$ must hold.

Consider the program segment S'' defined by

WHILE B $\neq$ 0 DO
c := c + a
b := b - 1
ENDWHILE

and let p be $(a^2 = ab + c)$, then, by the loop invariant rule and part 1 of the present example, we have that: $p \{S''\} (p \wedge (b = 0))$.

Combining parts 1, 2 and 3 of Example 14.5, with the composition rule, we conclude that program square is partially correct with respect to the initial assertion $integer(a)$ and the final assertion $(a^2 = ab + c) \wedge (b = 0)$. This shows that program square indeed computes a^2. In order to picture the whole proof at a glance, we annotate program square with the final intermediate assertions; each assertion is written to the right of the instruction after which it is true; we thus have

```
PROGRAM square
VAR a,b,c:  integer
BEGIN
   READ a                          integer(a)
   IF a < 0 THEN
      a:= -a
   ENDIF                           a ≥ 0
   b:= a
   c:= 0                           a² = ab + c
   WHILE b ≠ 0 DO
      c:= c + a
      b:= b - 1
   ENDWHILE                        (a² = ab + c) ∧ (b = 0)
   WRITE c                         (a² = c)
END
```

$integer(a)$

$a \geq 0$

$a^2 = ab + c$

$(a^2 = ab + c) \wedge (b = 0)$
$(a^2 = c)$

We can show the axioms and rules of our deductive system for proving partial correctness of iterative program are sound, so that any statement $p\{S\}q$ obtained by this deductive system will be valid. This system is, however, not complete, and it can be shown that there does not exist a complete deductive system for proving all valid partial correctness assertions, even for a toy programming language. (Hard.)

EXERCISE 14.8 Show that the program power terminates and writes the result a^k (with the convention $0^0 = 1$). ◇

```
PROGRAM power
VAR a,k,r:  integer
BEGIN
   READ a,k
   n:= k
   IF n < 0 and a = 0 THEN
      WRITE undefined result
   OTHERWISE
      r:= 1
      WHILE n < 0 DO
         r:= r / a
         n:= n+1
      ENDWHILE
      WHILE n > 0 DO
         r:= r * a
         n:= n-1
      ENDWHILE
      WRITE r
   ENDIF
END
```

The case of recursive programs is slightly more complex because several recursive calls can occur in the same program and several cases can also occur when the result is obtained without any recursive call. Recursive programs still can, however, be studied in a similar way. Consider the following procedure listing the inorder traversal of a binary tree (see Example 3.24):

```
PROCEDURE inorder(x:  BT)
BEGIN
IF x ≠ ∅ THEN
   inorder(LeftChild(x))
   WRITE root(x)
   inorder(RightChild(x))
ENDIF
END
```

In order to prove termination, we can consider the mapping $h: BT \longrightarrow \mathbb{N}$ giving the height of a binary tree. The value of $h(x)$ strictly decreases at each recursive call, namely, $h(LeftChild(x)) < h(x)$ and $h(RightChild(x)) < h(x)$. Since there can be no strictly decreasing infinite sequence in $\mathbb{N}$, we deduce that there are a finite number of recursive calls. Consequently, the procedure inorder always terminates.

Note that the choice of the mapping h is arbitrary. We could have chosen the mapping $n: BT \longrightarrow \mathbb{N}$ giving the number of nodes of a tree or even the mapping $id: BT \longrightarrow BT$. Indeed, the relation 'to be a subtree of' is a well-founded ordering on the set BT of binary trees. It is formally defined as the reflexive and transitive closure of the relation

$$\forall a \in A, \forall l, r \in BT, \quad l < (a, l, r) \text{ and } r < (a, l, r).$$

Moreover, it verifies $LeftChild(x) < x$ and $RightChild(x) < x$, which proves that the procedure inorder terminates.

Usually, to prove the termination of a recursive program we associate it with an expression that is given its values in a well-founded ordering. Most often, this expression depends only on the arguments of the recursive program. Let V be the value of the expression applied to the arguments of the program. ($V = h(x)$ in the above example.) We must then verify that, for each recursive call, the value of this expression for the parameters of the call ($h(LeftChild(x))$ and $h(RightChild(x))$ in the above example) is strictly less than V. As a result we can conclude that the program terminates.

Similarly, in order to prove a property of a recursive program, we associate with it a property P connecting the arguments of the program with its result. We then directly show that P holds in all the cases when the program terminates without recursive calls. Assuming that the property holds for all the recursive calls of the program, we prove that it still holds when the program terminates. Thus we have another inductive proof that P holds on exit of the program regardless of the values of its initial arguments. Consider, for instance, the following function:

```
FUNCTION power(a,n: integer): y: integer
BEGIN
IF n = 0 THEN
   y:=1
OTHERWISE
   IF n = 1 THEN
      y:=a
   OTHERWISE
      IF n is even THEN
         y:=power(a * a, n / 2)
      OTHERWISE
         y:=a * power(a * a, (n - 1) / 2)
      ENDIF
   ENDIF
ENDIF
RETURN(y)
END
```

Let us show that, $\forall n \geq 0$, this function computes a^n. We thus consider the property P: 'power$(a, n) = a^n$' and we prove that it holds when the function returns its value y:

• There are two cases where the function directly returns a value: if $n = 0$, the result is $1 = a^0$, and if $n = 1$, the result is $a = a^1$. The property P is thus verified in both cases.

• If n is even, the result is power$(a \times a, n/2)$. By the induction hypothesis we have power$(a \times a, n/2) = (a \times a)^{n/2} = a^n$. The property P still holds.

• Similarly, if n is odd, the result is $a \times$ power$(a \times a, (n-1)/2)$. By the induction hypothesis we have

$$\text{power}(a \times a, (n-1)/2) = (a \times a)^{(n-1)/2} = a^{n-1}.$$

We deduce that the result is $a \times a^{n-1} = a^n$ and property P still holds.

This can be formalized as an Hoare inference rule for recursion. Let

$$f(x : y) \quad \text{body}$$

be a recursive procedure with argument x, result y and defined by body; let p (resp. q) be an initial (resp. final) assertion; we extend Hoare's inference rules for iterative programs by an inference rule for recursion:

recursion $\quad \dfrac{(p\{f(x:y)\}q) \Longrightarrow (p\{\text{body}\}q)}{p\,\{f(x:y)\,\text{body}\}\,q}$,

which means: if, when we assume the partial correctness of all internal calls with respect to p and q we can prove the partial correctness of body with respect to p and q, then the recursive procedure $f(x:y)$ body is indeed partially correct with respect to p and q.

In order to apply this rule to the preceding program power, note that p is: $integer(a,n)$, and q is: $y = a^n$. power annotated with the final assertions is given below:

```
FUNCTION power(a,n: integer): y: integer
BEGIN
IF n = 0 THEN
   y:= 1                                    (n = 0) ∧ (y = 1)
OTHERWISE
   IF n = 1 THEN
      y:= a                                 (n = 1) ∧ (y = a)
   OTHERWISE
      IF n is even THEN
         y:= power(a*a, n/2)                (even(n)) ∧ (y = (a × a)^(n/2))
      OTHERWISE
         y:= a*power(a*a, (n-1)/2)          (odd(n)) ∧ (y = a × (a × a)^((n-1)/2))
      ENDIF
   ENDIF
ENDIF                                       (y = a^n)
RETURN(y)                                   (y = a^n)
END
```

EXERCISE 14.9   Show that $\forall n \in \mathbb{N}$, the call $\text{Fact}(n)$ of the Fact function defined below terminates and computes $n!$. ◇

```
FUNCTION Fact(n: integer): y: integer
BEGIN
IF n = 0 THEN y:= 1
OTHERWISE y:= n*Fact(n-1)
ENDIF
RETURN(y)
END
```

Let the Ackermann function be defined by

```
FUNCTION Ackermann(n,m:  integer): y:  integer
BEGIN
IF n = 0 THEN
   y:= (m + 1)
OTHERWISE
   IF m = 0 THEN
      y:= (Ackermann(n - 1,1))
   OTHERWISE
      y:= (Ackermann(n - 1, Ackermann(n, m - 1))
   ENDIF
ENDIF
RETURN(y)
END
```

EXERCISE 14.10 Show that $\forall n, m \in \mathbb{N}$, the call Ackermann$(n, m)$ terminates. Hint: use the lexicographic ordering on $\mathbb{N}^2$ (see Example 2.30). ◇

EXERCISE 14.11 Show that in the PROLOG program given below, the call Q(x) terminates with the result *true*, $\forall x \in \mathbb{N}$. ◇

$$Q(x) \longleftarrow (x = 0)$$
$$Q(x) \longleftarrow (y = x - 1) \wedge Q(y)$$

EXERCISE 14.12
1. Prove that Quicksort and pivot terminate.
2. Prove that Quicksort is partially correct with respect to the final assertion q
$i \leq k \leq l \leq j \Longrightarrow \bigl(T(k) \leq T(l)\bigr)$. ◇

CHAPTER 15

ANSWERS TO EXERCISES

Chapter 1

1.1. We have $A \cap \overline{B} = (A \cap \overline{A}) \cup (A \cap \overline{B}) = A \cap (\overline{A} \cup \overline{B}) = A \cap \overline{A \cap B}$. Similarly, $A \cap \overline{C} = A \cap \overline{A \cap C}$, and we deduce $A \cap B = A \cap C \Longrightarrow A \cap \overline{B} = A \cap \overline{C}$.
Since $\overline{\overline{X}} = X$ for any subset X of E, this also proves the converse:
$$A \cap \overline{B} = A \cap \overline{C} \Longrightarrow A \cap \overline{\overline{B}} = A \cap \overline{\overline{C}} \Longrightarrow A \cap B = A \cap C.$$

1.2. Assume $A \cup B \subseteq A \cup C$ and $A \cap B \subseteq A \cap C$. We have $B \subseteq A \cup B \subseteq A \cup C$. Thus
$$B \subseteq B \cap (A \cup C) = (B \cap A) \cup (B \cap C) \subseteq (A \cap C) \cup (B \cap C) = (A \cup B) \cap C \subseteq C.$$
Let us show that $B = C \iff (A \cup B = A \cup C$ and $A \cap B = A \cap C)$. The implication $\Longrightarrow$ is straightforward. Conversely, applying the preceding result twice we have $B \subseteq C$ and $C \subseteq B$, and hence $B = C$.

1.3. By the definition of the symmetric difference we have $A \triangle B = (A \cap \overline{B}) \cup (\overline{A} \cap B)$. Thus
$$(A \triangle B) \cap (A \triangle C) = \big((A \cap \overline{B}) \cup (\overline{A} \cap B)\big) \cap \big((A \cap \overline{C}) \cup (\overline{A} \cap C)\big)$$
$$= \Big((A \cap \overline{B}) \cap \big((A \cap \overline{C}) \cup (\overline{A} \cap C)\big)\Big)$$
$$\cup \Big((\overline{A} \cap B) \cap \big((A \cap \overline{C}) \cup (\overline{A} \cap C)\big)\Big)$$
$$= (A \cap \overline{B} \cap \overline{C}) \cup (\overline{A} \cap B \cap C)$$
$$= (A \cap \overline{B \cup C}) \cup (\overline{A} \cap (B \cap C)).$$
Because $\overline{B \cup C} \subseteq \overline{B \cap C}$, we have
$$(A \triangle B) \cap (A \triangle C) \subseteq (A \cap \overline{B \cap C}) \cup (\overline{A} \cap (B \cap C)) = A \triangle (B \cap C).$$
Because A and $\overline{A}$ are disjoint, equality holds if and only if $A \cap \overline{B \cup C} = A \cap \overline{B \cap C}$. We have
$$A \cap \overline{B \cup C} = A \cap \overline{B \cap C} \iff A \cap \overline{B} \cap \overline{C} = A \cap (\overline{B} \cup \overline{C})$$
$$\iff A \cap \overline{B} \cap \overline{C} = (A \cap \overline{B}) \cup (A \cap \overline{C})$$
$$\iff A \cap \overline{B} = A \cap \overline{C}.$$
Hence, using Exercise 1.1, equality holds if and only if $A \cap B = A \cap C$. Similarly, we prove that $A \triangle (B \cup C) \subseteq (A \triangle B) \cup (A \triangle C)$ and that equality holds if and only if $A \cap (B \cup C) = A \cap B \cap C$.

301

1.4. 1. Let $x \in A \bigcup \left(\bigcap_{i \in I} B_i \right)$. If $x \in A$ then $\forall i \in I, x \in A \cup B_i$, and thus $x \in \bigcap_{i \in I}(A \cup B_i)$. Otherwise, $\forall i \in I, x \in B_i$, and hence $x \in A \cup B_i$ and we again have $x \in \bigcap_{i \in I}(A \cup B_i)$. We have thus shown that $A \bigcup \left(\bigcap_{i \in I} B_i \right) \subseteq \bigcap_{i \in E}(A \cup B_i)$. Conversely, if $x \in \bigcap_{i \in I}(A \cup B_i)$, we have $\forall i \in I, x \in A \cup B_i$. Hence, either $x \in A$ or $\forall i \in I, x \in B_i$. We thus have $x \in A \bigcup \left(\bigcap_{i \in I} B_i \right)$, and this proves the reverse inclusion.

2.
$$x \in A \cap \left(\bigcup_{i \in I} B_i \right) \iff \exists i \in I : x \in A \cup B_i$$
$$\iff x \in \bigcup_{i \in I}(A \cup B_i).$$

1.5. $\forall x \in E, f(x) \in F = \bigcup_{i \in I} A_i$. Thus $\exists i \in I, f(x) \in A_i$. We then have $x \in f^{-1}(A_i)$ and, consequently, $E = \bigcup_{i \in I} f^{-1}(A_i)$.

Let $i, j \in I$ be such that $i \neq j$, $f^{-1}(A_i) \cap f^{-1}(A_j) = f^{-1}(A_i \cap A_j) = f^{-1}(\emptyset) = \emptyset$. The family $\left(f^{-1}(A_i) \right)_{i \in I}$ is hence *almost* a partition of E.

A necessary and sufficient condition for all the sets $f^{-1}(A_i)$ to be non-empty is that $\forall i, A_i \cap Im(f) \neq \emptyset$. This condition is verified if f is surjective. Thus, a sufficient condition for all the sets $f^{-1}(A_i)$ to be non-empty is that f is surjective.

1.6. Assume that the image by f of any partition of E is a partition of F.

Let (E) be the rough partition of E into a single set. $f(E)$ is a partition of F. Hence, $F = f(E)$, which means that f is surjective.

Let $(\{x\})_{x \in E}$ be the discrete partition of E into singleton sets. $(\{f(x)\})_{x \in E}$ is a partition of F, and hence $\forall x, y \in E, x \neq y \Longrightarrow \{f(x)\} \cap \{f(y)\} = \emptyset$. That is, $f(x) \neq f(y)$. Hence, f is injective.

Conversely, assume that f is bijective. Let $(A_i)_{i \in I}$ be any partition of E. We have

- $\bigcup_{i \in I} f(A_i) = f(\bigcup_{i \in I} A_i) = f(E) = F$ because f is surjective.
- Let $i, j \in I$ be such that $i \neq j$. Since $(A_i)_{i \in I}$ is a partition, $A_i \cap A_j = \emptyset$, and since f is injective, $f(A_i) \bigcap f(A_j) = f(A_i \bigcap A_j) = f(\emptyset) = \emptyset$.
- $\forall i \in I, f(A_i) \neq \emptyset$, because $A_i \neq \emptyset$ and f is a mapping. Hence, $\left(f(A_i) \right)_{i \in I}$ is a partition of F.

1.7. 1. It is clear that the inclusion $X \subseteq f^{-1}(f(X))$ always holds. Assume f is injective and let $X \subseteq A$. If $x \in f^{-1}(f(X))$ then $f(x) \in f(X)$, and hence $\exists y \in X$ such that $f(x) = f(y)$. Because f is injective, we have that $x = y$, and hence $x \in X$. Finally, $f^{-1}(f(x)) \subseteq X$, and thus $X = f^{-1}(f(X))$.

Conversely, let $x, y \in A$ be such that $f(x) = f(y)$. Letting $X = \{x\}$, we have

$$f(y) \in f(X) = \{f(x)\}.$$

Hence, $y \in f^{-1}(f(X)) = X = \{x\}$. Hence, $y = x$, and f is injective.

2. Here, too, the inclusion $f(f^{-1}(Y)) \subseteq Y$ always holds. Assume f is surjective and let $Y \subseteq B$. $\forall y \in Y, \exists x \in A$ such that $f(x) = y$. Hence, $x \in f^{-1}(Y)$ and $y = f(x) \in f(f^{-1}(Y))$. We thus have $Y \subseteq f(f^{-1}(Y))$, and therefore $Y = f(f^{-1}(Y))$.

Conversely, let $y \in B$ and $Y = \{y\}$. We have $y \in Y = f(f^{-1}(Y))$, and hence $\exists x \in f^{-1}(Y) \subseteq A$ such that $f(x) = y$. Hence, f is surjective.

Chapter 1

1.8. Note, first, that the inclusion $f(X \cap Y) \subseteq f(X) \cap f(Y)$ always holds.
Assume f is injective. Let $X, Y \subseteq A$ and let $z \in f(X) \cap f(Y)$. $\exists x \in X : f(x) = z$ and $\exists y \in Y : f(y) = z$. Because f is injective, $f(x) = z = f(y)$ implies $x = y$. Hence, $x = y \in X \cap Y$ and $z = f(x) \in f(X \cap Y)$. Thus, $f(X) \cap f(Y) \subseteq f(X \cap Y)$, and therefore $f(X) \cap f(Y) = f(X \cap Y)$.
Conversely, let $x, y \in A$ be such that $f(x) = f(y)$. Let $X = \{x\}$ and $Y = \{y\}$. We have $f(x) = f(y) \in f(X) \cap f(Y) = f(X \cap Y)$. Hence, $f(X \cap Y) \neq \emptyset$, and thus $X \cap Y \neq \emptyset$. This implies $x = y$.

1.9. 1. Let us show, first, that f is injective if and only if $A \cup B = E$.
Note, first, that because intersection distributes over union we have
$$X \cap (A \cup B) = (X \cap A) \cup (X \cap B).$$
Assume that $A \cup B = E$. Let $X, Y \subseteq E$ be such that $f(X) = f(Y)$. We have
$$X = X \cap E = X \cap (A \cup B) = (X \cap A) \cup (X \cap B) = (Y \cap A) \cup (Y \cap B) = Y.$$
f is thus injective.
To prove the converse, consider its contrapositive. Assume $A \cup B \subsetneq E$. Let $x \in E \setminus (A \cup B)$. We have $f(\{x\}) = (\emptyset, \emptyset) = f(\emptyset)$ and $\{x\} \neq \emptyset$. Hence, f is not injective.
2. Let us now show that f is surjective if and only if $A \cap B = \emptyset$.
Assume $A \cap B = \emptyset$. Let $(X, Y) \in \mathcal{P}(A) \times \mathcal{P}(B)$. Because $X \subseteq A$, $X \cap A = X$; because $Y \subseteq B$, $(Y \cap A) \subseteq (B \cap A) = \emptyset$. We thus have $(X \cup Y) \cap A = (X \cap A) \cup (Y \cap A) = X \cup \emptyset = X$. For the same reasons $(X \cup Y) \cap B = Y$. Hence, $f(X \cup Y) = (X, Y)$, and f is surjective.
We prove the converse by contradiction. Assume f is surjective and $A \cap B \neq \emptyset$. Let $x \in A \cap B$ and let X be a preimage by f of $(\{x\}, \emptyset)$; i.e. $X \cap A = \{x\}$ and $X \cap B = \emptyset$. We have $x \in X \cap A \subseteq X$ and $x \in B$, and hence $x \in X \cap B$, which contradicts $X \cap B = \emptyset$.
3. We deduce from the two preceding points that f is bijective if and only if $A \cup B = E$ and $A \cap B = \emptyset$. If, moreover, we assume that A and B are non-empty, this can be stated by saying that (A, B) is a partition of E.

1.10. (i) Let $x, y \in E$ be such that $(g \circ f)(x) = (g \circ f)(y)$. Because g and f are injective we deduce that $f(x) = f(y)$ and this implies that $x = y$. Hence, $g \circ f$ is injective.
(ii) Let $z \in G$. Because g is surjective, there exists $y \in F$ such that $g(y) = z$. Because f is surjective, there exists $x \in E$ such that $f(x) = y$. We deduce that $(g \circ f)(x) = z$, which proves that $g \circ f$ is surjective.

1.11. 1. (i) Assume that $g \circ f$ is injective. $\forall x, y \in A$, we have
$$f(x) = f(y) \implies g(f(x)) = g(f(y)) \implies x = y.$$
Hence, f is injective.
(ii) Assume that $g \circ f$ is surjective. $\forall z \in C, \exists x \in A$ such that $(g \circ f)(x) = z$. Hence, $\exists y \in B$ $(y = f(x))$ such that $g(y) = z$, and so g is surjective.
2. Let $f: \{1\} \longrightarrow \{1, 2\}$ be defined by $f(1) = 1$ and $g: \{1, 2\} \longrightarrow \{1\}$ be defined by $g(1) = g(2) = 1$. The mapping $g \circ f$ is a bijection (it is the identity), f is not surjective and g is not injective.
3. (i) Let $x, y \in B$. Since f is surjective, $\exists x', y' \in A$ such that $f(x') = x$ and $f(y') = y$. We have
$$g(x) = g(y) \implies g \circ f(x') = g \circ f(y')$$
$$\implies x' = y'$$
$$\implies x = y.$$
Hence, g is injective.
(ii) Let $y \in B$. Since $g \circ f$ is surjective, $\exists x \in A$ such that $(g \circ f)(x) = g(y)$. Because g is injective, we deduce that $f(x) = y$. Hence, f is surjective.

1.12. 1. Using Exercise 1.11, we deduce from the hypotheses that f, $g \circ f$, h, and $f \circ h$ are injective and that f and $f \circ h$ are surjective. Hence, f is bijective. From $f \circ h$ surjective and f injective we deduce that h is surjective. Consequently, h is bijective.

Lastly, from $g \circ f$ injective and f surjective we deduce that g is injective, and from $f \circ h \circ g$ surjective and $f \circ h$ injective we deduce that g is surjective. Hence, g is also bijective.

2. idem.

1.13. (i) ($\Longrightarrow$) Assume f is injective. Choose any element x_0 of E. (This is possible because $E \neq \emptyset$.) Let $y \in F$. If y has a preimage x by f then this preimage is unique since f is injective; then let $r(y) = x$. Otherwise, let $r(y) = x_0$. For any x of E, x is the unique preimage of $f(x)$. We thus have that $\forall x \in E, \quad r \circ f(x) = r(f(x)) = x$.

(ii) ($\Longrightarrow$) Let $y \in F$. Because f is surjective, there exist elements x in E such that $y = f(x)$. Choose an arbitrary element x among those and let $s(y) = x$. $\forall y \in F$ we have $f(s(y)) = y$, and thus $f \circ s = id_F$.

(iii) ($\Longrightarrow$) Assume f is bijective. It suffices to show that the mapping s defined above also verifies $s \circ f = id_E$. Let $x \in E$ and $y = f(x)$. Because f is injective, x is the unique preimage of y by f. We thus have $s(y) = x$ and, consequently, $s(f(x)) = x$.

The converses are deduced from Exercise 1.11, together with the fact that r is surjective, s is injective and f^{-1} is bijective.

1.14. Let us prove by induction on n that if there exists an injection from $[m]$ to $[n]$ then $m \leq n$.

If $n = 0$ then $[n] = \emptyset$, and if there exists a mapping from $[m]$ to $\emptyset$ then it must be true that $[m] = \emptyset$, and so $m = 0$.

Let n be a fixed integer. Assume that the existence of an injection from $[m]$ to $[n]$ implies $m \leq n$. Let $f: [m] \longrightarrow [n+1]$ be an injection. If $f^{-1}(n+1) = \emptyset$ then $f: [m] \longrightarrow [n]$ is an injection, and by the hypothesis we have that $m \leq n$, and hence $m \leq n+1$. Otherwise, let p be the unique preimage of $n+1$ by f. If $p = m$ then the mapping $g: [m-1] \longrightarrow [n]$ defined by $g(x) = f(x)$ is an injection. (g is the restriction of f to $[m-1]$.) Hence, $m - 1 \leq n$, and thus $m \leq n+1$. Lastly, if $p \neq m$ then the mapping $g: [m-1] \longrightarrow [n]$ defined by

$$g(x) = \begin{cases} f(x) & \text{if } x \neq p, \\ f(m) & \text{if } x = p, \end{cases}$$

is an injection. Here, too, we have $m - 1 \leq n$, and so $m \leq n+1$.

1.15. (i) Let $f: E \longrightarrow [n]$ and $g: F \longrightarrow [p]$ be two bijections. We can easily verify that the mapping $h: E \cup F \longrightarrow [n+p]$ defined by $\forall x \in E, h(x) = f(x)$ and $\forall x \in F, h(x) = g(x) + n$ is a bijection. We deduce that $|E \cup F| = n + p = |E| + |F|$.

(ii) Consequence of (i).

(iii) With the same notations as for (i), we define the mapping $h: E \times F \longrightarrow [n \times p]$ by letting $\forall (x, y) \in E \times F, h(x, y) = p \times (f(x) - 1) + g(y)$. We check that h is a bijection and hence $|E \times F| = |E| \times |F|$. We can also use (ii) to prove this result because the family $(E \times \{y\})_{y \in F}$ is a partition of $E \times F$.

(iv) If $E = \{a_1, \ldots, a_n\}$, then a mapping $f: E \longrightarrow F$ is fully specified by the images $b_1, \ldots, b_n$ of $a_1, \ldots, a_n$. Each b_i, where $i \in [n]$, can be chosen in $|F|$ possible ways. We thus have $|F|^n = |F|^{|E|}$ possible choices for f. So, there is a bijection from the set of mappings of E to F to the Cartesian product F^n, and this is the reason why the set of mappings from E to F is denoted by F^E.

(v) The mapping associating with each subset A of E its characteristic function χ_A is a bijection. Thus $|\mathcal{P}(E)| = |\{0,1\}^E| = 2^{|E|}$.

1.16. Let $A = [m]$ be the set of pigeons to be placed and let A_i be the set of pigeons nesting in the pigeonhole i, for $i \in [n]$. $A_1, \ldots, A_n$ form a partition of A, and thus

$$|A| = \sum_{i=1}^{n} |A_i|.$$

If we had that $\forall i \in [n]$, $|A_i| < m/n$, we would deduce $|A| = \sum_{i=1}^{n} |A_i| < n \times m/n = m$, a contradiction; we thus have that $|A_i| \geq m/n$ for at least one i.

1.17. Note, first, that the existence of a surjection from E to F is equivalent to the existence of an injection from F to E. So let $f \colon E \longrightarrow F$ and $g \colon F \longrightarrow E$ be two injections. We define the following sets by induction:

- $E_0 = E$ and $F_0 = F$.
- $F_{n+1} = f(E_n)$, $F'_{n+1} = F_n \setminus F_{n+1}$, $E'_{n+1} = g(F'_{n+1})$, and $E_{n+1} = E_n \setminus E'_{n+1}$.

We then verify that $X = \bigcap_{n \geq 0} E_n$ and $X' = \bigcup_{n \geq 1} E'_n$ almost form a partition of E, i.e. $X \cap X' = \emptyset$ and $E = X \cup X'$ with X or X' possibly empty. Indeed, let $x \in X'$. $\exists n \geq 1$ such that $x \in E'_n$. By the definition of E_n, $x \notin E_n$ and hence $x \notin X$. The sets X and X' are hence disjoint. Let $x \in E \setminus X$ and let $n \geq 0$ be such that $x \in E_n$ and $x \notin E_{n+1}$. We necessarily have $x \in E'_{n+1} \subseteq X'$. Thus $E = X \cup X'$.

Similarly, $Y = \bigcap_{n \geq 0} F_n$ and $Y' = \bigcup_{n \geq 1} F'_n$ almost form a partition of E.

Note that $X' \subseteq Im(g)$ because $\forall n \geq 1$, $E'_n = g(F'_n)$. Because g is injective, any element x of $Im(g)$, and hence of X', has a unique preimage by g denoted by $g^{-1}(x)$. The mapping $h \colon E \longrightarrow F$ is hence fully defined by

$$h(x) = \begin{cases} f(x) & \text{if } x \in X, \\ g^{-1}(x) & \text{if } x \in X'. \end{cases}$$

It remains to check that it is a bijection.

Let $x, y \in E$ and assume $h(x) = h(y)$. If $x, y \in X$ then $x = y$ because f is injective. If $x, y \in X'$ then $x = g(g^{-1}(x)) = g(h(x)) = g(h(y)) = y$. Finally, note that $f(X) = \bigcap_{n \geq 0} f(E_n) = Y$ and that $g(Y') = \bigcup_{n \geq 1} g(F'_n) = \bigcup_{n \geq 1} E'_n = X'$. Because g is injective, we deduce that $g^{-1}(X') = Y'$. Hence, $x \in X$ and $y \in X'$ cannot hold because in this case $h(x) = f(x) \in f(X) = Y$ and $h(y) = g^{-1}(y) \in g^{-1}(X') = Y'$. We would then have $h(x) = h(y) \in Y \cap Y' = \emptyset$. Hence, h is injective. We finally show that h is surjective. Let $y \in F$. If $y \in Y'$ then $g(y) \in g(Y') = X'$, and hence $h(g(y)) = g^{-1}(g(y)) = y$. If $y \in Y$ then $y \in F_1 = Im(f)$. Hence, y has a unique preimage x by f. We then show that $\forall n \geq 0$, $x \in E_n$. Indeed, we have $y \in Y \subseteq F_{n+1} = f(E_n)$, and because f is injective this implies that $x \in E_n$. Hence, $x \in X = \bigcap_{n \geq 0} E_n$, and $h(x) = f(x) = y$.

1.18. Consider the mapping $f \colon \mathbb{N}^2 \longrightarrow \mathbb{N}$ defined by

$$f(x, y) = y + \bigl(0 + 1 + 2 + \cdots + (x+y)\bigr).$$

This mapping gives a diagonal enumeration of the elements of $\mathbb{N}^2$ as indicated in Figure 15.1.

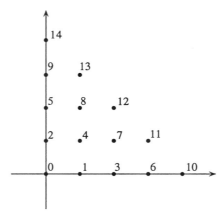

Figure 15.1

We show that f is bijective. Let $n \in \mathbb{N}$. There exists a unique $a \in \mathbb{N}$ such that

$$(0 + 1 + \cdots + a) \leq n < (0 + 1 + \cdots + (a+1)).$$

The unique preimage of n by f is given by $y = n - (0 + 1 + \cdots + a)$ and $x = a - y$.

A different bijection from $\mathbb{N}^2$ to $\mathbb{N}$ is given by the function $g: \mathbb{N}^2 \longrightarrow \mathbb{N}$ defined by $g(x, y) = 2^x(2y + 1) - 1$. The fact that it is a bijection follows from the remark that any strictly positive integer is the product of a power of two and an odd number.

1.19. By contradiction. Assume that U is countable and let $(x_n)_{n \in \mathbb{N}}$ be an enumeration of U. We can represent the values $(x_n)_{n \in \mathbb{N}}$ by the following table:

x	0	1	2	$\ldots$	n	$\ldots$
x_0	x_0^0	x_0^1	x_0^2	$\ldots$	x_0^n	$\ldots$
x_1	x_1^0	x_1^1	x_1^2	$\ldots$	x_1^n	$\ldots$
x_2	x_2^0	x_2^1	x_2^2	$\ldots$	x_2^n	$\ldots$
$\vdots$	$\vdots$	$\vdots$	$\vdots$	$\ddots$	$\vdots$	$\ddots$
x_n	x_n^0	x_n^1	x_n^2	$\ldots$	x_n^n	$\ldots$
$\vdots$	$\vdots$	$\vdots$	$\vdots$	$\ddots$	$\vdots$	$\ddots$

We then apply a so-called *diagonalization* argument: let u be the sequence defined by

$$\forall n \in \mathbb{N}, \quad u_n = \begin{cases} 0 & \text{if } x_n^n = 1, \\ 1 & \text{if } x_n^n = 0. \end{cases}$$

Then u cannot be equal to any x_n because $u_n \neq x_n^n$ for any n.

This also proves that $\mathcal{P}(\mathbb{N})$ is uncountable because $\mathcal{P}(\mathbb{N})$ is in bijection with U. To this end we associate with each subset A of $\mathbb{N}$ the sequence u defined by the characteristic function χ_A of A (i.e. $u_n = \chi_A(n)$ for any n in $\mathbb{N}$).

1.20. Let $\mathbb{1}$ and $\mathbb{1}'$ be two units. We have $\mathbb{1}' = \mathbb{1}' * \mathbb{1} = \mathbb{1}$.

1.21. 1. $(e, e') \in (\mathcal{R}_1 \cup \mathcal{R}_2)^{-1}$ if and only if $(e', e) \in \mathcal{R}_1 \cup \mathcal{R}_2$ if and only if $((e', e) \in \mathcal{R}_1$ or $(e', e) \in \mathcal{R}_2)$ if and only if $((e, e') \in (\mathcal{R}_1)^{-1}$ or $(e, e') \in (\mathcal{R}_2)^{-1})$ if and only if $(e, e') \in (\mathcal{R}_1)^{-1} \cup (\mathcal{R}_2)^{-1}$.

2. $(e, e') \in (\mathcal{R}_1 \cap \mathcal{R}_2)^{-1}$ if and only if $(e', e) \in \mathcal{R}_1 \cap \mathcal{R}_2$ if and only if $((e', e) \in \mathcal{R}_1$ and $(e', e) \in \mathcal{R}_2)$ if and only if $((e, e') \in (\mathcal{R}_1)^{-1}$ and $(e, e') \in (\mathcal{R}_2)^{-1})$ if and only if $(e, e') \in (\mathcal{R}_1)^{-1} \cap (\mathcal{R}_2)^{-1}$.

3. $(e, e') \in (\overline{\mathcal{R}})^{-1}$ if and only if $(e', e) \in \overline{\mathcal{R}}$ if and only if $(e', e) \notin \mathcal{R}$ if and only if $(e, e') \notin \mathcal{R}^{-1}$ if and only if $(e, e') \in \overline{\mathcal{R}^{-1}}$.

4. Assume $\mathcal{R}_1 \subseteq \mathcal{R}_2$. If $(e, e') \in \mathcal{R}_1^{-1}$, then $(e', e) \in \mathcal{R}_1$, and hence $(e', e) \in \mathcal{R}_2$ and $(e, e') \in \mathcal{R}_2^{-1}$. Conversely, assume $\mathcal{R}_1^{-1} \subseteq \mathcal{R}_2^{-1}$. If $(e, e') \in \mathcal{R}_1$, then $(e', e) \in \mathcal{R}_1^{-1}$, and hence $(e', e) \in \mathcal{R}_2^{-1}$ and $(e, e') \in \mathcal{R}_2$.

5. $(e, e') \in (\mathcal{R}^{-1})^{-1}$ if and only if $(e', e) \in \mathcal{R}^{-1}$ if and only if $(e, e') \in \mathcal{R}$.

6. We know that $\mathcal{R}^* = Id_E \cup \mathcal{R}^+$. But $\mathcal{R}^+ = \bigcup_{i>0} \mathcal{R}^i$, and as $Id_E \subseteq \mathcal{R}$, we have $Id_E \subseteq \mathcal{R}^+$, and hence $\mathcal{R}^* = \mathcal{R}^+$.

The converse is false. Let $\mathcal{R}$ be the relation defined by: $x \mathrel{\mathcal{R}} y \iff x \neq y$ on a set E with at least two distinct elements. Then $Id_E \not\subseteq \mathcal{R}$, but $Id_E \subseteq \mathcal{R}^2 \subseteq \mathcal{R}^+$.

1.22. 1.(a) $(e, e') \in (\mathcal{R}_1.\mathcal{R}_2)^{-1}$ if and only if $(e', e) \in \mathcal{R}_1.\mathcal{R}_2$ if and only if $(\exists e'' : (e', e'') \in \mathcal{R}_1$ and $(e'', e) \in \mathcal{R}_2)$ if and only if $(\exists e'' : (e'', e') \in \mathcal{R}_1^{-1}$ and $(e, e'') \in \mathcal{R}_2^{-1})$ if and only if $(e, e') \in (\mathcal{R}_2)^{-1}.(\mathcal{R}_1)^{-1}$.

1.(b) $(e, e') \in (\mathcal{R}_1 \cup \mathcal{R}_2).\mathcal{R}$ if and only if $(\exists e'' : (e, e'') \in \mathcal{R}_1 \cup \mathcal{R}_2$ and $(e'', e') \in \mathcal{R})$ if and only if $(\exists e'' : ((e, e'') \in \mathcal{R}_1$ or $(e, e'') \in \mathcal{R}_2)$ and $(e'', e') \in \mathcal{R})$ if and only if $(\exists e'' : ((e, e'') \in \mathcal{R}_1$ and $(e'', e') \in \mathcal{R})$ or $((e, e'') \in \mathcal{R}_2$ and $(e'', e') \in \mathcal{R}))$ if and only if $(\exists e_1 : ((e, e_1) \in \mathcal{R}_1$ and $(e_1, e') \in \mathcal{R})$ or $\exists e_2 : ((e, e_2) \in \mathcal{R}_2$ and $(e_2, e') \in \mathcal{R}))$ if and only if $(e, e') \in \mathcal{R}_1.\mathcal{R} \cup \mathcal{R}_2.\mathcal{R}$.

1.(c) is proved in the same way as 1.(b).

2. The inclusion is straightforward; the difference with case 1.(b) is that

$$\exists e'' : ((e, e'') \in \mathcal{R}_1 \text{ and } (e'', e') \in \mathcal{R}) \text{ and } ((e, e'') \in \mathcal{R}_2) \text{ and } (e'', e') \in \mathcal{R})$$

is not equivalent to

$$\exists e_1 : ((e, e_1) \in \mathcal{R}_1 \text{ and } (e_1, e') \in \mathcal{R}) \text{ and } \exists e_2 : ((e, e_2) \in \mathcal{R}_2 \text{ and } (e_2, e') \in \mathcal{R}).$$

Indeed, let $\mathcal{R}_1 = \{(a, b)\}$, $\mathcal{R}_2 = \{(a, c)\}$, and $\mathcal{R} = \{(b, d), (c, d)\}$. We have $\mathcal{R}_1 \cap \mathcal{R}_2 = \emptyset$, and hence $(\mathcal{R}_1 \cap \mathcal{R}_2).\mathcal{R} = \emptyset$ whilst $\mathcal{R}_1.\mathcal{R} = \mathcal{R}_2.\mathcal{R} = \{(a, d)\}$, and hence $\mathcal{R}_1.\mathcal{R} \cap \mathcal{R}_2.\mathcal{R} = \{(a, d)\}$.

1.23. 1. If $\mathcal{R}$ is left complete, then $\forall e, e' \in E, e \mathrel{\mathcal{R}} e'$. Hence, $\forall e, e' \in E, e' \mathcal{R}^{-1} e$, and so $\mathcal{R}^{-1}$ is right complete. The proof of the other case is symmetric.

2.(i) If $Id_E \subseteq \mathcal{R}.\mathcal{R}^{-1}$ we have $\forall e \in E, (e, e) \in \mathcal{R}.\mathcal{R}^{-1}$. As $(e, e) \in \mathcal{R}.\mathcal{R}^{-1} \implies \exists e' : (e, e') \in \mathcal{R}$, this shows that $\mathcal{R}$ is left complete. Conversely, if $\mathcal{R}$ is left complete, then for any e there exists an e' such that $(e, e') \in \mathcal{R}$. We thus have that $(e', e) \in \mathcal{R}^{-1}$, and hence $(e, e) \in \mathcal{R}.\mathcal{R}^{-1}$.

2.(ii) If $Id_E \subseteq \mathcal{R}^{-1}.\mathcal{R} = \mathcal{R}^{-1}.(\mathcal{R}^{-1})^{-1}$ then $\mathcal{R}^{-1}$ is left complete, and hence $\mathcal{R}$ is right complete (see point 1).

3. If $(e, e') \in \mathcal{R} \cap \mathcal{R}^{-1}$ then $(e, e') \in \mathcal{R}$ and $(e, e') \in \mathcal{R}^{-1}$. So $(e', e) \in \mathcal{R}^{-1}$ and $(e', e) \in \mathcal{R}$, and hence $(e', e) \in \mathcal{R} \cap \mathcal{R}^{-1}$. If $(e, e') \in \mathcal{R} \cup \mathcal{R}^{-1}$ then $(e, e') \in \mathcal{R}$ or $(e, e') \in \mathcal{R}^{-1}$, and hence $(e', e) \in \mathcal{R}^{-1}$ or $(e', e) \in \mathcal{R}$. Thus $(e', e) \in \mathcal{R} \cup \mathcal{R}^{-1}$.

4. If $(e, e') \in \mathcal{R}_1 \cap \mathcal{R}_2$ and if $(e', e'') \in \mathcal{R}_1 \cap \mathcal{R}_2$ then $(e, e') \in \mathcal{R}_1$, $(e, e') \in \mathcal{R}_2$, $(e', e'') \in \mathcal{R}_1$, and $(e', e'') \in \mathcal{R}_2$. Because $\mathcal{R}_1$ and $\mathcal{R}_2$ are transitive, $(e, e'') \in \mathcal{R}_1$ and $(e, e'') \in \mathcal{R}_2$. Hence, $(e, e'') \in \mathcal{R}_1 \cap \mathcal{R}_2$, which is thus indeed transitive.

The relation $\mathcal{R}_1 = \{(a,b)\}$ is transitive because $\mathcal{R}_1^2 = \emptyset \subseteq \mathcal{R}_1$. Similarly, $\mathcal{R}_2 = \{(b,a)\}$ is also transitive. Their union, $\mathcal{R} = \{(a,b),(b,a)\}$, is not transitive because $\mathcal{R}^2 = \{(a,a),(b,b)\}$ is not in $\mathcal{R}$.

5. Assume that $e\ \mathcal{R}^+\ e'$ and $e'\ \mathcal{R}^+\ e''$. Therefore, there exist two integers i and j such that $e\ \mathcal{R}^i\ e'$ and $e'\ \mathcal{R}^j\ e''$. We thus have $(e,e'') \in \mathcal{R}^{i+j} \subseteq \mathcal{R}^+$.

6. Because $\mathcal{R} \subseteq \mathcal{R}^+$, we have that if $\mathcal{R} \neq \mathcal{R}^+$ then $\mathcal{R}^+ \not\subseteq \mathcal{R}$. Therefore, there exists at least one integer i such that $\mathcal{R}^i \not\subseteq \mathcal{R}$. Let i_0 be the least such integer. This integer i_0 is strictly larger than 1 because $\mathcal{R}^0 = Id_E \subseteq \mathcal{R}^+$ and $\mathcal{R}^1 = \mathcal{R} \subseteq \mathcal{R}^+$. Moreover, $\mathcal{R}^{i_0-1} \subseteq \mathcal{R}$. Because $\mathcal{R}^{i_0} \not\subseteq \mathcal{R}$, there exist e_0 and e_{i_0} such that $e_0 \mathcal{R}^{i_0} e_{i_0}$ and $e_0\ \overline{\mathcal{R}}\ e_{i_0}$. Therefore, there exist $e_1, \ldots, e_{i_0-1}$ such that $e_j\ \mathcal{R}\ e_{j+1}$ for $0 \leq j < i_0$. We thus have $e_1\ \mathcal{R}^{i_0-1}\ e_{i_0}$, and hence $e_1\ \mathcal{R}\ e_{i_0}$. Because $\mathcal{R}$ is transitive we have that $e_0\ \mathcal{R}\ e_{i_0}$, a contradiction.

1.24. If $\mathcal{R}$ were symmetric, we would have $m = n+1 \iff n = m+1$, and that is impossible. If it were reflexive, we would have $n = n+1$. If it were transitive, we would have

$$m = n+1 \text{ and } p = m+1 \implies p = n+1.$$

We see that $n\ \mathcal{R}^i\ m$ if and only if $m = n+i$, and hence $\mathcal{R}^+$ is the strict ordering $<$ on the integers and $\mathcal{R}^*$ is the large ordering $\leq$ (see Chapter 2).

1.25. This relation is not transitive. For instance, 2 and 6 have a common divisor (2), 6 and 3 have a common divisor (3), but the only divisor common to 2 and 3 is 1.

1.26. 1. The relation $\mathcal{R}$ is symmetric if the matrix M is symmetric (i.e. if $\forall i,j,\ m_{i,j} = m_{j,i}$). It is reflexive if there are only 1s on the diagonal, and it is irreflexive if there are only 0s. It is antisymmetric if $\forall i,j, m_{i,j} \neq m_{j,i}$.

2. $M' = M_{\mathcal{R}^{-1}}$ is the transpose of matrix M: $m'_{i,j} = m_{j,i}$. $M' = M_{\overline{\mathcal{R}}}$ is the matrix defined by $m'_{i,j} = 1 - m_{i,j}$. If $M = M_{\mathcal{R}}$ and $M' = M_{\mathcal{R}'}$, $M'' = M_{\mathcal{R}.\mathcal{R}'}$ is the matrix defined by $m''_{i,j} = 1$ if and only if $\exists k : m_{i,k} m'_{k,j} = 1$.

1.27. By Proposition 1.23, the least equivalence relation containing $\mathcal{R}$ and $\mathcal{R}'$ is $((\mathcal{R} \cup \mathcal{R}') \cup (\mathcal{R} \cup \mathcal{R}')^{-1})^*$. Because $\mathcal{R}$ and $\mathcal{R}'$ are symmetric, $\mathcal{R} \cup \mathcal{R}'$ is symmetric and equal to $(\mathcal{R} \cup \mathcal{R}')^{-1}$. Because $\mathcal{R}$ and $\mathcal{R}'$ are reflexive, $\mathcal{R} \cup \mathcal{R}'$ contains the identity, and $(\mathcal{R} \cup \mathcal{R}')^* = (\mathcal{R} \cup \mathcal{R}')^+$ (see Exercise 1.21).

1.28. We first show that if $\mathcal{R} \subseteq \mathcal{R}'$ then $\forall e \in E, [e]_{\mathcal{R}} \subseteq [e]_{\mathcal{R}'}$. Indeed, $e' \in [e]_{\mathcal{R}} \implies e'\ \mathcal{R}\ e \implies e'\ \mathcal{R}'\ e \implies e' \in [e]_{\mathcal{R}'}$. Hence, $E/\mathcal{R}$ is a refinement of $E/\mathcal{R}'$.

Conversely, we let $e,e' \in E$ be such that $e\ \mathcal{R}\ e'$ and show that $e\ \mathcal{R}'\ e'$. As $e\ \mathcal{R}\ e'$, $e' \in [e]_{\mathcal{R}}$; since $E/\mathcal{R}$ is a refinement of $E/\mathcal{R}'$, there exists $e'' \in E$ such that $[e]_{\mathcal{R}} \subseteq [e'']_{\mathcal{R}'}$. We thus have $e' \in [e'']_{\mathcal{R}'}$ and $e \in [e'']_{\mathcal{R}'}$ (because $e,e' \in [e]_{\mathcal{R}}$), and hence $e\ \mathcal{R}'\ e''\ \mathcal{R}'\ e'$.

1.29. 1. Straightforward.

2. Assume $\mathcal{R}$ is antisymmetric, and let $x,y \in E$ be such that $x \neq y$. Because $x\ \mathcal{R}\ y$ and $y\ \mathcal{R}\ x$ cannot hold simultaneously, we can assume without loss of generality that $x\ \overline{\mathcal{R}}\ y$. There thus exists $X \in \mathcal{F}_x \setminus \mathcal{F}_y$ (i.e. $\exists X \in \mathcal{F}$ such that $x \in X$ and $y \notin X$), and hence $|X \cap \{x,y\}| = 1$.

Conversely, let x and y be two distinct elements of E and let $X \in \mathcal{F}$ be such that $|X \cap \{x,y\}| = 1$. For instance, $x \in X$ and $y \notin X$ (i.e. $X \in \mathcal{F}_x \setminus \mathcal{F}_y$). Hence, $x\ \overline{\mathcal{R}}\ y$.

3. Let $x,y \in E$ be such that $x\ \mathcal{R}\ y$ (i.e. $\mathcal{F}_x \subseteq \mathcal{F}_y$). Let $X \in \mathcal{F}_y$ and $\overline{X} \in (\mathcal{F} \setminus \mathcal{F}_y) \subseteq (\mathcal{F} \setminus \mathcal{F}_x)$. Then $X \in \mathcal{F}_x$ and we have that $\mathcal{F}_y \subseteq \mathcal{F}_x$, and hence $y\ \mathcal{R}\ x$.

4. If $\mathcal{R}$ is symmetric, it is an equivalence relation. We show that for all $x \in E$, we have $[x] = \bigcap_{X \in \mathcal{F}_x} X$. Let $y \in [x]$. We have $x\ \mathcal{R}\ y$, and hence $\forall X \in \mathcal{F}_x, y \in X$ (i.e. $y \in \bigcap_{X \in \mathcal{F}_x} X$). Conversely, let $y \in \bigcap_{X \in \mathcal{F}_x} X$. We have $\mathcal{F}_x \subseteq \mathcal{F}_y$, and hence $x\ \mathcal{R}\ y$ (i.e. $y \in [x]$).

Chapter 2

Because $\mathcal{F}$ is closed by intersection, we deduce that $\forall x \in E, [x] \in \mathcal{F}$.

Letting $X \in \mathcal{F}$, we show that X is a union of equivalence classes. Let $x \in X$. We have that $X \in \mathcal{F}_x$ and thus $[x] = \bigcap_{Y \in \mathcal{F}_x} Y \subseteq X$. We thus have $X = \bigcup_{x \in X} [x]$.

We deduce $\overline{X} = \bigcup_{x \notin X} [x]$. Since $\mathcal{F}$ is closed by union, we have $\overline{X} \in \mathcal{F}$.

1.30. 1. Commutativity of $+$ and $\times$ is clear and associativity is easily checked. The units of $+$ and $\times$ are $(0, 1)$ and $(1, 1)$.

2. Reflexivity and symmetry are straightforward. Transitivity follows immediately from the property: $(a, b) \sim (c, d)$ if and only if $\dfrac{a}{b} = \dfrac{c}{d}$ in $\mathbb{Q}$. But the goal of the exercise is the definition of $\mathbb{Q}$, and so a direct proof of this result follows.

Assume that $(a, b) \sim (c, d) \sim (e, f)$. We have $ad = bc$ and $cf = de$. Thus $adf = bcf = bde$, and because $d \neq 0$, $af = be$ (i.e. $(a, b) \sim (e, f)$). Let us show now that $\sim$ is a congruence. Assume that $(a, b) \sim (c, d)$ and let $(e, f) \in E$. We have $(a, b) + (e, f) = (af + be, bf)$ and $(c, d) + (e, f) = (cf + de, df)$. Therefore,

$$(af + be) \times df = adf^2 + bdef = bcf^2 + bdef = bf(cf + de).$$

Hence, $(a, b) + (e, f) \sim (c, d) + (e, f)$, which shows that $\sim$ is a congruence for $+$ because this operation is commutative. We show similarly that $\sim$ is a congruence for $\times$.

3. Let $(a, b) \in E$. We have that $(a, b) + (-a, b) = (ab - ab, bb) = (0, bb)$ and $(0, bb) \sim (0, 1)$. Hence, in $E/\sim$ the inverse of $[(a, b)]$ for $[+]$ is $[(-a, b)]$. The distributivity of $[\times]$ over $[+]$ is easily verified.

Let $[(a, b)]$ be an element of $E/\sim$ with $[(c, d)]$ as its inverse for $[\times]$. We have that $(a, b) \times (c, d) = (ac, bd) \sim (1, 1)$, and hence $ac = bd \neq 0$ and $[(a, b)] \neq [(0, 1)]$. Conversely, if $[(a, b)] \neq [(0, 1)]$ then $a \neq 0$ and $(b, a) \in E$. We then have that $(a, b) \times (b, a) = (ab, ab) \sim (1, 1)$, and hence $[(b, a)]$ is the inverse of $[(a, b)]$ for $[\times]$.

Chapter 2

2.1. Because $\mathcal{E} = Id_E \cup (\mathcal{R} \cap \mathcal{R}^{-1})$ is included in $Id_E \cup \mathcal{R}$ and because $\mathcal{R}$ is transitive, we indeed have $\mathcal{E}.\mathcal{R}.\mathcal{E} \subseteq \mathcal{R}$.

2.2. The relation $\mathcal{R}^\dagger$ is reflexive by definition.

If $x \mathcal{R}^\dagger y$ and $y \mathcal{R}^\dagger x$, and if $x \neq y$, then

$$x \mathcal{R} y \,,\, y \overline{\mathcal{R}} x \,,\, y \mathcal{R} x \,,\, x \overline{\mathcal{R}} y,$$

a contradiction. We must thus have that $x = y$, and the relation is antisymmetric.

To show the transitivity of $\mathcal{R}^\dagger$, it suffices to show that if $x \mathcal{R}^\dagger y$ and $y \mathcal{R}^\dagger z$, with $x \neq y$, $y \neq z$ and $x \neq z$, then $x \mathcal{R}^\dagger z$. By the definition of $\mathcal{R}^\dagger$, we have

$$x \mathcal{R} y \,,\, y \overline{\mathcal{R}} x \,,\, y \mathcal{R} z \,,\, z \overline{\mathcal{R}} y.$$

Since $\mathcal{R}$ is a transitive relation, $x \mathcal{R} z$. Assume that $z \mathcal{R} x$; we thus also have, by the transitivity of $\mathcal{R}$, $z \mathcal{R} y$, which is excluded. Hence, $z \overline{\mathcal{R}} x$ and thus $x \mathcal{R}^\dagger z$.

2.3. First, we show a slightly stronger result than the mere existence of a linear extension, namely:

> if we are given two incomparable elements e and e' in E, then there is a linear extension $\leq_t$ of $\leq$ such that $e \leq_t e'$.

Note that if E does not contain two incomparable elements, then it is totally ordered and the existence of a linear extension is clear.

We prove this result by induction on the number n of elements of E. If $n = 0$ or $n = 1$, the property holds vacuously because E does not contain two elements.

We assume $|E| = n + 1$, with $n > 0$. With each element x of E, we associate an integer $r(x)$ defined as the cardinality of the set $\{y \in E \,/\, y < x\}$. It is clear that $x < x' \Longrightarrow r(x) < r(x')$. Moreover, if $r(x) > 0$, there exists a y such that $y < x$. We easily deduce

$$\forall x \in E, r(x) > 0 \Longrightarrow \exists x' < x : r(x') = 0.$$

We then let

$$x_0 = \begin{cases} e & \text{if } r(e) = 0, \\ x & \text{otherwise.} \end{cases}$$

where x is an arbitrary element of E such that $x < e$ and $r(x) = 0$.

By the induction hypothesis, there is a linear extension $\leq'_t$ of the restriction of $\leq$ to $E \setminus \{x_0\}$, and it can be chosen such that $e \leq'_t e'$ if $e \neq x_0$. We then obtain the required result by defining $\leq_t = \leq'_t \cup \{(x_0, x) \,/\, x \in E\}$.

Let $\leq'$ be the intersection of the linear extensions of $\leq$. We have that $\leq \,\subseteq\, \leq'$. If the equality does not hold, then there must exist x and y (where $x \neq y$) such that $x \leq' y$ and $x \not\leq y$. We have that $y \not\leq x$, because if $y \leq x$ we would also have that $y \leq' x$ and thus $x = y$. We have just seen that it is possible to find a linear extension $\leq_t$ of $\leq$ such that $y \leq_t x$. As $\leq' \,\subseteq\, \leq_t$, we also have $x \leq_t y$ and thus $x = y$, a contradiction.

2.4. 1. We have $(x_1, x_2) \leq (y_1, y_2) \Longrightarrow \pi_i(x_1, x_2) = x_i \leq y_i = \pi_i(y_1, y_2)$.

2. Let $E = \{a, b\}$ with $a < b$. In $E \times E$, (a, b) and (b, a) are incomparable.

3. Let b be the bijection from $(E_1 \times E_2) \times E_3$ to $E_1 \times (E_2 \times E_3)$ associating $(x_1, (x_2, x_3))$ with $((x_1, x_2), x_3)$. It is an isomorphism because

$$((x_1, x_2), x_3) \leq ((y_1, y_2), y_3) \iff (x_1, x_2) \leq (y_1, y_2) \text{ and } x_3 \leq_3 y_3$$
$$\iff x_1 \leq_1 y_1, x_2 \leq_2 y_2 \text{ and } x_3 \leq_3 y_3$$
$$\iff x_1 \leq_1 y_1 \text{ and } (x_2, x_3) \leq (y_2, y_3)$$
$$\iff (x_1, (x_2, x_3)) \leq (y_1, (y_2, y_3)).$$

Similarly,

$$(x_1, x_2) \leq (y_1, y_2) \iff x_1 \leq_1 y_1 \text{ and } x_2 \leq_2 y_2 \iff (x_2, x_1) \leq (y_2, y_1).$$

2.5. 1. Obvious.

2. Let $E_1 = \{a, b\}$ with $a < b$ and $E_2 = \{c, d\}$, where c and d are incomparable. The lexicographic ordering on $E_2 \times E_1$ is $\{(ca, cb), (da, db)\}$. The lexicographic ordering on $E_1 \times E_2$ is $\{(ac, bc), (ac, bd), (ad, bc), (ad, bd)\}$.

3. Let (x_1, x_2) and (y_1, y_2) be in $E_1 \times E_2$. Because E_1 is totally ordered, three cases are possible:

- $x_1 < y_1$, and in this case $(x_1, x_2) < (y_1, y_2)$,
- $y_1 < x_1$, and in this case $(y_1, y_2) < (x_1, x_2)$,
- $x_1 = y_1$, and, because E_2 is totally ordered,
  - either $x_2 \leq y_2$, and then $(x_1, x_2) \leq (y_1, y_2)$,
  - or $y_2 \leq x_2$ and $(y_1, y_2) \leq (x_1, x_2)$.

Chapter 2

2.6. 1. Let E' be an antichain having at least two distinct elements x and y. Because E is totally ordered, assume that e.g. $x < y$. Then the relation $\leq \cap (E' \times E')$ contains the pair (x, y) and cannot be equal to the identity.

2. It follows that an antichain of an ordered set, or the intersection of a chain and an antichain, cannot contain two or more elements, because then we could find an antichain containing two distinct elements x and y with $x < y$.

2.7. If $x \leq y$, then $[x, y]$ contains x and y. If $[x, y]$ is non-empty, then there exists z such that $x \leq z$ and $z \leq y$, and hence $x \leq y$.

2.8. Assume that no element of $[x, y]$ covers x. We thus have

$$\forall z \in [x, y], \exists z' \in [x, z] : x \neq z' \text{ and } z \neq z'.$$

In other words, $\forall z \in [x, y], \exists z' : x < z' < z$.

We can thus construct a sequence $(z_i)_{i \geq 0}$ of elements of $[x, y]$ such that $z_0 = y$ and $\forall i \geq 0$, $x < z_{i+1} < z_i \leq y$, contradicting the fact that $[x, y]$ is finite.

2.9. By definition of the covering relation, $\prec \subseteq \leq$; because $\leq$ is a reflexive and transitive relation and $\prec^*$ is the least reflexive and transitive relation containing $\prec$, we have $\prec^* \subseteq \leq$.

Conversely, let us show by complete induction the property

$$\forall n, \forall x, y \in E, \quad x < y \text{ and } |[x, y]| \leq n \quad \Longrightarrow \quad x \prec^* y.$$

Assume that $x < y$ and $|[x, y]| = n$. By Exercise 2.8, there exists z such that $x \prec z$ and $z \leq y$. If $z = y$, we have $x \prec^* y$. Otherwise, $[z, y] \subseteq [x, y] \setminus \{x\}$ and $|[z, y]| < n$. By the induction hypothesis, $z \prec^* y$ and thus $x \prec^* y$.

2.10. 1. Consider the ordered set E obtained by adding to $\mathbb{N}$, equipped with the usual ordering on the integers, an element a, incomparable to any other element. Let $E' = E$. Element a is indeed the unique maximal element of E', but it is not the greatest element.

2. If E is totally ordered, then any finite subset has a unique maximal element which is also the greatest element. The converse is false: the set E of Example 2.25, 2, is such that for any subset having a unique maximal element, this maximal element is also the greatest element, but E is not totally ordered.

2.11. The same proof as in Example 2.30, 3, shows that $\preceq'$ is a well-founded ordering. This proof shows that in general the lexicographic product of two well-founded sets is also a well-founded set.

Moreover, $\preceq'$ is a total ordering because $<_1$ and $<_2$ are total orderings.

2.12. 1. The 'only if' direction is clear. 'If' direction: we prove by induction on n that any n-element subset has a least upper bound and a greatest lower bound.

2. The 'only if' direction is clear. 'If' direction: let F be any subset of E, and let $\text{Min}(F)$ be the set of lower bounds of F, then the least upper bound of $\text{Min}(F)$ is the greatest lower bound of F.

2.13. Let x and y be two elements of e such that $x \leq y$. Clearly, $\sup(\{x, y\}) = y$, and the continuity of f implies that $\sup(\{f(x), f(y)\}) = f(y)$, and hence $f(x) \leq f(y)$.

2.14. 1. $f(\bot) = \bot$ because $\sup(\emptyset) = \bot$ and because f preserves the least upper bound of $\emptyset$.

2. If $f(\bot) = \bot$, the least fixed point of f is $\bot$.

2.15. Let $(X_i)_{i \geq 0}$ be a sequence of subsets of $E \times E$. Let $\mathcal{R}_1 = Id_E \cup \mathcal{R}.(\bigcup_{i \geq 0} X_i)$ and $\mathcal{R}_2 = \bigcup_{i \geq 0}(Id_E \cup \mathcal{R}.X_i)$. Show that $\mathcal{R}_1 = \mathcal{R}_2$, which proves the continuity of f.

$$(x,y) \in \mathcal{R}_1 \iff x = y \text{ or } \exists z : x \mathcal{R} z \text{ and } (z,y) \in \bigcup_{i \geq 0} X_i$$

$$\iff x = y \text{ or } \exists z : x \mathcal{R} z \text{ and } \exists i : (z,y) \in X_i.$$

$$(x,y) \in \mathcal{R}_2 \iff \exists i : x = y \text{ or } \exists z : x \mathcal{R} z \text{ and } (z,y) \in X_i.$$

We see that these two conditions are equivalent.

The least element of $\mathcal{P}(E \times E)$ is the empty relation $\emptyset_E$. By Theorem 2.40, the least fixed point of f is $\bigcup_{k \geq 0} f^k(\emptyset_E)$. The result is then immediately deduced from the identity

$$f^{k+1}(\emptyset_E) = Id_E \cup \mathcal{R} \cup \mathcal{R}^2 \cup \cdots \cup \mathcal{R}^k.$$

This identity is easily proved by induction. Indeed,

$$f^0(\emptyset_E) = \emptyset_E,$$
$$f^1(\emptyset_E) = Id_E \cup \mathcal{R}.\emptyset_E = Id_E,$$
$$f^2(\emptyset_E) = f(Id_E) = Id_E \cup \mathcal{R}.Id_E = Id_E \cup \mathcal{R},$$
$$\vdots$$
$$f^{k+2}(\emptyset_E) = f(Id_E \cup \mathcal{R} \cup \mathcal{R}^2 \cup \cdots \cup \mathcal{R}^k)$$
$$= Id_E \cup \mathcal{R}.(Id_E \cup \mathcal{R} \cup \mathcal{R}^2 \cup \cdots \cup \mathcal{R}^k)$$
$$= Id_E \cup \mathcal{R} \cup \mathcal{R}^2 \cup \cdots \cup \mathcal{R}^k \cup \mathcal{R}^{k+1},$$
$$\vdots$$

2.16. Because $x \sqcap y \leq x$ and $x \sqcap y \leq y$, we also have that $x \sqcap y \leq x'$ and $x \sqcap y \leq y'$ and thus $x \sqcap y$ is less than or equal to the greatest lower bound $x' \sqcap y'$ of x' and y'.

Similarly, $x' \leq x' \sqcup y'$ and $y' \leq x' \sqcup y'$, and hence $x \leq x' \sqcup y'$ and $y \leq x' \sqcup y'$. Thus $x \sqcup y \leq x' \sqcup y'$.

2.17. 1. We must show the identity

$$\gcd(x, \text{lcm}(y,z)) = \text{lcm}(\gcd(x,y), \gcd(y,z)).$$

We decompose x, y and z into products of primes. We can thus write

$$x = p_0^{n_0} p_1^{n_1} \cdots p_k^{n_k},$$
$$y = p_0^{m_0} p_1^{m_1} \cdots p_k^{m_k},$$
$$x = p_0^{r_0} p_1^{r_1} \cdots p_k^{r_k},$$

where $p_0, p_1, \ldots, p_k$ are the $k+1$ first primes, and where their exponents are positive or null integers. The fact that $p_0^{a_0} p_1^{a_1} \cdots p_k^{a_k}$ divides $p_0^{b_0} p_1^{b_1} \cdots p_k^{b_k}$ can hence be written as: $\forall i, a_i \leq b_i$. We can thus also represent the number 0 in this form (even though it is not a product of primes) by assuming that all exponents are infinite.

Moreover, the lcm of the two numbers $p_0^{a_0} p_1^{a_1} \cdots p_k^{a_k}$ and $p_0^{b_0} p_1^{b_1} \cdots p_k^{b_k}$ is equal to

$$p_0^{\sup(a_0,b_0)} p_1^{\sup(a_1,b_1)} \cdots p_k^{\sup(a_k,b_k)}$$

Chapter 2   313

and their gcd is equal to
$$p_0^{\inf(a_0,b_0)} p_1^{\inf(a_1,b_1)} \ldots p_k^{\inf(a_k,b_k)}.$$

We thus have to show that for any three integers, possibly equal to ∞, we have
$$\inf(n, \sup(m, r)) = \sup(\inf(n, m), \inf(n, r)).$$

Without loss of generality we can assume $m \leq r$. In this case $\inf(n, m) \leq \inf(n, r)$ and
$$\inf(n, \sup(m, r)) = \inf(n, r) = \sup(\inf(n, m), \inf(n, r)).$$

2. If the lattice is complemented, then for any integer n there exists an integer m such that $\gcd(n, m) = 1$ and $\text{lcm}(n, m) = 0$. Let $n > 1$; then we must have that $m \neq 0$, because otherwise n would be a common divisor of n and 0. On the other hand, the lcm of two numbers must necessarily divide their product; thus nm divides 0, and this implies $m = 0$, a contradiction.

2.18. 1. If $\mathcal{R}_1$ and $\mathcal{R}_2$ are two equivalence relations, then $\mathcal{R}_1 \sqcap \mathcal{R}_2 = \mathcal{R}_1 \cap \mathcal{R}_2$ and $\mathcal{R}_1 \sqcup \mathcal{R}_2$ is the least equivalence relation containing $\mathcal{R}_1$ and $\mathcal{R}_2$ (see Exercise 1.27).

2. Consider the three equivalence relations $\equiv_1$, $\equiv_2$ and $\equiv_3$ on $E = \{a, b, c\}$ whose equivalence classes are, respectively:
$$\{a, c\}, \{b\}$$
$$\{a, b\}, \{c\}$$
$$\{b, c\}, \{a\}$$

It is easy to see that $(\equiv_1 \cap \equiv_2) = (\equiv_1 \cap \equiv_3) = Id_E$. The least upper bound of the two equivalence relations $\equiv_1 \cap \equiv_2$ and $\equiv_1 \cap \equiv_3$ is thus Id_E. Since $a \equiv_2 b \equiv_3 c$, the least upper bound of the relations $\equiv_2$ and $\equiv_3$ is the full relation, whose intersection with $\equiv_1$ is $\equiv_1$. Thus:
$$\equiv_1 \sqcap (\equiv_2 \sqcup \equiv_3) \neq (\equiv_1 \sqcap \equiv_2) \sqcup (\equiv_1 \sqcap \equiv_3)$$

and the lattice of equivalence relations is not distributive.

3. Although it is not distributive, the lattice of equivalence relations is nevertheless complemented (but the complement is not unique). Let $\equiv$ be an equivalence relation on E. With each equivalence class C we associate an arbitrary element x_C of that class. Let X be the set of these elements. We thus have $\forall e \in E, |X \cap [e]_\equiv| = 1$.

Let $\equiv'$ be the equivalence relation whose classes are X and $\{y\}$ for any $y \in E \setminus X$. Then $\equiv \cap \equiv' = Id_E$. Indeed, let e and e' be such that $e \equiv e'$ and $e \equiv' e'$. Because $e \equiv' e'$, if $e \notin X$ then $e = e'$, and if $e \in X$ then $e' \in X$ and $\{e, e'\} \subseteq X \cap [e]_\equiv$, and hence $e = e'$. On the other hand, the least upper bound of the relations $\equiv$ and $\equiv'$ is the full relation. Indeed, let y_1 and y_2 be any two elements of E. Let, for $i = 1, 2$,
$$y_i' = \begin{cases} y_i & \text{if } y_i \in X, \\ x_{[y_i]_\equiv} & \text{otherwise.} \end{cases}$$

We then have: $y_1 \equiv y_1' \equiv' y_2' \equiv y_2$.

2.19. $\nu(\bot) = \nu(\bot) \sqcup \bot = \top$, $\nu(\top) = \nu(\top) \sqcap \top = \bot$.

2.20. We assume that there exists an element x such that $x = \nu(x)$. We then have:
$$x = x \sqcap x = x \sqcap \nu(x) = \bot,$$
$$x = x \sqcup x = x \sqcup \nu(x) = \top.$$

Hence $\bot = \top$, a contradiction.

Chapter 3

3.1. The case $r = 1$ is clear for S_n and was studied in Example 3.3 for T_n. Let $r \neq 1$. We consider the property $P(n)$: '$S_n = \dfrac{r^{n+1} - 1}{r - 1}$'. We verify

$$P(0): \quad S_0 = r^0 = 1 = \frac{r - 1}{r - 1}.$$

Let $n \geq 0$. We assume that $P(n)$ is true and we verify

$$P(n+1): \quad S_{n+1} = S_n + r^{n+1} = \frac{r^{n+1} - 1}{r - 1} + r^{n+1} = \frac{r^{n+2} - 1}{r - 1}.$$

We thus deduce that $\forall n \geq 0$, $S_n = \dfrac{r^{n+1} - 1}{r - 1}$. The proof is similar for T_n.

3.2. 1. Let $P(n)$ be the property '$S_n = 2n^4 - n^2$'. $P(1)$ is true because $S_1 = 1 = 2 - 1$. Let $n \geq 1$. We assume $P(n)$. We have $S_{n+1} = S_n + (2n+1)^3 = 2n^4 - n^2 + (2n+1)^3 = 2n^4 + 8n^3 + 11n^2 + 6n + 1 = 2(n+1)^4 - (n+1)^2$. Thus $P(n+1)$ is true. We deduce $\forall n \geq 1$, $P(n)$.

2. We note that $T_1 = \frac{1}{3}, T_2 = \frac{1}{3} + \frac{1}{15} = \frac{2}{5}$ and $T_3 = \frac{1}{3} + \frac{1}{15} + \frac{1}{35} = \frac{3}{7}$. We guess the property $Q(n)$: '$T_n = \dfrac{n}{2n+1}$' that we prove by induction.

We have already verified $Q(1), Q(2)$ and $Q(3)$. Let $n \geq 1$ be such that $Q(n)$ is true. We have

$$T_{n+1} = T_n + \frac{1}{4(n+1)^2 - 1} = \frac{n}{2n+1} + \frac{1}{(2n+1)(2n+3)}$$
$$= \frac{2n^2 + 3n + 1}{(2n+1)(2n+3)} = \frac{n+1}{2n+3}.$$

Hence $Q(n+1)$ is true. We deduce that $\forall n \geq 1$, $Q(n)$.

3.3. 1. We have $P(x+1) - P(x) = x^2 + (2a+1)x + (a+b+1/3)$. The property is thus verified if and only if $2a + 1 = 0$ and $a + b + 1/3 = 0$, i.e. $a = -1/2$ and $b = 1/6$.

2. We prove this by induction on n. Let $Q(n)$ be the property '$P(n)$ is an integer'. $P(0) = 0$, and hence $Q(0)$ is true. Let $n \geq 0$ and assume $Q(n)$ is true. $P(n+1) = P(n) + n^2$, and hence $Q(n+1)$ is true. We deduce $\forall n \geq 0, Q(n)$.

3. We prove by induction the property $R(n)$: $S_n = P(n+1)$. $S_0 = 0 = P(1)$, and thus $R(0)$ is true. Let $n \geq 0$, and assume $R(n)$. We have $S_{n+1} = S_n + (n+1)^2 = P(n+1) + (n+1)^2 = P(n+2)$, and hence $R(n+1)$ is true. We deduce $\forall n \geq 0$, $R(n)$.

Finally, $\dfrac{n(n+1)(2n+1)}{6} = \dfrac{1}{3}n^3 + \dfrac{1}{2}n^2 + \dfrac{1}{6}n = P(n) + n^2 = P(n+1)$.

3.4. By induction on n. Let $P(n)$ be the property '$\forall A \subseteq \{1, 2, \ldots, 2n\}$ such that $|A| \geq n + 1$, $\exists a, b \in A$ such that $a < b$ and $a \mid b$', where the relation $\mid$ means 'divides', i.e. $a \mid b \iff (\exists c \in \mathbb{N}, b = ac)$.

We verify $P(1)$: if $A \subseteq \{1, 2\}$ and $|A| \geq 2$ then $A = \{1, 2\}$ and the pair $(a, b) = (1, 2)$ is suitable. For $n \geq 1$ we assume $P(n)$ and show $P(n+1)$. Let $A \subseteq \{1, 2, \ldots, 2n+2\}$ be such that $|A| \geq n + 2$. Let $B = A \cap \{1, 2, \ldots, 2n\}$. If $|B| \geq n + 1$ then by hypothesis $\exists a, b \in B \subseteq A$ such that $a < b$ and $a \mid b$. Otherwise, $\{2n+1, 2n+2\} \subseteq A$ must hold. If $n + 1 \in A$ then the pair $(n+1, 2n+2)$ is suitable. Otherwise, $|B \cup \{n+1\}| \geq n + 1$ and $\exists a, b \in B \cup \{n+1\}$ such that $a < b$ and $a \mid b$. Necessarily, $a \neq n + 1$. If $b \neq n + 1$ then $a, b \in A$ are suitable. Otherwise, $a, 2n+2 \in A$ are suitable because $a \mid b$ and $b \mid (2n+2)$. Finally, $P(n+1)$ is true and we deduce $\forall n \geq 1$, $P(n)$.

3.5. We show this result by induction on i. If $i = 0$ then, since $\mathcal{R}^0 = Id_E$ is the unit of the product of relations, $\mathcal{R}^j = \mathcal{R}^0.\mathcal{R}^j$. We assume that $\mathcal{R}^{i+j} = \mathcal{R}^i.\mathcal{R}^j$. Since the product of relations is associative, $\mathcal{R}^{i+1} = \mathcal{R}.\mathcal{R}^i$, we have

$$\mathcal{R}^{i+1}.\mathcal{R}^j = \mathcal{R}.\mathcal{R}^i.\mathcal{R}^j = \mathcal{R}.\mathcal{R}^{i+j} = \mathcal{R}^{i+1+j}.$$

3.6. 1. Let $n \in \mathbb{N}$. We assume $P(n)$: $\exists k \in \mathbb{N}$, $10^n - 1 = 9k$ and we show $P(n+1)$. We have $10^{n+1} - 1 = 10(10^n - 1) + 10 - 1 = 9(10k+1)$ and hence $9 \mid 10^{n+1} - 1$. The proof is similar for Q: we assume $\exists k \in \mathbb{N}$, $10^n + 1 = 9k$ and we deduce $10^{n+1} + 1 = 10(10^n + 1) - 10 + 1 = 9(10k - 1)$.
2. $P(0)$ is true because $10^0 - 1 = 0 = 9 \times 0$, and hence $\forall n \geq 0, P(n)$. $Q(0)$ is false because $10^0 + 1 = 2$. Hence, we can draw no conclusion. In fact we will show that $\forall n \geq 0$, $Q(n)$ is false. More precisely, we show that $R(n)$: 'the remainder of the division of $10^n + 1$ by 9 is 2'. $R(0)$ is true because $10^0 + 1 = 2$. Let $n \in \mathbb{N}$ be such that $R(n)$ is true: $\exists k \in \mathbb{N}, 10^n + 1 = 9k + 2$. We show $R(n+1) : 10^{n+1} + 1 = 10(10^n + 1) - 10 + 1 = 90k + 11 = 9(10k+1) + 2$. Thus $\forall n \geq 0$, $R(n)$.

3.7. In fact the inductive step (I) holds only for $n \geq 2$. An individual number n is indeed a member of G_1, but in order for it to also be a member of G_2, it is necessary (and sufficient) that n be greater than or equal to 2. We have thus shown $P(1)$ and $\forall n \geq 2$, $P(n) \Longrightarrow P(n+1)$. We can draw no conclusion.

3.8. Let P be a property depending on n. If P verifies (B) and (I), then P verifies (I'): indeed, (B) and (I) imply that $\forall n, P(n)$, and hence (I') is true.
Conversely, if P verifies (I'), then:

- P verifies (B) by Remark 3.5, 1, and hence $P(0)$ is true,
- P verifies (I): otherwise, $\exists n, \neg(P(n) \Longrightarrow P(n+1))$ (see Chapter 5). Let n_0 be the least n such that $(P(n_0) \Longrightarrow P(n_0 + 1))$ is false; we then have: $(\forall k < n_0 + 1, P(k))$ and $\neg P(n_0 + 1)$, which contradicts (I').

On structures more complex than $\mathbb{N}$, the second induction principle is more powerful than the first one. In fact, the first induction principle may even not hold, as is shown by the next example. Consider the set $2\omega = \{0, 1, 2, \ldots, \omega, \omega + 1, \omega + 2, \ldots\}$, consisting of two consecutive copies of $\mathbb{N}$. Let $P(x)$ be the property 'x is finite'. P verifies (B) and (I), but P is false on 2ω, because $P(\omega)$, $P(\omega + 1)$, etc., are false. On the other hand, P does not verify (I'), because $\forall k < \omega$, $P(k)$, whilst $P(\omega)$ does not hold.

3.9. 1. Here, induction is not needed; it suffices to expand and simplify.
2. We prove by induction the property

$$P(m): \text{`}\exists n \in \mathbb{N}, \exists \varepsilon_1, \varepsilon_2, \ldots, \varepsilon_n \in \{-1, 1\}, \ m = \varepsilon_1 1^2 + \varepsilon_2 2^2 + \cdots + \varepsilon_n n^2\text{'}.$$

$P(0)$: $0 = 1^2 + 2^2 - 3^2 + 4^2 - 5^2 - 6^2 + 7^2$,
$P(1)$: $1 = 1^2$,
$P(2)$: $2 = -1^2 - 2^2 - 3^2 + 4^2$,
$P(3)$: $3 = -1^2 + 2^2$.
Let $m \geq 4$; assume that $\forall k < m, P(k)$. $P(m-4)$ is thus true:

$$\exists n \in \mathbb{N}, \exists \varepsilon_1, \varepsilon_2, \ldots, \varepsilon_n \in \{-1, 1\}, \ (m-4) = \varepsilon_1 1^2 + \varepsilon_2 2^2 + \cdots + \varepsilon_n n^2.$$

We deduce $m = \varepsilon_1 1^2 + \varepsilon_2 2^2 + \cdots + \varepsilon_n n^2 + (n+1)^2 - (n+2)^2 - (n+3)^2 + (n+4)^2$ and hence $P(m)$ is true.
Using the second induction principle, we can deduce $\forall n \geq 0, P(n)$.

3.10. One implication is straightforward. If $p, q \in \mathbb{N}$ exist such that $u = w^p$ et $v = w^q$, we have $u \cdot v = w^{p+q} = v \cdot u$.

The converse implication is proved by induction on $|u| + |v|$. That is, we consider the property $P(n)$: '$\forall u, v \in A^*$ such that $|u| + |v| = n$ we have $u \cdot v = v \cdot u \Longrightarrow \exists w \in A^*, \exists p, q \in \mathbb{N}$: $u = w^p$ and $v = w^q$'. We use the second induction principle on $\mathbb{N}$. Let $n \in \mathbb{N}$. We assume that $\forall k < n, P(k)$ is true. Let $u, v \in A^*$ be such that $|u| + |v| = n$. By symmetry we can assume $|u| \leq |v|$, and hence Figure 15.2.

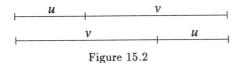

Figure 15.2

u is thus a prefix of v. If $u = \varepsilon$ or $u = v$, then we obtain the result by letting $w = v$. If u is a proper prefix of v, let $v' \in A^*$ be such that $v = uv'$. We easily verify that $v'u = v = uv'$. As $|v'| < |v|$, we apply the induction hypothesis to the pair (u, v'): $\exists w \in A^*, \exists p, q \in \mathbb{N}$ such that $u = w^p$ and $v' = w^q$. Since $v = u \cdot v'$ we deduce $v = w^{p+q}$, and this concludes the proof.

3.11. We show, first, that $L^* = L \cdot L^* \cup \{\varepsilon\}$. Indeed,

$$L^* = \cup_{n \geq 0} L^n = L^0 \cup \left(\cup_{n \geq 1} L^n\right) = \{\varepsilon\} \cup \left(\cup_{n \geq 0} L \cdot L^n\right) = \{\varepsilon\} \cup L \cdot L^*$$

We deduce that $L^* \cdot M = \{\varepsilon\} \cdot M \cup L \cdot L^* \cdot M$. Therefore, the language $L^* \cdot M$ is a solution of the equation $X = L \cdot X \cup M$.

We now show that this solution is unique. Let $X \subseteq A^*$ be such that $X = L \cdot X \cup M$. In order to show that $L^* \cdot M \subseteq X$ it suffices to prove that $\forall n \in \mathbb{N}, L^n \cdot M \subseteq X$ because $L^* \cdot M = \bigcup_{n \geq 0} L^n \cdot M$. Hence, we show by induction the property $P(n)$: '$L^n \cdot M \subseteq X$'. We have $L^0 \cdot M = \{\varepsilon\} \cdot M = M \subseteq M \cup L \cdot X = X$. Thus $P(0)$ is true. Let $n \in \mathbb{N}$ and assume that $P(n)$ holds. We have $L^{n+1} \cdot M = L \cdot L^n \cdot M \subseteq L \cdot X \subseteq L \cdot X \cup M = X$. Thus $P(n+1)$ is true.

Conversely, we prove by induction the property

$$Q(n): \text{'}(w \in X \text{ and } |w| = n) \Longrightarrow w \in L^* \cdot M\text{'},$$

which will prove the reverse inclusion. We use the second induction principle. Let $n \in \mathbb{N}$. We assume that $\forall k < n, Q(k)$ is true. Let $w \in X$ be such that $|w| = n$. Since $X = L \cdot X \cup M$, two cases may occur. Either $w \in M$ and we have directly $w \in L^* \cdot M$ since $M \subseteq L^* \cdot M$. Moreover, $w \in L \cdot X$ and $\exists (u, v) \in L \times X$ such that $w = u \cdot v$. Since $\varepsilon \notin L$, we obtain $|u| > 0$ and hence $|v| = |w| - |u| < |w|$. By the induction hypothesis, we have $v \in L^* \cdot M$. Therefore, $w = u \cdot v \in L \cdot L^* \cdot M \subseteq L^* \cdot M$. This proves that $Q(n)$ is true and concludes the proof.

3.12. In order to prove that $X \subseteq BT$, we show by induction the property $P(n)$: '$BT_n \subseteq X$'. BT_0 is reduced to the unique element of the basis of the inductive definition of BT, and hence $BT_0 \subseteq BT$ and $P(0)$ is verified. Let $n \in \mathbb{N}$, and assume $P(n)$ is true. Let $a \in A$ and let $l, r \in BT_n$. By the induction hypothesis, $BT_n \subseteq BT$. Thus applying the inductive step of the definition of BT we obtain $(a, l, r) \in BT$. Hence,

$$\{(a, l, r) \, / \, a \in A, l, r \in BT_n\} \subseteq BT.$$

We deduce $BT_{n+1} \subseteq BT$ by again using the inductive hypothesis: $BT_n \subseteq BT$. Therefore, $\forall n \in \mathbb{N}, BT_n \subseteq BT$ and thus $X = \bigcup_{n \in \mathbb{N}} BT_n \subseteq BT$.

Conversely, in order to show that $BT \subseteq X$, it suffices to prove that X verifies the conditions (B) and (I) of the inductive definition of BT.

(B) $\emptyset \in BT_0 \subseteq X$.
(I) Let $l, r \in X$, $n, m \in \mathbb{N}$ exist such that $g \in BT_n$ and $d \in BT_m$. By symmetry, we can assume that $m \leq n$. Since $BT_m \subseteq BT_n$ we deduce $l, r \in BT_n$, and thus $\forall a \in A, (a, l, r) \in BT_{n+1} \subseteq X$.

Finally, we have proved $X = BT$.

Chapter 3   317

3.13. We denote by $\leq$ the prefix ordering on A^*. We saw in Example 3.13 that $\forall x \in D$, $l(x) = r(x)$. We consider the property $P(x)$: '$\forall y \in A^*$, $y \leq x \implies l(y) \geq r(y)$'. We show by induction that the elements of D verify P. Clearly, $P(\varepsilon)$ is true. Let $x \in D$ be such that $P(x)$ holds, and let y be a prefix of (x). We distinguish three cases:

- $y = \varepsilon$ or $y = ($. We then have $l(y) \geq r(y)$.
- $y = (y'$ with y' prefix of x. We then have $l(y) = 1 + l(y') \geq 1 + r(y') > r(y') = r(y)$.
- $y = (x)$. We then have $l(y) = 1 + l(x) = 1 + r(x) = r(y)$.

We show similarly that if $x, y \in D$ verify P then $x \cdot y$ verifies P. We deduce that $D \subseteq L$.

Conversely, we show by induction on $|x|$ that $x \in L \implies x \in D$. We thus consider the property $Q(n)$: '$\forall x \in A^*$ such that $|x| = n$, we have $x \in L \implies x \in D$'. We use the second induction principle. Let $n \in \mathbb{N}$. We assume that $\forall k < n$, $Q(k)$ is true. Let $x \in L$ be such that $|x| = n$. If $n = 0$ then $x = \varepsilon \in D$. We thus assume $n > 0$. We denote by y the least non-empty prefix of x such that $l(y) = r(y)$; we thus have $y \in L$. We distinguish two cases:

- $|y| < |x|$. Let $z \in A^*$ be such that $x = y \cdot z$. We easily verify that $z \in L$. Moreover, because y is non-empty we also have that $|z| < |x| = n$. By the induction hypothesis we deduce $y, z \in D$. Using the inductive definition of D we obtain $x = y \cdot z \in D$.
- $|y| = |x|$. In this case, we show that $x = (z)$ with $z \in L$. Let x_1 and x_n be the first and last letters of x. We have $x_1 \leq x$ and thus $l(x_1) \geq r(x_1)$, which imply $x_1 = ($. As $l(x) = r(x)$, we necessarily have that $n \geq 2$. Let $z \in A^*$ be such that $x = x_1 z x_n$. We have $x_1 z \leq x$ and thus $l(x_1 z) \geq r(x_1 z)$. As $l(x) = r(x)$, we deduce $x_n =)$ and $l(x_1 z) = r(x_1 z) + 1$. Because $x_1 = ($ we also have that $l(z) = r(z)$. Finally, let z' be a prefix of z. $x_1 z'$ is a prefix of x, and hence $l(x_1 z') \geq r(x_1 z')$. Since $y = x$ we must have that $l(x_1 z') > r(x_1 z')$, and thus $l(z') = l(x_1 z') - 1 \geq r(x_1 z') = r(z')$. We have thus shown that $z' \in L$. Since $|z'| < n$, we deduce that $z' \in D$ and, consequently, $x = (z') \in D$.

Finally, we have shown $\forall n \in \mathbb{N}$, $P(n)$, which means that $L \subseteq D$.

3.14. 1. We proceed of course by induction. The desired property is

$$P(x): \text{'}n(x) \leq 2^{h(x)} - 1\text{'}.$$

The unique element of the basis is the empty tree and we have $2^{n(\emptyset)} - 1 = 2^0 - 1 = 0 \geq 0 = n(\emptyset)$. Let $x = (a, l, r)$ and assume $P(l)$ and $P(r)$. We have

$$n(x) = 1 + n(l) + n(r) \leq 2^{h(l)} + 2^{h(r)} - 1 \leq 2 \times 2^{max(h(l), h(r))} - 1 = 2^{h(x)} - 1.$$

2. We proceed in the same way as for the property $Q(x)$: '$f(x) \leq 2^{h(x)-1}$'. Because of the definition of the function f, we must directly verify Q for the empty tree and for the tree reduced to a single element.

- $f(\emptyset) = 0 \leq 2^{-1} = 2^{h(\emptyset)-1}$,
- $f((a, \emptyset, \emptyset)) = 1 = 2^0 = 2^{h((a,\emptyset,\emptyset))-1}$.

We then verify the inductive step. Let $x = (a, l, r) \in BT$ with $g \neq \emptyset$ or $d \neq \emptyset$ and assume $Q(l)$ and $Q(r)$. $f(x) = f(l) + f(r) \leq 2^{h(l)-1} + 2^{h(r)-1} \leq 2 \times 2^{max(h(l), h(r))-1} = 2^{h(x)} - 1$.

3.15. 1. The inductive definition of the set SBT is

(B) $\forall a \in A, (a, \emptyset, \emptyset) \in SBT$,

(I) $l, r \in SBT \implies \forall a \in A, (a, l, r) \in SBT$.

2. Let $P(x)$ be the property '$n(x) = 2f(x) - 1$'.

- $\forall a \in A$, if $x = (a, \emptyset, \emptyset)$ then $n(x) = 1 = 2 \times 1 - 1 = 2f(x) - 1$.
- Let $l, r \in SBT$ be such that $P(l)$ and $P(r)$ hold. Let $a \in A$ and $x = (a, l, r)$. We have $n(x) = 1 + n(l) + n(r) = 1 + 2f(l) - 1 + 2f(r) - 1 = 2(f(l) + f(r)) - 1 = 2f(x) - 1$. Hence $\forall x \in SBT$, $P(x)$ is true.

3.16. 1. The inductive definition of the set BBT is

(B) $\emptyset \in BBT$,

(I) if $l, r \in BBT$ verify $|h(l) - h(r)| \leq 1$ then $\forall a \in A, (a, l, r) \in BBT$.

2. We show by induction the property $P(x)$: '$n(x) \geq u_{h(x)}$'.

(B) $n(\emptyset) = 0 = u_0 = u_{h(\emptyset)}$.

(I) Let $l, r \in BBT$ be such that $|h(l) - h(r)| \leq 1$ and $P(l)$ and $P(r)$ hold. Let $a \in A$ and $x = (a, l, r) \in BBT$. We thus have $n(x) = 1 + n(l) + n(r) \geq 1 + u_{h(l)} + u_{h(r)}$.

- If $h(l) < h(r)$ then we have $h(x) = h(r) + 1$ and $h(l) = h(r) - 1 = h(x) - 2$. Thus $n(x) \geq 1 + u_{h(x)-2} + u_{h(x)-1} = u_{h(x)}$.
- Similar proof if $h(l) > h(r)$.
- If $h(l) = h(r)$ then $h(x) = h(l) + 1 = h(r) + 1$. Thus

$$n(x) \geq 1 + 2u_{h(x)-1} \geq 1 + u_{h(x)-1} + u_{h(x)-2} = u_{h(x)}$$

because the sequence $(u_n)_{n \in \mathbb{N}}$ is monotone increasing.

In all three cases we have thus verified $P(x)$. We deduce that $\forall x \in BBT, P(x)$.

3.17. 1. The proof is by induction. Denote by $P(L)$ the property '$\widetilde{L} \in Rat$'. We show that $\forall L \in Rat, P(L)$ is true.

(B) $\widetilde{\emptyset} = \emptyset \in Rat$ and $\forall a \in A, \widetilde{\{a\}} = \{a\} \in Rat$. Thus $P(L)$ is verified for any language L of the basis of the inductive definition of Rat.

(I_1) Let $L, M \in Rat$ be such that $P(L)$ and $P(M)$ hold. We have $\widetilde{L \cup M} = \widetilde{L} \cup \widetilde{M}$. By the hypothesis, $\widetilde{L}, \widetilde{M} \in Rat$, and applying ($I_1$) we obtain $\widetilde{L \cup M} \in Rat$.

(I_2) Let $L, M \in Rat$ be such that $P(L)$ and $P(M)$ hold. We easily verify that $\widetilde{L \cdot M} = \widetilde{M} \cdot \widetilde{L}$. Applying the hypothesis and (I_2) we obtain $\widetilde{L \cdot M} \in Rat$.

(I_3) Let $L \in Rat$ be such that $P(L)$ is true. We have $\widetilde{L^*} = \widetilde{L}^* \in Rat$ (hypothesis and (I_3)).

We thus have proved: $\forall L \in Rat, \widetilde{L} \in Rat$.

2. The proof is similar. Let $Q(L)$ be the property '$LF(L) \in Rat$'. We show that $\forall L \in Rat, Q(L)$ is true.

(B) $LF(\emptyset) = \emptyset \in Rat$ and $\forall a \in A, LF(\{a\}) = \{\varepsilon, a\} \in Rat$ since $\{\varepsilon\} = \emptyset^* \in Rat$ (I_3) and $\{\varepsilon, a\} = \{\varepsilon\} \cup \{a\} \in Rat$ (B and I_1).

(I) Let $L, M \in Rat$ be such that $Q(L)$ and $Q(M)$ hold. We have

- $LF(L \cup M) = LF(L) \cup LF(M) \in Rat$ (induction hypothesis and I_1),
- $LF(L \cdot M) = LF(L) \cup (L \cdot LF(M)) \in Rat$ (induction hypothesis, I_1 and I_2),
- $LF(L^*) = L^* \cdot LF(L) \in Rat$ (induction hypothesis, I_2 and I_3).

We have thus proved: $\forall L \in Rat, LF(L) \in Rat$.

3.18. The mirror image is a mapping from A^* to A^*. In order to define it inductively, we use the inductive definition of A^* given in Example 3.9. We obtain

$$\widetilde{u} = \begin{cases} \varepsilon & \text{if } u = \varepsilon, \\ a \cdot \widetilde{v} & \text{if } u = v \cdot a. \end{cases}$$

3.19. 1. $Q(\varepsilon)$ holds and $Q(x)$ holds implies that $Q(ax)$ holds.

Chapter 3

2. $g((a_1), y) = (a_1 y)$.

3. We check by induction that

$$g\bigl((a_n(a_{n-1}(\ldots(a_1)\ldots))), y\bigr) = g\bigl(\varepsilon, (a_1(\ldots(a_{n-1}(a_n y))\ldots))\bigr).$$

This is clear for $n = 0$. Moreover, assuming it holds for n,

$$g\bigl((a_{n+1}(a_n(\ldots(a_2(a_1))\ldots))), y\bigr) = g\bigl((a_n(\ldots(a_2(a_1))\ldots)), (a_{n+1} y)\bigr)$$
$$= g\bigl(\varepsilon, (a_1(a_2(\ldots(a_n(a_{n+1} y))\ldots)))\bigr).$$

4. Let $rev(x) = g(x, \varepsilon)$ with

$$g(\varepsilon, y) = y$$
$$g((al), y) = g(l, (ay)).$$

Hence,

$$rev\bigl((a_1(a_2(\ldots(a_n)\ldots)))\bigr) = g\bigl((a_1(a_2(\ldots(a_n)\ldots))), \varepsilon\bigr)$$
$$= g\bigl(\varepsilon, (a_n(\ldots(a_2(a_1 \varepsilon))\ldots))\bigr) = (a_n(\ldots(a_2(a_1))\ldots)).$$

In Exercise 3.18, even though the strings of X were generated in the form $((\ldots(a_1(a_2))\ldots a_{n-1})a_n)$, where the parentheses are added for outlining the structure implicit in Example 3.9 part 2, A^* is endowed with an associative concatenation and the mirror is defined as a mapping from X into A^*. Here, the lists of L are generated in the form $(a_1(a_2(\ldots(a_n(a_{n+1}))\ldots)))$ by prefixing with letters of A, and, since rev is a mapping $L \longrightarrow L$, the only operations available in L are the prefixings by a letter of A. Thus rev must be defined in terms of these prefixings only, and hence the more complex definition.

3.20. $\mathbb{N} \times \mathbb{N}^*$ coincides with the subset X of $\mathbb{N} \times \mathbb{N}$ inductively defined by

(B) $\forall (n, m) \in \mathbb{N}^2, n < m \Longrightarrow (n, m) \in X$,

(I) $(n, m) \in X \Longrightarrow (n + m, m) \in X$.

This definition is induced by the inductive definition given for the modulo function. It is obviously not the most natural definition of $\mathbb{N} \times \mathbb{N}^*$.

3.21. The function gcd is defined by

$$\gcd(n, m) = \begin{cases} n & \text{if } m = 0, \\ m & \text{if } n = 0, \\ \gcd(n - m, m) & \text{if } 0 < m \leq n, \\ \gcd(n, m - n) & \text{if } 0 < n < m. \end{cases}$$

The corresponding definition of X is

(B) $\forall n \in \mathbb{N} \setminus \{0\}, (0, n) \in X$ and $(n, 0) \in X$,

(I) $(n, m) \in X$ and $m \neq 0 \Longrightarrow (n + m, m) \in X$,

$(n, m) \in X$ and $n \neq 0 \Longrightarrow (n, m + n) \in X$.

3.22. We have
$$n(x) = \begin{cases} 0 & \text{if } x = \emptyset, \\ 1 + n(l) + n(r) & \text{if } x = (a, l, r). \end{cases}$$

$$l(x) = \begin{cases} 0 & \text{if } x = \emptyset, \\ 1 & \text{if } x = (a, \emptyset, \emptyset), \\ l(g) + l(d) & \text{if } x = (a, g, d) \text{ with } g \neq \emptyset \text{ or } d \neq \emptyset. \end{cases}$$

3.23. The inductive definition of the preorder traversal of a binary tree is
$$\text{Pref}(x) = \begin{cases} \varepsilon & \text{if } x = \emptyset, \\ a \cdot \text{Pref}(l) \cdot \text{Pref}(r) & \text{if } x = (a, l, r). \end{cases}$$

3.24. By Proposition 3.27, it suffices to show that $C(k\mathbb{N}) \subseteq k\mathbb{N}$. If $n \in C(k\mathbb{N})$, p and q exist in $k\mathbb{N}$ such that $n = p + q$. But $p = kp'$ and $q = kq'$, and hence $n = k(p' + q') \in k\mathbb{N}$.

3.25. $\hat{C}(X)$ is the convex hull of X.

1. C is clearly monotone increasing; it is finitary by definition, because $x \in C(X)$ implies that there exist $a, b \in X$ such that $x \in [a, b] = C(\{a, b\})$.

2. $\hat{C}(A \cup B)$ is C-closed and contains A, and hence $\hat{C}(A \cup B)$ contains $\hat{C}(A)$; symmetrically, $\hat{C}(A \cup B)$ contains $\hat{C}(B)$; and thus, again because of the closure property, $\hat{C}(A \cup B)$ contains $C(\hat{C}(A) \cup \hat{C}(B))$.

3. By 2, $\forall [a, b] \subseteq C(\bigcup_{F \in \text{fin}(X)} \hat{C}(F))$, $\exists F_1 \in \text{fin}(X)$, with
$$[a, b] \subseteq \hat{C}(F_1) \subseteq \bigcup_{F \in \text{fin}(X)} \hat{C}(F)$$
which is thus C-closed.

4. $\hat{C}$ is monotone increasing; it is also finitary, since by (3) $\hat{C}(X) = \bigcup_{F \in \text{fin}(X)} \hat{C}(F)$ and because

- $\forall F \in \text{fin}(X)$, $\hat{C}(F) \subseteq \hat{C}(X)$ and
- $\bigcup_{F \in \text{fin}(X)} \hat{C}(F)$ is a C-closed set containing X.

5. Yes, because the proofs of 2 – 4 do not depend on the explicit definition of C.

3.26. $C'(\hat{C}(F_0)) = (C(\hat{C}(F_0)) \cup F_0) \subseteq \hat{C}(F_0)$ because $\hat{C}(F_0)$ is C-closed and contains F_0. Moreover, $\emptyset \subseteq \hat{C}(F_0)$ hence $\hat{C}'(\emptyset) \subseteq \hat{C}(F_0)$ by Proposition 3.27.

We verify that $F_0 \subseteq \hat{C}'(F_0)$ and $C(\hat{C}'(F_0)) \subseteq \hat{C}'(F_0)$ similarly, hence, by the same Proposition, 3.27, $\hat{C}(F_0) \subseteq \hat{C}'(\emptyset)$.

3.27. Since C is finitary, $\hat{E} = \cup_{n \in \mathbb{N}} E_n$, and we verify by induction on n that $E_n \subseteq T$, $\forall n \in \mathbb{N}$.

We might also note that $C(T) \subseteq T$ (by the definition of T), and hence T is C-closed; because T contains F_0, T also contains its closure $\hat{E}$.

Moreover, since T is the least set verifying conditions (B) and (I) of Definition 3.14, and $\hat{E}$ also verifies these conditions, $T \subseteq \hat{E}$.

3.28. Let $\mathcal{R}$ be a subset of $U \times V$. We define

$$\Gamma(R) = \bigcup_{i>0}\{(f(\sigma_1,\ldots,\sigma_i), h_f(v_1,\ldots,v_i)) \,/\, (\sigma_j, v_j) \in \mathcal{R}, f \in F_i\}.$$

Then Γ is finitary.

Let $R_0 = \{(f, h(f)) \,/\, f \in F_0\}$ and let $\hat{R} = \hat{\Gamma}(T_0)$. Let $P = \{\sigma \,/\,$ there exists a unique $v\colon (\sigma, v) \in \hat{R}\}$. Then $\hat{C}(F_0) \subseteq P$. By the universal induction principle it suffices to show that $F_0 \subseteq P$ and $C(P) \subseteq P$.

1. If $F \in F_0$, $(f, v) \in \hat{R} \Longrightarrow (f, v) \in R_0$. Let $R_{i+1} = R_i \bigcup \Gamma(R_i)$ and let i be the least integer such that $(f, v) \in R_i$. If $i = 0$ the proof is complete. Otherwise $(f, v) \in R_i$ and $(f, v) \notin R_{i-1}$, and hence $(f, v) \in \Gamma(R_{i-1})$. But if $(\sigma, v) \in \Gamma(R_{i-1})$, then $\sigma = f_n(\sigma_1, \ldots, \sigma_n)$ with $f_n \in F_n$ and $f \neq \sigma$, a contradiction.

2. If $f \in F_0$ then $(f, h(f)) \in R_0$, and if $(f, v) \in \hat{R}$ then $(f, v) \in R_0$; hence $v = h(f)$ and $F_0 \subseteq P$.

3. Let $e \in C(P)$. Then $e = f(e_1, \ldots, e_n)$ with $e_j \in P$ and hence there exists a unique v_j such that $(e_j, v_j) \in \hat{R}$; but then $(e, h_f(v_1, \ldots, v_n)) \in \Gamma(\hat{R}) \subseteq \hat{R}$. If $(e, v') \in \hat{R}$, we again consider the least i such that $(e, v') \in R_i$. Because $e \notin F_0, i > 0$ and $(e, v') \notin R_{i-1}$, and hence $(e, v') \in \Gamma(R_{i-1})(e, v') = (f'(e'_1, \ldots, e'_{n'}), h_f(v'_1, \ldots, v'_{n'}))$ with $(e'_j, v'_j) \in R_j$. We deduce that $f = f'$, $n = n'$, and $e_j = e'_j$; hence, since $e_j \in P$, we have that $v_j = v'_j$ and $v' = h_f(v_1, \ldots, v_n)$.

Then let $h^*(e)$ be equal to the unique v such that $(e, v) \in \hat{R}$, and it is easy to check that h^* verifies the required properties.

3.29. 1. We show by induction that $E_i = \{1, \ldots, i\}$. $E_0 = \{1\}$, and if $E_i = \{1, \ldots, i\}$ then $E_{i+1} = E_i \cup C(E_i) = E_i \cup \{2, \ldots, i+1\} = \{1, \ldots, i+1\}$.

2. Let E be its limit. We thus have, by Proposition 3.30, $E \subseteq \hat{C}(\{1\})$ and hence $C(E) \subseteq \hat{C}(\{1\})$. As E is infinite, $0 \in C(E)$, and hence $\mathbb{N} = \{0\} \cup E \subseteq \hat{C}(\{1\})$.

3. Since $E \subsetneq \hat{C}(\{1\})$, C is not finitary. Indeed, let X be infinite, then $0 \in C(X)$ and for any finite subset X' of X, $0 \notin C(X')$.

Chapter 4

4.1. Let us recall our hypotheses.

$N_0 : x \sqcup x = x$,
$N_1 : x \sqcap (y \sqcup z) = (x \sqcap y) \sqcup (x \sqcap z)$,
$N'_1 : (x \sqcup y) \sqcap z = (x \sqcap z) \sqcup (y \sqcap z)$,
$N_{2a} : x \sqcap \top = x$,
$N_3 : x \sqcup \bot = x = \bot \sqcup x$,
$N_4 : x \sqcap \bar{x} = \bot$ and $x \sqcup \bar{x} = \top$.

1. Let us first show that $\sqcap$ is idempotent. Using N_3, N_4, N_1 then again N_4, and N_2, we obtain

$$x \sqcap x = (x \sqcap x) \sqcup \bot = (x \sqcap x) \sqcup (x \sqcap \bar{x}) = x \sqcap (x \sqcup \bar{x}) = x \sqcap \top = x.$$

$\top \sqcap x = x$ also holds since, using N_4, N_1', N_0', N_4 and N_3,

$$\top \sqcap x = (x \sqcup \overline{x}) \sqcap x = (x \sqcap x) \sqcup (x \sqcap \overline{x}) = x \sqcup \bot = x.$$

Using N_{2a}, N_4, N_1, N_1', we obtain

$$\top \sqcup x = (\top \sqcup x) \sqcap \top = (\top \sqcup x) \sqcap (x \sqcup \overline{x})$$
$$= ((\top \sqcup x) \sqcap x) \sqcup ((\top \sqcup x) \sqcap \overline{x}) = ((x \sqcup (x \sqcap x)) \sqcup (\overline{x} \sqcup (x \sqcap \overline{x})).$$

Using N_0', N_0, N_4 and N_3, we obtain

$$((x \sqcup (x \sqcap x)) \sqcup (\overline{x} \sqcup (x \sqcap \overline{x})) = x \sqcup (\overline{x} \sqcup \bot) = x \sqcup \overline{x} = \top.$$

Similarly,

$$(x \sqcup \top) = (x \sqcup \top) \sqcap (x \sqcup \overline{x}) = x \sqcup \overline{x} = \top.$$

2. Let us assume that $y \sqcap x = \bot$ and $y \sqcup x = \top$. Using N_{2b}, N_4, N_1', N_4 and N_3, we obtain

$$\overline{x} = \top \sqcap \overline{x} = (y \sqcup x) \sqcap \overline{x} = (y \sqcap \overline{x}) \sqcup (x \sqcap \overline{x}) = y \sqcap \overline{x},$$

and, using N_{2a}, N_4, N_1 and N_3,

$$y = y \sqcap \top = y \sqcap (x \sqcup \overline{x}) = (y \sqcap x) \sqcup (y \sqcap \overline{x}) = \bot \sqcup (y \sqcap \overline{x}) = y \sqcap \overline{x}.$$

By N_4, it follows that $x = \overline{\overline{x}}$.

3. Using N_4, N_3, N_1, N_4 and N_3, we have

$$\bot = x \sqcap \overline{x} = x \sqcap (\overline{x} \sqcup \bot) = (x \sqcap \overline{x}) \sqcup (x \sqcap \bot) = \bot \sqcup (x \sqcap \bot) = x \sqcap \bot.$$

As $x = \overline{\overline{x}}$, we also have $\overline{x} \sqcap x = \bot$ and

$$\bot = \overline{x} \sqcap x = (\overline{x} \sqcup \bot) \sqcap x = (\overline{x} \sqcap x) \sqcup (\bot \sqcap x) = \bot \sqcup (\bot \sqcap x) = \bot \sqcap x.$$

4. These four equalities can be proved in similar ways, so we check only one of them:

$$x = x \sqcap \top = x \sqcap (\top \sqcup y) = (x \sqcap \top) \sqcup (x \sqcap y) = x \sqcup (x \sqcap y).$$

The associativity of $\sqcup$ then follows:

$$(x \sqcup y) \sqcup z = ((x \sqcup y) \sqcup z) \sqcap (x \sqcup \overline{x}) = ((x \sqcup y) \sqcup z) \sqcap x) \sqcup ((x \sqcup y) \sqcup z) \sqcap \overline{x})$$
$$= (((x \sqcup y) \sqcap x) \sqcup (z \sqcap x)) \sqcup (((x \sqcup y) \sqcap \overline{x}) \sqcup (z \sqcap \overline{x})) = x \sqcup ((y \sqcap \overline{x}) \sqcup (z \sqcap \overline{x})),$$
$$x \sqcup (y \sqcup z) = (x \sqcup (y \sqcup z)) \sqcap (x \sqcup \overline{x}) = ((x \sqcup (y \sqcup z)) \sqcap x) \sqcup ((x \sqcup (y \sqcup z)) \sqcap \overline{x})$$
$$= x \sqcup ((x \sqcap \overline{x}) \sqcup ((y \sqcup z) \sqcap \overline{x})) = x \sqcup ((y \sqcup z) \sqcap \overline{x}) = x \sqcup ((y \sqcap \overline{x}) \sqcup (z \sqcap \overline{x})).$$

and also the commutativity:

$$(x \sqcup y) \sqcap (y \sqcup x) = ((x \sqcup y) \sqcap y) \sqcup ((x \sqcup y) \sqcap x) = y \sqcup x,$$
$$(x \sqcup y) \sqcap (y \sqcup x) = (x \sqcap (y \sqcup x)) \sqcup (y \sqcap (y \sqcup x)) = x \sqcup y.$$

Using N_1 and N_1', together with the associativity and commutativity of $\sqcup$, we obtain
$$(x \sqcup y) \sqcap (x \sqcup z) = x \sqcup (x \sqcap y) \sqcup (x \sqcap z) \sqcup (y \sqcap z)$$
and this is equal to $x \sqcup (y \sqcap z)$ by N_5.

5. In order to show that $\overline{x \sqcap y} = \overline{x} \sqcup \overline{y}$, it is enough to show that $(\overline{x} \sqcup \overline{y}) \sqcap (x \sqcap y) = \bot$ and $(\overline{x} \sqcup \overline{y}) \sqcup (x \sqcap y) = \top$. The following holds:
$$(\overline{x} \sqcup \overline{y}) \sqcap (x \sqcap y) = (\overline{x} \sqcap (x \sqcap y)) \sqcup (\overline{y} \sqcap (x \sqcap y)).$$
Since $x = x \sqcup (x \sqcap y)$, and $y = y \sqcup (x \sqcap y)$ we obtain
$$\bot = \overline{x} \sqcap x = \overline{x} \sqcap (x \sqcup (x \sqcap y)) = \overline{x} \sqcap (x \sqcap y),$$
$$\bot = \overline{y} \sqcap y = \overline{y} \sqcap (y \sqcup (x \sqcap y)) = \overline{y} \sqcap (x \sqcap y),$$
and therefore
$$(\overline{x} \sqcup \overline{y}) \sqcap (x \sqcap y) = \bot \sqcup \bot = \bot.$$
The following also holds: $\overline{y} = \top \sqcap \overline{y} = (x \sqcup \overline{x}) \sqcap \overline{y} = (x \sqcap \overline{y}) \sqcup (\overline{x} \sqcap \overline{y})$, and so
$$\overline{y} \sqcup (x \sqcap y) = (x \sqcap \overline{y}) \sqcup (\overline{x} \sqcap \overline{y}) \sqcup (x \sqcap y) = (x \sqcap (\overline{y} \sqcup y)) \sqcup (\overline{x} \sqcap \overline{y}) = x \sqcup (\overline{x} \sqcap \overline{y})$$
and therefore
$$\overline{x} \sqcup \overline{y} \sqcup (x \sqcap y) = \overline{x} \sqcup x \sqcup (\overline{x} \sqcap \overline{y}) = \top \sqcup (\overline{x} \sqcap \overline{y}) = \top.$$
We can easily deduce that $x \sqcup y = \overline{\overline{x}} \sqcup \overline{\overline{y}} = \overline{\overline{x} \sqcap \overline{y}}$ hence $\overline{x \sqcup y} = \overline{x} \sqcap \overline{y}$.
We thus obtain $y \sqcap x = \overline{\overline{y} \sqcup \overline{x}} = \overline{\overline{x} \sqcup \overline{y}} = x \sqcap y$ and
$$(x \sqcap y) \sqcap z = \overline{\overline{(x \sqcap y)} \sqcup \overline{z}} = \overline{\overline{x} \sqcup \overline{y} \sqcup \overline{z}} = \overline{\overline{x} \sqcup \overline{(y \sqcap z)}} = x \sqcap (y \sqcap z).$$

4.2. 1. Multiplying $x = ax + b\overline{x}$ by x, we obtain $xx = axx + b\overline{x}x$, i.e. $x = ax$, whence $x \leq a$. Multiplying by $\overline{x}$, we obtain $0 = b\overline{x}$, and since $b = b(x + \overline{x}) = bx + b\overline{x}$, $b = bx$ holds, therefore $b \leq x$ also holds.
Conversely, If $b \leq x \leq a$, we have $x = ax$ and $b = bx$, whence $b\overline{x} = bx\overline{x} = 0$, and therefore $ax + b\overline{x} = x + 0 = x$.

2. Multiplying $ax + b\overline{x} = 0$ by $\overline{x}$, we obtain $b\overline{x} = 0$ and we have just shown that this implies $b \leq x$. Multiplying by x, we obtain $ax = 0$, and this implies $x \leq \overline{a}$ for the same reasons.
Conversely, $b \leq x$ implies $b\overline{x} = 0$ and $x \leq \overline{a}$ implies $ax = 0$, whence $ax + b\overline{x} = 0$.

3.(i) Multiplying $x = au + b\overline{u}$ by $\overline{a}\overline{b}$, we obtain $\overline{a}\overline{b}x = \overline{a}\overline{b}au + \overline{a}\overline{b}b\overline{u} = 0$. As $x = au + b\overline{u}$, $\overline{x} = (\overline{a} + \overline{u})(\overline{b} + u) = \overline{a}\overline{b} + \overline{a}u + \overline{u}\overline{b}$, and multiplying by ab we obtain $ab\overline{x} = 0$. Therefore, $\overline{a}\overline{b}x + ab\overline{x} = 0$, and, applying the Schröder formula, $ab \leq x \leq \overline{\overline{a}\overline{b}} = a + b$. As $b \leq a$, $ab = b$ and $a + b = a$, the result follows.

3.(ii) Let us assume that $b \leq x \leq a$ and $\overline{b}x \leq u \leq \overline{a} + x$. Since $u = u(\overline{b}x + \overline{\overline{b}x}) = u(\overline{b}x + b + \overline{x})$ and $\overline{b}x \leq u$, we obtain $u = \overline{b}x + ub + u\overline{x}$. Since $u \leq \overline{a} + x$, $u = u(\overline{a}+x) = u\overline{a} + ux$ and $u\overline{x} = u\overline{a}\,\overline{x}$. Letting $y = u$ and $z = u\overline{x}$, $u = \overline{b}x + by + \overline{a}z$ holds. It follows that $\overline{u} = (b + \overline{x})(\overline{b} + \overline{y})(a + \overline{z})$ and therefore
$$au + b\overline{u} = a(\overline{b}x + by + \overline{a}z) + b(b + \overline{x})(\overline{b} + \overline{y})(a + \overline{z})$$
$$= a\overline{b}x + aby + b\overline{y}(a + \overline{z}).$$
Since $b \leq x \leq a$, we have in particular $b(a + \overline{z}) = b$, and we can simplify:
$$au + b\overline{u} = a\overline{b}x + aby + b\overline{y}(a + \overline{z})$$
$$= \overline{b}x + by + b\overline{y}$$
$$= \overline{b}x + b.$$
Since $b \leq x$, we conclude $b = bx$ and $\overline{b}x + b = \overline{b}x + bx = x$.

4.3. By definition $x \leq y$ if and only if $x \sqcup y = y$, and $h(x) \leq' h(y)$ if and only if $h(x) \sqcup' h(y) = h(y)$. Now $h(x \sqcup y) = h(x) \sqcup' h(y)$.

Let E be a set. The mapping h from $\mathcal{P}(E)$ to itself defined by

$$h(X) = \begin{cases} \emptyset & \text{if } X = \emptyset, \\ E & \text{otherwise,} \end{cases}$$

is a monotone mapping for the inclusion ordering. It is not a homomorphism of Boolean algebras, because, if X is different from $\emptyset$ and E, then $h(X) = h(\overline{X}) = E$, which contradicts the equality $\emptyset = h(\emptyset) = h(X \cap \overline{X}) = h(X) \cap \overline{(X)}$.

4.4. Let us first show that E' is a distributive lattice. To this end, it suffices to show that E' is closed under the sum and product operations, and that e is the maximal element of E'. We have

$$x = xe, y = ye \implies x + y = xe + ye = (x+y)e,$$
$$x = xe, y = ye \implies xy = xeye = xye,$$
$$x = xe \implies x \leq e.$$

Let us next show that E' is complemented.

- $x\hat{x} = x\overline{x}e = 0e = 0$,
- $x + \hat{x} = xe + \overline{x}e = (x + \overline{x})e = 1e = e$.

Last, it is easy to check that h is a homomorphism by just writing down the definitions.

4.5. We first prove the lemma: if X is a subset of F, then

$$\sum_{Y \subseteq X} \left(\prod_{x \in Y} g(x) \prod_{x \in X \setminus Y} \overline{g(x)} \right) = \prod_{x \in X} (g(x) + \overline{g(x)}) = 1.$$

The proof is by induction on the cardinality of X. If X is empty, the least upper bound

$$\sum_{Y \subseteq X} \left(\prod_{x \in Y} g(x) \prod_{x \in X \setminus Y} \overline{g(x)} \right)$$

coincides with the least upper bound of the empty set which is equal to 1.

Let y be an element of F which is not in X. We have

$$\prod_{x \in X \cup \{y\}} (g(x) + \overline{g(x)}) = (g(y) + \overline{g(y)}) \prod_{x \in X} (g(x) + \overline{g(x)})$$

$$= (g(y) + \overline{g(y)}) \sum_{Y \subseteq X} \left(\prod_{x \in Y} g(x) \prod_{x \in X \setminus Y} \overline{g(x)} \right)$$

$$= \sum_{Y \subseteq X} (g(y) \prod_{x \in Y} g(x) \prod_{x \in X \setminus Y} \overline{g(x)}) + \sum_{Y \subseteq X} (\overline{g(y)} \prod_{x \in Y} g(x) \prod_{x \in X \setminus Y} \overline{g(x)})$$

$$= \sum_{Y \subseteq X} \left(\prod_{x \in Y \cup \{y\}} g(x) \prod_{x \in X \setminus Y} \overline{g(x)} \right) + \sum_{Y \subseteq X} \left(\prod_{x \in Y} g(x) \prod_{x \in (X \cup \{y\}) \setminus Y} \overline{g(x)} \right)$$

$$= \sum_{Y \subseteq X \cup \{y\}} \left(\prod_{x \in Y} g(x) \prod_{x \in X \setminus Y} \overline{g(x)} \right).$$

Chapter 4

Let us now define the mapping h. Let $h(\emptyset) = 0$ and, for $\mathcal{X} \in \mathcal{F}$, $\mathcal{X} \neq \emptyset$, let

$$h(\mathcal{X}) = \sum_{X \in \mathcal{X}} \left(\prod_{x \in X} g(x) \prod_{x \notin X} \overline{g(x)} \right)$$

and this is an element of B.

By the lemma, we have $h(\mathcal{P}(F)) = \sum_{Y \subseteq F} \left(\prod_{x \in Y} g(x) \prod_{x \in F \setminus Y} \overline{g(x)} \right) = 1$.

We also have

$$h(i(y)) = \sum_{X : y \in X} \left(\prod_{x \in X} g(x) \prod_{x \notin X} \overline{g(x)} \right)$$

$$= g(y) \sum_{X \subseteq F \setminus \{y\}} \left(\prod_{x \in X} g(x) \prod_{x \in (F \setminus \{y\}) \setminus X} \overline{g(x)} \right)$$

$$= g(y) 1 = g(y).$$

Let us show that h is a homomorphism.

It is clear, by definition, that $h(\emptyset) = 0$ and we have just proved that $h(\mathcal{P}(F)) = 1$. It immediately follows, by the definition of h, that $h(\mathcal{X} \cup \mathcal{Y}) = h(\mathcal{X}) + h(\mathcal{Y})$.

By the De Morgan laws, $h(\mathcal{X} \cap \mathcal{Y}) = h(\mathcal{X}) h(\mathcal{Y})$ will hold if $h(\overline{\mathcal{X}}) = \overline{h(\mathcal{X})}$, and this must be shown anyway. In order to show that $h(\overline{\mathcal{X}}) = \overline{h(\mathcal{X})}$, it suffices to show that $h(\overline{\mathcal{X}}) + h(\mathcal{X}) = 1$ and $h(\overline{\mathcal{X}}) h(\mathcal{X}) = 0$.

The first of these two equalities is a consequence of $h(\mathcal{X} \cup \mathcal{Y}) = h(\mathcal{X}) + h(\mathcal{Y})$.

Let us compute $h(\overline{\mathcal{X}}) h(\mathcal{X})$. By definition of h this is equal to

$$\sum_{X_1 \notin \mathcal{X}} \left(\prod_{x \in X_1} g(x) \prod_{x \notin X_1} \overline{g(x)} \right) \sum_{X_2 \in \mathcal{X}} \left(\prod_{x \in X_2} g(x) \prod_{x \notin X_2} \overline{g(x)} \right)$$

and hence also equal to

$$\sum_{X_1 \in \mathcal{X}, X_2 \notin \mathcal{X}} \left(\prod_{x \in X_1 \cup X_2} g(x) \prod_{x \notin X_1 \cap X_2} \overline{g(x)} \right).$$

If $X_1 \notin \mathcal{X}$ and if $X_2 \in \mathcal{X}$, then necessarily $X_1 \neq X_2$. Hence it is enough to show that in this case $\prod_{x \in X_1 \cup X_2} g(x) \prod_{x \in \overline{X_1 \cup X_2}} \overline{g(x)} = 0$. Now, if $X_1 \neq X_2$, there exists an element y belonging to X_1 but not to X_2, or belonging to X_2 but not to X_1. In either case, $\prod_{x \in X_1 \cup X_2} g(x) \prod_{x \in \overline{X_1 \cup X_2}} \overline{g(x)}$ can be written $g(y) \overline{g(y)} z$ and is therefore equal to 0.

4.6. The sixteen functions are given by the following table:

x y	f_0	f_1	f_2	f_3	f_4	f_5	f_6	f_7	f_8	f_9	f_{10}	f_{11}	f_{12}	f_{13}	f_{14}	f_{15}
0 0	0	0	0	0	0	0	0	0	1	1	1	1	1	1	1	1
0 1	0	0	0	0	1	1	1	1	0	0	0	0	1	1	1	1
1 0	0	0	1	1	0	0	1	1	0	0	1	1	0	0	1	1
1 1	0	1	0	1	0	1	0	1	0	1	0	1	0	1	0	1

Their polynomial form follows

$$
\begin{array}{rcccl}
f_0 & = & \tilde{f}_0 & = & 0 \\
f_1 & = & \tilde{f}_7 & = & xy \\
f_2 & = & \tilde{f}_{11} & = & x\overline{y} \\
f_3 & = & \tilde{f}_3 & = & x \\
f_4 & = & \tilde{f}_{13} & = & \overline{x}y \\
f_5 & = & \tilde{f}_5 & = & y \\
f_6 & = & \tilde{f}_9 & = & \overline{x}y + x\overline{y} \\
f_7 & = & \tilde{f}_2 & = & x + y \\
f_8 & = & \tilde{f}_{14} & = & \overline{x}\,\overline{y} \\
f_9 & = & \tilde{f}_6 & = & xy + \overline{x}\,\overline{y} \\
f_{10} & = & \tilde{f}_{10} & = & \overline{y} \\
f_{11} & = & \tilde{f}_2 & = & x + \overline{y} \\
f_{12} & = & \tilde{f}_{12} & = & \overline{x} \\
f_{13} & = & \tilde{f}_4 & = & \overline{x} + y \\
f_{14} & = & \tilde{f}_8 & = & \overline{x} + \overline{y} \\
f_{15} & = & \tilde{f}_{15} & = & 1
\end{array}
$$

Chapter 5

5.1. We note that $p \Longrightarrow q$ is equivalent to $\neg q \Longrightarrow \neg p$. Thus, $[\neg q$ and $(\neg q \Longrightarrow \neg p)] \Longrightarrow \neg p$ is equivalent to the *modus tollens* rule. On the other hand, substituting $\neg q$ for p and $\neg p$ for q in the *modus ponens* rule, we have $[\neg q$ and $(\neg q \Longrightarrow \neg p)] \Longrightarrow \neg p$, which can thus be deduced from the *modus ponens* rule.

The converse can be proved similarly. We notice that $p \Longrightarrow q$ is equivalent to $\neg p$ or q, i.e. $(\neg p \vee q)$; then the *modus ponens* rule that is written $[p \wedge (p \Longrightarrow q)] \Longrightarrow q$ is equivalent to $p \Longrightarrow [(p \Longrightarrow q) \Longrightarrow q]$, and also to $p \Longrightarrow [\neg(p \Longrightarrow q) \vee q]$, whose contrapositive $[(p \Longrightarrow q) \wedge \neg q] \Longrightarrow \neg p$ is the *modus tollens* rule.

5.2. 1. $p \Longrightarrow q$ and its contrapositive implication $\neg q \Longrightarrow \neg p$ are both true. The converse $q \Longrightarrow p$, and its contrapositive $\neg p \Longrightarrow \neg q$, are usually false.

2. $p \Longrightarrow q$ is clearly false; its converse is, however, true. This example also shows that the quantifiers $\forall$ and $\exists$ may not be permuted (see also Exercise 5.14).

5.3. 1. The identities given above are immediately deduced from the truth table given below:

p	q	$\neg p$	$p \wedge q$	$p \vee q$	$p \supset q$	$p \supset \neg q$	$\neg p \supset q$	$\neg(p \supset \neg q)$
1	1	0	1	1	1	0	1	1
1	0	0	0	1	0	1	1	0
0	1	1	0	1	1	1	1	0
0	0	1	0	0	1	1	0	0

2. By induction on n.

In order to prove the last identity, first show that

$$\overline{I(F_n \wedge (F_{n-1} \wedge (\cdots \wedge (F_2 \wedge F_1))\cdots))} = \overline{I(F_n)I(F_{n-1})\cdots I(F_2)I(F_1)}$$
$$= \overline{I(F_n)} + \overline{I(F_{n-1})} + \cdots + \overline{I(F_1)}$$

(See Proposition 4.3.)

5.4. If F is unsatisfiable, then $\forall I$, $I(F) = 0$, and thus $\forall I$, $I(\neg F) = \overline{I(F)} = 1$, hence $\neg F$ is valid. The converse can be proved similarly.

5.5. 1. The sequent $(\emptyset, G)$ is true in I if and only if

$$(\forall F \in \emptyset, I(F) = 1) \implies I(G) = 1. \qquad (I)$$

But assertion $\forall F \in \emptyset$, $I(F) = 1$ is trivially true, because it can be rewritten as: $\forall F$, $[F \in \emptyset \implies I(F) = 1]$; $F \in \emptyset$ is always false because the set $\emptyset$ is always empty, but then, since 'false implies anything', the implication $F \in \emptyset \implies I(F) = 1$, which can be also written as $0 \implies I(F) = 1$, is always true, and thus $(\forall F \in \emptyset, I(F) = 1) = 1$. The implication (I) is thus reduced to $1 \implies I(G) = 1$, which is true if and only if $I(G) = 1$, i.e. if and only if G is true in I. Validity can be similarly verified.

⚡ Such arguments about the empty set and the satisfaction of implications $0 \implies G$ or $1 \implies G$ are tricky, but alas often useful, and should be handled with care in order to avoid errors.

2. Sequent $(\emptyset, (p \supset q))$ is true in I if and only if $p \supset q$ is true in I, i.e. if and only if $I(p) = 0$ or $I(q) = 1$.

Sequent $(\{p, (p \supset q)\}, q)$ is valid because $I(p) = 1$ and $I(p \supset q) = \overline{I(p)} + I(q) = 1$ implies that $I(q) = 1$.

5.6. We assume $S = (\mathcal{F}, G) = (\{F_n, \ldots, F_1\}, F)$; then $\phi(S) = F_n \supset (F_{n-1} \supset (\cdots (F_1 \supset F) \cdots))$ and the result is shown exactly as in Proposition 5.9.

5.7. $\iff$ is a congruence. It suffices to verify

$$\forall * \in \{\wedge, \vee, \supset\}, \quad \text{if } F_1 \iff F_1' \text{ and } F_2 \iff F_2', \text{ then}$$

$$(F_1 * F_2) \iff (F_1' * F_2') \quad \text{and}$$
$$\neg F_1 \iff \neg F_1'.$$

For instance: $F_1 \iff F_1'$ and $F_2 \iff F_2'$ imply that: $I(F_1) = I(F_1')$ and $I(F_2) = I(F_2')$, hence, for instance for $* = \vee$, $I((F_1 * F_2)) = I(F_1) + I(F_2) = I(F_1') + I(F_2') = I((F_1' * F_2'))$. The other cases can be verified similarly, and follow from the fact that I is a homomorphism from the set of formulas, equipped with the binary operations $\wedge, \vee, \supset$ and the unary operation $\neg$, to the Boolean algebra $\mathbb{B}$, equipped with the operations $\cdot, +, \circ$ and $^-$, where $\circ$ is defined by $x \circ y = \overline{x} + y$.

5.8. He mixed $p \supset q$ and $\neg p \supset \neg q$ (that is, the contrapositive of the converse $q \supset p$ of $p \supset q$).

5.9. Let $A = $ 'I love Anne', $M = $ 'I love Mary', and $P = M \supset A = $ 'If I love Mary, then I love Anne'. In 1 we have $(M \vee A) \wedge (M \supset A) = (M \vee A) \wedge (\neg M \vee A) = A$. We can thus conclude with certainty that he loves Anne, but his feelings for Mary no longer fall within the scope of logic. (He may love or may not love Mary; neither of these two assertions follows from (a) and (b).)

On the other hand, in 2, we know that our logician certainly loves Anne **and** Mary. Indeed, both assertions (a) and (b) hold: (a): $P \supset M$, and (b): $M \supset P$. Assume that P is false. Then, because 'false implies anything', it follows from (a) that M is true. But then P is also true by (b), and thus also A; if P is true, then it also follows from (a) that M is true, and thus also A because P is true. Our logician is thus, with high probability, bigamous (unless his fiancée is called Mary–Anne).

5.10. There are neither free variables nor free occurrences in the first formula; in the second formula, variable x is free: the first occurrence of x is bound and the second is free.

5.11. (i) $\forall x\, (P(x) \supset Q(x))$.

(ii) $\exists x\, (P(x) \wedge Q(x))$. This is not to be confused with $\exists x\, (P(x) \supset Q(x))$, which *is not* the translation of (ii), but which means 'some individuals either are not Ps or else are Qs'.

(iii) $\forall x\, (P(x) \supset \neg Q(x))$, or also, $\neg \exists x\, (P(x) \wedge Q(x))$.

(iv) $\exists x\, (P(x) \wedge \neg Q(x))$.

5.12. If x is not free in F, then for any valuation v' such that $v' =_{X-\{x\}} v$, we have, by Proposition 5.32, $\bar{v}(F) = \bar{v}'(F)$. Thus:

$$\bar{v}(\forall x F) = 1 \iff \text{for all } v' =_{X-\{x\}} v : \bar{v}'(F) = 1 \iff \bar{v}(F) = 1 \iff$$
$$\iff \text{there is } v' =_{X-\{x\}} v \text{ such that } \bar{v}'(F) = 1 \iff \bar{v}(\exists x F) = 1$$

5.13. We have, allowing for some notational flexibility in formulas and writing xRy instead of $R(x,y)$,

$$F = (\forall x\, xRx) \wedge (\forall x\, \forall y\, (xRy \wedge yRx \supset x = y))$$
$$\wedge\ (\forall x\, \forall y\, \forall z\, (xRy \wedge yRz \supset xRz)).$$

We add the formula $\forall x\, \forall y\, (xRy \vee yRx)$ in order to obtain a total ordering.

5.14. 1. Clearly, for any r, $\exists y \forall x\, r(x,y) \supset \forall x \exists y\, r(x,y)$ is true, but the converse implication is false; for instance in $\mathbb{N}$, $\forall x \exists y\ x < y$, but $\neg \exists y \forall x\ x < y$. Similarly, $\exists y (p(y) \wedge q(y)) \supset (\exists y p(y)) \wedge (\exists y q(y))$, but the converse is false; for instance in $\mathbb{N}^*$, $(\exists y\, \text{even}(y)) \wedge (\exists y\, \text{odd}(y))$ is true, but $\exists y\, (\text{even}(y) \wedge \text{odd}(y))$ is false.

2. By the preceding question, it thus suffices to verify the converse implication when $r(x,y) = p(x) \wedge q(y)$. This implication follows from the fact that, as $q(y)$ does not depend on x, we can choose the same y for all xs verifying $\forall x \exists y\, r(x,y)$. The formal proof is given below. We assume that $\bar{v}(\forall x \exists y [p(x) \wedge q(y)]) = 1$. Then, $\forall v' =_{X-\{x\}} v$, there exists v'' such that $v'' =_{X-\{y\}} v'$: $\bar{v}''(p(x)) = 1$ and $\bar{v}''(q(y)) = 1$. Note that $\bar{v}''(q(y))$ depends only on $y_0 = v''(y)$; we thus have $\bar{v}''(q(y)) = q_S(y_0) = 1$ and $\bar{v}''(p(x)) = \bar{v}'(p(x)) = 1$, and thus $\forall v' =_{X-\{x\}} v$, $\bar{v}'(p(x)) = 1$. Hence, finally, $\forall v' =_{X-\{x\}} v : \bar{v}'(p(x)) = 1$ and $q_S(y_0) = 1$. Because $\bar{v}'(p(x))$ depends only on $v'(x)$, we can transform this assertion into $\forall v' =_{X-\{x\}} v'' : \bar{v}'(p(x)) = 1$ and $\bar{v}''(q(y)) = 1$, which shows that $\exists w = v'' =_{X-\{y\}} v$, $\forall v' =_{X-\{x\}} v'' : \bar{v}'(p(x)) = 1$ and $\bar{v}'(q(y)) = \bar{v}''(q(y)) = 1$. Hence the required result, $\bar{v}(\exists y \forall x\, [p(x) \wedge q(y)]) = 1$.

See Proposition 5.43 for a generalization of this result, i.e. if x is not free in G and y is not free in F, then:

$$(\forall x\, F) \wedge (\exists y\, G) \approx \forall x \exists y\, (F \wedge G) \approx \exists y \forall x\, (F \wedge G)$$

5.15. We obtain one of the equivalent prenex forms:

$$\exists x \forall x' \forall y\, \Big(P(x) \wedge (Q(y) \supset R(x'))\Big)$$
$$\forall x \forall y \exists x'\, \Big(P(x') \wedge (Q(y) \supset R(x))\Big)$$
$$\forall x \exists x' \forall y\, \Big(P(x') \wedge (Q(y) \supset R(x))\Big)$$

5.16. We prove that there is no programmer. Let us define the following predicates:

$$\begin{cases} p(x) & \text{means '}x\text{ is a programmer'} \\ e(x,y) & \text{means '}x\text{ writes programs for }y\text{'} \end{cases}$$

Then:

- Rule (a) becomes

$$\forall x \left(p(x) \supset \forall y \left(\neg e(y,y) \supset e(x,y) \right) \right)$$

and also

$$\forall x \left(\neg p(x) \vee \forall y \left(e(y,y) \vee e(x,y) \right) \right),$$

and by Proposition 5.43

$$\forall x \, \forall y \left(\neg p(x) \vee e(y,y) \vee e(x,y) \right). \tag{F}$$

- Rule (b) becomes

$$\neg \Big(\exists x \, \exists y \left(p(x) \wedge e(y,y) \wedge e(x,y) \right) \Big)$$

and also

$$\forall x \, \forall y \left(\neg p(x) \vee \neg e(y,y) \vee \neg e(x,y) \right). \tag{G}$$

By Proposition 5.42 (i), $F \wedge G$ can be written

$$\forall x \, \forall y \left(\left(\neg p(x) \vee e(y,y) \vee e(x,y) \right) \bigwedge \left(\neg p(x) \vee \neg e(y,y) \vee \neg e(x,y) \right) \right)$$

and, by distributivity,

$$\forall x \, \forall y \left(\neg p(x) \bigvee \left(e(y,y) \vee e(x,y) \right) \bigwedge \left(\neg e(y,y) \vee \neg e(x,y) \right) \right).$$

Let S be a $\{p, e\}$-structure in which this formula is valid, i.e. $\emptyset \models_S F \wedge G$. This implies that for any element a of S, the valuation v_a such that $v_a(x) = v_a(y) = a$ verifies

$$\bar{v}_a \left(\neg p(x) \bigvee \left(e(y,y) \vee e(x,y) \right) \bigwedge \left(\neg e(y,y) \vee \neg e(x,y) \right) \right) = 1$$

i.e.

$$\overline{p_S(a)} \bigvee \left(e_S(a,a) \vee e_S(a,a) \right) \bigwedge \left(\neg e_S(a,a) \vee \neg e_S(a,a) \right) = 1$$

and thus $p_S(a) = 0$, meaning that there is no programmer.

More formally, we have proved that for any structure S, if $F \wedge G$ is valid in S, then $\forall x \, \neg p(x)$ is also valid in S, i.e. $F \wedge G \models \forall x \, \neg p(x)$.

As there is no programmer, it is then normal to obtain contradictory conclusions if we assume that x is a programmer, because in that case the hypothesis $p(x)$ is false, and we can deduce anything from false (see Exercise 5.5).

5.17. As in Theorem 5.17, it suffices to verify by induction on the length of proofs that each use of a proof rule generates only valid sequents from valid sequents. We verified in Theorem 5.17 that each proof rule of propositional logic given in Definition 5.11 is valid; it thus suffices to verify that the three rules given in Definition 5.49 are valid.

For instance, validity of the instantiation rule immediately follows from Proposition 5.41, and validity of the generalization rule immediately follows from Proposition 5.38.

5.18. 1. We introduce the predicates

$R(x, y)$, for denoting 'x likes y',

$A(x)$, for 'x is an alpinist',

$S(x)$, for 'x is a skier',

and the constants

s for 'snow', r for 'rain', b for 'Bernard', and c for 'Christopher'.

The hypotheses of the exercise can be written:

$$F_1 = \forall x(A(x) \supset \neg R(x,r))$$
$$F_2 = \forall x(S(x) \supset R(x,s))$$
$$F_3 = \forall x(A(x) \vee S(x))$$
$$F_4 = \forall y(R(b,y) \vee R(c,y))$$
$$F_5 = \exists y(R(b,y) \wedge \neg R(c,y)).$$

Let $\mathcal{F} = \{F_1, F_2, F_3, F_4, F_5\}$ be the five formulas given above; $\mathcal{F}$ expresses the requirements of the Alpine Club.

2. We want to determine the models of the five formulas of $\mathcal{F}$.

We first show that it is not possible for Bernard to be an alpinist. Assume that there exists a model H such that

$$\emptyset \models_H A(b).$$

Applying instantiation and *modus ponens* to F_1, we have

$$\emptyset \models_H \neg R(b,r).$$

Applying instantiation and *modus ponens* to F_4, we have

(i) $\emptyset \models_H R(c,r)$, hence $\emptyset \not\models_H R(b,r) \wedge \neg R(c,r)$.

By F_5, we must thus have

(ii) $\emptyset \models_H R(b,s) \wedge \neg R(c,s)$.

If $\emptyset \models_H A(c)$, then, by F_1, $\emptyset \models_H \neg R(c,r)$, which contradicts (i), and if $\emptyset \models_H S(c)$, then, by F_2, $\emptyset \models_H R(c,s)$, which contradicts (ii). We thus necessarily have $\emptyset \models_H \neg A(c) \wedge \neg S(c)$, which contradicts F_3.

We thus cannot have $\emptyset \models_H A(b)$, and because of F_3, we have

$$\emptyset \models_H S(b).$$

By F_2, this implies

(iii) $\emptyset \models_H R(b,s)$.

And hence, because false implies anything, $\emptyset \models_H \neg R(b,s) \supset R(c,s)$. In order for H to be a model of F_4, we must also have

$$\emptyset \models_H \neg R(b,r) \supset R(c,r).$$

On the other hand, in order for F_5 to be satisfied in H, we must have

$$\emptyset \models_H \bigl(R(b,r) \wedge \neg R(c,r)\bigr) \vee \bigl(R(b,s) \wedge \neg R(c,s)\bigr),$$

which, in view of (iii), can be written
$$\emptyset \models_H \big(R(b,r) \land \neg R(c,r)\big) \lor \neg R(c,s).$$
If we then let $\emptyset \models_H R(b,r)$, $\emptyset \models_H \neg R(c,r)$ and $\emptyset \models_H R(c,s)$, we see that all formulas are satisfied, and Christopher may be an alpinist, a skier or even both. Thus let $E = \{b, c\}$.
- Let H be defined by: $S(b)$, $S(c)$, $\neg A(b)$, $\neg A(c)$, $R(b,r)$, $\neg R(c,r)$, $R(b,s)$ and $R(c,s)$ are true in H; H is a model of $\mathcal{F}$.
- Let H_1 be defined by: $S(b)$, $S(c)$, $\neg A(b)$, $A(c)$, $R(b,r)$, $\neg R(c,r)$, $R(b,s)$ and $R(c,s)$ are true in H_1; H_1 is also a model of $\mathcal{F}$.
- Let H_2 be defined by: $S(b)$, $\neg S(c)$, $\neg A(b)$, $A(c)$, $R(b,r)$, $\neg R(c,r)$, $R(b,s)$ and $R(c,s)$ are true in H_2; H_2 is also a model of $\mathcal{F}$.
- Let H_3 be defined by: $S(b)$, $\neg S(c)$, $\neg A(b)$, $A(c)$, $R(b,r)$, $\neg R(c,r)$, $R(b,s)$ and $\neg R(c,s)$ are true in H_3; H_3 is also a model of $\mathcal{F}$.

In fact H, H_1, H_2, H_3 are Herbrand models of $\mathcal{F}$. (See Section 5.4.2.)

3. Let F be the formula $\exists x\big((A(x) \land \neg S(x)) \lor (\neg A(x) \land S(x))\big)$. We want to prove that $\mathcal{F} \vdash F$. We reason by contradiction and assume that $\mathcal{F} \cup \{\neg F\}$ is satisfiable; let H be a model of $\mathcal{F} \cup \{\neg F\}$. Note first that
$$\neg F = \forall x\big((\neg A(x) \lor S(x)) \land (A(x) \lor \neg S(x))\big)$$
$$\iff \forall x\Big((\neg A(x) \land (A(x) \lor \neg S(x))) \lor (S(x) \land (A(x) \lor \neg S(x)))\Big)$$
$$\iff \forall x\big((\neg A(x) \land \neg S(x)) \lor (A(x) \land S(x))\big)$$

Let $F' = \forall x\big((\neg A(x) \land \neg S(x)) \lor (A(x) \land S(x))\big)$. Because H is a model of $\neg F$, H is a model of F'. Applying instantiation and *modus ponens* to F' and F_3, we have
$$\emptyset \models_H A(b) \land S(b), \text{ and } \emptyset \models_H A(c) \land S(c); \text{ and hence,}$$
$$\emptyset \models_H A(b) \text{ and } \emptyset \models_H A(c).$$
Because H is a model of F_1, this implies
$$\emptyset \models_H \neg R(b,r) \text{ and } \emptyset \models_H \neg R(c,r).$$
But H is also a model of F_4, namely, $\forall y(R(b,y) \lor R(c,y))$: a contradiction.

Note that, in fact, we have proved the stronger result that $\{F_1, F_3, F_4\} \vdash F$, because formulas F_2 and F_5 have not been used in the proof.

5.19. $I_0 = \{edge(a,b), edge(b,c), path(a,b), path(b,c), path(a,c)\}$, and also, for any $K \subset \{a,b,c\}^2$,
$$I_K = I_0 \cup \{path(l,l') \,/\, (l,l') \in K\}$$
$$J_K = I_0 \cup \{edge(l,l') \,/\, (l,l') \in K\} \cup \{path(l,l') \,/\, (l,l') \in K\}$$
$$\cup \{path(l,l') \,/\, (l,l_1) \in K \text{ and } (l_1,l') \in K\}$$
$$\cup \{path(l,l') \,/\, (l,l_1) \in K \text{ and } (l_1,l_2) \in K \text{ and } (l_2,l') \in K\}$$

In other words, on a graph having exactly three vertices a, b and c, Herbrand models of $\mathcal{F}$ yield any relation *path* such that

(i)   *path* is a transitive relation (because of formula r_4);
(ii) *path* contains the relation *edge* (because of formula r_3);
(iii) the relation *edge* contains at least one edge from a to b and one edge from b to c (because of formulas r_1 and r_2).

(i) and (ii) mean that *path* is the transitive closure of *edge*.

5.20. Let $\mathcal{L}' = \mathcal{L} \cup \{f\}$ and let S' be an $\mathcal{L}'$-structure satisfying F'; the $\mathcal{L}$-structure S that is deduced from S' by omitting the function f_S interpreting f is a model of F, i.e. $\emptyset \models_S F$. Conversely, let S be an $\mathcal{L}$-structure modelling F; because $\emptyset \models_S F$, for every $a_1, \ldots, a_n$ in the domain E of S, there exists an a in E such that $\emptyset \models_S G[x_1 := a_1]\ldots[x_n := a_n][y := a]$. Defining $f_{S'}(a_1, \ldots, a_n) = a$ yields an expansion S' of S such that $\emptyset \models_{S'} F'$.

5.21. The possible prenex forms of F are

$$F_1 = \forall u \exists v [R(u) \vee R'(v)] \quad \text{and} \quad F_2 = \exists v \forall u [R(u) \vee R'(v)].$$

Note that F, F_1 and F_2 are equivalent. The corresponding Skolemizations are

$$F_1' = \forall u [R(u) \vee R'(f(u))] \quad \text{and} \quad F_2' = \forall u [R(u) \vee R'(a)].$$

Note that F_1 and F_1' are not equivalent, nor are F_2 and F_2'.

5.22. A possible prenex form of F is $\forall u \exists v \forall w \exists z [R(u,v) \vee \neg R'(w,z)]$, yielding the Skolemization $\forall u \forall w [R(u, f_1(u)) \vee \neg R'(w, f_2(u,w))]$.
Another possible prenex form of F is $\forall u \forall w \exists v \exists z [R(u,v) \vee \neg R'(w,z)]$, yielding the Skolemization $\forall u \forall w [R(u, f_1(u,w)) \vee \neg R'(w, f_2(u,w))]$.

5.23. H defined by $\{S(b), S(c), R(b,r), R(b,s), R(c,s)\}$ and H_3 defined by $\{S(b), A(c), R(b,r), R(b,s)\}$ (see Exercise 5.18, 2) are the minimal Herbrand models for the Alpine Club. H and H_3 are incomparable, and their intersection $I = \{S(b), R(b,r), R(b,s)\}$ is not a model of $\mathcal{F}$.

5.24. The least Herbrand model of P is defined by $I_M = \emptyset$, i.e. the *path* relation is the empty relation.

5.25. Here the Herbrand universe is $U_H = \{s^n(a) \,/\, n \in \mathbb{N}\}$; the least Herbrand model M of the program P is defined by the whole Herbrand basis, i.e. $I_M = B_H = \{i(s^n(a)) \,/\, n \in \mathbb{N}\}$. M can be seen as modelling the set of integers with a successor funtion.

5.26. Let $M = \langle U_H, I_M \rangle$ be the least Herbrand model of P.
1. If $A \in Th(P) \cap B_H$, then $A \in I_M$ because A must be true in any model of P.
2. Conversely, if $A \in I_M$, then A must be true in all Herbrand models of P, and as a result, because A is a universal formula, A must be true in all models of P, thus $A \in Th(P)$.

5.27. Any set $\mathcal{F}$ of Horn clauses which is satisfiable has a least Herbrand model. Let $P \subset \mathcal{F}$ be the set of program clauses in $\mathcal{F}$; P has a least Herbrand model M defined by $I_M = B_H \cap Th(P)$ (see Exercise 5.26). We can prove that M is also the least Herbrand model of $\mathcal{F}$: let $C = \forall x_1 \cdots \forall x_p (\neg A_1 \vee \cdots \vee \neg A_n)$ be a negative clause in $\mathcal{F}$, because $P \cup \{C\}$ is satisfiable, $Th(P \cup \{C\})$ is not contradictory, and hence, by Theorem 5.57, $P \cup \{C\}$ has a model; by Theorem 5.64 $P \cup \{C\}$ has a Herbrand model $H = \langle U_H, I_H \rangle$. Because $C = \forall x_1 \cdots \forall x_p (\neg A_1 \vee \cdots \vee \neg A_n)$, and $\emptyset \models_H C$, for any valuation $v: \{x_1, \ldots, x_p\} \longrightarrow U_H$, $\bar{v}(C) = 1$, hence for any valuation there is an i such that $\bar{v}(\neg A_i) = 1$, which means that the ground instance $A_i[x_1 := v(x_1)]\cdots[x_p := v(x_p)]$ of A_i is not in I_H. Because $I_M = \cap \{I_H \,/\, \emptyset \models_H P\}$, and $\emptyset \models_H P$, $A_i[x_1 := v(x_1)]\cdots[x_p := v(x_p)]$ is not in I_M either; whence: for any valuation $v: \{x_1, \ldots, x_p\} \longrightarrow U_H$, there is an i such that $A_i[x_1 := v(x_1)]\cdots[x_p := v(x_p)] \notin I_M$, and thus $\emptyset \models_M C$. C is true in the least Herbrand model M of P.

Chapter 5

5.28. 1. Straightforward.

2. I is a model of P if and only if for any valuation s, and any rule r of P, $\emptyset \models_I s^*(r)$, where $s^*(r)$ is obtained by substituting $s(x)$ for x in r, for any variable x. In other words, I is a model of P if and only if for any ground instance $A_1, \ldots, A_n \Longrightarrow A$ of a clause r of P, $\emptyset \models_I [A_1, \ldots, A_n \Longrightarrow A]$, i.e. if and only if for any ground instance $A_1, \ldots, A_n \Longrightarrow A$ of a clause r of P, $\emptyset \models_I A_1, \ldots$, $\emptyset \models_I A_n$ imply $\emptyset \models_I A$, that is if and only if $T_P(I) \subset I$.

3. Immediate by the two preceding questions, and the fact that the least fixed point of T_P is defined by $\inf\{I \in \mathcal{P}(B_H) \,/\, T_P(I) \subset I\}$.

4. Let $\{K_i\}_{i \in \mathbb{N}}$ be any increasing sequence of subsets of $\mathcal{P}(B_H)$. Because T_P is monotone: $\sup_i T_P(K_i) \subset T_P(\sup_i K_i)$.
Let us prove the reverse inclusion: $\sup_i T_P(K_i) \supset T_P(\sup_i K_i)$; let $A \in T_P(\sup_i K_i)$, then there is a ground instance $(A_1, \ldots, A_n \Longrightarrow A)$ of a clause r of P, with $A_1 \in \sup_i K_i, \ldots, A_n \in \sup_i K_i$. We assume that $A_1 \in K_{i_1}, \ldots, A_n \in K_{i_n}$; because K_i is an increasing sequence, if we let $l = \sup\{i_1, \ldots, i_n\}$, we have $K_{i_1} \subset K_l, \ldots, K_{i_n} \subset K_l$ and hence $A_1 \in K_l, \ldots, A_n \in K_l$; thus $A \in T_P(K_l) \subset \sup_i T_P(K_i)$.

5. As demonstrated in question 3, the least Herbrand model of P is the least fixpoint of T_P. By Theorem 2.40, the least fixpoint of T_P is $\sup(\{T_P^n(\emptyset) \,/\, n \in \mathbb{N}\})$.

6. By induction on n, we can see that the sequence $T_P^n(B_H)$ is decreasing, hence $T_P(T_P^n(B_H)) = T_P^{n+1}(B_H) \subset T_P^n(B_H)$; in light of question 2, $T_P^n(B_H)$ is a model of P.

7. K is a model of P because $T_P(K) \subset K$.

8. The following example shows that K is not a fixpoint of P. Let P be defined by:

$r_1:$ $\quad\quad\quad\quad\quad\quad\quad\quad p(x) \Longrightarrow q(a)$,
$r_2:$ $\quad\quad\quad\quad\quad\quad\quad\quad p(x) \Longrightarrow p(f(x))$,
$r_3:$ $\quad\quad\quad\quad\quad\quad\quad\quad q(x) \Longrightarrow q(f(x))$,
$r_4:$ $\quad\quad\quad\quad\quad\quad\quad\quad q(x) \Longrightarrow q(b)$.

Then $\quad I_M = \emptyset \quad$ and
$$B_H = \left\{ p(X), q(X) \,/\, X \in \{f^n(a), f^n(b), n \in \mathbb{N}\} \right\}$$
$$T_P(B_H) = \left\{ q(X), p(Y) \,/\, X \in \{f^n(a), f^n(b), n \in \mathbb{N}\}, Y \in \{f^k(a), f^k(b), k \geq 1\} \right\}$$
$$T_P^2(B_H) = \left\{ q(X), p(Y) \,/\, X \in \{f^n(a), f^n(b), n \in \mathbb{N}\}, Y \in \{f^k(a), f^k(b), k \geq 2\} \right\}$$
$$\ldots$$
$$K = \left\{ q(X) \,/\, X \in \{f^n(a), f^n(b), n \in \mathbb{N}\} \right\}$$
$$T_P(K) = \{q(X), q(Y) \,/\, X = f^n(a), Y = f^p(b), n \geq 1, p \geq 0\}$$
$$T_P^2(K) = \{q(X), q(Y) \,/\, X = f^n(a), Y = f^p(b), n \geq 2, p \geq 0\}$$
$$\ldots$$
$$\nu(T_P) = \{q(Y) \,/\, Y = f^p(b), p \geq 0\}$$

Then $T_P(K) \neq K$, and the greatest fixpoint of T_P is $\nu(T_P) \subsetneq K$. Note that the least fixpoint of T_P is defined by $I_M = \emptyset$.
More generally, it can be shown that if $\nu(T_P)$ is the greatest fixpoint of T_P, then $\nu(T_P) \subset K$.

Chapter 6

6.1. We have n possible choices for the first element, then $n-1$ possible choices for the second element, etc. Hence $A_n^k = n(n-1)(n-2)\cdots(n-k+1) = \dfrac{n!}{(n-k)!}$.

6.2. Consider a partition of a set E with $a+b$ elements in $E = A \cup B$, where $A \cap B = \emptyset$, $|A| = a$, and $|B| = b$. Then $\binom{a+b}{p}$ is the number of subsets with p elements of E; any subset with p elements of E can be obtained by choosing k elements in A and $n-k$ elements in B, for $0 \leq k \leq \inf(p, a)$.

6.3. 1. Apply Exercise 6.2 with $a = b = p = n$:
$$\binom{2n}{n} = \sum_{k=0}^{n} \binom{n}{k}\binom{n}{n-k}.$$
Hence, since $\binom{n}{k} = \binom{n}{n-k}$,
$$\binom{2n}{n} = \sum_{k=0}^{n} (\binom{n}{k})^2.$$

2. We apply the same method as in Exercise 6.2. $\binom{3n}{n}$ is the number of subsets with n elements of $E = A \cup B \cup C$, where $|A| = |B| = |C| = n$ and A, B, C are pairwise disjoint. We have $\sum_{i+j+k=n} \binom{n}{k}\binom{n}{i}\binom{n}{j} = \binom{3n}{n} = \sum_{k=0}^{n} \sum_{i=0}^{n-k} \binom{n}{k}\binom{n}{i}\binom{n}{n-i-k} = \sum_{k=0}^{n} \sum_{i=0}^{n-k} \binom{n}{k}\binom{n}{i}\binom{n}{i+k}$.
We might also directly apply the result of Exercise 6.2 by saying that a subset with n elements of $E = A \cup B \cup C$ can be obtained by choosing a subset with p elements in A and $n-p$ elements in $B \cup C$; i.e. $\sum_{k=0}^{n} \binom{n}{k} \sum_{i=0}^{n-k} \binom{n}{i}\binom{n}{n-i-k} = \sum_{k=0}^{n} \binom{n}{k} \sum_{i=0}^{n-k} \binom{n}{i}\binom{n}{i+k}$.

6.4. The number of n-tuples of disjoint subsets with p elements of a set with np elements is $\binom{np}{p} \times \binom{(n-1)p}{p} \times \binom{(n-2)p}{p} \times \cdots \times \binom{p}{p}$. Because the ordering of the subsets in the partition does not matter, $n!$ n-tuples of subsets correspond to a single partition, and we have
$$N = \frac{\binom{np}{p} \times \binom{(n-1)p}{p} \times \binom{(n-2)p}{p} \times \cdots \times \binom{p}{p}}{n!}.$$

6.5. $S = \binom{n}{p} \sum_{q=0}^{p} (-1)^q \binom{p}{q} = \binom{n}{p} \sum_{q=0}^{p} \binom{p}{q}(-1)^q 1^{p-q} = \binom{n}{p}(1-1)^p = 0$.

6.6. 1. By induction on k. For $k = 0$, the identity is clearly true. Assume it holds for k; i.e. $\sum_{p=0}^{k} \binom{n+p}{p} = \binom{n+k+1}{k}$. Then, for $k+1$, $\sum_{p=0}^{k+1} \binom{n+p}{p} = \sum_{p=0}^{k} \binom{n+p}{p} + \binom{n+k+1}{k+1} = \binom{n+k+1}{k} + \binom{n+k+1}{k+1} = \binom{n+k+2}{k+1}$, by the identity (6.1) of Proposition 6.5. To intuitively motivate this identity, consider Pascal's triangle: the sum of the values at the black dots is to be found in the white square (see Figure 15.3).

Figure 15.3

Chapter 6

2. By induction on n. For $n = p$ the identity is trivially true. Assume it is true for n. Then $\binom{p}{p} + \binom{p+1}{p} + \cdots + \binom{n}{p} + \binom{n+1}{p} = \binom{n+1}{p+1} + \binom{n+1}{p} = \binom{n+2}{p+1}$. A look at Pascal's triangle will support intuition here as well.

Let $S_p = \sum_{k=1}^{n} k^p$.

- For $p = 1$, we have $S_1 = \sum_{k=1}^{n} k = \sum_{k=1}^{n} \binom{k}{1} = \binom{n+1}{2} = \dfrac{n(n+1)}{2}$.

- For $p = 2$, we have $\binom{k}{2} = \dfrac{k(k-1)}{2} = \dfrac{k^2}{2} - \dfrac{k}{2}$; hence

$$2\binom{n+1}{3} = 2\left(\sum_{k=2}^{n}\binom{k}{2}\right) = (S_2 - 1) - (S_1 - 1) \quad \text{and}$$

$$S_2 = S_1 + 2\binom{n+1}{3} = \dfrac{n(n+1)(n-1)}{3} + \dfrac{n(n+1)}{2} = \dfrac{n(n+1)(2n+1)}{6}$$

- For $p = 3$, we have $\binom{k}{3} = \dfrac{k(k-1)(k-2)}{6} = \dfrac{k^3 - 3k^2 + 2k}{6}$, and thus

$$6\binom{n+1}{4} = (S_3 - 9) - 3(S_2 - 5) + 2(S_1 - 3) = S_3 - 3S_2 + 2S_1;$$

hence, finally,

$$S_3 = \dfrac{(n+1)n(n-1)(n-2)}{4} + \dfrac{n(n+1)(2n+1)}{2} - n(n+1) = \dfrac{n^2(n+1)^2}{4} = S_1^2.$$

6.7. By induction on the degree of P.

- If P has degree 0, (i.e. if $\forall x$, $P(x) = c$), then: $\sum_{i=0}^{n+1}(-1)^i\binom{1}{i}P(x+i) = c\sum_{i=0}^{n+1}(-1)^i\binom{n+1}{i} = 0$.

- If P has degree $d+1$, then P can be written $P = xQ + c$, where Q has degree d. Let $n \geq d+1$. Hence,

$$\sum_{i=0}^{n+2}(-1)^i\binom{n+2}{i}P(x+i) = x\sum_{i=0}^{n+2}(-1)^i\binom{n+2}{i}Q(x+i)$$
$$+ \sum_{i=0}^{n+2}(-1)^i\binom{n+2}{i}iQ(x+i) + c\sum_{i=0}^{n+2}(-1)^i\binom{n+2}{i}$$

By the induction hypothesis, the first summand of this sum is zero; the third summand is trivially zero. We only have to compute $\sum_{i=0}^{n+2}(-1)^i i\binom{n+2}{i}Q(x+i)$, which can also be written

$$\sum_{i=0}^{n+1}(-1)^{i+1}(i+1)\binom{n+2}{i+1}Q(x+i+1).$$

Since $(i+1)\binom{n+2}{i+1} = (n+2)\binom{n+1}{i}$, we have

$$\sum_{i=0}^{n+1}(-1)^{i+1}(i+1)\binom{n+2}{i+1}Q(x+i+1) = -(n+2)\sum_{i=0}^{n+1}(-1)^i\binom{n+1}{i}Q(x+i+1) = 0.$$

6.8. The required probability distribution is

$$p = \frac{\text{number of favourable cases}}{\text{number of possible cases}} = \frac{n_f}{n}$$

There are twenty-seven cubes, each one having twenty-four possible orientations in the space, and they can be set in $(27)!$ possible ways. Thus, there are altogether $n = 24^{27}(27)!$ rebuilding possibilities.

In order to compute n_f, we consider the various types of cube and the ways in which they can be set up in order to rebuild a big red cube:

- There are eight cubes with three red sides; each one can be oriented in three different ways in the space, and they can be permuted in 8! ways, and hence there are $3^8!$ 'right ways' to set up these eight cubes.
- There are twelve cubes with two red sides; each one can be oriented in two different ways in the space, and they can be permuted in $(12)!$ ways, and hence there are $2^{12}(12)!$ 'right ways' to set up these eight cubes.
- There are six cubes with one red side; each one can be oriented in four different ways in the space, and they can be permuted in 6! ways, and hence there are $4^6 6!$ 'right ways' to set up these eight cubes.
- There is one white cube (the central cube), it can be oriented in twenty-four different ways, and hence, finally, $n_f = 3^8 \times 2^{12} \times 4^6 \times 24 \times 8! \times (12)! \times 6!$.

For the inquiring reader, $p \simeq 1,83 \times 10^{-37}$.

6.9. 1. There are eight possible choices for the face value of four cards of the same kind, and, as all four cards of the chosen face value are in the hand, it only remains to choose the fifth card of the hand among the twenty-eight remaining cards. Hence: 8×28 hands containing a four-of-a-kind.

2. There are eight possible choices for the face value of the three cards of the same kind and $\binom{4}{3}$ possible choices for the cards of the three-of-a-kind among the four cards of the chosen face value in the deck. We then have to choose the fourth and fifth cards in the hand. These fourth and fifth cards must have face values different from the face value of the three-of-a-kind (otherwise we would have a four-of-a-kind) and must not be of equal face values (otherwise we would have a three-of-a-kind **and** a pair). They can thus be chosen in $\binom{7}{2} \times 4 \times 4$ possible ways ($\binom{7}{2}$ choices for two face values and four choices for the face values of the fourth and fifth cards). Hence, finally: $8 \times \binom{4}{3} \times \binom{7}{2} \times 4^2$ hands containing a three-of-a-kind.

3. There are eight possible choices for the strength of the pair, then $\binom{4}{2}$ possible choices for the cards of the pair, and finally $\binom{7}{3} \times 4 \times 4 \times 4$ possible choices for the remaining three cards. Hence: $8 \times \binom{4}{2} \times \binom{7}{3} \times 4^3$ hands containing a pair.

6.10. $\binom{16}{8}$. It suffices to choose the place of the eight bits equal to 1.

6.11. 1. There corresponds such a generalized characteristic function for each pair (A_1, A_2) verifying (6.2); conversely, for any function $f: E \to \{0, 1, 2\}$, if we let $A_1 = f^{-1}(1)$ and $A_2 = f^{-1}(\{1, 2\})$, then (A_1, A_2) verifies (6.2). There are 3^n such functions; thus $N_1 = 3^n$.

2. Similarly, we define a function $f: E \to \{0, 1, 2, 3\}$ for each triple (A_1, A_2, A_3) verifying (6.3), and we deduce that $N_2 = 4^n$.

Chapter 6

6.12. 1. There are n_1 possible choices for the element of A and n_2 possible choices for the element of B. After that we can choose $p-2$ elements in $\overline{A \cup B}$; hence

$$N = n_1 n_2 \binom{n-(n_1+n_2)}{p-2}, \quad \text{letting } \binom{m}{q} = 0 \text{ if } q > m.$$

2. There are altogether $\binom{n}{p}$ subsets with p elements. Let C (resp. D) be the subsets with p elements having no element of A (resp. B). We have

$$|C \cup D| = |C| + |D| - |C \cap D| = \binom{n-n_1}{p} + \binom{n-n_2}{p} - \binom{n-(n_1+n_2)}{p},$$

since $C \cap D$ is the set of subsets with p elements having no element from $A \cup B$. Thus $N = \binom{n}{p} - \binom{n-n_1}{p} - \binom{n-n_2}{p} + \binom{n-(n_1+n_2)}{p}$.

6.13. 1. 4^n.

2. Let A_1 (resp. A_2, A_3, A_4) be the set of strings of length n where the letter a (resp. b, c, d) does not occur. We want to count the number of elements in the complement of $A_1 \cup A_2 \cup A_3 \cup A_4$, i.e. $N = |\overline{A_1 \cup A_2 \cup A_3 \cup A_4}|$. We have $N = 4^n - |A_1 \cup A_2 \cup A_3 \cup A_4|$, and, by Proposition 6.13,

$$|A_1 \cup A_2 \cup A_3 \cup A_4| = \sum |A_1| - \sum_{i<j} |A_i \cap A_j|$$
$$+ \sum_{i<j<k} |A_i \cap A_j \cap A_k| - |A_1 \cap A_2 \cap A_3 \cap A_4|$$
$$= 4 \times 3^n - 6 \times 2^n + 4, \quad \text{since } |A_1 \cap A_2 \cap A_3 \cap A_4| = 0.$$

Thus $N = 4^n - 4 \times 3^n + 6 \times 2^n - 4$.

6.14. Let A (resp. B) be the permutations of $\{a,b,c,d,e,f\}$ containing 'ac' (resp. 'bde'). We want to count $\overline{A} \cap \overline{B} = \overline{A \cup B}$. By Proposition 6.13, $|A \cup B| = |A| + |B| - |A \cap B|$, and

- $|A| = 4! \times 5$, since in a permutation containing 'ac' there are 5 possible positions of 'ac' and $4!$ permutations of $b, d, e,$ and f,
- $|B| = 3! \times 4$, since there are $3!$ permutations of $a, c,$ and f and 4 possible positions of 'bde', and
- $|A \cap B| = 3! = 6$, since we have to permute the three subsequences 'ac', 'bde', and f.

Hence, finally: $N = |\overline{A \cup B}| = 6! - 4! \times 5 - 3! \times 4 + 6 = 582$.

6.15. $u_0 = 1$, $u_1 = 2$, $u_2 = 3$ and $u_n = 4$, for $n \geq 3$. The only possible strings are thus

$$\underbrace{10\ldots0}_{n-2}\begin{Bmatrix}0\\1\end{Bmatrix} \quad \text{and} \quad \underbrace{00\ldots0}_{n-2}\begin{Bmatrix}0\\1\end{Bmatrix}.$$

6.16. 1. An increasing mapping is necessarily injective; thus, we must have $n \leq m$. With each increasing function, we can associate a unique set with n distinct values between 1 and m, namely, $\{f(1), \ldots, f(n)\}$. Conversely, with each set of n distinct values between 1 and m, we can associate a unique increasing sequence $f(1) < \cdots < f(n)$. There is thus a one-to-one correspondence between the set of subsets consisting of n elements between 1 and m and the set of increasing mappings from $\{1, \ldots, n\}$ to $\{1, \ldots, m\}$. Hence, the number of increasing mappings from $\{1, \ldots, n\}$ to $\{1, \ldots, m\}$ is $\binom{m}{n}$.

2. We want to count the number of increasing mappings such that the element $k+1$ has a preimage. This boils down to determining the number of subsets with n elements between 1 and m containing $k+1$. There are $\binom{n-1}{m-1}$ such subsets. (It remains, after choosing $k+1$, to choose $n-1$ elements among the $m-1$ elements left.)

3. We wish to determine the number of increasing functions such that

$$|\{a \,/\, f(a) < k\}| = |\{a \,/\, f(a) > k\}|.$$

We must distinguish two cases here, according to the parity of n.

- If n is even, k has no preimage. The elements between 1 and $n/2$ have their images between 1 and $k-1$, and the elements between $(n/2)+1$ and n have their images between $k+1$ and m. It is thus sufficient to choose $n/2$ values in $\{1,\ldots,k-1\}$ and $n/2$ values in $\{k+1,\ldots,m\}$. This is possible only if $(n+1)/2 \le k \le m - n/2$ holds. The required number of functions is then $\binom{k-1}{n/2}\binom{m-k}{n/2}$.

- If $n = 2p+1$ is odd, then the p first values of f are in $\{1,\ldots,k-1\}$, the $(p+1)$th value is k, and the p last values of f are in $\{k+1,\ldots,m\}$. The required number of functions is thus $\binom{k-1}{p}\binom{m-k}{p} = \binom{k-1}{E\lfloor n/2 \rfloor}\binom{m-k}{E\lfloor n/2 \rfloor}$, where $E\lfloor n/2 \rfloor$ denotes the largest integer less than or equal to $n/2$.

4. We now wish to compute the number of injections verifying

$$|\{a \,/\, f(a) < k\}| = |\{a \,/\, f(a) > k\}|.$$

To cover all of them, it suffices to consider all possible permutations of increasing functions verifying the condition. We thus have $n!\binom{k-1}{p}\binom{m-k}{p}$ functions, with $n=2p$ or $n=2p+1$.

6.17. 1. We will construct strings of length n. We thus have n positions, q_1 letters $a_1,\ldots$, and q_p letters a_p. Let us first place the letters a_1 in the string. There are $\binom{n}{q_1}$ ways of placing them. After all the letters a_1 are placed, there are $n-q_1$ places left. We then place the letters a_2. There are $\binom{n-q_1}{q_2}$ ways of positioning them, and there are $n - q_1 - q_2$ available positions afterwards. We place all the letters successively. The number of solutions is thus $\binom{n}{q_1}\binom{n-q_1}{q_2}\cdots\binom{n-(q_1+\cdots+q_{p-1})}{q_p}$.

(a) The formal polynomial $(X_1+X_2+\cdots+X_p)^n$ can be written, assuming that the variables $X_1,\ldots,X_n$ commute, as

$$(X_1+X_2+\cdots+X_p)^n = \sum_{q_1+q_2+\cdots+q_p=n} \binom{n}{q_1}\cdot\binom{n-q_1}{q_2}\cdots\binom{n-q_1-\cdots-q_{p-1}}{q_p} X_1^{q_1} X_2^{q_2}\cdots X_p^{q_p}.$$

(b) We could have done a direct computation by noticing that, among the $n!$ ways of placing n letters in n places, the q_i letters a_i can be permuted in $q_i!$ ways. We thus obtain $\dfrac{n!}{q_1!\cdots q_p!}$ and therefore deduce $\binom{n}{q_1}\binom{n-q_1}{q_2}\cdots\binom{n-(q_1+\cdots+q_{p-1})}{q_p} = \dfrac{n!}{q_1!\cdots q_p!} = \binom{n}{q_1,\ldots,q_p}$.

2. We apply the result of question 1 (b) with $n = k!$, $p = (k-1)!$, and $\forall i = 1,\ldots,p$, $q_i = k$. We obtain $\dfrac{(k!)!}{k!(k-1)!}$, which is an integer.

3. 1287.

Chapter 6   339

6.18. Let $f(x) = (1+x)^{2n} + (1-x)^{2n}$. Note that

$$(1+x)^{2n} = \sum_{p=0}^{2n} \binom{2n}{p} x^p, \quad \text{and}$$

$$(1-x)^{2n} = \sum_{p=0}^{2n} (-1)^p \binom{2n}{p} x^p; \quad \text{thus}$$

$$f(x) = (1+x)^{2n} + (1-x)^{2n} = 2\sum_{p=0}^{n} \binom{2n}{2p} x^{2p}.$$

Taking derivatives, we have

$$f'(x) = 4\sum_{p=1}^{n} p\binom{2n}{2p} x^{2p-1}, \quad \text{and}$$

$$f''(x) = 4\sum_{p=1}^{n} p(2p-1)\binom{2n}{2p} x^{2p-2} = 8\sum_{p=1}^{n} p^2 \binom{2n}{2p} x^{2p-2} - 4p\sum_{p=1}^{n} \binom{2n}{2p} x^{2p-2},$$

$$= 8g(x) - \frac{1}{x} f'(x).$$

Because we defined $g(x) = \sum_{p=1}^{n} p^2 \binom{2n}{2p} x^{2p-2}$, we thus have

$$g(x) = \frac{1}{8}\left(f''(x) + \frac{1}{x} f'(x)\right).$$

Since

$$f'(x) = 2n\left((1+x)^{2n-1} - (1-x)^{2n-1}\right), \quad \text{and}$$

$$f''(x) = 2n(2n-1)\left((1+x)^{2n-2} + (1-x)^{2n-2}\right),$$

we have

$$g(1) = \frac{1}{8}\left(2n(2n-1)2^{2n-2} + 2n\cdot 2^{2n-1}\right)$$

$$= n(2n+1)2^{2n-4} = \sum_{p=0}^{n} p^2 \binom{2n}{2p}.$$

But

$$g(1) = \sum_{p=1}^{n} p^2 \binom{2n}{2p} = \sum_{p=0}^{n} p^2 \binom{2n}{2p}.$$

6.19. Method 1
(a) x_{p+1} can take all the values from 0 to n. For each fixed value of x_{p+1}, $x_1 + \cdots + x_p = n - x_{p+1}$, and there are thus $F(n - x_{p+1}, p)$ solutions. Hence $F(n, p+1) = \sum_{k=0}^{n} F(k, p)$.
(b) By induction on n. If $n = 0$, then $\binom{p}{p} = \binom{p-1}{p-1} = 1$; and if $\binom{n+p}{p} = \sum_{k=0}^{n} \binom{k+p-1}{p-1}$, then:

$$\sum_{k=0}^{n+1} \binom{k+p-1}{p-1} = \sum_{k=0}^{n} \binom{k+p-1}{p-1} + \binom{n+1+p-1}{p-1}$$

$$= \binom{n+p}{p} + \binom{n+p}{p-1} \quad \text{(by the induction hypothesis)}$$

$$= \binom{n+1+p}{p} \quad \text{(by equation (6.1))}.$$

We can also write $\sum_{k=0}^{n} \binom{k+p-1}{p-1} = \sum_{k=p-1}^{n+p-1} \binom{k}{p-1}$, and then apply the result of Exercise 6.6, 2.

(c) We show that $F(n,p) = \binom{n+p-1}{p-1}$ by induction on p. If $p = 1$, then $F(n,p) = 1 = \binom{n}{0}$; the verification of the recurrence immediately follows from questions (a) and (b).

Method 2

(a) Let $E_{n,p}$ be the set of p-tuples $(x_1, \ldots, x_p) \in \mathbb{N}^p$ such that $x_1 + \cdots + x_p = n$; so $F(n,p) = |E_{n,p}|$. We can decompose $E_{n,p+1}$ in a partition: $E_{n,p+1} = A \cup B$, with $A = \{(x_1, \ldots, x_p, 0) \in \mathbb{N}^{p+1} \ / \ x_1 + \cdots + x_p = n\}$ and $B = \{(x_1, \ldots, x_p, x_{p+1}) \in \mathbb{N}^{p+1} \ / \ x_{p+1} > 0 \text{ and } x_1 + \cdots + x_p + x_{p+1} = n\}$. Therefore,

- $|A| = |E_{n,p}|$, since any element of A can be obtained by adding a 0 component to a p-tuple in $E_{n,p}$.
- $|B| = |E_{n-1,p+1}|$, since B is in a one-to-one correspondence with $E_{n-1,p+1}$ as follows: if $(x_1, \ldots, x_p, x_{p+1}) \in B$, subtract 1 from x_{p+1} in order to obtain $(x_1, \ldots, x_p, x_{p+1} - 1) \in E_{n-1,p+1}$.

We therefore deduce that $F(n, p+1) = F(n, p) + F(n-1, p+1)$.

(b) By induction on $k = n + p$. If $k = 1$, then $n = 0$, $p = 1$, and $F(n,p) = 1 = \binom{0}{0}$. Assume the result is true for k, and let us compute $F(n,p)$ with $n+p = k+1$. By (a) we have $F(n,p) = F(n,p-1) + F(n-1,p)$, and by the recurrence, which can be applied because $n+p-1 = n-1+p = k$, we have $F(n,p) = \binom{n+p-2}{n} + \binom{n+p-2}{n-1} = \binom{n+p-1}{n}$.

Chapter 7

7.1. 1. $b_0 = 2$ (the empty tree $\emptyset$ and the tree $(a, \emptyset, \emptyset)$), $b_1 = 3$, $b_2 = 12$.
2. For $n \geq 2$, $b_n = \sum_{k=0}^{n-1} b_k \times b_{n-1-k}$.

7.2. 1.
- $u_0 = 1$ (the only binary tree of depth 0 is $\emptyset$),
- and for $n \geq 1$, $u_n = 1 + ku_{n-1}^2$: indeed a binary tree of depth less than or equal to n is of the form $\emptyset$ or (x, b_l, b_r) with $x \in \Sigma$ and with b_l, b_r binary trees of depth less than or equal to $n - 1$.

2.
- $v_0 = 1$,
- for $n \geq 1$, $v_n = k(v_{n-1}^2 + 2v_{n-1}u_{n-2})$; indeed a binary tree of depth exactly n with $n \geq 1$, is of the form (x, b_l, b_r) with $x \in \Sigma$ and
  - either b_l, b_r binary trees of depth exactly $n - 1$,
  - or $b_l \in U_{n-2}$ and $b_r \in AB_{n-1}$ (b_l is of depth strictly less than $n - 1$, namely, b_l is of depth less than or equal to $n - 2$, and b_r is of depth exactly $n - 1$),
  - or, symmetrically, $b_r \in U_{n-2}$ and $b_l \in AB_{n-1}$.

7.3. If the nth line intersects the preceding lines in i distinct points, it will determine $i + 1$ new regions, the first and the $(i+1)$th region will be infinite, the $i - 1$ intermediate regions will be bounded. We advise the reader to draw a picture to support his/her intuition. The maximum possible number of bounded regions determined is thus, assuming that each new line is not parallel to any of the others,

$$\text{for} \quad n > 2: \quad r_n = (n-2) + r_{n-1} = (n-2) + (n-3) + \cdots + 1 = (n-1)(n-2)/2.$$

7.4. If there is a single circle it will obviously define two regions in the plane. Assume that $n - 1$ circles define r_{n-1} regions, and add a new circle; it will intersect each of the old $n - 1$ circles in two points, and each crossing of an intersection point will define a new region, hence $2(n - 1)$ new regions. We deduce, summing equations (7.3) for $i = 2, \ldots, n - 1$, that $r_n = n(n-1) + 2$.

Chapter 7

7.5. For $n = 2^k$, $k = 1, \ldots, l$, we write

$$(\times 2^{k-1}) \qquad t_2 = 1$$
$$(\times 2^{k-2}) \qquad t_4 = 2t_2 + 1$$
$$\vdots$$
$$(\times 2) \qquad t_{2^{k-1}} = 2t_{2^{k-2}} + 1$$
$$t_{2^k} = 2t_{2^{k-1}} + 1$$

and we sum the above equalities multiplied by the indicated summation factors, hence:

$$t_{2^n k} = 2^k - 1$$

We can also show by induction that $t_{2^k} = 2^k - 1$.

(B) $t_2 = 2^1 - 1 = 1$,

(I) $t_{2^{k+1}} = 2t_{2^k} + 1 = 2(2^k - 1) + 1 = 2^{k+1} - 1$.

For $n \neq 2^k$, the solution t_n is not uniquely defined (see Proposition 7.8).

7.6. $s_p = -p$, $\forall p \in \mathbb{N}$, since $s_1 = 1 - 2$ and $s_{p+1} = s_p + 2p + 1 - (2p + 2) = s_p - 1$.

7.7. $u_n = u_{n-1} + u_{n-2}$, with $u_0 = u_1 = u_2 = 1$.

Similarly, with other types of recurrence, it suffices to impose initial conditions that are too stringent.

7.8. In general, it is enough to have initial conditions that are too weak.
$u_n = u_{n-1} + u_{n-2}$, with $u_0 = 1$.
$u_n = 2u_{n/2}$ with $u_1 = 1$; we then have the solutions $u_{2^n} = 2^n$ and $u_{(2k+1)2^n} = f(2k+1)2^n$, where f can be an arbitrary function.

7.9. Notice that: $u_n - u_{n-1} = 3u_{n-1}$, hence :

$$u_n = 4u_{n-1} = 4^2 u_{n-2} = \ldots = 4^n u_0.$$

7.10. The characteristic polynomial of the recurrence is $r^3 - 5r^2 + 8r - 4 = 0$; it has a simple root 1 and a double root 2; the solution of the recurrence is thus of the form: $u_n = a + (b + cn)2^n$, the initial conditions allow us to determine a, b, c, since for $n = 0, 1, 2$, we obtain

$$a + b = 0,$$
$$a + 2(b + c) = 1,$$
$$a + 4(b + 2c) = 2.$$

hence: $a = -b = -2$ and $c = -1/2$ and, finally,

$$u_n = 2^{n+1} - n2^{n-1} - 2.$$

7.11. 1. a, b, aba, abb, baa, bab.

2. By induction on $|w|$. If $|w| \leq 1$, the result is true. Assume $w \in B$ and $|w| \leq n \implies |w|$ is odd. Let $w' \in B$ and $|w'| = n + 1$; then $w' = abw$ or $w' = baw$, with $w \in B$ and $|w| = |w'| - 2 \leq n - 1$, thus $|w|$ is odd by the recurrence, and thus $|w'| = |w| + 2$ is also odd.
The converse is not true; for instance $w = aaa \notin B$.

3. $u_1 = 2$ and $u_2 = 0$; the recurrence relation is $u_n = 2u_{n-2}$.
Solving it directly, we obtain $u_{2n} = 0$, $u_{2n+1} = 2^{n+1}$. The characteristic polynomial method gives the characteristic polynomial: $r^2 - 2 = 0$, hence $u_n = \lambda(\sqrt{2})^n + \mu(-\sqrt{2})^n$, with $\lambda = -\mu = 1/\sqrt{2}$.

7.12. 1. $L_0 = \{\varepsilon\}$, $L_1 = \{a,b,c,d\}$, $L_2 = \{xy \,/\, x, y \in \Sigma, xy \neq ab\} = \Sigma^2 - \{ab\}$. $u_0 = 1$, $u_1 = 4$, $u_2 = 15$.

2. $w' \in L_{n-1}$, and, moreover, $w' \notin bL_{n-2}$ if $x = a$.

3. $L_n = \{a,b,c,d\}L_{n-1} - \{ab\}L_{n-2}$. Since $bL_{n-2} \subsetneq L_{n-1}$,

$$|L_n| = |\{a,b,c,d\}L_{n-1}| - |\{ab\}L_{n-2}| = 4|L_{n-1}| - |L_{n-2}|.$$

4. Characteristic polynomial: $r^2 - 4r + 1 = 0$, roots $r = 2 \pm \sqrt{3}$, hence:

with
$$u_n = \lambda(2+\sqrt{3})^n + \mu(2-\sqrt{3})^n$$
$$\lambda + \mu = u_0 = 1$$
$$\lambda(2+\sqrt{3}) + \mu(2-\sqrt{3}) = u_1 = 4$$

$$\lambda = \frac{\sqrt{3}+2}{2\sqrt{3}}, \qquad \mu = \frac{\sqrt{3}-2}{2\sqrt{3}}$$

5. We obtain $u_n = \dfrac{\sqrt{2}+1}{2}(2+\sqrt{2})^n + \dfrac{1-\sqrt{2}}{2}(2-\sqrt{2})^n$.

7.13. The general solution of (7.11) (page 131) is of the form $u_n = ac^n e^{nit} + bc^n e^{-nit}$. Recall that
$$e^{iz} = \cos z + i\sin z \qquad \text{and} \qquad e^{-iz} = \cos z - i\sin z$$
We thus have
$$u_n = ac^n(\cos(nt) + i\sin(nt)) + bc^n(\cos(nt) - i\sin(nt))$$
$$= (a+b)c^n \cos(nt) + i(a-b)c^n \sin(nt))$$
hence the result with $\lambda = a+b$ and $\mu = i(a-b)$; if the initial values are real, then the solution of (7.11) will be of the form $u_n = ac^n e^{nit} + bc^n e^{-nit}$, with a and b conjugate, and we will thus finally obtain $u_n = \mathrm{Re}\,(a)c^n \cos(nt) + \mathrm{Im}\,(a)c^n \sin(nt)$.

7.14. 1. The associated characteristic polynomial $r^2 - r + 2$ has the following two conjugate complex roots: $(1 \pm i\sqrt{7})/2$. We thus have

$$u_n = \alpha\left(\frac{1+i\sqrt{7}}{2}\right)^n + \beta\left(\frac{1-i\sqrt{7}}{2}\right)^n$$

If u_0 and u_1 are real, α and β are conjugate; indeed in this case, if $\alpha = a + ia'$ and $\beta = b + ib'$,

$$\alpha + \beta = u_0,$$
$$\alpha\left(\frac{1+i\sqrt{7}}{2}\right) + \beta\left(\frac{1-i\sqrt{7}}{2}\right) = u_1;$$

hence $a + b = u_0$, $a' = -b'$, and finally, after solving the second equation,

$$a = b = \frac{u_0}{2} \qquad a' = -b' = -\frac{2u_1 - u_0}{2\sqrt{7}}$$

In this case we thus have
$$u_n = \mathrm{Re}\left(\alpha\left(\frac{1+i\sqrt{7}}{2}\right)^n\right)$$

2. Letting $v_n = u_n + 2$, v_n verifies the recurrence: $v_n = v_{n-1} - 2v_{n-2}$, and we are back to case 1.

7.15. 1. The characteristic polynomial of the recurrence is: $2r^2 - 3r + 1 = 0$; it has the roots $r = 1$ and $r = 1/2$; the general solution is thus of the form $u_n = a + b/2^n$.

2. The characteristic polynomial of the recurrence is: $r^2 - 4r + 4 = 0$; it has the double root $r = 2$; the general solution is thus of the form $u_n = (an + b)2^n$.

7.16. 1. We have

$$\begin{pmatrix} u_n \\ v_n \end{pmatrix} = M \times \begin{pmatrix} u_{n-1} \\ v_{n-1} \end{pmatrix} = \begin{pmatrix} 4 & 2 \\ -3 & -1 \end{pmatrix} \times \begin{pmatrix} u_{n-1} \\ v_{n-1} \end{pmatrix}.$$

The eigenvalues of M are 1 and 2, and the associated eigenvectors are

$$V_1 = \begin{pmatrix} -2 \\ 3 \end{pmatrix} \quad \text{and} \quad V_2 = \begin{pmatrix} 1 \\ -1 \end{pmatrix}$$

we deduce that

$$M = N \times \Delta \times N^{-1} = N \times \begin{pmatrix} 1 & 0 \\ 0 & 2 \end{pmatrix} \times N^{-1},$$

with

$$N = \begin{pmatrix} -2 & 1 \\ 3 & -1 \end{pmatrix} \quad \text{and} \quad N^{-1} = \begin{pmatrix} 1 & 1 \\ 3 & 2 \end{pmatrix}$$

hence

$$\begin{pmatrix} u_n \\ v_n \end{pmatrix} = N \times \Delta^n \times N^{-1} \times \begin{pmatrix} a \\ b \end{pmatrix} = N \times \begin{pmatrix} 1 & 0 \\ 0 & 2^n \end{pmatrix} \times \begin{pmatrix} a+b \\ 3a+2b \end{pmatrix}$$
$$= \begin{pmatrix} -2a - 2b + 2^n(3a + 2b) \\ 3a + 3b - 2^n(3a + 2b) \end{pmatrix}$$

2. Letting $u'_n = \log u_n$, $v'_n = \log v_n$, we are back to case 1.

7.17. 1. $u_1 = 0$, $v_1 = 1$, $w_1 = 0$, $u_2 = 1$, $v_2 = 0$, $w_2 = 1$.

2. For $x \in \Sigma$, and $f \in \Sigma^n$, with $n \geq 1$, we have

$$fx \in L_0 \quad \Longleftrightarrow \quad \begin{cases} \text{either } f \in L_0 \text{ and } x = 0, \\ \text{or } f \in L_1 \quad\;\; \text{and } x = 1. \end{cases}$$

we thus have $u_{n+1} = u_n + v_n$; similarly $v_{n+1} = u_n + w_n$ and $w_{n+1} = w_n + v_n$.

3. Note that $\forall n \geq 1$, $u_{n+1} - w_{n+1} = u_n - w_n$; as, moreover, $u_1 = w_1 = 0$, we deduce $\forall n \geq 1$, $u_n = w_n$.

Thus,

$$\forall n \geq 1, \quad v_{n+1} = 2u_n,$$
$$\forall n \geq 2, \quad u_{n+1} = u_n + 2u_{n-1}.$$

We can then compute u_n; the characteristic polynomial is $r^2 - r - 2 = (r-2)(r+1) = 0$; hence $u_n = a2^n + b(-1)^n$; from the initial conditions $u_1 = 0$ and $u_2 = 1$ we deduce $a = 1/6$ and $b = 1/3$; we obtain finally:

$$\forall n \geq 1, \quad u_n = w_n = \frac{2^{n-1} + (-1)^n}{3},$$
$$v_n = \frac{2^{n-1} - 2(-1)^n}{3}.$$

The matrix method here would give

$$\begin{pmatrix} u_n \\ v_n \\ w_n \end{pmatrix} = M \times \begin{pmatrix} u_{n-1} \\ v_{n-1} \\ w_{n-1} \end{pmatrix} = \begin{pmatrix} 1 & 1 & 0 \\ 1 & 0 & 1 \\ 0 & 1 & 1 \end{pmatrix} \times \begin{pmatrix} u_{n-1} \\ v_{n-1} \\ w_{n-1} \end{pmatrix}.$$

The eigenvalues of M are -1, 1 and 2, and associated eigenvectors are

$$V_{-1} = \begin{pmatrix} 1 \\ -2 \\ 1 \end{pmatrix}, \quad V_1 = \begin{pmatrix} 1 \\ 0 \\ -1 \end{pmatrix} \quad \text{and} \quad V_2 = \begin{pmatrix} 1 \\ 1 \\ 1 \end{pmatrix}.$$

We deduce that

$$M = N \times \Delta \times N^{-1} = N \times \begin{pmatrix} 1 & 0 & 0 \\ 0 & -1 & 0 \\ 0 & 0 & 2 \end{pmatrix} \times N^{-1}, \quad \text{with}$$

$$N = \begin{pmatrix} 1 & 1 & 1 \\ 0 & -2 & 1 \\ -1 & 1 & 1 \end{pmatrix} \quad \text{and} \quad N^{-1} = \frac{1}{6}\begin{pmatrix} 3 & 0 & -3 \\ 1 & -2 & 1 \\ 2 & 2 & 2 \end{pmatrix}.$$

7.18. Let $u_n = |L_n|$, where

$$L_n = \{w \,/\, w \in L \text{ and } |w| = n\}$$

and construct L_n from L_{n-1}. Then we can write:

$$L_n = \{w = w'a \,/\, w' \in \{a,c,d\}^{n-1}\} \cup \{w = w'x \,/\, w' \in L_{n-1}, x \in \{b,c,d\}\},$$

thus $u_n = 3u_{n-1} + 3^{n-1}$, with $u_0 = 1$, $u_1 = 4$. The characteristic polynomial is $(r-3)^2 = 0$, and thus $u_n = \lambda 3^n + \mu n 3^n$, with (because of the initial conditions) $\lambda = 1$ and $\mu = 1/3$, i.e. $u_n = 3^n + n3^{n-1}$.

7.19. This is a recurrence of the form (7.14) with $l = 1$, $b_1 = -1$ and $P_1(n) = 1$. The characteristic equation is $(r^2 - r - 2)(r+1) = (r-2)(r+1)^2 = 0$; the general solution of the recurrence is thus $u_n = a2^n + (bn+c)(-1)^n$. Letting $n = 0, 1, 2$, we obtain the linear system in a, b, c:

$$a + c = 1,$$
$$2a - b - c = 1,$$
$$4a + 2b + c = 4,$$

giving the same result (fortunately!).

7.20. The characteristic polynomial is $(x^2 - 3x + 2)(x-1)^2 = 0$, or $(x-2)(x-1)^3 = 0$. The general solution is thus of the form $u_n = \lambda 2^n + a + bn + cn^2$; to determine a, b, c, we first use the recurrence for computing $u_2 = 0$ and $u_3 = 4$; we then deduce: $\lambda = 4$, $a = -4$, $b = -2$, $c = -2$.

7.21. Letting $v_k = u_{2^k}$, we have $v_k = 4v_{k-1} + 2^{2k}$, hence $u_{2^k} = v_k = (u_1 + k)2^{2k}$.

7.22. 1. Let $v_n = 1/u_n$, hence $2v_n = v_{n-1} + v_{n-2}$, and we obtain a linear recurrence. The characteristic polynomial is $(2r+1)(r-1) = 0$, hence $v_n = \lambda(-1/2)^n + \mu$; we find $\lambda = 2/3(1/a - 1/b)$ and $\mu = 1/3(1/a + 2/b)$.

2. Let $v_n = \log u_n$, hence $2v_n = v_{n-1} + v_{n-2}$, $v_0 = 0$, $v_1 = \log 2$. We finally obtain $u_n = 2^{2/3(1-1/2^n)}$.

3. Let a_n be the solution of $v_n = v_{n-1} + v_{n-2}$, with $v_0 = a_0$ and $v_1 = a_1$, and let b_n be the solution of $v_n = v_{n-1} + v_{n-2}$, with $v_0 = b_0$ and $v_1 = b_1$. We check that $u_n = a_n/b_n$ verifies recurrence 3.

7.23. $(\Delta u)_n = \dfrac{-k}{(n-1)n(n+1)\cdots(n+k-1)} = \dfrac{-k}{(n-1)}u_n$, for $n > 0$.

Chapter 8

8.1. The sequence $u_0 = 1$, $u_1 = 0$, $u_2 = 0$, ..., $u_n = 0$, ..., represented by the polynomial $U(X) = 1$.

8.2. $\mathbf{u}(X) = (1 + X)^r \times (1 + X^2)^r = (1 + X + X^2 + X^3)^r$. We have

$$(1+X)^r = 1 + rX + \cdots + r(r-1)\cdots(r-k+1)\frac{X^k}{k!} + \cdots + X^r$$

$$= \sum_{p=0}^{r} \binom{r}{p} X^p$$

$$(1+X^2)^r = 1 + rX^2 + \cdots + r(r-1)\cdots(r-k+1)\frac{X^{2k}}{k!} + \cdots + X^{2r}$$

$$= \sum_{p=0}^{r} \binom{r}{p} X^{2p}$$

hence
$$(1+X)^r \times (1+X^2)^r = (1+X+X^2+X^3)^r$$
$$= \sum_n \Big(\sum_{k'+2k=n} \binom{r}{k'}\binom{r}{k} \Big) X^n$$
$$= \sum_n \Big(\sum_{k=0}^{n} \binom{r}{k}\binom{r}{n-2k} \Big) X^n$$

No simpler form is known for the coefficients of this series.

8.3. Let: $\mathbf{v}_1 = 1$, and $\mathbf{v}_n = 0$ for all $n \neq 1$, namely, $\mathbf{v} = X$.

8.4. 1. Let $\mathbf{u}$ be an invertible series, $\exists \mathbf{w} \in \mathbb{C}[\![X]\!]$ with $\mathbf{uw} = \mathbb{1}$; hence $a_0 w_0 = 1$ and a_0 is invertible.

2. Conversely, assume a_0 invertible with inverse w_0, i.e. $w_0 a_0 = 1$. Then, as $\mathbf{u} = a_0 + X \mathbf{u}_l$, $w_0 \mathbf{u} = 1 + X w_0 \mathbf{u}_l = 1 + \mathbf{v}$ with $\mathbf{v} = w_0 X \mathbf{u}_l$. As in Lemma 8.8, we can show that

$$w_0 \mathbf{u}(\mathbb{1} - \mathbf{v} + \cdots + (-1)^n \mathbf{v}^n + \cdots) = \mathbb{1}.$$

Hence $\mathbf{u}$ is invertible.

8.5. Assuming that all the series considered converge for value x, lines (2) – (7) are consequences of the rules giving the power series expansions of derivatives, integrals, $\dfrac{1}{1-x}$, and of the definition convolution product of series.

8.6. 1. By induction on n.
(B) For $n = 0$, $\int_0^\infty e^{-t} dt = -\big[e^{-t}\big]_0^\infty = 1 = 0!$.
(I) Assuming the result is true for $n - 1$, an integration by parts gives

$$\int_0^\infty t^n e^{-t} dt = \int_0^\infty n t^{n-1} e^{-t} dt - \big[t^n e^{-t}\big]_0^\infty = n \int_0^\infty t^{n-1} e^{-t} dt = n!$$

2. It is an immediate consequence: $\hat{u}(xt) = \sum_{n\geq 0} \frac{u_n}{n!} x^n t^n$, whence

$$\int_0^\infty \hat{u}(xt) e^{-t} dt = \sum_{n\geq 0} \left(\frac{u_n}{n!} x^n \int_0^\infty t^n e^{-t} dt \right) = \sum_{n\geq 0} u_n x^n.$$

8.7. The restriction $deg(U) < deg(V)$ can easily be deleted. If $deg(U) \geq deg(V)$, we can divide polynomial U by polynomial V and obtain: $g(x) = P_1(x) + \dfrac{U_1(x)}{V(x)}$ with $deg(U_1) < deg(V)$.

8.8. 1. Note that

$$\frac{x^{2^k}}{1 - x^{2^{k+1}}} = \sum_{p=0}^\infty x^{2^k + p2^{k+1}} = \sum_{p=0}^\infty x^{2^k(1+2p)}.$$

Furthermore, note that if $n \geq 1$, then there are unique integers $k \geq 0$ and $p \geq 0$, such that $n = 2^k(1+2p)$. Thus in the sum

$$\sum_{k=0}^\infty \frac{x^{2^k}}{1 - x^{2^{k+1}}} = \sum_{k=0}^\infty \sum_{p=0}^\infty x^{2^k(1+2p)},$$

each formal product x^n appears exactly once for each $n \geq 1$. Thus, we have

$$\sum_{k=0}^\infty \frac{x^{2^k}}{1 - x^{2^{k+1}}} = \frac{x}{1-x}.$$

2. Recall that the Fibonacci numbers F_n are defined by: $F_n = \dfrac{r_1^n - r_2^n}{\sqrt{5}}$, where $r_1 = \dfrac{1+\sqrt{5}}{2}$ and $r_2 = \dfrac{1 - \sqrt{5}}{2}$ are the roots of the equation $r^2 = r + 1$, and satisfy $r_1 r_2 = -1$. Thus

$$\sum_{k=0}^\infty \frac{1}{F_{2^k}} = \sqrt{5} \left(\sum_{k=0}^\infty \frac{1}{r_1^{2^k} - r_2^{2^k}} \right) = \sqrt{5} \left(\sum_{k=0}^\infty \frac{r_2^{2^k}}{(r_1 r_2)^{2^k} - r_2^{2^{k+1}}} \right)$$

$$= \sqrt{5} \left(\frac{r_2}{-1 - r_2^2} + \sum_{k=1}^\infty \frac{r_2^{2^k}}{1 - r_2^{2^{k+1}}} \right) = \sqrt{5} \left(\frac{r_2}{-1 - r_2^2} + \frac{r_2}{1 - r_2} - \frac{r_2}{1 - r_2^2} \right)$$

$$= \sqrt{5} \left(\frac{r_2}{-2 - r_2} + \frac{r_2}{1 - r_2} + 1 \right) \qquad \text{(since } r_2^2 = r_2 + 1\text{)}$$

$$= \sqrt{5} \left(\frac{\sqrt{5}}{5} + \frac{\sqrt{5} - 3}{2} + 1 \right) = \frac{7 - \sqrt{5}}{2}.$$

8.9. The sequence of Fibonacci numbers has generating series

$$F(z) = \frac{P_1(z) - z P_0(z)}{1 - z - z^2} = \frac{z}{1 - z - z^2};$$

hence

$$\sum_{n\geq 0} F_{2n} z^{2n} = \frac{1}{2} \left(\frac{z}{1 - z - z^2} + \frac{-z}{1 + z - z^2} \right)$$

$$= \frac{1}{2} \left(\frac{2z^2}{1 - 3z^2 + z^4} \right).$$

Chapter 8

and
$$\sum_{n\geq 0} F_{2n} z^n = \frac{z}{1 - 3z + z^2}$$

8.10. 1. Let $u(z) = \sum_{n\geq 0} u_n z^n$; then the recurrence equation implies
$$2(u(z) - u_0 - u_1 z) = 3z(u(z) - u_0) - z^2 u(z) \;;$$
hence
$$u(z) = \frac{pz + q}{2 - 3z + z^2} = \frac{pz + q}{(1-z)(2-z)}$$
$$= \frac{a}{(1-z)} + \frac{b}{(2-z)} = a \sum_{n\geq 0} z^n + \frac{b}{2} \sum_{n\geq 0} \frac{z^n}{2^n} \;,$$
with $u_0 = a + b/2$ and $u_1 = a + b/4$.

2. If $u(z) = \sum_{n\geq 0} u_n z^n$, then
$$u(z) = \frac{pz + q}{1 - 4z + 4z^2} = \frac{pz + q}{(1 - 2z)^2}$$
$$= \frac{a}{(1-2z)^2} + \frac{b}{(1-2z)} = 2a \sum_{n\geq 0} n 2^n z^{n-1} + b \sum_{n\geq 0} 2^n z^n$$
with $u_0 = b$ and $u_1 = 4a + 2b$.

8.11. Let $u(z) = \sum_{n\geq 0} u_n z^n$; then the recurrence equation gives: $\forall n \geq 2$, $u_n z^n = 4 u_{n-1} z^n - 4 u_{n-2} z^n + (n-1) z^n$. As
$$\sum_{n\geq 2} (n-1) z^n = \sum_{n\geq 1} n z^{n+1} = \sum_{n\geq 0} n z^{n+1} = z^2 \left(\sum_{n\geq 0} n z^{n-1} \right) = z^2/(1-z)^2,$$
we deduce $u(z) - 1 - z = 4z(u(z) - 1) - 4z^2 u(z) + z^2/(1-z)^2$, and
$$u(z) = \frac{z^2 + (1 - 3z)(1 - z)^2}{(1-z)^2 (1-2z)^2}$$
$$= \frac{1}{(1-z)^2} + \frac{2}{(1-z)} + \frac{1}{2(1-2z)^2} - \frac{5}{2(1-2z)}$$
$$= \sum_{n\geq 0} (n+1) z^n + 2 \sum_{n\geq 0} z^n + \frac{1}{2} \sum_{n\geq 0} (n+1) 2^n z^n - \frac{5}{2} \sum_{n\geq 0} 2^n z^n$$

Let $u_n = n + 3 + (n+1) 2^{n-1} - 5 \times 2^{n-1}$. The reader can also solve the recurrence equation using the characteristic polynomial method and check that the same result is obtained.

8.12. We have $u(x) = 2xv(x) + x^2 u(x) + 1$, and $v(x) = xu(x) + x^2 v(x)$. The second equation gives $v(x) = xu(x)/(1 - x^2)$; hence
$$u(x) = \frac{1 - x^2}{1 - 4x^2 + x^4} \;, \qquad v(x) = \frac{x}{1 - 4x^2 + x^4} \;.$$
Noting that the common denominator is a function of x^2 we introduce
$$w(z) = \frac{1}{1 - 4z + z^2}$$
$$= \sum_{n\geq 0} \left(\frac{3 + 2\sqrt{3}}{6} (2 + \sqrt{3})^n + \frac{3 - 2\sqrt{3}}{6} (2 - \sqrt{3})^n \right) z^n,$$
and we have $u(x) = (1 - x^2) w(x^2)$ and $v(x) = xw(x^2)$. Finally: $u_{2n+1} = v_{2n} = 0$, $v_{2n+1} = w_n$ and $u_{2n} = w_n - w_{n-1} = (2 + \sqrt{3})^n/(3 - \sqrt{3}) + (2 - \sqrt{3})^n/(3 + \sqrt{3})$.

8.13. Let $u(x) = \sum_{n\geq 0} u_n x^n$; the recurrence equation yields $u(x) - 1 = 3x(u(x)-1) - 2x^2 u(x) + 2x^2/(2-x)^2$, i.e.

$$u(x) = \frac{(1-3x)(2-x)^2 + 2x^2}{(2-x)^2(1-x)(1-2x)}$$

$$= \frac{(1-3x)(2-x)^2 + 2x^2}{2(2-x)^2(x-1)(x-1/2)},$$

which could be expanded to

$$u(x) = \frac{\alpha}{(2-x)^2} + \frac{\beta}{(2-x)} + \frac{\gamma}{(1-x)} + \frac{\delta}{(1-2x)}.$$

Instead we will apply Proposition 8.13 directly, which enables us to conclude

$$u(x) = \frac{(1-3x)(2-x)^2 + 2x^2}{2(2-x)^2(x-1)(x-1/2)} = \sum_{n\geq 0}\left(a + b2^n + \frac{cn+d}{2^n}\right)x^n, \qquad (E)$$

with $a = 0$, $b = -5/9$, and $c = 2/3$. In order to determine d, let $x = 0$ in equation (E); we have $b + d = 4/4 = 1$ (also equal to u_0), and thus $d = 14/9$; hence, finally,

$$u_n = -\frac{5}{9}2^n + \left(\frac{2}{3}n + \frac{14}{9}\right)\frac{1}{2^n}.$$

One could also compute the characteristic polynomial $(x-1)(x-2)(x-1/2)^2 = 0$, and this gives a general solution of the form $u_n = a + b2^n + (cn+d)/2^n$. But then, one is left with the problem of finding a, b, c, d by solving a system of linear equations. Note that the constants a, b, c, d are not equal to $\alpha, \beta, \gamma, \delta$.

8.14. Let p be the number of tokens of value 2 and q the number of tokens of value 3. The problem amounts to finding the number of solutions of the equation $2p + 3q = n$. This is the same type of problem as given in Section 8.2.3. We will, however, solve the equation $2p + 3q = n$ directly, without computing the partial fraction expansion. Indeed, here,

$$u(x) = 1 + x^2 + x^4 + \cdots + x^{2p} + \cdots.$$

$$w(x) = 1 + x^3 + x^6 + \cdots + x^{3q} + \cdots.$$

$$v(x) = u(x) \times w(x) = \sum_{n\geq 0}\left(\sum_{i+j=n} u_i w_j\right) x^n.$$

Since $u_i = \begin{cases} 1 & \text{if } i = 2k, \\ 0 & \text{otherwise,} \end{cases}$ and $w_j = \begin{cases} 1 & \text{if } j = 3k, \\ 0 & \text{otherwise,} \end{cases}$ it follows that the number v_n of ways of bringing up a total of n with tokens of value 2 and 3 is given by

$$v_n = \sum_{i+j=n} u_i w_j = \sum_{i=0}^{n} u_i w_{n-i} = \sum_{k=0}^{\lfloor n/2 \rfloor} w_{n-2k},$$

where $\lfloor n/2 \rfloor$ is the largest integer less than or equal to $n/2$. Finally,

- for n even, $w_{n-2k} = 1 \iff n - 2k$ is a multiple of 6,
- for n odd, $w_{n-2k} = 1 \iff n - 2k$ is an odd multiple of 3 $\iff$ the remainder of the division of $n - 2k$ by 6 is 3.

Hence,

- for n even, v_n is the number of multiples of 6 between 0 and n,
- for n odd, v_n is the number of odd multiples of 3 between 0 and n.

8.15. Let us first find the recurrence equation defining the number u_n of Morse code words taking n time units: the last letter of the word is

- either a dot, and there are then u_{n-2} possibilities for the beginning of the word,
- or a dash, and there are then u_{n-3} possibilities for the beginning of the word,

hence $u_n = u_{n-2} + u_{n-3}$ for $n \geq 4$, with $u_0 = u_1 = 0$, $u_2 = u_3 = 1$.
The characteristic polynomial is given by $r^3 - r - 1 = 0$ and the generating series is given by

$$u(z) = \frac{z^2 + z^3}{1 - z^2 - z^3} = -1 + \frac{1}{1 - z^2 - z^3}.$$

We thus must find the roots of $V(z) = z^3 + z^2 - 1$. To this end, we will use the so-called Cardan method: letting $r = 1/z$ gives us the characteristic polynomial (see also Remark 8.15). We then look for a solution of the form $r = u + v$, which yields $u^3 + v^3 + 3u^2v + 3uv^2 - (u+v) - 1 = 0$, or $u^3 + v^3 + (u+v)(3uv - 1) - 1 = 0$. Assuming that $3uv - 1 = 0$ we have to solve

$$\begin{cases} 3uv = 1, \\ u^3 + v^3 = 1, \end{cases} \tag{15.1}$$

hence $v = 1/3u$ and $u^3 + 1/27u^3 = 1$. Hence u^3 and v^3 are roots of $u^6 - u^3 + 1/27 = 0$, with the conditions (15.1). We thus have $u^3 = 1/6(3 \pm \sqrt{23/3})$ and the roots r_1, r_2, r_3 of $r^3 - r - 1 = 0$ are, letting $\psi = \sqrt[3]{1/6(3 + \sqrt{23/3})}$ and $\psi' = \frac{1}{3}\psi^{-1} = \sqrt[3]{1/6(3 - \sqrt{23/3})}$,

$$r_1 = \psi + \psi', \qquad r_2 = j\psi + j^2\psi', r_3 = j^2\psi + j\psi'$$

where j and j^2 are the cubic roots of 1.
Finally, let us apply Proposition 8.11 and write

$$\frac{1}{1 - z^2 - z^3} = \sum_{n \geq 0} \left(\sum_{i=1}^{3} a_i r_i^{n+1} \right)$$

with for $i = 1, 2, 3$

$$a_i = \frac{1}{V'(1/r_i)} = \frac{1}{3/r_i^2 + 2/r_i} = \frac{r_i^2}{3 + 2r_i}.$$

Z The solution is different from the one of the preceding exercise because different sequences of dots and dashes taking the same total amount of time result in different words u_n.

8.16. Reasoning as in Section 8.2.3, we can check that the number of ways of changing $n\$$ with \$1, \$2 and \$5 bills is the coefficient of x^n in the series

$$v(x) = \frac{\alpha_3}{(1-x)^3} + \frac{\alpha_2}{(1-x)^2} + \frac{\alpha_1}{1-x} + \frac{\beta}{1+x}$$

$$+ \frac{\gamma}{1 - e^{i\alpha}x} + \frac{\bar{\gamma}}{1 - e^{-i\alpha}x} + \frac{\delta}{1 - e^{2i\alpha}x} + \frac{\bar{\delta}}{1 - e^{-2i\alpha}x}$$

$$= \sum_{n \geq 0} \left((a_2 n^2 + a_1 n + a_0) + (-1)^n b \right.$$

$$\left. + (c e^{ni\alpha} + \bar{c} e^{-ni\alpha}) + (d e^{2ni\alpha} + \bar{d} e^{-2ni\alpha}) \right) x^n .$$

We can find the coefficients a_2, b, c, d of the above partial fraction expansion by using Proposition 8.11 and Proposition 8.13. We find $a_2 = 1/20$, $b = 1/8$, $c = e^{2i\alpha}/(5(1-e^{-i\alpha})(1-e^{-2i\alpha}))$, $d = e^{-i\alpha}/(5(1-e^{-2i\alpha})(1-e^{i\alpha}))$. Finally a_1 and a_0 can be determined by assigning values to x.

The coefficient v_n of x^n can, however, be determined using slightly simpler computations; writing

$$v(x) = (u \times w)(x) = \frac{1}{(1-x)(1-x^2)(1-x^5)} = \frac{1}{(1-x)(1-x^2)} \times \frac{1}{(1-x^5)}$$

$$= (1+x) \times \frac{1}{(1-x^2)^2} \times \frac{1}{(1-x^5)}$$

$$= \sum_{n \geq 0} ((n+1)x^{2n}(1+x) \times \sum_{n \geq 0} x^{5n}) = \sum_{n \geq 0} \left(\sum_{j=0}^{n} u_j w_{n-j} \right),$$

we deduce that $v_n = \sum_{k=0}^{\lfloor n/5 \rfloor} w_{5k} u_{n-5k}$; as $u_{2n} = u_{2n+1} = n+1$. We have, after computations,

$$v_n = \sum_{0 \leq i \leq \lfloor n/5 \rfloor} (1+5i) + \sum_{0 \leq i < \lfloor n/5 \rfloor} (3+5i),$$

which immediately yields

$$v_n = 5(\lfloor n/5 \rfloor)^2 + 4\lfloor n/5 \rfloor - 3.$$

For $n = 100$, we find $v_{100} = 2077$.

8.17. Let $u(x) = \sum_{n \geq 0} u_n x^n$; then $u'(x) = \sum_{n \geq 0} n u_n x^{n-1}$ hence $2u'(x) = u(x) + e^x$; solving the differential equation $2y' - y = e^x$ yields $y = e^x + ce^{x/2}$, and the initial condition $y(0) = 2$ implies $c = 1$; thus, finally, $u(x) = e^x + e^{x/2}$, and $u_n = \frac{1}{n!}\left(1 + \frac{1}{2^n}\right)$.

We could also have applied the summation factors technique: multiplying the equation giving u_k by $2^{k-1}(k-1)!$ and summing for k ranging from 0 to n, we find $2^n n! u_n = \sum_{k=0}^{n-1} 2^k + 2 = 2^n + 1$.

8.18. 1. By structural induction.

2. It immediately follows, from the inductive definition of the words in the Dyck language, that $u_n = \sum_{i=0}^{n-1} u_i u_{n-i-1}$ for $n \geq 1$, and $u_0 = 1$. u_n is thus the nth Catalan number.

3. The words of the Dyck language on the alphabet A_k always have an even length; if v_n is the number of words of length $2n$ in the Dyck language on A_k, we obtain the recurrence equation $v_n = k \sum_{i=0}^{n-1} v_i v_{n-i-1}$ for $n \geq 1$, and $v_0 = 1$. This recurrence equation can be solved in two different ways:

- Noting that $v_1 = ku_1$ we can show by induction on n that $\forall n \geq 1$, $v_n = k^n u_n$.
- Computing as for the sequence b_n we will obtain the equation

$$kxv^2 - v + 1 = 0$$

defining the generating series $v = v(x)$ of the sequence v_n. Solving, we find

$$v(x) = \frac{1 - \sqrt{1 - 4kx}}{2kx} = b(kx),$$

wherefrom it follows that

$$\forall n \geq 0, \quad v_n = \frac{k^n}{n+1}\binom{2n}{n} = \frac{k^n}{n+1} \frac{(2n)!}{n!n!}.$$

Chapter 8   351

8.19. Multiplying the equation $u_n = u_{n-1} + 2u_{n-2} + \cdots + nu_0$ by x^n for each $n > 0$ in $\mathbb{N}$, and summing the equalities thus obtained we have

$$u(x) - 1 = xu(x) + 2x^2 u(x) + \cdots + nx^n u(x) + \cdots$$
$$= x\frac{1}{(1-x)^2}u(x)$$

and

$$u(x) = \frac{xu(x)}{(1-x)^2} + 1,$$

hence

$$u(x) = \frac{1 - 2x + x^2}{1 - 3x + x^2} = 1 + \frac{x}{1 - 3x + x^2},$$

and thus (see Exercise 8.9), $u_0 = 1$, and $u_n = F_{2n}$ for $n > 0$.

8.20. For all $n \geq 2$ in $\mathbb{N}$, let us multiply the equation $u_n = -2nu_{n-1} + \sum_{k=0}^{n} \binom{n}{k} u_k u_{n-k}$ by $\frac{x^n}{n!}$, and then sum the equalities thus obtained; we have:

$$\hat{u}(x) - x = -2x\hat{u}(x) + \sum_n \left(\sum_k \binom{n}{k} u_k u_{n-k} \right) \frac{x^n}{n!}$$
$$= -2x\hat{u}(x) + \sum_n \left(\sum_k \frac{n!}{k!(n-k)!} u_k u_{n-k} \right) \frac{x^n}{n!}$$
$$= -2x\hat{u}(x) + \sum_n \left(\sum_k \frac{u_k x^k}{k!} \frac{u_{n-k} x^{n-k}}{(n-k)!} \right)$$
$$= -2x\hat{u}(x) + (\hat{u}(x))^2$$

Hence $\hat{u}(x) = -2x\hat{u}(x) + (\hat{u}(x))^2 + x$; and finally, $\hat{u}(x) = 1/2(1 + 2x - \sqrt{1 + 4x^2})$, and thus: $\forall n \geq 1$, $u_{2n+1} = 0$ and $u_{2n} = (-1)^n (2n)! b_{n-1}$, where b_{n-1} is the $(n-1)$th Catalan number.

8.21. 1.(a) $t_1 = 3$ (no constraint); $t_2 = 3^2 = 9$ (no constraint); $t_3 = 3.3.2 = 18$ (we choose the first two colours in A: 3×3 choices, we then choose the third colour in $A \setminus \{a\}$: two choices.

1.(b) Recurrence equation; in order to form a size n solution, with $n \geq 3$:

- we first form a size $(n-1)$ solution; we can do this in t_{n-1} different ways;
- we then choose the last colour in $A \setminus \{a_{n-2}\}$; this can be done in two different ways.

We thus have, for $n \geq 3$, $t_n = 2t_{n-1}$.

1.(c) We have $t_1 = 3$. We deduce from (b) that $t_n = 2^{n-2} t_2$, hence, since $t_2 = 9$,

$$t_n = 9.2^{n-2} \text{ if } n \geq 2.$$

2.(a) $s_1 = 3$ (no constraint); $s_2 = 9$ (no constraint); for s_3, if $a_1 \neq a_3$ there are eighteen possible choices, and if $a_1 = a_2 = a_3$, three possible choices, hence $s_3 = 18 + 3 = 21$; for s_4, there are six solutions of the form $aaab$, six solutions of the form $abbb$, with $a \neq b$ and three solutions of the form $aaaa$; hence $s_4 = t_4 + 6 + 6 + 3 = 51$.

2.(b) Recurrence equation; consider a size n solution $a_1, a_2, \ldots, a_n$ with $n \geq 3$.

Two cases are possible:

(α) $a_{n-2} \neq a_n$. In this case, we form a size $(n-1)$ solution: s_{n-1} choices are possible for doing so, we then choose the last colour in $A\setminus\{a_{n-2}\}$ and two choices are possible for this last colour.

(β) $a_{n-2} = a_{n-1} = a_n$. In this case, we form a size $(n-2)$ solution; s_{n-2} choices are possible for doing so. We then complete by two occurrences of the last letter a_{n-2}.

We thus obtain the recurrence equation

$$s_n = 2s_{n-1} + s_{n-2}, \qquad \text{for } n \geq 3.$$

2.(c) Associated generating series. Let $s(z) = \sum_{n \geq 1} s_n z^n$. By the above formula, we have

$$\sum_{n \geq 3} s_n z^n = 2 \sum_{n \geq 3} s_{n-1} z^n + \sum_{n \geq 3} s_{n-2} z^n,$$

thus

$$s(z) - s_1 z - s_2 z^2 = 2z(s(z) - s_1 z) + z^2 s(z)$$

or

$$s(z)(z^2 + 2z - 1) = -3(z^2 + z)$$

and

$$s(z) = \frac{-3z(z+1)}{z^2 + 2z - 1}.$$

Then find the partial fraction expansion of $s(z)$:

$$s(z) = z\left(-1 + \frac{1+\sqrt{2}}{2(z+1+\sqrt{2})} + \frac{1-\sqrt{2}}{2(z+1-\sqrt{2})}\right)$$

$$= z\left(-1 + \frac{1}{2} \times \frac{1}{1-z(1-\sqrt{2})} + \frac{1}{2} \times \frac{1}{1-z(1+\sqrt{2})}\right),$$

we find

$$s_n = \frac{3}{2}[(1-\sqrt{2})^n + (1+\sqrt{2})^n] \quad \text{for } n \geq 1.$$

8.22. 1. $b_1 = 0$, $b_2 = 1$ (transposition), $b_3 = 2$.

2. Let f be defined by $f(e_i) = c_j$ if and only if pupil e_i receives the exercise c_j of e_j.

$$\begin{pmatrix} e_1 \\ \vdots \\ e_n \end{pmatrix} \xrightarrow{f} \begin{pmatrix} c_1 \\ \vdots \\ c_n \end{pmatrix}.$$

Let $c_i = f(e_1)$ be the exercise given to pupil e_1: we have $(n-1)$ choices for c_i. Two disjoint cases are possible: $f(e_i) = c_1$, or $f(e_i) = c_j$ with $j \neq 1, i$.

• In the first case, pupils $\{e_1, e_i\}$ interchange the exercises $\{c_1, c_i\}$; hence e_1 and e_i interchange their exercises and $n - 1$ choices are possible for e_i; b_{n-2} choices remain to redistribute the $n-2$ remaining exercises amongst the $n - 2$ remaining pupils. Hence, $(n-1)b_{n-2}$ possible choices altogether.

• In the second case, we delete the pair $\{e_i, c_i\}$. The $(n-1)$ remaining pupils distribute the $(n-1)$ remaining exercises: we have b_{n-1} possible choices. Pupil e_1 then obtains exercise c_j with $j \neq 1, i$. This exercise is in fact for pupil e_i. e_1 thus gives exercise c_j to e_i, and obtains exercise c_i in exchange ($i \neq 1$). There are thus $(n-1)b_{n-1}$ possible choices.

Chapter 9

Finally,
$$b_n = (n-1)(b_{n-1} + b_{n-2}), \quad \text{for } n \geq 2. \tag{I}$$
(with $b_0 = 1$ and $b_1 = 0$).

3. Let us prove the equation
$$b_n - nb_{n-1} = (-1)^n, \quad \text{for } n \geq 2. \tag{II}$$
by an inductive proof.

- If $n = 2$, $b_2 - 2b_1 = b_2 = 1 = (-1)^2$. The formula is true in this case.
- Assuming the formula is true for n, let us form $b_{n+1} - (n+1)b_n$. Applying formula (I), which is possible because $n + 1 \geq 2$, we have

$$n(b_n + b_{n-1}) - nb_n - b_n = nb_{n-1} - b_n$$
$$= -(-1)^n \quad \text{(by the induction hypothesis)}$$
$$= (-1)^{n+1}$$

The formula is thus indeed proved for all $n > 1$. We can note that, letting $b_0 = 1$, the formula is also true for $n = 1$.

4. Let $b(z) = \sum_{n \geq 0} \dfrac{b_n z^n}{n!}$.

Applying formula (II), true for $n \geq 1$,

$$\sum_{n \geq 1} \frac{b_n z^n}{n!} - \sum_{n \geq 1} nb_{n-1}\frac{z^n}{n!} = \sum_{n \geq 1} \frac{(-1)^n z^n}{n!},$$

i.e. $b(z) - b_0 - zb(z) = e^{-z} - 1$, or $b_0 = 0$; hence $b(z) = \dfrac{e^{-z}}{1-z}$.

Chapter 9

9.1. The constant c such that $kn \leq cn$ is different for each term kn; on the other hand, when writing: $\sum_{k=1}^{n} O(n) = nO(n)$, we implicitly assume that **all** the $O(n)$s refer to the same constant c. In other words, when $k = n$, $kn \neq O(n)$.
A correct argument would be: $\sum_{k=1}^{n} kn = \sum_{k=1}^{n} O(n^2) = O(n^3)$.

9.2. 1. Yes.

2. No, for instance: $f(n) = n + \log_2 n = O(n)$, but $2^{f(n)} = 2^{n+\log_2 n} = n2^n$ is not an $O(2^n)$ since $2^{f(n)}/2^n$ goes to infinity when n goes to infinity.

9.3. No: if $g_1(n) = n^2$ and $g_2(n) = 1$, $n = O(g_1(n) + g_2(n))$, but $n \notin g_1(n) + O(g_2(n))$.

9.4. Similar to the proof of case (i).

9.5. 1. By recurrence on n it can be checked that $u_n \geq 1$ for all n. Let, for $n \geq 1$,

$$u_n^2 - u_{n-1}^2 = (u_n - u_{n-1})(u_n + u_{n-1}),$$
$$= \frac{1}{u_{n-1}}(2u_{n-1} + \frac{1}{u_{n-1}}) = 2 + \frac{1}{u_{n-1}^2} \geq 2.$$

As $u_n \geq 1$,
$$\frac{1}{u_{n-1}^2} \leq \frac{1}{u_{n-1}}.$$

2. From 1 it follows that
$$2 \leq u_n^2 - u_{n-1}^2 \leq 2 + \frac{1}{u_{n-1}} = 2 + u_n - u_{n-1},$$

whence the inequalities (9.1).

3. Finally, summing the inequalities (9.1), we have $2n \leq u_n^2 - c^2 \leq 2n + u_n - c$. We deduce that:
- $\lim_{n\to\infty} u_n^2 = \infty$,
- hence, $u_n = o(u_n^2)$ and $u_n^2 = \Omega(2n)$,
- from $2n \leq u_n^2 - c^2 \leq 2n + u_n - c$ we deduce, since $u_n = o(u_n^2)$ and $c = o(u_n^2)$, that

$$2n + o(u_n^2) \leq u_n^2 \leq 2n + o(u_n^2),$$

i.e. $u_n \sim \sqrt{2n}$.

9.6. $\left(e^{an^b} \times n^c \times (\log n)^d\right) \prec \left(e^{a'n^{b'}} \times n^{c'} \times (\log n)^{d'}\right)$ if and only if
- either $a > 0, a' > 0$, and $bacd < b'a'c'd'$ in the lexicographic ordering,
- or $a < 0, a' < 0$, and
  - $b > b'$,
  - or $b = b'$ and $acd < a'c'd'$, in the lexicographic ordering,
- or $a = 0$ or $a' = 0$, and $a < a'$ (car $b > 0$),
- or $a = 0$ and $a' = 0$, and $cd < c'd'$ in the lexicographic ordering,
- or $a < 0$ and $a' > 0$.

9.7. By induction on k; let $k = 1$, and assume there are two principal parts of f with respect to E. We thus have: $f = a_1 g_1 + o(g_1)$ and $f = a_2 g_2 + o(g_2)$; hence: $\lim_{n\to\infty} \frac{f(n)}{a_1 g_1(n)} = 1 = \lim_{n\to\infty} \frac{f(n)}{a_2 g_2(n)}$ and thus, $\lim_{n\to\infty} \frac{a_2 g_2(n)}{a_1 g_1(n)} = 1$. Hence $\lim_{n\to\infty} \frac{g_2(n)}{g_1(n)} = \frac{a_1}{a_2}$, with $0 \neq \frac{a_1}{a_2} \neq \infty$. Henceforth g_1 and g_2 have the same order of magnitude; by condition (ii) of Definition 9.19, we deduce that $g_1 = g_2$ and $a_1 = a_2$. The inductive step of the induction is proved in a similar way.

9.8. Note that
$$(1+n)^{1/n} = e^{\frac{\log(1+n)}{n}},$$
and that
$$\log(1+n) = \log\left(n\left(1 + \frac{1}{n}\right)\right) = \log n + \log\left(1 + \frac{1}{n}\right)$$
$$= \log n + \frac{1}{n} - \frac{1}{2n^2} + O(n^{-2}).$$

Let
$$u(n) = \frac{\log n}{n} + \frac{1}{n^2} - \frac{1}{2n^3} + O(n^{-3}),$$
then $\lim_{n\to\infty} u(n) = 0$, and squaring, then cubing, the asymptotic power series expansion of $u(n)$, we have
$$(u(n))^2 = \frac{(\log n)^2}{n^2} + \frac{2\log n}{n^3} + o\left(\frac{\log n}{n^3}\right),$$
$$(u(n))^3 = \frac{(\log n)^3}{n^3} + o\left(\frac{(\log n)^3}{n^3}\right).$$
We deduce
$$f(n) = e^{u(n)} = 1 + u(n) + \frac{(u(n))^2}{2!} + \frac{(u(n))^3}{6} + o((u(n))^3)$$
$$= 1 + \frac{\log n}{n} + \frac{1}{2}\frac{(\log n)^2}{n^2} + \frac{1}{n^2} + \frac{1}{6}\frac{(\log n)^3}{n^3} + o\left(\frac{(\log n)^3}{n^3}\right).$$

9.9. 1. Nothing valuable; we would obtain a term of the form
$$\sum_{0<k<n} \frac{1}{k^2(n-k)} = \frac{1}{n}H_{n-1}^2 + \frac{2}{n^2}H_{n-1},$$
implying only that $u_n = \Omega(1/n)$.

2. Note, first, that
$$u(1) = \exp\left(\sum_{k\geq 1}\frac{1}{k^2}\right) = e^{\pi^2/6};$$
Writing
$$nu_n = \sum_{0\leq k<n}\frac{u_k}{n} + \sum_{0\leq k<n}u_k\left(\frac{1}{n-k} - \frac{1}{n}\right),$$
i.e.
$$nu_n = \frac{1}{n}\sum_{0\leq k}u_k - \frac{1}{n}\sum_{n\leq k}u_k + \frac{1}{n}\sum_{0\leq k<n}\frac{ku_k}{n-k}. \qquad (S)$$
The first sum in (S) returns $u(1)$, and we can show that:
$$\sum_{n\leq k}u_k = O\left(\frac{(\log n)^2}{n}\right) \quad \text{and} \quad O\left(\sum_{0\leq k<n}\frac{(\log n)^2}{k(n-k)}\right) = O\left(\frac{(\log n)^3}{n}\right);$$
this enables us to bound the second and third terms of the sum (S) by $O\left(\frac{(\log n)^3}{n^3}\right)$; bootstrapping once more we obtain
$$u_n = \frac{e^{\pi^2/6}}{n^2} + O\left(\frac{\log n}{n^3}\right).$$

9.10. It is easy to see that $u_n < \sum_0^n n = n^2 = O(n^2)$. Assuming
$$u_n = an^2 + bn + c \qquad (I)$$
and plugging in the recurrence defining u_n, we have
$$u_n = n + u_{n-1} = an^2 + (1+b-2a)n + a - b + c,$$
wherefrom we deduce, identifying with (I),
- that $a = b = 1/2$,
- and that, if (I) is true for u_{n-1} with $a = b = 1/2$, then (I) is also true for u_n; writing that (I) is true for u_0, we have $c = 0$ and thus $u_n = n(n+1)/2$.

Chapter 10

10.1. We prove this result for the case of undirected graphs. The case of directed graphs is quite similar.
Let $G = (V, E, \delta)$ and $G' = (V', E', \delta')$. It is easy to see that if
- either G' is a subpartial graph of G,
- or G' is a partial graph of a subgraph of G,

then
$$V' \subseteq V, E' \subseteq E \quad \text{and} \quad \forall e \in E', \delta'(e) = \delta(e).$$

It thus suffices to prove the converse of this property.
(a) Let H be the graph (V', E'', δ'') with $A'' = \{e \in E \mid \delta(e) \subseteq V'\}$ and $\forall e \in A'', \delta''(e) = \delta(e)$. Then, by definition, H is a subgraph of G. Because $\forall a \in A', \delta'(e) = \delta(e) \subseteq V'$, $A' \subseteq A''$ and G' is a partial graph of H.
(b) Let H' be the graph (V, E', δ') which clearly is a partial graph of G. Then, $\forall e \in E'$, $\delta'(e) \subseteq V'$ and thus G' is a subgraph of H'.

10.2. Each edge of a directed graph is counted once in the outdegree of its origin and once in the outdegree of its target. The sum of all indegrees is thus equal to the sum of all outdegrees and is also equal to the number of edges.

10.3. 1. Let V_k, for $0 \leq k \leq K$, be the set of the vertices of degree k. By definition, $|V_k| = n_k$. As $V = \bigcup_{k=0}^{K} V_k$ and as the V_ks are pairwise disjoint, $n = |V| = \sum_{k=0}^{K} n_k$.
The sum of the degrees of the vertices is

$$p = \sum_{v \in V} d(v) = \sum_{k=0}^{K} \sum_{v \in V_k} d(v) = \sum_{k=0}^{K} k|V_k| = \sum_{k=0}^{K} kn_k.$$

Moreover, we know that $p = 2m$.

2. Since $n_K \neq 0$, $K \leq Kn_K \leq \sum_{k=0}^{K} kn_k = 2m$. This bound is reached for the graph with one vertex and m edges that all are loops.

3. Let v be any fixed vertex. For all vertices v' of the graph (including vertex v), denote by $E_{v'}$ the set of edges connecting v and v', i.e.

$$\begin{cases} E_{v'} = \{e \mid \delta(e) = \{v, v'\}\} & \text{if } v \neq v', \\ E_v = \{e \mid \delta(e) = \{v\}\}. \end{cases}$$

The degree of v is thus equal to $2|E_v| + \sum_{v' \neq v} |E_{v'}|$. Because the graph has no loop, $E_v = \emptyset$, and because the graph has no multiple edges, $E_{v'}$ has at most one element for $v' \neq v$. Hence, $d(v) \leq \sum_{v' \neq v} 1 = n - 1$. The bound is reached for complete graphs, i.e. graphs such that $\forall v, v' \in V, v \neq v' \implies \exists e \in E : \delta(e) = \{v, v'\}$.

10.4. Because the graph is simple, the degree of a vertex v is the number of vertices $t \neq v$ that are adjacent to v, i.e. connected to v by an edge.

1. Let v be a vertex. There are $n - 1$ other vertices; at most $n - 1$ among them are adjacent to v, and hence $d(v) \leq n - 1$.

2. If a vertex v has degree $n - 1$, it is adjacent to $n - 1$ other vertices, i.e. to all other vertices; consequently, all other vertices (each of which is adjacent to v) are of degree ≥ 1.

3. Assume that the degrees are pairwise different. They form a sequence $d_1 < d_2 < \cdots < d_n$. We have that $d_1 \geq 0$, and 1 implies that $d_n < n$. Because the d_is are integers, this gives $d_1 = 0, d_2 = 1, \ldots, d_n = n - 1$ which contradicts 2. Hence, there are at least two equal degrees.

Chapter 10

10.5. 1. Let V_x be the set of vertices of $V - \{x, y\}$ which are adjacent to x, and let V_y be the set of the vertices of $V - \{x, y\}$ which are adjacent to y. We assume that $z \in V_x \cap V_y$. Then x, y, z, which are mutually adjacent, form a triangle, a contradiction. Thus $V_x \cap V_y = \emptyset$. We also have $V_x \cup V_y \subseteq S - \{x, y\}$ by definition. Hence,

$$n_x + n_y = |V_x| + |V_y| = |V_x \cap V_y| \leq |V - \{x, y\}| = |V| - 2.$$

2. By induction on $n = |V|$

- Initialization: It holds for $n = 1, 2$:

$$n = 1 : |E| = 0 \leq \frac{1}{4} = \frac{|V|^2}{4},$$
$$n = 2 : |E| \leq 1 \leq \frac{4}{4} = \frac{|V|^2}{4}.$$

- We assume that the property is true for n. We show that it is true for $n+2$. Let $|V| = n+2$. If there are no edges, then $|E| = 0 \leq |V|^2/4$. If there is at least one edge, consider two adjacent vertices x and y. Let $V' = V - \{x, y\}$ and $A' = \{a \in A \mid \delta(e) \subseteq V'\}$. The graph (V', E', δ) has no triangle, and $|V'| = n$. By the induction hypothesis we have $|E'| \leq |V'|^2/4 = \dfrac{n^2}{4}$. The edges of E are

  - those of E',
  - the edge between x and y,
  - those between x and the vertices of V', and
  - those between y and the vertices of V'.

Thus $|E| = |E'| + 1 + n_x + n_y$. Since $|E'| \leq \dfrac{n^2}{4}$ and $n_x + n_y \leq |V| - 2 = |V'|$ (by 1), we have

$$|E| \leq \frac{n^2}{4} + 1 + n = \frac{n^2 + 4n + 4}{4} = \frac{(n+2)^2}{4}.$$

The property is also true for $n + 2$.
By induction it is thus true for all $n \in \mathbb{N}$.

10.6. Let $c = v_0, e_1, v_1, \ldots, v_{n-1}, e_n, v_n$ be a chain of minimal length connecting v_0 and v_n with $v_0 \neq v_n$. If this chain were not elementary, there would exist $i < j$ such that $v_i = v_j$. Consider the sequence c' defined by

$$c' = \begin{cases} v_j, e_{j+1}, v_{j+1}, \ldots, v_{n-1}, e_n, v_n & \text{if } v_i = v_0, \\ v_0, e_1, v_1, \ldots, v_{i-1}, e_i, v_i & \text{if } v_j = v_n, \\ v_0, e_1, v_1, \ldots, v_i, e_{j+1}, v_{j+1}, \ldots, v_{n-1}, e_n, v_n & \text{if } v_0 \neq v_i = v_j \neq v_n. \end{cases}$$

This sequence is indeed a chain connecting v_0 and v_n and it has a length strictly less than c, a contradiction.
The same result can be obtained, with the same proof, for paths in directed graphs.

10.7. If $v = v''$, then $d(v, v'') = 0$, and the inequality is verified. If $v = v'$ or if $v' = v''$, the result is straightforward. Consider now the case when $v \neq v''$, $v \neq v'$, and $v' \neq v''$.
If $d(v, v') = \infty$ or if $d(v', v'') = \infty$, the result is clear. Otherwise, there exists a chain of length $d(v, v')$ connecting v to v' and a chain of length $d(v', v'')$ connecting v' to v''. Concatenating these two chains yields a chain of length $d(v, v') + d(v', v'')$ connecting v to v''. The shortest chain connecting v to v'' thus has length less than or equal to $d(v, v') + d(v', v'')$.

10.8. 1. The number of vertices is the number of ways of choosing two elements among five. It is $\binom{5}{2} = \dfrac{5!}{2!3!} = 10$. Each vertex $v = \{n, m\}$ is connected to three others by an edge; these three other vertices are those obtained by choosing two elements in $\{0, 1, 2, 3, 4\} \setminus \{n, m\}$. The number of edges is equal to the half-sum of the degrees, i.e. $\frac{1}{2}(10 \times 3) = 15$. Lastly, let $\{n, m\}$ and $\{p, q\}$ be two distinct vertices; if $\{n, m\} \cap \{p, q\} = \emptyset$, these two vertices are at distance 1; otherwise their intersection is reduced to a single element and we may write $\{a, b\}$ and $\{b, c\}$. Let d and e be the two remaining elements. $\{d, e\}$ is disjoint from $\{a, b\}$ and from $\{b, c\}$ and is thus at distance 1 from each one of them. The diameter of the graph is thus 2.

2. To increase the readability we have labelled only a subset of the graph (Figure 15.4).

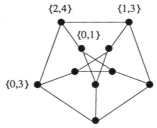

Figure 15.4

10.9. 1. The result is true for $k = 1$ by the definition of M. Assume that the $m_{ij}^{(k)}$ entry of M^k is the number of paths of length k from v_i to v_j. Let C_{ij}^k be the set of paths of length k from v_i to v_j, and let P_{ilj}^{k+1} be the subset of C_{ij}^{k+1} consisting of the paths whose last edge has origin v_l. It is easy to see that $C_{ij}^{k+1} = \bigcup_{l=1}^{n} P_{ilj}^{k+1}$, and since the sets P_{ilj}^{k+1} are disjoint, $|C_{ij}^{k+1}| = \sum_{l=1}^{n} |P_{ilj}^{k+1}|$. Moreover, every path of P_{ilj}^{k+1} is obtained by extending a path of C_{il}^k by one edge with origin v_l and target v_j. Hence, $|P_{ilj}^{k+1}| = |C_{il}^k| \times m_{lj}$. Since by the induction hypothesis $|C_{il}^k| = m_{il}^{(k)}$ holds, we have

$$|C_{ij}^{k+1}| = \sum_{l=1}^{n} m_{il}^{(k)} \times m_{lj},$$

i.e. $|C_{ij}^{k+1}| = m_{ij}^{(k+1)}$.

2. Yes. We associate with (V, E, δ) having the n vertices $v_1, \ldots, v_n$ the matrix M defined in the following way. Each entry $M_{i,j} = m_{ij}$ of M (for $1 \leq i \leq n$ and $1 \leq j \leq n$) is the number of edges e of E such that $\delta(e) = \{v_i, v_j\}$. We note that the matrix M is symmetrical, i.e. $m_{ij} = m_{ji}$, and we reason by induction as in case 1.

10.10. 1. The set of the edges of G_4 is the union of a set H of 'horizontal' edges and of a set V of 'vertical' edges. With each pair $\{(x, y), (x + 1, y)\}$ of elements of $\mathbb{Z}^2$ is associated a horizontal edge connecting them. Similarly, a vertical edge connects all pairs of elements $\{(x, y), (x, y + 1)\}$. If c is any chain connecting (x, y) and (x', y'), it is easy to see that $|x - x'|$ is less than or equal to the number of horizontal edges of c and that $|y - y'|$ is less than or equal to the number of vertical edges of c. Hence, $|x - x'| + |y - y'| \leq d_{G_4}((x, y), (x', y'))$.

It is also easy to see that there exists a 'horizontal' chain of length $|x - x'|$ connecting (x, y) and (x', y), and a 'vertical' chain of length $|y - y'|$ connecting (x', y) and (x', y'). There thus exists a chain of length $|x - x'| + |y - y'|$ connecting (x, y) and (x', y'). Hence, $d_{G_4}((x, y), (x', y')) \leq |x - x'| + |y - y'|$.

2. In order to obtain G_8, we add to G_4 the 'oblique' edges connecting (x, y) and $(x + 1, y + 1)$ and those connecting (x, y) and $(x + 1, y - 1)$. If c is any chain of length n connecting (x, y) and

(x',y'), we easily show that $|x-x'| \leq n$ and $|y-y'| \leq n$, and hence $\max(|x-x'|,|y-y'|) \leq d_{G_8}((x,y),(x',y'))$.

Let (x,y) and (x',y') be two elements of $\mathbb{Z}^2$. We assume that $|x-x'| \leq |y-y'|$ and let $n = |x-x'|$. The proof would be similar if $|y-y'| \leq |x-x'|$. Through oblique chains of length n we may connect (x,y) to $(x+n,y+n)$, to $(x-n,y-n)$, to $(x+n,y-n)$ and to $(x-n,y+n)$. We now assume that $x \leq x'$, i.e. $x' = x+n$. The symmetrical case could be dealt with similarly. Then we may connect (x',y') to $(x',y+n)$ by a vertical chain of length $|y'-y-n|$ and to $(x',y-n)$ by a vertical chain of length $|y'-y+n|$. Because the least of the two numbers $|y'-y+n|$ and $|y'-y-n|$ is $|y-y'|-n$ we may connect (x,y) to (x',y') by a chain of length $n + |y-y'| - n = \max(|x-x'|,|y-y'|)$. Hence, $\max(|x-x'|,|y-y'|) \geq d_{G_8}((x,y),(x',y'))$ (Figure 15.5).

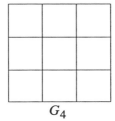
G_4
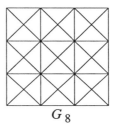
G_8

Figure 15.5

10.11. 1. Each vertex v_i is indeed the endpoint of two edges, which are

$$\begin{cases} \{e_i, e_{i+1}\} & \text{if } i < n, \\ \{e_1, e_n\} & \text{if } i = n. \end{cases}$$

2. By induction on n. Recall that the sum of the degrees of an undirected graph is equal to twice its number of edges. If each one of the n vertices of a graph is of degree 2, this graph thus has n edges.

- If $n = 1$, G has a single edge e and a single vertex v with $\delta(e) = \{v\}$. G is thus isomorphic to C_1.
- If $n = 2$, G has two vertices v and v' and two edges e and e'. Since G is connected, v is connected to v' by one of the two edges. Assume that it is edge e. We then have $\delta(e) = \{v,v'\}$. If $\delta(e') = \{v\}$ then v is of degree 3; if $\delta(e') = \{v'\}$ then v' is of degree 3. We thus have $\delta(e') = \{v,v'\}$, and G is isomorphic to C_2.
- Assume that G has $n+1$ vertices, with $n \geq 2$. Let v be a vertex of G and e' an edge with endpoint v.
  - If $\delta(e') = \{v\}$, then since v is of degree 2, v is neither the origin nor the target of any other edge, and $\{v\}$ is a connected component of G, which is excluded.
  - If $\delta(e') = \{v,v'\}$ with $v \neq v'$, there thus exists another edge e'' with endpoint v'. Hence $\delta(e'') = \{v''\}$ cannot hold for the same reasons as previously. Thus $\delta(e'') = \{v,v''\}$, with $v \neq v''$.
  - Thus $\delta(e') = \{v,v'\}$ with $v \neq v'$, and $\delta(e'') = \{v,v''\}$, with $v \neq v''$. If $v' = v''$, then v' can be neither the origin nor the target of any edge other than e' and e''. The set $\{v,v'\}$ is a connected component of G which is not equal to the whole of G because G has strictly more than two vertices, and this is excluded.
  - The remaining case is that $\delta(e') = \{v,v'\}$ and $\delta(e'') = \{v,v''\}$, with $v \neq v'$, $v \neq v''$ and $v' \neq v''$. Consider the graph G' obtained from G by deleting vertex v and edges e' and e''

and by adding an edge e with $\delta(e) = \{v', v''\}$. It is a graph with n vertices each of which is of degree 2. By the induction hypothesis, G' is isomorphic to C_n. We then easily see that G is obtained from G' by inserting v between v' and v'', and thus that G is isomorphic to C_{n+1}.

3. Let G be a graph and let $V_1, V_2, \ldots, V_k$ be its connected components. Let G_i be the subgraph of G whose set of vertices is V_i. Then G is the disjoint union of the G_is. Moreover, each G_i is a connected graph all of whose vertices have degree 2; it is thus isomorphic to some C_{n_i}.

10.12. The chromatic number of the complete graph K_4 with four vertices, and where all distinct pairs of vertices are connected by an edge, is 4.

This graph is indeed planar. Assume that it can be coloured with strictly less than four colours. There would then be two distinct vertices of the same colour, which is impossible since these two vertices are connected by an edge.

10.13. If one of the two connected components were not a tree, it would necessarily contain a simple cycle which would also be in the initial graph. This is impossible because the initial graph is a tree.

10.14. Let s be the sum of the degrees of the n vertices of G. Because G is a tree, it has $n-1$ edges and $s = 2(n-1)$. Let k be the number of vertices of degree 2 and let k' be the number of vertices of degree greater than or equal to 3. Then $n = k + k' + 2$ and $s \geq 2 + 2k + 3k'$. As $k' = n - 2 - k$ and $s = 2n - 2$, we have $2n - 2 \geq 2 + 2k + 3n - 6 - 3k$, i.e. $k \geq n - 2$. As $k \leq n - 2$, we have $k = n - 2$.

Because G is connected, there exists an elementary chain connecting the two vertices of degree 1 (see Exercise 10.6). Assume that this chain does not contain all the vertices of the graph. Since the vertices in this chain are of degree 2 and since they already belong to two edges and since the endpoints belong to one edge, no vertex of this chain can be connected by an edge to a vertex which is not in the chain. In this case the graph would not be connected, which is excluded.

10.15. Let $V = \{v_1, \ldots, v_n\}$ be the set of vertices of tree G. We show by induction on the number n of vertices the following more precise result: $\forall v_i \in V$, $\forall \alpha > 0$, $\forall R > 0$; we may always draw G on a circular segment with angle α and radius R in such a way that the edges consist of linear segments without cross-sections except at the endpoints, and that vertex v_i is the origin of the circular segment.

• If $n \leq 2$ the result is clear, because for $n = 1$, G is reduced to a point, and for $n = 2$, G is reduced to a single edge. (In fact it suffices to consider the case in which $n = 1$ as the basis case.)

• If $n > 2$, let e be an edge such that $\delta(e) = \{v_i, v_j\}$. Delete e from tree G. We obtain two trees G_1 and G_2 such that G_i has $n_i < n$ vertices for $i = 1, 2$, $v_i \in G_1$ and $v_j \in G_2$. By the induction hypothesis, we may thus draw G_1 (resp. G_2) in a circular segment with origin v_i (resp. v_j), angle $\alpha/2$, and radius $R/2$ in such a way that the edges consist of linear segments without cross-sections except at the endpoints. It then suffices to first draw G_1 and to then draw the edge e connecting v_i and v_j in such a way that it forms an $\alpha/2$ angle with the picture of G_1 and that it has a length of $R/2$, and to finally draw G_2 starting from v_j (Figure 15.6).

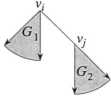

Figure 15.6

Planarity is an immediate consequence.

Chapter 10

10.16. Let Pref(L) be the set $\{u \in A^* \mid \exists v \in A^* : uv \in L\}$ (see Exercise 11.10). We construct a directed graph G whose vertices are the strings of Pref(L) and whose edges connect all the pairs (u, ua) with $a \in A$ and both u and $ua \in$ Pref(L).

The outdegree of each vertex of this graph is indeed finite since it is always less than or equal to the number of elements of A.

By the definition of Pref(L), any string $a_0 \cdots a_n$ of L is the target of the path ε, a_0, $a_0 a_1, \ldots, a_0 a_1 \cdots a_n$ with origin ε. Since L is infinite, ε is the origin of infinitely many paths of G.

Applying Proposition 10.25, we deduce the existence in G of an infinite path $\varepsilon, a_0, a_0 a_1$, $\ldots, a_0 a_1 \cdots a_n, \ldots$ with origin ε. Because each string $a_0 a_1 \cdots a_n$ is in Pref(L), the infinite string $a_0 a_1 \cdots a_n \cdots$ does indeed have the required property.

10.17. Let F_n be the set of the injections $f_n : \{0, 1, \ldots, n\} \to \mathbb{N}$ such that $\forall i \in \{0, 1, \ldots, n\}$, $(i, f_n(i)) \in S$. By (i), F_n is non-empty, and it is clear that if $n \neq m$ then $F_n \cap F_m = \emptyset$.

On the other hand, each F_n is finite. If k_n is the cardinality of the finite set $\{m \in \mathbb{N} \mid (n, m) \in V\}$, we easily show by induction on n that the number of elements of F_n is less than or equal to $k_0 \times k_1 \times \cdots \times k_n$.

We define the relation R on $\cup_{n \geq 0} F_n$ by $f\ R\ g$ if and only if there is an $n \geq 0$ such that $f \in F_n$, $g \in F_{n+1}$, and $\forall i \in \{0, 1, \ldots, n\}$, $f(i) = g(i)$.

If g is in F_{n+1}, its restriction g' to $\{0, 1, \ldots, n\}$ is indeed such that $g'\ R\ g$. Proposition 10.26 can thus be applied: there exists a sequence of injections $f_0, f_1, \ldots, f_n, \ldots$ such that $f_n \in F_n$ and $f_n\ R\ f_{n+1}$.

We then define $f : \mathbb{N} \to \mathbb{N}$ by $f(n) = f_n(n)$. Thus $(n, f(n)) = (n, f_n(n)) \in V$. Finally, f is an injection. Indeed, we easily show by induction on m that $\forall n, m \in \mathbb{N}, \forall i \leq n, f_n(i) = f_{n+m}(i)$. If there were $n < m$ such that $f(n) = f(m)$, then $f_m(n) = f_n(n) = f(n) = f(m) = f_m(m)$ and thus f_m would not be injective, a contradiction.

10.18. Both questions can be proved by induction on the number of vertices of the tree.

1. If the complete binary tree G has only one vertex, it has indeed an odd number of vertices. If it has $n + 1$ vertices, its root r must necessarily have two children v and v'. Graphs $G(v)$ and $G(v')$ are again complete binary trees having, respectively, p and p' vertices, with $p + p' = n$. We thus have $1 \leq p \leq n - 1$ and $1 \leq p' \leq n - 1$. By the induction hypothesis, p and p' are odd numbers. Hence $n = p + p'$ is even and $n + 1$ is odd.

2. If a complete binary tree has a single vertex, it has a single leaf, and the property is true. For a tree with $2n - 1$ vertices, we proceed as previously: $G(v)$ is a tree with $2p - 1$ vertices and p leaves, $G(v')$ is a tree with $2p' - 1$ vertices and p' leaves and G is a tree with $2p - 1 + 2p' - 1 + 1 = 2(p + p') - 1$ vertices and with $p + p'$ leaves.

Chapter 11

11.1. 1. YES. The composition of mappings is indeed associative and has a unit that is the identity mapping.

2. NO. The 'power' operation is not associative: $(2^1)^2 = 2^2 = 4$, $2^{(1^2)} = 2^1 = 2$.

3. YES. The empty string is an even length string, and the (associative) product of two even length strings is again an even length string.

4. YES. Let $|u|_a$ and $|u|_b$ be the number of occurrences of a and b in u. Indeed $|\varepsilon|_a = |\varepsilon|_b = 0$; if $|u|_a = |u|_b$ and $|v|_a = |v|_b$, then $|uv|_a = |u|_a + |v|_a = |u|_b + |v|_b = |uv|_b$.

5. YES. Union is associative and its unit is the empty set.

6. YES and NO. The intersection of the subsets of a set E is an associative operation. Its unit is is E; indeed, if Z is this unit, $\forall e \in E, \{e\} \cap Z = \{e\}$ should hold (i.e. $\forall e \in E, e \in Z$). Since the unit must be a finite subset, we have a monoid if and only if E is finite.

11.2. 1. The matrices I, A and B have determinant $+1$, and a product of matrices with determinant $+1$ is again a matrix with determinant $+1$. If $M = \begin{pmatrix} x & y \\ z & w \end{pmatrix}$ is a matrix with non-negative integral coefficients, then $MA = \begin{pmatrix} x & x+y \\ z & z+w \end{pmatrix}$ and $MB = \begin{pmatrix} x+y & y \\ z+w & w \end{pmatrix}$ are matrices with non-negative integral coefficients.

2. Let $M = \begin{pmatrix} x & y \\ z & w \end{pmatrix}$. Its determinant Δ is equal to $xw - yz$. Let $y = x + p$ and $w = z + q$, with $p, q \in \mathbb{Z}$. Then $\Delta = x(z+q) - (x+p)z = xq - pz$. If $\Delta = 1$, we have $xq = 1 + pz$. If $y \geq x$, then $p \geq 0$ and, since x and z are non-negative, $q > 0$ and thus $w > z$. If $w \leq z$, then $q \leq 0$, and hence $p < 0$ and $y < x$.

3. It is easy to see that

$$A^{-1} = \begin{pmatrix} 1 & -1 \\ 0 & 1 \end{pmatrix} \text{ and } B^{-1} = \begin{pmatrix} 1 & 0 \\ -1 & 1 \end{pmatrix}.$$

Let $M = \begin{pmatrix} x & y \\ z & w \end{pmatrix}$. If the first column is greater that the second column, the matrix $MB^{-1} = \begin{pmatrix} x-y & y \\ z-w & w \end{pmatrix}$ has determinant $+1$ and non-negative integral coefficients. The sum of its coefficients is $x + z$, strictly less than $x + y + z + w$. (If $y + w = 0$, M has determinant zero.) $M' = MB^{-1}$ is thus an element of $\mathcal{M}_1$, and so is $M = M'B$. The uniqueness of the decomposition of M in a product of matrices A and B follows from the fact that M cannnot be written as $M''A$, with $M'' \in \mathcal{M}_1$, because $M'' = MA^{-1} = \begin{pmatrix} x & y-x \\ z & w-z \end{pmatrix}$ with $y - x \leq 0$ and $w - z \leq 0$, and if $M'' \in \mathcal{M}_1$ we have $y - x = 0$ and $w - z = 0$, which is impossible if the determinant of M'' is 1.

If the second column of M is greater than the first column, the argument is similar modulo the interchange of A and B.

4. The equality of $\mathcal{M}_1$ and $\mathcal{M}_2$ easily follows from the above.

11.3. Let $h(a) = A$ and $h(b) = B$. This homomorphism is surjective by the definition of $\mathcal{M}_1$. It is injective because the decomposition of any matrix M of $\mathcal{M}_1 = \mathcal{M}_2$ into a product of matrices A and B is unique.

11.4. 1. It is necessary and sufficient that $\forall x \in A$, $h(x)$ is a length 2 string.

2. It is necessary and sufficient that $\forall x \in A$, $h(x)$ is a string containing no bs.

3. Let, for instance, the two homomorphisms

$$h\colon (A\setminus\{a\})^* \longrightarrow (A\setminus\{b\})^* \text{ and } g\colon (A\setminus\{b\})^* \longrightarrow (A\setminus\{a\})^*$$

be defined as follows:
$$h(b) = a\,,$$
$$h(x) = x, \quad \forall x \notin \{a,b\}\,,$$
$$g(a) = b\,,$$
$$g(x) = x, \quad \forall x \notin \{a,b\}\,.$$

Then $gh\colon (A\setminus\{a\})^* \longrightarrow (A\setminus\{a\})^*$ is the identity, and h is thus an isomorphism.

11.5. It is clear that if one of the following three cases holds then $uv = xy$. See Figure 15.7.

(i) If $u = x$ and $v = y$ then $uv = xy$.

(ii) If $ut = x$ and $v = ty$ then $uv = uty = xy$.

(iii) If $u = xt$ and $tv = y$ then $uv = xtv = xy$.

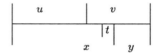

Figure 15.7

The converse is proved by complete induction on $n = |u| + |x|$:

- If $n = 0$ then $|u| = |x| = 0$. Hence $u = x = \varepsilon$, and thus $u = y$.
- Assume that $|u| + |x| = n + 1$.
  - If $|u| = 0$ then case (ii) holds and we let $t = x$.
  - If $|x| = 0$ then case (iii) holds and we let $t = u$.
  - If both $|u|$ and $|x|$ are strictly positive, and $uv = xy$, we have $u = au'$, $x = ax'$, and $u'v = x'y$. Since $|u'| + |x'| = n - 1$, we can apply the induction hypothesis:

    (i) If $|u| = |x|$, then $|u'| = |x'|$. Hence $u' = x'$, $v = y$, and thus $u = x$.

    (ii) If $|u| < |x|$, then $|u'| < |x'|$. Hence $u't = x'$ and $v = ty$, and also $ut = au't = ax' = x$.

    (iii) If $|u| > |x|$, then $|u'| > |x'|$. Hence $u' = x't$ and $tv = t$, and also $u = au' = ax't = xt$.

11.6. 1. If $u = w^m$ and $v = w^n$, it is clear that $uv = vu$. To show the converse, we reason by complete induction on the length of uv.

- If $|uv| = 0$, then $u = v = \varepsilon$, and we let $w = \varepsilon, m = n = 1$.
- If $|u| = 0$, then $u = \varepsilon$, and we let $w = v, m = 1$, and $n = 0$. (Recall that $w^0 = \varepsilon$.)
- If $|v| = 0$, then $v = \varepsilon$, and we let $w = u$, $m = 0$, and $n = 1$.
- If $|uv| = n+1$ (with $|u|$ and $|v|$ strictly positive) then, by Levi's lemma, one of the following three cases holds.

- If $|u| = |v|$, then $u = v$; we then let $w = u = v$, and $p = q = 1$.
- If $|u| < |v|$, then $ut = v$ and $v = tu$, and hence $ut = tu$. As $|u| > 0$, $|u| + |t| = |v| < |u| + |v|$, and we can apply the induction hypothesis: $u = w^m$ and $t = w^n$, and thus $v = ut = w^{m+n}$.
- If $|u| > |v|$, then $u = tv$ and $tv = u$, and hence $vt = tv$. For the same reasons as the above, we can apply the induction hypothesis: $t = w^m$ and $v = w^n$, and thus $u = vt = w^{m+n}$.

2. Here again the necessary condition is straightforward. Thus, assume $u^p = v^q$.
- If $|u| = |v|$, we let $w = u = v$ and $m = n = 1$.
- If $|u| \neq |v|$, we can assume $|u| < |v|$ because the equation is symmetrical in u and v. Let k and r be integers such that $|v| = k|u| + r$, with $r < |u|$. We easily deduce that $v = u^k t$ and $u = tt'$ (with $|t| = r$, and thus $|t'| \neq 0$), and thus that $(tt')^p = ((tt')^k t)^q$. Deleting the $(tt')^k$ prefix common to both strings, we have $(tt')^{p-k} = t((tt')^k t)^{q-1}$. Noting also that $(|t| + |t'|)(p - k) = |t| + (q-1)(|t| + k(|t| + |t'|))$, we have that $q - 1 > 0$, because otherwise $(|t| + |t'|)(p - k) = |t|$, which is impossible since both $|t|$ and $|t'|$ are non-zero. The string $t((tt')^k t)^{q-1}$ thus starts with ttt', and the string $(tt')^{p-k}$, which is equal, starts with $tt't$. Because these two prefixes have the same length, they are equal and thus $tt' = t't$. We can then apply the preceding result: $t = w^m$ and $t' = w^n$, and hence $u = w^{m+n}$ and $v = w^{k(m+n)+m}$.

3. By studying the conditions that the length of w satisfies, we find that $2|u| = |u|$, namely, $|u| = 0$ and thus $u = \varepsilon$.

4. Applying 1, we have $u = w^m$ and $a = w^n$. We must thus have $1 = |a| = n|w|$; hence $n = |w| = 1$, and thus $w = a$.

5. Since the condition $0 \neq 0$ is always false, this boils down to the condition that $ua = bu$ never holds (with $a \neq b$). Let n (resp. p, q) be the number of occurrences of a in u (resp. ua, bu). We have that $p = n + 1$ and $q = n$, and if ua were equal to bu, we would have that $p = q$, which would be impossible.

11.7. 1. The product of languages is associative:
$$(L \cdot L') \cdot L'' = L \cdot (L' \cdot L'') = \{uu'u'' \ / \ u \in L, u' \in L', u'' \in L''\}.$$
The language $\{\varepsilon\}$ is indeed the unit because
$$L \cdot \{\varepsilon\} = \{uv \ / \ u \in L, v \in \{\varepsilon\}\} = \{u\varepsilon \ / \ u \in L\} = L.$$
For the same reasons, $L = \{\varepsilon\} \cdot L$.

2. $u \in (\bigcup_{i \in I} L_i) \cdot L$ if and only if $(\exists v \in \bigcup_{i \in I} L_i, \exists w \in L: u = vw)$ if and only if $(\exists i \in I, \exists u \in L_i, \exists w \in L: u = vw)$ if and only if $\exists i \in I: u \in L_i \cdot L$.

3. It is easy to prove by induction that $(L + \{\varepsilon\})^n = \bigcup_{i=0}^{n} L^i$. It holds for $n = 0$. Applying the induction hypothesis, we have $(L + \{\varepsilon\})^{n+1} = (\bigcup_{i=0}^{n} L^i) \cdot (L \cup \{\varepsilon\})$, which is equal to $\bigcup_{i=0}^{n+1} L^i$ by the result of the preceding question and the fact that $\{\varepsilon\}$ is the unit of the product.

4. $\emptyset^* = \{\varepsilon\} + \emptyset \cdot L^* = \{\varepsilon\}$.

11.8. 1. X^* is the set of strings (including the empty string) consisting only of bs.
Y is the set of non-empty strings whose first letter is not a b and whose other letters are all bs. Indeed, $u \in Y$ if and only if $u = vw$ with $u \in A \setminus \{b\}$ and $v \in \{b\}^*$.
Y^* is the set consisting of the empty string and of all the non-empty strings which do not start with b.

2. Let u be a string of A^* whose first letter is not a b. It can thus be written
$$x_1 b^{p_1} x_2 b^{p_2} \cdots b^{p_{n-1}} x_n b^{p_n},$$
with $x_i \in A \setminus \{b\}$. Each of the strings $x_i b^{p_i}$ is in Y, and thus $u \in Y^*$.

3. Let u be a string in A^*. If it contains at least one letter different from b, it can be written $b^p x u'$. By 1, $b^p \in X^*$, and by 2, $xu' \in Y^*$. If u contains only bs, then it is in X^*, and as $\varepsilon \in Y^*$, we have that $u \in X^* \cdot Y^*$. This decomposition is unique: because u contains only bs, the only string of Y^* that can occur in u is the empty string.

11.9. The only paths with trace ε are the empty paths, which are in the set of paths whose source and target are equal.

11.10. Let $\mathcal{A} = (S, T, I, F)$ be a finite-state automaton recognizing L. We will obtain a finite-state automaton $\mathcal{A}' = (S, T, I, F')$ recognizing $\mathrm{Pref}(L)$ by including in F' any state s of $\mathcal{A}$ such that there exists a path c whose source is a state of I and whose target is s, and a path c' whose source is s and whose target is a state of F.

11.11. Let L be a finite non-empty language. Let $P(L)$ be the set $\{u \in A^* \,/\, \exists v \in A^* \colon uv \in L\}$. This set is finite because any string of L has but a finite number of prefixes, and it is non-empty because it contains L and the empty string ε. Define the finite-state automaton $\mathcal{A} = (S, T, \{i\}, F)$ as follows:

- $S = P(L)$,
- $T = \{(u, a, ua) \,/\, u, ua \in P(L)\}$,
- $i = \varepsilon$,
- $F = L$.

It is clear that this automaton is deterministic. It is also easy to see that there exists a path with source ε and with trace u if and only if $u \in P(L)$, and then the target state of this path is u. We thus have $L(\mathcal{A}) = L$.

If L is the empty language, then it is recognized by the empty automaton, which is a deterministic finite-state automaton.

11.12. If there exists a circuit going through a state which is on a path going from an initial state to a final state, then there exist three paths: c' going from an initial state s_i to a state s, c, non-empty, going from s to s, and c'' going from s to a final state s_f. Let u, v and w be the traces of these three paths. Since for any $n \geq 0$, $c_1 c^n c_2$ is a path with trace $uv^n w$ in the automaton, and since v is not the empty string, the language recognized by the automaton contains infinitely many strings.

Conversely, if there is no circuit going through a state which is on a path going from an initial state to a final state, then no path from an initial state to a final state can go twice through the same state. Such paths thus have, by the pigeonhole principle (see Proposition 1.8 and Exercise 1.16), a length strictly less than the number of states of the finite-state automaton.

If n is the number of states of the automaton, the recognized strings will have length strictly less than n. If k is the number of letters in the alphabet, the number of strings of length strictly less than n is $1 + k + k^2 + \cdots + k^{n-1}$, which is equal to

$$\begin{cases} n & \text{if } k = 1, \\ \dfrac{k^n - 1}{k - 1} & \text{if } k \geq 2. \end{cases}$$

This bound is the same for deterministic finite-state automata, and it is indeed reached, as the next example shows.

Let $S = \{0, 1, \ldots, n-1\}$, with initial state 0, final states $F = \{0, 1, \ldots, n-2\}$ and $T = \{(i-1, a, i) \,/\, 1 \leq i \leq n, a \in A\}$. It is easy to see that this automaton is deterministic and recognizes all the strings of length strictly less than n. Conversely, if a finite-state automaton with n states recognizes a number of strings equal to this bound, it must then recognize all the strings of length strictly less than n and it has the above-given form. Complete finite-state automata recognizing finite languages must have at least a non-final state. In this case the bound is

$$\begin{cases} n - 1 & \text{if } k = 1, \\ \dfrac{k^n - 2}{k - 1} & \text{if } k \geq 2. \end{cases}$$

The deterministic finite-state automaton reaching this bound is

$$S = \{0, 1, \ldots, n-1\}, \quad F = \{0, 1, \ldots, n-2\} \quad i = 0 \quad \text{and}$$

$$T = \{(i-1, a, i) \,/\, 1 \leq i \leq n-1, a \in A\} \cup \{(n-1, a, n-1) \,/\, a \in A\}.$$

11.13. 1. Because we add to $\mathcal{A}$ transitions (s, a, s) when there are no transitions (s, a, s') in $\mathcal{A}$, then we have that if $\mathcal{A}$ is deterministic, $\mathcal{A}'$ is also deterministic.

2. Any path of $\mathcal{A}$ is also a path of $\mathcal{A}'$, and hence $L(\mathcal{A}) \subseteq L(\mathcal{A}')$. This inclusion is strict. If $\mathcal{A}$ is the finite-state automaton over the alphabet $\{a, b\}$, whose only transition is (i, b, f), it recognizes the language $\{b\}$. Completing it as indicated, we add the transitions (i, a, i), (f, a, f) and (f, b, f), and the recognized language becomes $a^*b(a+b)^*$.

11.14.

$$\{i\} \xrightarrow{a} \emptyset, \quad \{i\} \xrightarrow{b} \{x, y\}, \quad \{i\} \xrightarrow{c} \emptyset$$

$$\emptyset \xrightarrow{a} \emptyset, \quad \emptyset \xrightarrow{b} \emptyset, \quad \emptyset \xrightarrow{c} \emptyset$$

$$\{x, y\} \xrightarrow{a} \{f\}, \quad \{x, y\} \xrightarrow{b} \{i, z\}, \quad \{x, y\} \xrightarrow{c} \{f, z\}$$

$$\{f\} \xrightarrow{a} \emptyset, \quad \{f\} \xrightarrow{b} \emptyset, \quad \{f\} \xrightarrow{c} \emptyset$$

$$\{i, z\} \xrightarrow{a} \{f'\}, \quad \{i, z\} \xrightarrow{b} \{x, y\}, \quad \{i, z\} \xrightarrow{c} \{f'\}$$

$$\{f, z\} \xrightarrow{a} \{f'\}, \quad \{f, z\} \xrightarrow{b} \emptyset, \quad \{f, z\} \xrightarrow{c} \{f'\}$$

$$\{f'\} \xrightarrow{a} \{z\}, \quad \{f'\} \xrightarrow{b} \emptyset, \quad \{f'\} \xrightarrow{c} \{z\}$$

$$\{z\} \xrightarrow{a} \{f'\}, \quad \{z\} \xrightarrow{b} \emptyset, \quad \{z\} \xrightarrow{c} \{f'\}$$

The initial state is $\{i\}$, and the final states are $\{f\}, \{f'\}$ and $\{f, z\}$.

11.15. Automata for questions 1, 2 and 8 are shown in Figure 15.8.

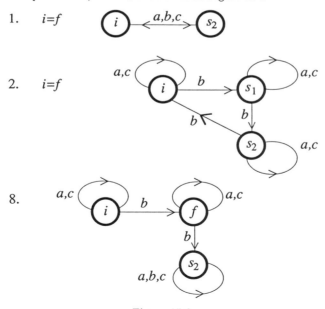

Figure 15.8

11.16. 1. It is again an application of the pigeonhole principle (see Proposition 1.8 and Exercise 11.12).

2. If $z \in L_{q,q'}$ and $|z| \geq n$, there exists a path γ with trace z from q to q'. Because this path has a length greater than or equal to n, it must be true that the path consisting of its n first transitions contains a circuit and γ can thus be written $c'cc''$ with $|c'c| \leq n$, and c such that its source is equal to its target. We have the required result by taking u to be the trace of c', v to be the trace of c, and w to be the trace of c''.

3. This immediately follows from the preceding point, by using the fact that $z \in L(\mathcal{A})$ if and only if $\exists q \in I, q' \in F \colon z \in L_{q,q'}$.

11.17. We will reason by contradiction in all cases. We will assume that the given language is recognized by a finite-state automaton with k states, and, applying the iteration lemma, we will show that it must also contain strings other than the desired ones.

1. The string $a^k b^k$ can be written uvw with $|uv| \leq k$ and $v \neq \varepsilon$. We thus have $u = a^p, v = a^q$, and $w = a^r b^k$ with $q \neq 0$ and $p + q + r = k$. We deduce that the automaton also recognizes $a^{k+q} b^k$, which is not in the language.

2. The string a^{k^2} can be written as $a^p a^q a^r$ with $0 < q \leq k$. The string a^{k^2+q} is also recognized by the automaton, but $k^2 + q$ is not a square. (The least square strictly greater than k^2 is $(k+1)^2 = k^2 + 2k + 1$.)

3. Let m be a prime number greater than k. $p + q + r$ with $0 < q \leq k$ and a^{m+qm} is again recognized by the automaton even though $m + qm$ is obviously not a prime number.

4. Let $w = a^k b$. The string ww is thus decomposed as $a^p, a^q \ a^r bw$, with $q > 0$, and the string $a^{k+q} b a^k b$ is accepted by the automaton even though it is not of the form ww.

5. The string $a^k b a^k$ is a palindrome. For the same reasons as given above, the string $a^{k+q} b a^k$ will be accepted by the automaton even though it is not a palindrome.

11.18. Answers to questions 1 and 5 are given in Figure 15.9 and to question 2 in Figure 15.10.

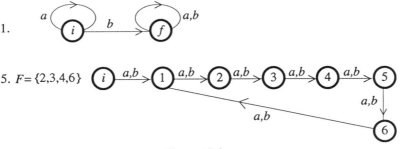

Figure 15.9

2. The completion of the following finite-state automaton, where $F = \{1, 2, 4, 6\}$.

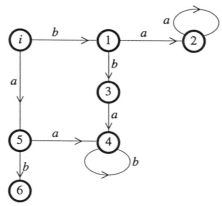

Figure 15.10

11.19. Let $A = \{a, b\}$.

Consider the complete non-deterministic finite-state automaton whose transitions are

$$(q, a, q), (q, b, q), (q', a, q') \text{ and } (q', b, q'),$$

where q is final, and where both q and q' are initial. This automaton recognizes A^*. The finite-state automaton $\mathcal{A}'$, whose final state is q', also recognizes A^*.

Consider the incomplete deterministic finite-state automaton whose transitions are (q, a, q'), and (q', b, q'), whose initial state is q, and whose only final state is q'. This automaton recognizes ab^* even though the automaton $\mathcal{A}'$, whose final state is q, recognizes $\{\varepsilon\}$, which is not the complement of ab^*.

11.20. Let $\{D_i \,/\, i \in I\}$ be an arbitrary set of mappings in $\mathcal{D}$ whose least upper bound is defined by $D(s, s') = \bigcup_{i \in I} D_i(s, s')$. Show that $\widehat{\mathcal{S}}(D)$ is the least upper bound of $\{\widehat{\mathcal{S}}(D_i) \,/\, i \in I\}$. To this end it is enough to verify that if $t = (s'', a, s')$, then $a \cdot \bigcup_{i \in I} D_i(s'', s') = \bigcup_{i \in I} a \cdot D_i(s'', s')$, and this is a straightforward consequence of Exercise 11.7 2.

11.21. $(0, a, 0), (0, b, 1), (1, a, 1), (1, b, 2), (2, a, 2), (2, b, 0)$ with initial state: 0, and with final state: 0.

$$x_{0,0} = ax_{0,0} + bx_{1,0} + \varepsilon$$
$$x_{0,1} = ax_{0,1} + bx_{1,1}$$
$$x_{0,2} = ax_{0,2} + bx_{1,2}$$
$$x_{1,0} = ax_{1,0} + bx_{2,0}$$
$$x_{1,1} = ax_{1,1} + bx_{2,1} + \varepsilon$$
$$x_{1,2} = ax_{1,2} + bx_{2,2}$$
$$x_{2,0} = ax_{2,0} + bx_{0,0}$$
$$x_{2,1} = ax_{2,1} + bx_{0,1}$$
$$x_{2,2} = ax_{2,2} + bx_{0,2} + \varepsilon$$

We deduce $x_{0,0} = a^*(bx_{1,0} + \varepsilon)$, $x_{1,0} = a^*bx_{2,0}$, $x_{2,0} = a^*bx_{0,0}$; hence

$$x_{0,0} = a^* + a^*ba^*ba^*bx_{0,0} \text{ and thus } x_{0,0} = (a^*ba^*ba^*b)^*a^*.$$

11.22. L is recognized by the finite-state automaton

$$(0, a, 1), (1, a, 2), (2, a, 3), (3, a, 3), (0, b, 0), (1, b, 0), (2, b, 0), (3, b, 3)$$

with initial state: 0, and with final states: 0,1,2. See Figure 15.11.

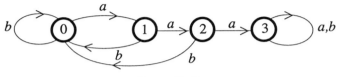

Figure 15.11

The equation system associated with this automaton has sixteen equations. We will thus write only those equations that are useful for determining the language recognized by this automaton:

$$x_{0,0} = ax_{1,0} + bx_{0,0} + \varepsilon$$
$$x_{0,1} = ax_{1,1} + bx_{0,1}$$
$$x_{0,2} = ax_{1,2} + bx_{0,2}$$
$$x_{1,0} = ax_{2,0} + bx_{0,0}$$
$$x_{1,1} = ax_{2,1} + bx_{0,1} + \varepsilon$$
$$x_{1,2} = ax_{2,2} + bx_{0,2}$$
$$x_{2,0} = ax_{3,0} + bx_{0,0}$$
$$x_{2,1} = ax_{3,1} + bx_{0,1}$$
$$x_{2,2} = ax_{3,2} + bx_{0,2} + \varepsilon$$
$$x_{3,0} = ax_{3,0} + bx_{3,0}$$
$$x_{3,1} = ax_{3,1} + bx_{3,1}$$
$$x_{3,2} = ax_{3,2} + bx_{3,2}$$

Consider the system consisting of the last three equations. Its least solution is $(\emptyset, \emptyset, \emptyset)$. This system can thus be simplified to

$$x_{0,0} = ax_{1,0} + bx_{0,0} + \varepsilon$$
$$x_{0,1} = ax_{1,1} + bx_{0,1}$$
$$x_{0,2} = ax_{1,2} + bx_{0,2}$$
$$x_{1,0} = ax_{2,0} + bx_{0,0}$$
$$x_{1,1} = ax_{2,1} + bx_{0,1} + \varepsilon$$
$$x_{1,2} = ax_{2,2} + bx_{0,2}$$
$$x_{2,0} = bx_{0,0}$$
$$x_{2,1} = bx_{0,1}$$
$$x_{2,2} = bx_{0,2} + \varepsilon$$

which can also be written

$$x_{0,0} = ax_{1,0} + bx_{0,0} + \varepsilon$$
$$x_{0,1} = ax_{1,1} + bx_{0,1}$$
$$x_{0,2} = ax_{1,2} + bx_{0,2}$$
$$x_{1,0} = (ab + b)x_{0,0}$$
$$x_{1,1} = (ab + b)x_{0,1} + \varepsilon$$
$$x_{1,2} = a + (ab + b)x_{0,2}$$

or

$$x_{0,0} = (aab + ab + b)x_{0,0} + \varepsilon$$
$$x_{0,1} = a + (aab + ab + b)x_{0,1}$$
$$x_{0,2} = aa + (aab + ab + b)x_{0,2}$$

we have

$$x_{0,0} = (aab + ab + b)^*$$
$$x_{0,1} = (aab + ab + b)^*a$$
$$x_{0,2} = (aab + ab + b)^*aa$$

Chapter 12

12.1. (ii) is true by hypothesis; it is enough to verify (i) and (iii); for (i), $\forall A \in \mathcal{T}$, $0 \leq P(A) \leq \sum_{\omega \in \Omega} P(\omega) = P(\Omega) = 1$. For (iii), $\forall n$, $P(A_n) = \sum_{\omega \in A_n} P(\omega)$, and thus $\sum_{n \in \mathbb{N}} P(A_n) = \sum_{n \in \mathbb{N}} \left(\sum_{\omega \in A_n} P(\omega) \right) = \sum_{\omega \in (\cup_{n \in \mathbb{N}} A_n)} P(\omega) = P(\cup_{n \in \mathbb{N}} A_n)$.

12.2. Let $a = $ 'A hits the target' and $b = $ 'B hits the target'. We have: $P(a \cup b) = P(a) + P(b) - P(a \cap b)$, and we obtain, assuming that a and b are independent (i.e. assuming that a and b are such that $P(a \cap b) = P(a)P(b)$): $P(a \cup b) = P(a) + P(b) - P(a)P(b) = \dfrac{11}{20}$.

12.3. Let $X_n \in \{B, R\} = \Omega$ be the colour of the ball obtained at the nth drawing. We have

$$P(X_1 = R) = \frac{r}{b+r}$$
$$P(X_2 = R\,/\,X_1 = R) = \frac{r-1}{b+r-1+c}$$
$$\vdots$$
$$P(X_{n+1} = R\,/\,X_n = R, \ldots, X_2 = R, X_1 = R) = \frac{r-n}{b+r-n+nc}$$
$$\vdots$$

By Proposition 12.19 we deduce that

$$P(X_1 = X_2 = \cdots = X_k = R) = \frac{r(r-1)\cdots(r-k+1)}{(b+r)(b+r+c-1)\cdots(b+r+(k-1)c-(k-1))}.$$

Chapter 12

12.4. Let $p_n = P(X_n = \text{'yes'})$. We have

$$p_1 = 1, \quad \text{and, for } n \geq 2,$$

$$p_n = P(X_{n-1} = \text{'yes' and } X_n = \text{'yes'}) + P(X_{n-1} = \text{'no' and } X_n = \text{'yes'})$$
$$= P(X_n = \text{'yes'} \,/\, X_{n-1} = \text{'yes'}) P(X_{n-1} = \text{'yes'})$$
$$\quad + P(X_n = \text{'no'} \,/\, X_{n-1} = \text{'yes'}) P(X_{n-1} = \text{'no'})$$
$$= p p_{n-1} + q(1 - p_{n-1})$$
$$= q + (p - q) p_{n-1} \,.$$

The recurrence equation $p_n = q + (p-q)p_{n-1}$ has the characteristic polynomial: $(r-(p-q))(r-1) = 0$. Assuming $p \neq 1$, we thus have a general solution $p_n = \lambda + \mu(p-q)^n$; $p_1 = 1$ and $p_2 = p$ give us

$$p_n = \frac{1}{2}(1 + (p-q)^{n-1}) \,.$$

If $p = 1$, then straightforwardly $p_n = 1$.

12.5. For all $n \in \mathbb{N}$, let B_n be the event $X^n = B$, and R_n be the event $X^n = R$.

(B) We indeed have: $P(B_1) = b/(b+r)$ and $P(R_1) = r/(b+r)$.

(I) Assume by induction that, **regardless of the initial values** of b, r, c, $P(B_n) = b/(b+r)$, and prove that $P(B_{n+1}) = b/(b+r)$. We have $P(B_{n+1}) = P(B_{n+1} \cap B_1) + P(B_{n+1} \cap R_1) = P(B_{n+1}/B_1)P(B_1) + P(B_{n+1}/R_1)P(R_1)$. By the induction hypothesis applied to the sequence $(X^n)_{n \geq 2}$, we have:

- $P(B_{n+1}/B_1) = (b+c)/(b+r+c)$, because in this case the sequence $(X^n)_{n \geq 2}$ corresponds to a Polya urn model with initial values $b+c, r, c$, and
- $P(B_{n+1}/R_1) = b/(b+r+c)$, because in this case the sequence $(X^n)_{n \geq 2}$ corresponds to a Polya urn model with initial values $b, r+c, c$.

Thus

$$P(B_{n+1}) = \frac{b+c}{b+r+c} \times \frac{b}{b+r} + \frac{b}{b+r+c} \times \frac{r}{b+r} = \frac{b}{b+r} \,.$$

12.6. Assuming both possible choices of the urn are equally probable,

$$\frac{r_1(b_2 + r_2)}{r_1(b_2 + r_2) + r_2(b_1 + r_1)} \,.$$

Indeed, letting the events: $V_i = $'urn U_i was chosen', for $i = 1, 2$, $R = $'a red ball was drawn': (V_1, V_2) is a partition, thus by Theorem 12.22

$$P(V_1/R) = \frac{P(V_1)P(R/V_1)}{P(V_1)P(R/V_1) + P(V_2)P(R/V_2)} \,.$$

$P(R/V_i) = \dfrac{r_i}{b_i + r_i}$, for $i = 1, 2$; moreover, both possible choices of the urn are equally probable, $P(V_1) = P(V_2) = 1/2$. Thus

$$P(V_1/R) = \frac{P(V_1)P(R/V_1)}{P(V_1)P(R/V_1) + P(V_2)P(R/V_2)}$$

$$= \frac{P(V_1)\dfrac{r_1}{b_1 + r_1}}{P(V_1)\dfrac{r_1}{b_1 + r_1} + P(V_2)\dfrac{r_2}{b_2 + r_2}}$$

$$= \frac{r_1(b_2 + r_2)}{r_1(b_2 + r_2) + r_2(b_1 + r_1)} \,.$$

12.7. This is a straightforward consequence of Theorem 12.22. Let the events: $d =$ 'the sample object is flawed', $a =$ 'the sample object comes from A', $b =$ 'the sample object comes from B'. Then,
$$P(a) = \frac{100}{100+200} = \frac{1}{3}, \quad P(b) = \frac{2}{3}, \quad P(d/a) = \frac{5}{100}, \quad P(d/b) = \frac{6}{100};$$
therefore
$$P(a/d) = \frac{P(a)P(d/a)}{P(b)P(d/b) + P(a)P(d/a)} = \frac{5}{17}.$$

12.8. Define the events
$i = \{\text{voter } e \text{ is from area } i\}$,
$c = \{\text{voter } e \text{ voted for candidate } C\}$.
1. $P(c) = \sum_{i=1}^{3} P(i \cap c) = \sum_{i=1}^{3} P(i)P(c/i) = 3/10 \times 2/5 + 1/2 \times 24/50 + 1/5 \times 3/5 = 48\%$.
2. We have
$$P(3/c) = \frac{P(3)P(c/3)}{\sum_{i=1}^{3} P(i)P(c/i)}$$
$$= \frac{1/5 \times 3/5}{3/10 \times 2/5 + 1/2 \times 24/50 + 1/5 \times 3/5} = \frac{12}{12 + 24 + 12} = 1/4.$$

12.9. We have $\Omega_2 = \{MM, MF, FM, FF\}$, with the uniform probability $P(\omega) = 1/4, \forall \omega \in \Omega_2$; $A = \{MM, MF, FM\}$ and $B = \{MF, FM\} = A \cap B$, hence $P(A \cap B) = 1/2$, but $P(B) = 1/2$ and $P(A) = 3/4$, thus $P(A \cap B) \neq P(A)P(B)$.
On the other hand, in the set of families with three children, we have
$$\Omega_3 = \{MMM, MMF, MFM, FMM, MFF, FMF, FFM, FFF\}$$
together with the uniform probability $P(\omega) = 1/8, \forall \omega \in \Omega_3$; hence $A = \{MMM, MMF, MFM, FMM\}$ and $B = \{MMF, MFM, FMM, MFF, FMF, FFM\}$, and thus $P(B) = 3/4$ and $P(A) = 1/2$, $P(A \cap B) = P(MMF, MFM, FMM) = 3/8 = P(A)P(B)$.

12.10. 1. $P(M) = 2/3, P(\overline{M}) = 1/3$;
$$P(D/M) = 0.7 \Longrightarrow P(\overline{D}/M) = 0.3 \qquad P(D/\overline{M}) = 0.2 \Longrightarrow P(\overline{D}/\overline{M}) = 0.8$$
$$P(F/M) = 0.2 \Longrightarrow P(\overline{F}/M) = 0.8 \qquad P(F/\overline{M}) = 0.9 \Longrightarrow P(\overline{F}/\overline{M}) = 0.1$$

2. We must compute $P(M/D \cap F)$. This conditional probability is equal to
$$P(M/D \cap F) = \frac{P(M \cap D \cap F)}{P(D \cap F)} = \frac{P(M \cap D \cap F)}{P(M \cap D \cap F) + P(\overline{M} \cap D \cap F)}.$$
But $P(M \cap D \cap F) = P(M)P(D \cap F/M)$. The fact that, given that a student likes maths, events D and F are independent is expressed by: $P(D \cap F/M) = P(D/M)P(F/M)$. Hence, $P(M \cap D \cap F) = P(M)P(D \cap F/M) = P(M)P(D/M)P(F/M)$. Similarly, substituting $\overline{M}$ for M, $P(\overline{M} \cap D \cap F) = P(\overline{M})P(D/\overline{M})P(F/\overline{M})$. Hence,
$$P(M/D \cap F) = \frac{P(M)P(D/M)P(F/M)}{P(M)P(D/M)P(F/M) + P(\overline{M})P(D/\overline{M})P(F/\overline{M})}$$
$$= \frac{2/3 \times 0.7 \times 0.2}{2/3 \times 0.7 \times 0.2 + 1/3 \times 0.2 \times 0.9} = 0.936.$$
Furthermore, the independence of the conditional events D and F on the hypothesis that a student likes maths implies
$$P(F/D \cap M) = P((F/M)/(D/M)) = P(F/M)$$
which is the equation corresponding to the independence of the conditional events D/M and F/M.

Chapter 12

12.11. 1. Let Ω be the sample space constituted by the six permutations of the letters a, b, c, together with the three tuples (a, a, a), (b, b, b), (c, c, c), and the uniform probability; each sample point thus has the probability $1/9$. Let A_k be the event: 'letter a appears at the kth place, for $k = 1, 2, 3$'. $P(A_k) = 3/9 = 1/3$, for $k = 1, 2, 3$; moreover $P(A_1 \cap A_2) = P(A_1 \cap A_3) = P(A_3 \cap A_2) = 1/9$; the events A_1, A_2, A_3 are thus pairwise independent, but they are not independent because $P(A_1 \cap A_2 \cap A_3) = 1/9$; thus A_3 is not independent of $A_1 \cap A_2$.

2. Consider the experiment consisting of tossing two dice; $\Omega = \{1, 2, \ldots, 6\}^2$ with uniform probability. Let A_1: 'the first die is a one', A_2: 'the second die is even', A_3: 'the total score of both dice is 7'. $P(A_1) = 1/6$, $P(A_2) = 1/2$, $P(A_3) = 1/6$, $P(A_1 \cap A_2) = 1/12$, $P(A_1 \cap A_3) = 1/36$, $P(A_3 \cap A_2) = 1/12$; the events A_1, A_2, A_3 are thus pairwise independent, but $P(A_1 \cap A_2 \cap A_3) = 1/36 \neq 1/6 \times 1/2 \times 1/6$.

12.12. 1. It is enough to verify that

$$\sum_{\omega = (i, j, k) \in \Omega} P(\omega) = 1.$$

The event '$i + j + k$ is even' consists of the tuples $(0, 0, 0)$, $(0, 1, 1)$, $(1, 0, 1)$ and $(1, 1, 0)$; we are thus left with four tuples each having probability $1/4$.

2.
$$A = \{i = 0\} = \{(0, j, k) \,/\, (j, k) \in \{0, 1\}^2\}$$
$$= \{(0, 0, 0), (0, 0, 1), (0, 1, 0), (0, 1, 1)\};$$

thus $P(A) = 1/2$. Similarly $P(B) = P(C) = 1/2$. Finally, it is easy to see that $P(A \cap B) = P(A \cap C) = P(B \cap C) = 1/4$; the three events $A = \{i = 0\}$, $B = \{j = 0\}$, $C = \{k = 0\}$ are thus pairwise independent.

3. No, because $A \cap B \cap C = (0, 0, 0)$ and thus $P(A \cap B \cap C) = 0 \neq P(A \cap B)P(C) = 1/4 \times 1/2 = 1/8$.

12.13. It is enough to prove that, $\forall \omega = (\omega_1, \ldots, \omega_n) \in \Omega$, $0 \leq P(\omega) \leq P(\Omega)$, which is straightforward, and that $\sum_{\omega \in \Omega} P(\omega) = 1$; but

$$\sum_{\omega \in \Omega} P(\omega) = \sum_{\omega_1 \in \Omega_1} \sum_{\omega_2 \in \Omega_2} \cdots \sum_{\omega_n \in \Omega_n} P_1(\omega_1) \times P_2(\omega_2) \cdots \times P_n(\omega_n)$$
$$= \sum_{\omega_2 \in \Omega_2} \cdots \sum_{\omega_n \in \Omega_n} P_2(\omega_2) \times \cdots \times P_n(\omega_n) \Big(\sum_{\omega_1 \in \Omega_1} P_1(\omega_1) \Big)$$
$$= \sum_{\omega_2 \in \Omega_2} \cdots \sum_{\omega_n \in \Omega_n} P_2(\omega_2) \times \cdots \times P_n(\omega_n) = \cdots$$
$$= \sum_{\omega_n \in \Omega_n} P_n(\omega_n) = 1.$$

12.14. We have $\Omega_k = \{1, 2, \ldots, k\}^n$, $\mathcal{T} = \mathcal{P}(\Omega_k)$ and for $\omega \in \Omega_k$ such that $|\omega_1| = i_1, \ldots, |\omega_k| = i_k$, $P(\omega) = p_1^{i_1} \cdots p_k^{i_k}$.

12.15. 1. Obvious since for $x \leq x'$, $F(x') = F(x) + \sum_{x \leq d < x'} P_X(d)$; since, moreover, $\sum_{x \leq d < x'} P_X(d) \geq 0$ (recall that probabilities are always non-negative), $F(x') \geq F(x)$.

2. We have: $\lim_{x \to \infty} F(x) = \lim_{x \to \infty} \sum_{d < x} P_X(d) = \sum_{d \in D} P_X(d) = P_X(D) = 1$. Suppose to simplify that X assumes values in $\mathbb{Z}$. Then, $\forall \varepsilon$, $\exists N$, $\sum_{d = -N}^{N-1} P_X(d) > 1 - \varepsilon$, and thus $F(-N) = \sum_{d < -N} P_X(d) < \varepsilon$, hence $\lim_{x \to -\infty} F(x) = 0$.

12.16. It suffices to check that $\sum_{(d,d') \in \{0,1\}^2} P_{(X,Y)}(d,d') = 4/4 = 1$, and similarly for Q, and that $\sum_{d \in \{0,1\}} P_X(d) = 1$. But, for $d \in \{0,1\}$, $P_X(d) = \sum_{d' \in \{0,1\}} P_{(X,Y)}(d,d') = P_{(X,Y)}(d,0) + P_{(X,Y)}(d,1) = P(X=d, Y=0) + P(X=d, Y=1) = 1/2$. Thus $\sum_{d \in \{0,1\}} P_X(d) = 2/2 = 1$. Similarly for Q_X.

12.17. Yes, provided that $f(X)$ and $g(Y)$ be defined, and to this end it suffices that f (resp. g) be defined on the image of X (resp. Y). We indeed have

$$P(f(X) = i, g(Y) = j) = \sum_{\substack{x \in f^{-1}(i) \\ y \in g^{-1}(j)}} P(X = x, Y = y)$$

$$= \sum_{\substack{x \in f^{-1}(i) \\ y \in g^{-1}(j)}} P(X = x) P(Y = y)$$

$$= P(g(X) = i) P(g(Y) = j).$$

12.18. 1. The distribution of W can be represented as follows:

x	1	2	3
$P_W(x)$	1/3	1/3	1/3

U can assume the values 2, 3, 4, 5, 6; in order to find $P(U = k)$, we look for the elementary events $(X = i) \cap (Y = j)$ satisfying $U = k$, i.e. such that $i + j = k$; for instance

$$P(U = 3) = P((X = 1) \cap (Y = 2)) + P((X = 2) \cap (Y = 1))$$

(since we have a subdivision into mutually exclusive events)

$$= P(X = 1)P(Y = 2) + P(X = 2)P(Y = 1)$$

(since X and Y are independent)

$$= 2/9.$$

We have the table

x	2	3	4	5	6
$P_U(x)$	1/9	2/9	1/3	2/9	1/9

Similarly, we have

x	-2	-1	0	1	2
$P_V(x)$	1/9	2/9	1/3	2/9	1/9

2. The values of (U, V) depend upon those of the tuple (X, Y, Z) which can assume $27 = 3^3$ distinct values; we find the table:

$P_{(U,V)}$ v	u	2	3	4	5	6
-2		a	a	a	0	0
-1		a	$2a$	$2a$	a	0
0		a	$2a$	$3a$	$2a$	a
1		0	a	$2a$	$2a$	a
2		0	0	a	a	a

where $a = 1/27$.

For instance, to compute $P\big((U = 3) \cap (V = -1)\big)$, note that

$$(U = 3) = (X = 1 \text{ and } Y = 2) \cup (X = 2 \text{ and } Y = 1),$$

and

$$(V = -1) = (X = 1 \text{ and } Z = 2) \cup (X = 2 \text{ and } Z = 3);$$

hence:

$$P\big((U = 3) \cap (V = -1)\big) = P\big((X = 1) \cap (Y = 2) \cap (Z = 2)\big)$$
$$+ P\big((X = 2) \cap (Y = 1) \cap (Z = 3)\big) = 1/27 + 1/27.$$

U and V are not independent, because, e.g.,

$$0 = P\big((U = 5) \cap (V = -2)\big) \neq P(U = 5)P(V = -2) = \frac{2}{9} \times \frac{1}{9}.$$

12.19. Let the r.v.'s $X_1, \ldots, X_n$ assume values in $D_1, \ldots, D_n$; we will say that $X_1, \ldots, X_n$ are independent if and only if for all $A_1 \subseteq D_1, \ldots, A_n \subseteq D_n$, the events $X_1 \in A_1, \ldots, X_n \in A_n$ are independent.

12.20. 1. $A \in \mathcal{T}$; this will occur if $\mathcal{T} = \mathcal{P}(\Omega)$.

2. We then have $E(\chi_A) = \sum_{d \in \mathbb{B}} dP(\chi_A = d) = P(\chi_A = 1) = P(A)$.

12.21. 1. We have $E(U) = 4$, and $E(V) = 0$.

2. $E(UV) = \sum_i \sum_j ij P\big((U = i) \cap (V = j)\big) = 2/3$. On the other hand,

$$\sigma(U) = \sqrt{E(U^2) - (E(U))^2} = \sqrt{156/9 - 16} = (2/3)\sqrt{3},$$
$$\sigma(V) = \sqrt{E(V^2) - (E(V))^2} = \sqrt{E(V^2)} = (2/3)\sqrt{3};$$

hence $\rho(U, V) = 1/2$, and this indeed confirms that U and V are not independent.

12.22. 1.(a) $E(Z) = (1-a)\sum_{k=1}^{\infty} k a^{k-1} = (1-a)\big(\sum_{k=0}^{\infty} a^k\big)' = \dfrac{1-a}{(1-a)^2} = \dfrac{1}{1-a}$. (see Chapter 8.) We can also have a direct computation: $(1-a)\sum_{k=1}^{\infty} ka^{k-1} = \sum_{k=1}^{\infty} ka^{k-1} - \sum_{k=1}^{\infty} ka^k = 1 + \sum_{k=1}^{\infty}(k+1)a^k - \sum_{k=1}^{\infty} ka^k = \sum_{k=1}^{\infty} a^k = \dfrac{1}{1-a}$.

(b) $P(Z \geq k) = a^{k-1}(1-a)\big(\sum_{i=0}^{\infty} a^i\big) = a^{k-1}$.

2.(a) Note that the event $(T \geq k)$ is the disjoint union of the mutually exclusive events $(T = k)$ and $(T \geq k+1)$; we deduce that $P(T \geq k) = P(T = k) + P(T \geq k+1)$.

It suffices to note that
- $(T \geq k) \iff (\inf(X,Y) \geq k) \iff (X \geq k \text{ and } Y \geq k)$,
- X and Y being independent, $P(X \geq u \text{ and } Y \geq v) = P(X \geq u)P(Y \geq v)$; and thus

$$P(T = k) = P(T \geq k) - P(T \geq k+1)$$
$$= P(X \geq k \text{ and } Y \geq k) - P(X \geq k+1 \text{ and } Y \geq k+1)$$
$$= P(X \geq k)P(Y \geq k) - P(X \geq k+1)P(Y \geq k+1)$$

(b) We deduce then from 2 (a) that $p(T = k) = p^{k-1}q^{k-1} - p^k q^k = (pq)^{k-1}(1 - pq)$.

3.(a) Firstly, note that
$$T_n = \inf(X_1, \ldots, X_n) \geq k \iff T_n = \inf(T_{n-1}, X_n) \geq k$$
$$\iff (T_{n-1} \geq k \text{ and } X_n \geq k)$$
Then check that, if $X_1, \ldots, X_n$ are independent, then T_{n-1} and X_n are independent; indeed
$$P(T_{n-1} \geq u \text{ and } X_n \geq v) = P(X_1 \geq u \text{ and } \ldots \text{ and } X_{n-1} \geq u \text{ and } X_n \geq v)$$
$$= P(X_1 \geq u) \cdots P(X_{n-1} \geq u) P(X_n \geq v)$$
$$= P(T_{n-1} \geq u) P(X_n \geq v) .$$

We can then apply 2 and generalize it by induction on n:
- for $n = 1$, $T_1 = X_1$ has a geometric distribution of ratio p;
- for the induction, use 2 and the above two remarks.

(b) By 1 (b) we know that $P(T_n > 1) = P(T_n \geq 2) = p^n$, hence the result.

12.23. Let X be the r.v. which, to each individual in the population, associates its height. Then $E(X) = 1.65$ and $\sigma(X) = 0.04$.

Markov's inequality will give us the rather rough upper bound $P(X \geq 1.80) \leq \dfrac{1.65}{1.80} = 0.916$.

Chebyshev's inequality will gives us the much better upper bound
$$P(X \geq 1.80) \leq P(|X - 1.65| \geq 0.15)$$
$$\leq \left(\frac{0.04}{0.15}\right)^2 = 0.071$$

12.24. We want to determine integers n such that
$$P\left(S_n \leq -\frac{1}{2}\right) \leq \frac{1}{100} .$$
The r.v.'s $X_1, \ldots, X_n$ are independent, and X_i has the same distribution as $-X_i$; thus S_n has the same distribution as $-S_n$, hence, for all $a > 0$
$$P(S_n \leq -a) = P(S_n \geq a) = \frac{1}{2} P(|S_n| \geq a)$$
Moreover, we have
$$E(X_i) = 0 \implies E(S_n) = 0 ,$$
$$var(X_i) = E(X_i^2) - \left(E(X_i)\right)^2 = 1 ,$$
$$X_1, \ldots, X_n \text{ independent} \implies var(X_1 + \cdots + X_n) = n$$
$$\text{by Proposition 12.51} \implies var(S_n) = \frac{1}{n^2} var(X_1 + \cdots + X_n) = \frac{1}{n} .$$
Chebyshev's inequality then gives us the upper bound
$$\forall a > 0, \quad P(|S_n| \geq a) \leq \frac{1}{na^2} ,$$
$$\text{hence} \quad P(S_n \leq -a) \leq \frac{1}{2na^2} ,$$
$$\text{and for } a = \frac{1}{2}, \quad P\left(S_n \leq -\frac{1}{2}\right) \leq \frac{2}{n} ,$$
$$\text{and lastly} \quad P\left(S_n \leq -\frac{1}{2}\right) \leq \frac{1}{100}, \quad \forall n \geq 100 .$$

12.25. $E(X) = d(-1)$; $var(X) = d(-2) - d(-1)^2$; $E(\log X) = -d'(0)$.

Chapter 12

12.26.
$$g_{X+Y}(z) = \sum_{n=0}^{\infty} P(X+Y=n)z^n$$
$$= \sum_{n=0}^{\infty} \left(\sum_{i+j=n} P(X=i, Y=j)z^n \right)$$
$$= \sum_{n=0}^{\infty} \left(\sum_{i+j=n} P(X=i)P(Y=j)z^n \right)$$
(since the r.v.'s X and Y are independent)
$$= \sum_{n=0}^{\infty} \left(\sum_{i+j=n} P(X=i)z^i P(Y=j)z^j \right)$$
$$= \left(\sum_{i=0}^{\infty} P(X=i)z^i \right) \left(\sum_{j=0}^{\infty} P(Y=j)z^j \right)$$
(by the definition of the product of series)
$$= g_X(z)g_Y(z) \ .$$

12.27. 1. $P(S_1 = k) = pq^{k-1}$, $g_{S_1}(z) = \sum_{k=1}^{\infty} pq^{k-1}z^k = \frac{p}{q} \sum_{k=1}^{\infty} (qz)^k = \frac{p}{q} \times \frac{1}{1-qz}$.

2. $S_r = X_1 + \cdots + X_r$, the r.v.'s X_i are mutually independent with the distribution of S_1 as the common distribution, thus $g_{S_r}(z) = (g_{S_1}(z))^r = \left(\frac{p}{q}\right)^r \times \left(\frac{1}{1-qz}\right)^r$.

3. We deduce from 2 that
$$P(S_r = k) = \left(\frac{p}{q}\right)^r \frac{r(r+1)\cdots(r+k-1)}{k!} q^k \ ,$$
$$E(S_r) = g'_{S_r}(1) = rq\left(\frac{p}{q}\right)^r \left(\frac{1}{1-q}\right)^{r+1}$$
$$= \frac{r}{pq^{r-1}} \ ,$$
$$var(S_r) = \frac{r(r+1)}{p^2q^{r-2}} + \frac{r}{pq^{r-1}} - \left(\frac{r}{pq^{r-1}}\right)^2 \ .$$

12.28. 1. The event $V = k$ can be subdivided into the union $\bigcup_{j=1}^{n} \left((U=j) \cap ((\sum_{i=1}^{j} X_i) = k) \right)$. These events being disjoint, we have
$$P(V=k) = \sum_{j=1}^{n} P\left((U=j) \cap \left(\sum_{i=1}^{j} X_i \right) = k \right)$$

U and X_i being independent,
$$P\left((U=j) \cap \left(\sum_{i=1}^{j} X_i \right) = k \right) = P(U=j) \times P\left(\left(\sum_{i=1}^{j} X_i \right) = k \right)$$

2. First note that, by Proposition 12.62 and the independence of the X_is, we can prove by induction on j that $\forall j = 1, \ldots, n$, the generating function of the r.v. $\sum_{i=1}^{j} X_i$ is given by: $g_{\sum_{i=1}^{j} X_i}(z) = (g(z))^j$. Compute the generating function of V:

$$g_V(z) = \sum_{k \in \mathbb{N}} P(V = k) z^k$$

$$= \sum_{k \in \mathbb{N}} \left(\sum_{j=1}^{n} P(U = j) \times P\left(\left(\sum_{i=1}^{j} X_i \right) = k \right) z^k \right)$$

$$= \sum_{j=1}^{n} P(U = j) \left(\sum_{k \in \mathbb{N}} P\left(\left(\sum_{i=1}^{j} X_i \right) = k \right) z^k \right)$$

$$= \sum_{j=1}^{n} P(U = j) g_{\sum_{i=1}^{j} X_i}(z)$$

$$= \sum_{j=1}^{n} P(U = j)(g(z))^j = f(g(z)).$$

3. Since $g(1) = 1$, $E(V) = g'_V(1) = g'(1) f'(g(1)) = E(X_i) E(U)$;
$$g''_V(z) = g''(z) f'(g(z)) + (g'(z))^2 f''(g(z));$$
hence $\text{var}(V) = \text{var}(U)(E(X_i))^2 + E(U) \text{var}(X_i)$.

12.29. 1. The generating functions of X and Y are given by

$$g_X(z) = \sum_{k=0}^{m} \binom{m}{k} p^k q^{m-k} z^k = (pz + q)^m,$$

$$g_Y(z) = \sum_{k=0}^{n} \binom{n}{k} p^k q^{n-k} z^k = (pz + q)^n.$$

and, since X and Y are independent,

$$g_{X+Y}(z) = g_X(z) g_Y(z) = (pz + q)^{n+m}.$$

$X + Y$ thus has a binomial distribution $b(p, m+n)$ of parameters $(p, m+n)$.

2. For $x \in \{1, 2, \ldots, s\}$, we have

$$P(X = x \,/\, S = s) = \frac{P(X = x, X + Y = s)}{P(X + Y = s)} = \frac{P(X = x, Y = s - x)}{P(X + Y = s)}$$

$$= \frac{P(X = x) P(Y = s - x)}{P(X + Y = s)}$$

(since X and Y are independent)

$$= \frac{\binom{m}{x} p^x q^{m-x} \binom{n}{s-x} p^{s-x} q^{n-s+x}}{\binom{m+n}{s} p^s q^{m+n-s}} = \frac{\binom{m}{x} \binom{n}{s-x}}{\binom{m+n}{s}}$$

for $x > s$, $P(X = x \,/\, S = s) = 0$.

The conditional distribution of X given S is thus a hypergeometric distribution (see Section 12.6.4).

12.30. Each sequence $(x_1, \ldots, x_n)$ containing k people of type 1 has probability $p^k q^{n-k}$ of being obtained. Moreover, there are $n!$ ways of permuting $(x_1, \ldots, x_n)$, but among the $n!$ ways, the $k!$ permutations of the k people of type 1 and the $(n-k)!$ permutations of the $n-k$ people of type 0 give the same result, hence: $P(S = k) = \dfrac{n!}{k!(n-k)!} p^k q^{n-k} = \binom{n}{k} p^k q^{n-k}$.

12.31. 1. To study the distribution of S_i, we can group together all the values $j = 1, \ldots, n$, $j \neq i$, in a single value d_i; then S_i has the binomial distribution $b(p_i, n)$.
$\Gamma(S_1, S_2) = E(S_1 S_2) - E(S_1)E(S_2)$, where S_1 and S_2 have binomial distributions with respective parameters p_1 and $p_2 = 1 - p_1$; we want to compute $E(S_1 S_2)$. Consider, for $i = 1, \ldots, n$, the characteristic functions $\chi_i^1 = \chi_{(X_i = 1)}$ and $\chi_i^2 = \chi_{(X_i = 2)}$, and note that $S_1 = \sum_{i=1}^n \chi_i^1$ and $S_2 = \sum_{j=1}^n \chi_j^2$; hence

$$E(S_1 S_2) = E\left(\left(\sum_{i=1}^n \chi_i^1\right)\left(\sum_{j=1}^n \chi_j^2\right)\right)$$

$$= E\left(\sum_{i=1}^n \sum_{j=1}^n \chi_i^1 \chi_j^2\right)$$

$$= n(n-1) p_1 p_2$$

and $\quad \Gamma(S_1, S_2) = n(n-1) p_1 p_2 - n^2 p_1 p_2 = -n p_1 p_2$

2. Let $(X_1, \ldots, X_p)$ be discrete r.v.'s; we can generalize the generating functions in order to represent the distribution of $(X_1, \ldots, X_p)$. Let $\overline{X} = (X_1, \ldots, X_p)$ and

$$g_{\overline{X}}(z_1, \ldots, z_p) = \sum_{i_1, \ldots, i_p} P(X_1 = i_1, \ldots, X_p = i_p) z_1^{i_1} \cdots z_p^{i_p} = E\left((z_1^{X_1} \cdots z_p^{X_p})\right).$$

Therefore,

$$g_{(S_1, \ldots, S_r)}(z_1, \ldots, z_r) = \sum_{i_1, \ldots, i_r} P(S_1 = i_1, \ldots, S_r = i_r) z_1^{i_1} \cdots z_r^{i_r}$$

$$= E\left((z_1^{S_1} \cdots z_r^{S_r})\right)$$

and, noting that $S_k = \sum_{i=1}^n \chi_{(X_i = k)}$,

$$g_{(S_1, \ldots, S_r)}(z_1, \ldots, z_r) = E\left((z_1^{S_1} \cdots z_r^{S_r})\right)$$

$$= E\left(\left(z_1^{\sum_{i=1}^n \chi(X_i=1)} \cdots z_r^{\sum_{i=1}^n \chi(X_i=r)}\right)\right)$$

$$= E\left(\prod_{j=1}^n (z_1^{\chi(X_j=1)} \cdots z_r^{\chi(X_j=r)})\right)$$

$$= \prod_{j=1}^n \left(E(z_1^{\chi(X_j=1)} \cdots z_r^{\chi(X_j=r)})\right)$$

(since the X_js are independent)

$$= \prod_{j=1}^n (p_1 z_1 + \cdots + p_r z_r) = (p_1 z_1 + \cdots + p_r z_r)^n$$

We can easily prove the following properties of generating functions depending on many variables:

- $g_{X_i}(z) = g_{\overline{X}}(1, \ldots, 1, z, 1, \ldots, 1)$, where $z_1 = \cdots = z_{i-1} = z_{i+1} = \cdots = z_p = 1$,
- $g_{X_1 + \cdots + X_p}(z) = g_{\overline{X}}(z, \ldots, z)$,
- $(X_1, \ldots, X_p)$ are independent if and only if $g_{\overline{X}}(z_1, \ldots, z_p) = \Pi_{k=1}^p g_{X_k}(z_k)$.

12.32. $X: \Omega \longrightarrow \mathbb{N}$, where Ω is the set of sequences $\{1^n 0 \, / \, n \in \mathbb{N}\} \cup \{111\cdots 11\cdots\}$, with the convention $1^0 0 = 0$, and with $\mathcal{T} = \mathcal{P}(\Omega)$, $\forall n \in \mathbb{N}$, $P(1^n 0) = p^n q$ and $P(111\cdots 11\cdots) = 0$.

12.33. 1. $1/2^{n-1}$.

2.
$$\sum_{i=2}^{5} \frac{1}{2^{i-1}} = \sum_{i=1}^{4} \frac{1}{2^i} = \frac{1/2(1-(1/2)^4)}{1/2} = \frac{15}{16}.$$

3. Let $A =$ 'an even number of tosses is necessary' and $B =$ 'an odd number of tosses is necessary'. We have $P(A) + P(B) = 1$, and

$$P(A) = \frac{1}{2} + \frac{1}{2^3} + \frac{1}{2^5} + \cdots = \sum_{n=1}^{\infty} \frac{1}{2^{2n-1}},$$

$$P(B) = \frac{1}{2^2} + \frac{1}{2^4} + \frac{1}{2^6} + \cdots = \frac{1}{2} P(A) = \sum_{n=1}^{\infty} \frac{1}{2^{2n}}.$$

Hence $P(A) = 2/3$.

12.34. The methods are similar to the ones used in Exercise 12.31. Note, however, that here the χ_i are not independent.

1. We will apply Proposition 12.56. We have

(i) $\qquad E(S_1) = E(\chi_1) + \cdots + E(\chi_n)$

(ii) $\qquad var(S_1) = var(\chi_1) + \cdots + var(\chi_n) + 2 \sum_{1 \leq i < j \leq n} \Gamma(\chi_i, \chi_j)$

Noting that

$$\forall i = 1, \ldots, n, \qquad E(\chi_i) = P(\chi_i = 1) = \frac{n_1}{N} = p_1,$$
$$\forall i = 1, \ldots, n, \qquad var(\chi_i) = E((\chi_i)^2) - E(\chi_i)^2 = p_1 - p_1^2 = p_1 p_2,$$
$$E(\chi_i, \chi_j) = P(\chi_i = 1, \chi_j = 1) = \frac{n_1(n_1 - 1)}{N(N-1)},$$
$$\Gamma(\chi_i, \chi_j) = \frac{n_1(n_1-1)}{N(N-1)} - \left(\frac{n_1}{N}\right)^2 = \frac{-n_1 n_2}{N^2(N-1)}.$$

Then applying (i) and (ii) we deduce

$$E(S_1) = np_1, \quad var(S_1) = np_1 p_2 \left(\frac{N-n}{N-1}\right).$$

2. Grouping together all types which are different from j reduces the problem to case 1.

12.35. Let

$$F(k) = P(S_1 = k) = \binom{n}{k} \frac{\binom{N-n}{n_1 - k}}{\binom{N}{n_1}}.$$

Chapter 12

We have
$$\frac{n_1}{N} \to 0, \ \frac{n}{N} \to 0, \ \text{and} \ F(0) = \frac{(N-n_1)(N-n_1-1)\cdots(N-n_1-n+1)}{N(N-1)\cdots(N-n+1)};$$
hence
$$\log F(0) = \log\left(1 - \frac{n_1}{N}\right) + \log\left(1 - \frac{n_1}{N-1}\right) + \cdots + \log\left(1 - \frac{n_1}{N-n+1}\right),$$
$$n \log\left(1 - \frac{n_1}{N}\right) \leq \log F(0) \leq n \log\left(1 - \frac{n_1}{N-n+1}\right).$$

Taking the limits, the right-hand side and left-hand side of the inequality go to $-\lambda$; thus $\log F(0) \to -\lambda$ and $F(0) \to e^{-\lambda}$. For $k \in \mathbb{N}$, $k \neq 0$, the ratio
$$\frac{F(k+1)}{F(k)} = \frac{(n-k)(n_1-k)}{(k+1)(N-n_1-n+k)}$$
is equivalent to $\dfrac{nn_1}{(k+1)N} \sim \dfrac{\lambda}{k+1}$; thus $F(k) \to e^{-\lambda}\dfrac{\lambda^k}{k!}$.

12.36.

1.
$$\sum_{n \leq m} P(X = n, Y = m) = \sum_{m=0}^{\infty} \left(\sum_{n=0}^{m} \frac{m!}{n!(m-n)!}\right) \frac{\lambda^m}{m!} e^{-2\lambda}$$
$$= \sum_{m=0}^{\infty} 2^m \frac{\lambda^m}{m!} e^{-2\lambda}$$
$$= e^{2\lambda} e^{-2\lambda} = 1.$$

The distribution of (X, Y) is thus entirely determined by the given probabilities.

2.
$$P(X = n) = \sum_{m=0}^{\infty} P(X = n, Y = m)$$
$$= \sum_{m \geq n} \frac{\lambda^m}{n!(m-n)!} e^{-2\lambda}$$
$$= \frac{\lambda^n}{n!} e^{-2\lambda} \left(\sum_{i=0}^{\infty} \frac{\lambda^i}{i!}\right)$$
$$= \frac{\lambda^n}{n!} e^{-\lambda}.$$

X thus has a Poisson distribution with mean λ.
$$P(Y = m) = \sum_{n=0}^{\infty} P(X = n, Y = m)$$
$$= \sum_{n=0}^{m} \frac{\lambda^m}{n!(m-n)!} e^{-2\lambda}$$
$$= \frac{\lambda^m}{m!} e^{-2\lambda} \left(\sum_{n=0}^{m} \frac{m!}{n!(m-n)!}\right)$$
$$= \frac{(2\lambda)^m}{m!} e^{-2\lambda}.$$

Y thus has a Poisson distribution with mean 2λ.

$$P(X=0, Y=0) = e^{-2\lambda},$$
$$P(X=0)P(Y=0) = e^{-\lambda}e^{-2\lambda} = e^{-3\lambda}.$$

Thus X and Y are not independent.

3. Note that $Y \geq X$, therefore $Y - X$ is integral-valued.

$$P(Y - X = r) = \sum_{n=0}^{\infty} P(X = n, Y = n + r)$$
$$= \left(\sum_{n=0}^{\infty} \frac{\lambda^{n+r}}{n!r!}\right) e^{-2\lambda}$$
$$= \frac{\lambda^r}{r!} e^{-\lambda}.$$

$Y - X$ thus has a Poisson distribution with mean λ.

$$P(X = n, Y - X = r) = P(X = n, Y = n + r)$$
$$= \frac{\lambda^{n+r}}{n!r!} e^{-2\lambda}$$
$$= P(X = n)P(Y - X = r).$$

Hence, the r.v.'s X and $Y - X$ thus are independent.

4.
$$\Gamma(X, Y) = \Gamma(X, X + Y - X)$$
$$= \Gamma(X, X) + \Gamma(X, Y - X)$$
$$= \Gamma(X, X) \quad \text{(because } X \text{ and } Y - X \text{ are independent)}$$
$$= \sigma^2(X) = \lambda.$$
$$\rho(X, Y) = \frac{\Gamma(X, Y)}{\sigma(X)\sigma(Y)} = \frac{\lambda}{\sqrt{\lambda}\sqrt{2\lambda}} = \frac{\sqrt{2}}{2}.$$

12.37. 1. $P(Z = z, N = n) = P(N = n)P(Z = z \,/\, N = n)$. Moreover the r.v. Z conditioned by N has a binomial distribution, i.e. $P(Z = z \,/\, N = n) = \binom{n}{z}p^z(1-p)^{n-z}$; hence

$$P(Z = z, N = n) = e^{-\lambda}\frac{\lambda^n}{n!}\binom{n}{z}p^z(1-p)^{n-z} = A(n, z).$$

2.
$$P(Z = z) = \sum_{n=z}^{\infty} P(Z = z, N = n) = \sum_{n=z}^{\infty} A(n, z)$$
$$= e^{-\lambda}\frac{(\lambda p)^z}{z!}\left(\sum_{n=0}^{\infty} \lambda^{n-z}\frac{(1-p)^{n-z}}{(n-z)!}\right)$$
$$= e^{-\lambda}\frac{(\lambda p)^z}{z!}e^{\lambda(1-p)}.$$

Thus Z has a Poisson distribution with mean λp.

3. $P(N = n \,/\, Z = z) = \dfrac{P(N = n, Z = z)}{P(Z = z)}$. We deduce that the distribution of N conditioned by Z is a Poisson distribution with mean $\lambda(1 - p)$.

4.
$$\rho(N, Z) = \frac{\Gamma(N, Z)}{\sigma(N)\sigma(Z)} = \frac{E(NZ) - E(N)E(Z)}{\sigma(N)\sigma(Z)}.$$

All the factors are known, except for $E(NZ)$.

$$E(NZ) = \sum_{n=1}^{\infty} \sum_{z=0}^{n} nz P(N = n, Z = z)$$

$$= \sum_{n=1}^{\infty} n e^{-\lambda} \frac{\lambda^n}{n!} \underbrace{\left(\sum_{z=0}^{n} z \frac{n!}{z!(n-z)!} p^z (1-p)^{n-z} \right)}_{\text{mean of a binomial distribution}}$$

$$= \sum_{n=1}^{\infty} n e^{-\lambda} \frac{\lambda^n}{n!} np$$

$$= p \sum_{n=1}^{\infty} n^2 e^{-\lambda} \frac{\lambda^n}{n!} = pE(N^2)$$

Since N has a Poisson distribution with mean λ, we have

$$\sigma^2(N) = E(N^2) - E(N)^2 = \lambda, \quad \text{hence}$$
$$E(N^2) = \lambda + \lambda^2, \quad \text{thus}$$
$$E(NZ) = \lambda p(1 + \lambda);$$
$$E(N) = \lambda; \quad E(Z) = \lambda p;$$
$$\sigma(N) = \sqrt{\lambda}; \quad \sigma(Z) = \sqrt{\lambda p}; \quad \text{hence}$$
$$\rho(N, Z) = \frac{p\lambda + p\lambda^2 - p\lambda^2}{\lambda \sqrt{p}} = \sqrt{p}.$$

Chapter 13

13.1. We check by induction on n that the distribution of each X_n is entirely determined by conditions 1 and 2.

13.2. 1. Let $n \geq 1$. Assume that for $k \leq n - 1$ the probability distribution of X_k is given by

$$P(X_k = \text{'yes'}) = P(X_k = \text{'no'}) = p = \frac{1}{2}.$$

Then
$$P(X_n = \text{'yes'}) = P(X_n = \text{'yes' and } X_{n-1} = \text{'yes'})$$
$$+ P(X_n = \text{'yes' and } X_{n-1} = \text{'no'})$$
$$= P(X_n = \text{'yes'}/X_{n-1} = \text{'yes'})P(X_{n-1} = \text{'yes'})$$
$$+ P(X_n = \text{'yes'}/X_{n-1} = \text{'no'})P(X_{n-1} = \text{'no'})$$
$$= \frac{1}{2}p_{11} + \frac{1}{2}p_{21} = \frac{1}{2}p + \frac{1}{2}(1-p) = \frac{1}{2}.$$

Hence we also have $P(X_n = \text{'no'}) = 1 - 1/2 = 1/2$.

We deduce, applying complete induction (Chapter 3, rule (I'_{n_0}) with $n_0 = 1$) that all the X_ns have the same distribution as X_0.

2. No: we have $P(X_n = \text{'yes'}) = 1/2$ whilst $P(X_n = \text{'yes'}/X_{n-1} = \text{'yes'}) = p$. In order for X_n and X_{n-1} to be independent $p = 1/2$ must hold. This necessary condition is also sufficient (see Exercise 13.6).

13.3. 1. $c = 0$. The condition is necessary (except if $b = 0$ or $r = 0$) by Example 13.3. It is clearly sufficient.

2. Yes. Indeed, we have, by Exercise 12.5: $P(X^{n+1} = B) = (b/(b+r))$, and in this case we will have $Y_{n+1} = Y_n + c$ and $P(X^{n+1} = R) = (r/(b+r))$, and in that case we will have $Y_{n+1} = Y_n$. Letting $Y_n = k$, we deduce:

$$P(Y_{n+1} = m / Y_n = k) = \begin{cases} (b/(b+r)) & \text{if } m = k + c, \\ (r/(b+r)) & \text{if } m = k, \\ 0 & \text{otherwise.} \end{cases}$$

13.4. 1. Assume that the turrets are numbered from 1 to 4. The system is in state i at time n, i.e. $X_n = i$ if the nth move of the sentry led him to the ith turret ($i = 1, \ldots, 4$).

2. X_n is a Markov chain because, for all $n > 0$,

$$P(X_n = i) = \begin{cases} 1/2 & \text{if } X_{n-1} \equiv (i-1)\,[4] \text{ or } X_{n-1} \equiv (i+1)\,[4]\,. \\ 0 & \text{otherwise.} \end{cases}$$

Its transition matrix is
$$P = \begin{pmatrix} 0 & 1/2 & 0 & 1/2 \\ 1/2 & 0 & 1/2 & 0 \\ 0 & 1/2 & 0 & 1/2 \\ 1/2 & 0 & 1/2 & 0 \end{pmatrix}.$$

13.5.

1. $$P(X_{n+1} = i) = \sum_{j=1}^{r} P(X_{n+1} = i \text{ and } X_n = j)$$
$$= \sum_{j=1}^{r} P(X_{n+1} = i/X_n = j)P(X_n = j)$$
$$= \sum_{j=1}^{r} p_{ji} P(X_n = j)\,.$$

Hence, $L_{n+1} = P^t \times L_n$, where P^t is the transpose matrix of P.

2. We deduce by induction on n that $L_{n+1} = (P^t)^{n+1} L_0$, $\forall n \geq 0$.

3. If all the X_ns have the same distribution, we have in particular $L_1 = P^t L_0 = L_0$. L_0 is thus an eigenvector of the matrix P^t for the eigenvalue 1. Conversely, if the probability distribution L_0 of X_0 is an eigenvector associated with the eigenvalue 1 of the matrix P^t, we have $L_1 = P^t L_0 = L_0$ and, by induction on n, $L_n = L_0$ for all n.

13.6. 1. (i) $\implies$ (ii) Let j be fixed column. X_k and X_{k+1} being independent, $\forall i$, $p_{ij} = P(X_{k+1} = j/X_k = i) = P(X_{k+1} = j)$.

(ii) $\implies$ (iii) For any k, $\forall i,j : P(X_{k+1} = j/X_k = i) = p_{ij} = p_j$. Moreover,

$$P(X_{k+1} = j) = \sum_i P(X_{k+1} = j \text{ and } X_k = i)$$

$$= \sum_i P(X_{k+1} = j/X_k = i) \times P(X_k = i)$$

$$= \sum_i p_{ij} P(X_k = i) = \sum_i p_j P(X_k = i) = p_j \sum_i P(X_k = i)$$

$$= p_j.$$

Thus, for all k,
$$P(X_{k+1} = j/X_k = i) = P(X_{k+1} = j),$$
i.e. X_{k+1} and X_k are independent.

(iii) $\implies$ (iv) Assume that for all k, X_k and X_{k+1} are independent and let us show that $\forall n \geq 1$, $(X_0, \ldots, X_n)$ are independent.

(B) The result holds for $n = 1$.

(I) Assume $(X_0, \ldots, X_n)$ are independent, and let us compute

$$P(X_0 = i_0, \ldots, X_n = i_n, X_{n+1} = i_{n+1})$$
$$= P(X_{n+1} = i_{n+1}/X_0 = i_0, \ldots, X_n = i_n)$$
$$\times P(X_0 = i_0, \ldots, X_n = i_n)$$
$$= P(X_{n+1} = i_{n+1}/X_n = i_n) P(X_0 = i_0, \ldots, X_n = i_n)$$
by the Markov property
$$= P(X_{n+1} = i_{n+1}/X_n = i_n) P(X_n = i_n) \cdots P(X_0 = i_0)$$
by the independence of $(X_0, \ldots, X_n)$
$$= P(X_{n+1} = i_{n+1}) P(X_n = i_n) \cdots P(X_0 = i_0)$$
by the independence of X_n and X_{n+1}.

$(X_0, \ldots, X_n, X_{n+1})$ are thus independent, hence the induction hypothesis and the result.

(iv) $\implies$ (i) Straightforward.

2. The distribution of X_n is given by $\forall j$, $P(X_n = j) = p_j$. All the X_ns thus have the same distribution, given by the row of the matrix P and independent of the probability distribution of X_0.

13.7. 1 and 2 are quite easy to verify by computations using the Markov property. Let us check the last equality, which is slightly more complex.

3. Let $B_n = (X_{n+1} \in A_1, \ldots, X_{n+k} \in A_k)$, $C_n = (X_n = E_{i_k})$, and $A = (X_0 \in A'_0, \ldots, X_{n-1} \in A'_{n-1})$; note that we can decompose A in a partition $A = \sum_{i \in I} A_i$, where each A_i is of the form $(X_0 = E_{i_0}, \ldots, X_{n-1} = E_{i_{n-1}})$. We then have

$$P(X_{n+1} \in A_1, \ldots, X_{n+k} \in A_k \, / \, X_0 \in A'_0, \ldots, X_{n-1} \in A'_{n-1}, X_n = E_{i_k})$$
$$= P(B_n \, / \, C_n \cap A) = P(B_n \, / \, C_n \cap \sum_{i \in I} A_i)$$
$$= \frac{\sum_{i \in I} P(B_n \cap C_n \cap A_i)}{\sum_{i \in I} P(C_n \cap A_i)}$$
$$= \frac{\sum_{i \in I} P(B_n \, / \, C_n \cap A_i) \times P(C_n \cap A_i)}{\sum_{i \in I} P(C_n \cap A_i)}.$$

By 2, we have $P(B_n \,/\, C_n \cap A_i) = P(B_0 \,/\, C_0)$, and hence

$$P(X_{n+1} \in A_1, \ldots, X_{n+k} \in A_k \,/\, X_0 \in A'_0, \ldots, X_{n-1} \in A'_{n-1}, X_n = E_{i_k})$$
$$= P(B_0 \,/\, C_0) \times \frac{\sum_{i \in I} P(C_n \cap A_i)}{\sum_{i \in I} P(C_n \cap A_i)}$$
$$= P(B_0 \,/\, C_0)$$
$$= P(X_1 \in A_1, \ldots, X_k \in A_k \,/\, X_0 = E_{i_k}).$$

13.8. Decompose the event $E_{ijn} =$ 'the system starting from E_i is in E_j after n steps' in the k disjoint events, $E_{ijn}^k = A_{ij}^k \cap E_{jj(n-k)}$, $k = 1, \ldots, n$, where

- $A_{ij}^k =$'the system starting from E_i reaches E_j for the first time after k steps' and
- $E_{jj(n-k)} =$'the system then goes E_j to E_j in $n-k$ steps'.

We have $p_{ij}^{(n)} = P(E_{ijn}) = \sum_{k=1}^n P(E_{ijn}^k)$ with

$$P(E_{ijn}^k) = P(A_{ij}^k \cap E_{jj(n-k)}) = P(E_{jj(n-k)} \,/\, A_{ij}^k) P(A_{ij}^k),$$

and by the Markov chain property, $P(E_{ijn}^k) = p_{jj}^{(n-k)} f_{ij}^{(k)}$.

13.9. Straightforward by noting that $P(X_0 = X_1 = \cdots = X_n = E_i \,/\, X_0 = E_i) = p_{ii}^n$.

13.10. If E_i is absorbing, then $p_{ii} = f_{ii}^{(1)} = f_{ii} = 1$, and thus E_i is persistent.

13.11. 1.

$$E(N_j \,/\, X_0 = i) = E\left(\sum_{n=0}^\infty \delta_{nj} \,/\, X_0 = i\right)$$
$$= \sum_{n=0}^\infty E(\delta_{nj} \,/\, X_0 = i)$$
$$= \sum_{n=0}^\infty \left(0 \times P(\delta_{nj} = 0 \,/\, X_0 = i) + 1 \times P(\delta_{nj} = 1 \,/\, X_0 = i)\right)$$
$$= \sum_{n=0}^\infty \left(0 \times (1 - p_{ij}^{(n)}) + 1 \times p_{ij}^{(n)}\right)$$
$$= \sum_{n=0}^\infty p_{ij}^{(n)}$$
$$= u_{ij}$$

2.

$$u_{ij} = E(N_j \,/\, X_0 = i)$$
$$= \sum_{k=0}^\infty P(N_j = k \,/\, X_0 = i)$$
$$= \sum_{k=0}^\infty f_{ij} P(N_j = k \,/\, X_0 = j) \qquad (15.2)$$
$$= f_{ij} \sum_{k=0}^\infty P(N_j = k \,/\, X_0 = j)$$
$$= f_{ij} E(N_j \,/\, X_0 = j) = f_{ij} u_{jj},$$

where (15.2) follows from (13.6). A direct computation using the distribution of N_j given that $X_0 = j$ yields $E(N_j \,/\, X_0 = j) = \sum_{k=0}^{\infty} k(1-f_{jj})f_{jj}^{k-1} = \dfrac{1}{1-f_{jj}} = u_{jj}$.

13.12. 1. X_n ranges over the values 0, 1, and 2.
Let B_1^l (resp. B_2^l) be the event: 'draw a black ball from U_1 (resp. U_2) given that U_1 (resp. U_2) contains l black balls'. Let $\overline{B}_1^l$ (resp. $\overline{B}_2^l$) be the complementary event, i.e. 'draw a non-black ball from U_1 (resp. U_2) given that U_1 (resp. U_2) contains l black balls'. We have, for $k, l \in \{0, 1, 2\}$:

$$P(X_n = k \,/\, X_{n-1} = l) = 0 \quad \text{if } |k-l| \geq 2,$$
$$P(X_n = l+1 \,/\, X_{n-1} = l) = P(\overline{B}_1^l) \times P(B_2^{2-l}) = \frac{5-l}{5} \times \frac{2-l}{5},$$
$$P(X_n = l \,/\, X_{n-1} = l) = P(B_1^l) \times P(B_2^{2-l}) + P(\overline{B}_1^l) \times P(\overline{B}_2^{2-l})$$
$$= \frac{l}{5} \times \frac{2-l}{5} + \frac{5-l}{5} \times \frac{5-(2-l)}{5},$$
$$P(X_n = l-1 \,/\, X_{n-1} = l) = P(B_1^l) \times P(\overline{B}_2^{2-l}) = \frac{l}{5} \times \frac{5-(2-l)}{5}.$$

The X_ns form a Markov chain because the above probabilities are independent of n.
2. We obtain the matrix

$$P = \begin{array}{c} 0 \\ 1 \\ 2 \end{array} \begin{pmatrix} \overset{0}{15/25} & \overset{1}{10/25} & \overset{2}{0} \\ 4/25 & 17/25 & 4/25 \\ 0 & 10/25 & 15/25 \end{pmatrix}$$

and the graph (Figure 15.12)

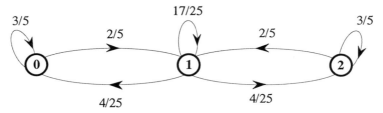

Figure 15.12

The chain is thus irreducible and all its states are persistent.
3. Returning to the result of Exercise 13.5, we deduce that all the X_ns have the same probability distribution if the vector

$$L_0 = \begin{pmatrix} P(X_0 = 0) \\ P(X_0 = 1) \\ P(X_0 = 2) \end{pmatrix}$$

is an eigenvector of the matrix P^t associated with the eigenvalue 1, i.e. if

$$P(X_0 = 0) = 2/9, \ P(X_0 = 1) = 5/9 \ \text{and} \ P(X_0 = 2) = 2/9.$$

13.13. 1. Each urn contains a fixed number (nine) of balls. The total number of black (resp. red) balls is also invariant. Assume that at step n urn U_1 contains k ($0 \leq k \leq 4$) black balls. Then at step n, U_2 will contain $l = 4 - k$ black balls and for all $n \geq 0$, we will have

$$P(X_{n+1} = j \,/\, X_n = k) = \begin{cases} \dfrac{9-k}{9} \times \dfrac{4-k}{9} & \text{if } j = k+1, \\ \dfrac{k}{9} \times \dfrac{4-k}{9} + \dfrac{9-k}{9} \times \dfrac{5+k}{9} & \text{if } j = k, \\ \dfrac{k}{9} \times \dfrac{5+k}{9} & \text{if } j = k-1, \\ 0 & \text{otherwise.} \end{cases}$$

Thus the X_ns indeed form a Markov chain because the same values will be obtained when computing $P(X_{n+1} = j \,/\, X_n = k, X_{n-1} = l_{n-1}, \ldots, X_0 = l_0)$.

The states of the chain are $\{0, 1, 2, 3, 4\}$ and the initial state satisfies

$$q_i = P(X_0 = i) = \begin{cases} 1 & \text{if } i = 2, \\ 0 & \text{if } i \neq 2. \end{cases}$$

2.
$$P = \frac{1}{81} \begin{pmatrix} 45 & 36 & 0 & 0 & 0 \\ 6 & 51 & 24 & 0 & 0 \\ 0 & 14 & 53 & 14 & 0 \\ 0 & 0 & 24 & 51 & 6 \\ 0 & 0 & 0 & 36 & 45 \end{pmatrix}$$

3. The graph is shown in Figure 15.13.

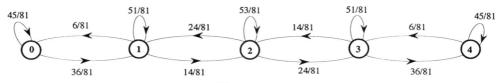

Figure 15.13

The chain is irreducible.

4.
- $L(0) = (0, 0, 1, 0, 0)$.
- $L(1) = \dfrac{1}{81}(0, 14, 53, 14, 0)$.
- $L(2) = \dfrac{1}{(81)^2}(6 \times 14, 14(51 + 53), 2 \times 24 \times 14 + 53^2, 14(53 + 51), 6 \times 14)$.

13.14. The graph is shown in Figure 15.14.

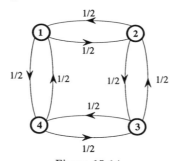

Figure 15.14

Chapter 13 389

The chain is irreducible and all its states are persistent; hence there can be no transient or absorbing states.

13.15. Let $T = \{E_1, \ldots, E_k\}$ be the set of transient states, and let $C_1 = \{E_{k+1}, \ldots, E_r\}$ be the irreducible closed set of persistent states. We note that the transition matrix P can be written as

$$P = \begin{pmatrix} M & P_1 \\ 0 & V_1 \end{pmatrix},$$

where M is a square $k \times k$ matrix representing the transition probabilities restricted to T, and P_1 is the matrix with k rows and $r - k$ columns defined by

$$M = \begin{pmatrix} p_{1,k+1} & p_{1,k+2} & \cdots & p_{1,k+r} \\ p_{2,k+1} & p_{2,k+2} & \cdots & p_{2,k+r} \\ \vdots & \vdots & \ddots & \vdots \\ p_{k,k+1} & p_{k,k+2} & \cdots & p_{k,k+r} \end{pmatrix}.$$

Note that the matrix M is not stochastic, but verifies

$$\forall i, j, \quad p_{ij} \geq 0 \quad \text{and} \quad \forall i, \quad \sum_{j=1}^{k} p_{ij} \leq 1. \tag{15.3}$$

We say that M is substochastic.

The equation system (13.7) is equivalent to

$$\begin{pmatrix} \lambda_1 \\ \lambda_2 \\ \vdots \\ \lambda_k \end{pmatrix} = M \times \begin{pmatrix} \lambda_1 \\ \lambda_2 \\ \vdots \\ \lambda_k \end{pmatrix} + \begin{pmatrix} p_{1,k+1} + p_{1,k+2} + \cdots + p_{1,k+r} \\ p_{2,k+1} + p_{2,k+2} + \cdots + p_{2,k+r} \\ \vdots \\ p_{k,k+1} + p_{k,k+2} + \cdots + p_{k,k+r} \end{pmatrix} = M \times \begin{pmatrix} \lambda_1 \\ \lambda_2 \\ \vdots \\ \lambda_k \end{pmatrix} + P_1 \times \begin{pmatrix} 1 \\ 1 \\ \vdots \\ 1 \end{pmatrix}. \tag{15.4}$$

Equations (13.1) tell us that

$$M \times \begin{pmatrix} 1 \\ 1 \\ \vdots \\ 1 \end{pmatrix} + P_1 \times \begin{pmatrix} 1 \\ 1 \\ \vdots \\ 1 \end{pmatrix} = (M \quad P_1) \times \begin{pmatrix} 1 \\ 1 \\ \vdots \\ 1 \end{pmatrix} = \begin{pmatrix} 1 \\ 1 \\ \vdots \\ 1 \end{pmatrix}$$

and hence that $\lambda_1 = \lambda_2 = \cdots = \lambda_k = 1$ is a solution of (15.4).

13.16. 1.

$$P = \begin{pmatrix} & 1 & 2 & a & s \\ 1 & q & p & r & 0 \\ 2 & 0 & q & r & p \\ a & 0 & 0 & 1 & 0 \\ s & 0 & 0 & 0 & 1 \end{pmatrix}$$

The graph is shown in Figure 15.15 below.

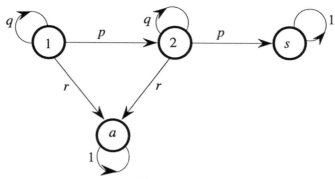

Figure 15.15

2. The chain is not irreducible; the transient states are 1 and 2 and the persistent states are a and s. The persistent states are also absorbing.

3. $\lambda_{i,t}^{n+1} = \sum_{j=1,2} p_{ij} \lambda_{j,t}^{n}$, and hence:

$$\lambda_{1,a}^{n+1} = q\lambda_{1,a}^{n} + p\lambda_{2,a}^{n},$$
$$\lambda_{2,a}^{n+1} = q\lambda_{2,a}^{n}.$$

Similarly:

$$\lambda_{1,s}^{n+1} = q\lambda_{1,s}^{n} + p\lambda_{2,s}^{n},$$
$$\lambda_{2,s}^{n+1} = q\lambda_{2,s}^{n}.$$

4. $\lambda_{i,t} = \sum_{n \in \mathbb{N}} \lambda_{i,t}^{n}$, and hence, summing the above equalities, $\lambda_{i,t} - \lambda_{i,t}^{1} = \sum_{j=1,2} p_{ij} \lambda_{j,t}$, or, noting that $\lambda_{i,t}^{1} = p_{it}$,

$$\begin{pmatrix} \lambda_{1,a} \\ \lambda_{2,a} \end{pmatrix} = \begin{pmatrix} q & p \\ 0 & q \end{pmatrix} \times \begin{pmatrix} \lambda_{1,a} \\ \lambda_{2,a} \end{pmatrix} + \begin{pmatrix} r \\ r \end{pmatrix}$$

and

$$\begin{pmatrix} \lambda_{1,s} \\ \lambda_{2,s} \end{pmatrix} = \begin{pmatrix} q & p \\ 0 & q \end{pmatrix} \times \begin{pmatrix} \lambda_{1,s} \\ \lambda_{2,s} \end{pmatrix} + \begin{pmatrix} 0 \\ p \end{pmatrix},$$

and hence

$$\begin{pmatrix} 1-q & -p \\ 0 & 1-q \end{pmatrix} \times \begin{pmatrix} \lambda_{1,a} \\ \lambda_{2,a} \end{pmatrix} = \begin{pmatrix} r \\ r \end{pmatrix} \qquad \begin{pmatrix} 1-q & -p \\ 0 & 1-q \end{pmatrix} \times \begin{pmatrix} \lambda_{1,s} \\ \lambda_{2,s} \end{pmatrix} = \begin{pmatrix} 0 \\ p \end{pmatrix},$$

thus

$$\begin{pmatrix} \lambda_{1,a} \\ \lambda_{2,a} \end{pmatrix} = \begin{pmatrix} 1-q & -p \\ 0 & 1-q \end{pmatrix}^{-1} \times \begin{pmatrix} r \\ r \end{pmatrix} \qquad \begin{pmatrix} \lambda_{1,s} \\ \lambda_{2,s} \end{pmatrix} = \begin{pmatrix} 1-q & -p \\ 0 & 1-q \end{pmatrix}^{-1} \times \begin{pmatrix} 0 \\ p \end{pmatrix}$$

and, finally,

$$\begin{pmatrix} \lambda_{1,a} \\ \lambda_{2,a} \end{pmatrix} = \begin{pmatrix} \frac{1}{1-q} & \frac{p}{(1-q)^2} \\ 0 & \frac{1}{1-q} \end{pmatrix} \times \begin{pmatrix} r \\ r \end{pmatrix} \qquad \begin{pmatrix} \lambda_{1,s} \\ \lambda_{2,s} \end{pmatrix} = \begin{pmatrix} \frac{1}{1-q} & \frac{p}{(1-q)^2} \\ 0 & \frac{1}{1-q} \end{pmatrix} \times \begin{pmatrix} 0 \\ p \end{pmatrix}.$$

We can check that $\lambda_{1,a} + \lambda_{1,s} = 1$, and similarly that $\lambda_{2,a} + \lambda_{2,s} = 1$.

Chapter 13

For $p = 0.6$, $q = 0.3$ and $r = 0.1$, we obtain $\lambda_{1,a} = 0.27$, $\lambda_{1,s} = 0.73$, $\lambda_{2,a} = 0.14$ and $\lambda_{2,s} = 0.86$.

5. The result is clear for $n = 0$; we can easily check the inductive step. We note that

$$\begin{pmatrix} \lambda_{1,t}^{n+1} \\ \lambda_{2,t}^{n+1} \end{pmatrix} = \begin{pmatrix} q & p \\ 0 & q \end{pmatrix} \times \begin{pmatrix} \lambda_{1,t}^{n} \\ \lambda_{2,t}^{n} \end{pmatrix} = M \times \begin{pmatrix} \lambda_{1,t}^{n} \\ \lambda_{2,t}^{n} \end{pmatrix} = M^n \times \begin{pmatrix} \lambda_{1,t}^{1} \\ \lambda_{2,t}^{1} \end{pmatrix},$$

or, for $n \geq 1$,

$$\lambda_{2,a}^n = rq^{n-1},$$
$$\lambda_{2,s}^n = pq^{n-1},$$
$$\lambda_{1,a}^1 = r,$$

and, for $n \geq 2$,

$$\lambda_{1,a}^n = r(q^{n-1} + (n-1)pq^{n-2}),$$
$$\lambda_{1,s}^n = p^2(n-1)q^{n-2}.$$

6. We have $P(N_i = n) = \lambda_{i,s}^n + \lambda_{i,a}^n$, thus

$$G_i(Z) = \sum_{n \geq 1} (\lambda_{i,s}^n + \lambda_{i,a}^n) Z^n,$$

$$G_2(Z) = \sum_{n \geq 1} (p+r) q^{n-1} Z^n = \frac{(p+r)Z}{1 - qZ},$$

$$G_1(Z) = \sum_{n \geq 1} rq^{n-1} Z^n + \sum_{n \geq 2} p(p+r)(n-1) q^{n-2} Z^n = \frac{rZ}{1 - qZ} + \frac{p(p+r)Z^2}{(1 - qZ)^2}.$$

N_i represents the waiting time of the chain in the transient states, given that it started from the initial state i. We verify that

$$P(N_i < \infty) = \sum_{n \in \mathbb{N}} P(N_i = n) = \sum_{n \geq 1} (\lambda_{i,s}^n + \lambda_{i,a}^n)$$
$$= G_i(1) = 1.$$

We have $G_2(1) = \frac{p+r}{1-q} = 1$ and $G_1(1) = \frac{(1-q)(p+r)}{(1-q)^2} = 1$.

Lastly, we have $m_i = E(N_i) = (G_i(Z))'(1)$, or by a direct computation,

$$m_2 = E(N_2) = \sum_{n \geq 1} n(p+r) q^{n-1} = \frac{(p+r)}{(1-q)^2} = \frac{1}{1-q},$$

$$m_1 = E(N_1) = \sum_{n \geq 1} nrq^{n-1} + \sum_{n \geq 2} n(n-1)p(p+r)q^{n-2} = \frac{r}{(1-q)^2} + \frac{2p(p+r)}{(1-q)^3} = \frac{r+2p}{(1-q)^2}.$$

m_i represents the average waiting time of the chain in the transient states before ultimate absorption at a or s, given that it started from initial state i.

For $p = 0.6$, $q = 0.3$ and $r = 0.1$, we obtain: $m_1 = 2.65$ and $m_2 = 1.43$.

7. The average number of years of study, m_i^s, represents the mean of the r.v. N_i when final success is assumed. Defining event S by: 'the chain, given that it started from the initial state i,

eventually reaches state s', we have $m_i^s = E(N_i/S)$, i.e. m_i^s is the mean of the r.v. N_i conditioned by S, or the mean of the r.v. N_i conditioned by final success.

We thus obtain

$$m_i^s = E(N_i/S) = \sum_{n \geq 1} nP(N_i = n/S) = \sum_{n \geq 1} \frac{nP(N_i = n \text{ and } S)}{P(S)}$$

$$= \sum_{n \geq 1} \frac{nP(N_i = n \text{ and } X_n = s)}{P(S)} = \sum_{n \geq 1} \frac{n\lambda_{i,s}^n}{\lambda_{i,s}}.$$

Similarly, defining event A by: 'the chain, given that it started from the initial state i, ends up in state a', we have

$$m_i^a = E(N_i/A) = \sum_{n \geq 1} \frac{n\lambda_{i,a}^n}{\lambda_{i,a}}.$$

m_i^t represents the average waiting time of the chain in the transient states before ultimate absorption at state t, with $t \in \{a, s\}$, given that it started from initial state i.

$$m_2^s = E(N_2/S) = \sum_{n \geq 1} npq^{n-1}/\lambda_{2,s} = \frac{p/(1-q)^2}{p/(1-q)} = \frac{1}{1-q},$$

$$m_1^s = E(N_1/S) = \sum_{n \geq 2} n(n-1)p^2 q^{n-2}/\lambda_{1,s} = \frac{2p^2/(1-q)^3}{p^2/(1-q)^2} = \frac{2}{1-q},$$

and, similarly,

$$m_2^a = E(N_2/A) = \sum_{n \geq 1} nrq^{n-1}/\lambda_{2,a} = \frac{r/(1-q)^2}{r/(1-q)} = \frac{1}{1-q}$$

and

$$m_1^a = E(N_1/A) = \sum_{n \geq 1} nr(q^{n-1} + (n-1)pq^{n-2})/\lambda_{1,a}$$

$$= \frac{r(1-q+2p)/(1-q)^3}{r(1-q+p)/(1-q)^2} = \frac{1-q+2p}{(1-q)(1-q+p)}.$$

for $p = 0.6$, $q = 0.3$ and $r = 0.1$, we obtain: $m_1^s = 2.86$, $m_1^a = 2.09$ and $m_2^s = m_2^a = 1.43$.

8. The Markov chain hypothesis means that the probability of success or failure neither improves nor worsens as years go by; thus the generic student remains equable whatever the circumstances.

Chapter 14

14.1. We will give only one case of 'optimistic' or minimalist termination (after $n-1$ comparisons) and a case of 'pessimistic' or maximalist termination (after $n+1$ comparisons), but we strongly advise the reader to study other possible cases.

Case 1 ('optimistic' ending, after $n-1$ comparisons): assume that, just before the two counters k and l meet, we have the following configuration (Figure 15.16):

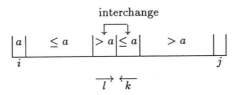

Figure 15.16

in that case the `pivot` procedure will stop after $n-1$ comparisons.

Case 2 ('pessimistic' ending, after $n+1$ comparisons): Assume that, before counters k and l meet, we have the following configuration (Figure 15.17):

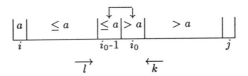

Figure 15.17

in this case the `pivot` procedure will stop after $n+1$ comparisons. Indeed, for $k = i_0$ and $k = i_0 - 1$, $T(k)$ will be compared to the pivot; then, for $l = i_0 - 1$ and $l = i_0$, $T(l)$ will be compared to the pivot; each of the elements $T(i_0)$ and $T(i_0 - 1)$ will thus be compared to the pivot **twice** before the procedure `pivot` finally stops.

14.2. 1. First check the case when $T[1,2] = (11,22)$; procedure `Quicksort` will call the `pivot` procedure with $i = 1$, $j = 2$; the `pivot` procedure will set $l := 2$, $k := 2$, $p := 11$; then it will perform the comparisons

- $T(k) > 11$ hence $k := 1$, and
- $T(l) \leq 11$ hence $l := 3$;

and it will stop after these two comparisons, since $l > k$.

Then check the case when $T[1,2] = (22,11)$; the `pivot` procedure will set $l := 2$, $k := 2$, $p := 22$; then it will perform the comparisons

- $T(k) \not> 22$ hence $k := 2$, and
- $T(l) \leq 22$ hence $l := 3$;

as $l < k$, `pivot` will then perform $\mathtt{interchange}(T(l), T(k))$, and it will stop, since $l > k$; `pivot` also stops here after two comparisons.

2. Similar.

14.3. Writing equations (14.1) for $i = 2, \ldots, n$, and summing these equations, we deduce

$$\forall n \geq 2, \quad p_n = \frac{(n+1)(n+2)}{2} - 3.$$

14.4. Computations are almost identical (albeit very slightly simpler).

14.5. Similar computations.

14.6. In order to prove termination, we associate with the loop the integral-valued expression v, which strictly decreases each time we execute the body of the WHILE loop.

In order to prove that the program computes the greatest common divisor of its arguments, let us introduce the loop invariant I: $\gcd(u, v) = k$, where k is a constant which does not change when we go through the loop. k is thus equal to $gcd(u_0, v_0)$ when the program begins, where u_0 and v_0 are the values read by the program, and k is equal to u_n when exiting the program, where u_n is the value of u when the loop is exited with $v = 0$.

14.7. We have the successive divisions

$$29 = 3 \times 8 + 5$$
$$8 = 5 \times 8 + 1$$
$$5 = 1 \times 3 + 2$$
$$3 = 1 \times 2 + 1$$
$$2 = 1 \times 1 + 1$$

hence

$$\frac{8}{29} = \cfrac{1}{3 + \cfrac{1}{1 + \cfrac{1}{1 + \cfrac{1}{1 + \cfrac{1}{2}}}}}$$

$n = 5$.

14.8. We associate with the first loop the value $-n$, which is a strictly positive integer and which strictly decreases at each execution of the WHILE loop; this first loop thus terminates. Similarly, in order to prove that the second WHILE loop terminates, we associate with it the value n.

In order to prove that the program indeed computes the power, we associate with it the loop invariant

$$I(a, n, r): r \times a^n = a^k ,$$

whose value remains constant at each execution of the WHILE loops; its value when entering the WHILE loops is $i = a^k$ and its value when exiting the WHILE loops is $i = r$. We thus have that

$r = a^k$. The annotated program is as follows:

```
PROGRAM power
VAR a,k,r:  integer
BEGIN
READ a,n                                    integer(a, k)
n:= k
IF n < 0 and a = 0 THEN
   WRITE undefined result
OTHERWISE
   r:= 1                                    (n ≠ 0) ∧ (r = 1) ∧ (a^n r = a^k)
   WHILE n < 0 DO
      r:= r / a
      n:= n+1
   ENDWHILE                                 (n = 0) ∧ (a^n r = a^k)
   WHILE n > 0 DO
      r:= r * a
      n:= n-1
   ENDWHILE                                 (n = 0) ∧ (a^n r = a^k)
   WRITE r                                  (r = a^k)
ENDIF
END
```

14.9. For proving termination, we use the expression $V = n$. For showing that the program indeed computes the factorial function, we prove the property $p(n) = (y = n!)$. **Fact** annotated with the final assertions is

```
FUNCTION Fact(n:  integer):  y:  integer
BEGIN
IF n = 0 THEN y:= 1                         (n = 0) ∧ (y = 1)
OTHERWISE y:= n*Fact(n-1)                   y = n × (n − 1)!
ENDIF                                       y = n!
RETURN(y)                                   y = n!
END
```

14.10. $V = (n, m)$, and $\mathbb{N}^2$ is endowed with the lexicographic ordering:

- If $n = 0$, the call **Ackermann**(n, m) terminates.
- If $n \neq 0$, **Ackermann**(n, m) calls:
  - Ackermann$(n-1, 1)$,
  - Ackermann$(n, m-1)$ and
  - Ackermann$(n-1, \text{Ackermann}(n, m-1))$.

We have in the lexicographic ordering,

$$(n-1, 1) < (n, m)$$
$$(n, m-1) < (n, m)$$
$$\text{and} \quad (n-1, \text{Ackermann}(n, m-1)) < (n, m)$$

14.11. $V = x$.

- If $x = 0$, the first clause of the program terminates with result Q(x)=*true*.
- Assume that: $\forall y \leq n-1$, Q(y) terminates with the result *true*, then for $x = n$, Q(x) has for result the result returned by the recursive call of Q(y), with $y = n-1$. Thus Q(x) also terminates with the result *true*.

14.12. 1. **pivot** terminates: with the first (resp. second) inner WHILE loop we associate the integer k (resp. l), which strictly decreases (resp. strictly increases) at each execution of the loop and is in the well-ordered set $\{i, i+1, \ldots, \sup(i+1, j)\}$, and with the outer WHILE loop we associate the integer $k - l$, which strictly decreases at each execution of the loop and is in the well-ordered set $\{-1\} \cup \mathbb{N}$.

Quicksort terminates: the length of the list to be sorted strictly decreases at each recursive call.

2. For the partial correctness, we simply give the programs annotated with the final assertions; each final assertion is written at the right-hand of the instruction after which it is true.

```
PROGRAM pivot
VAR i, j, k, l: integer
VAR T: integer list
BEGIN
  READ i,j,T
  l:= i+1
  k:= j                                    integer(i,j,k)
  p:= T(i)                                 q
  WHILE l ≤ k DO
    WHILE T(k) > p DO k:= k-1 ENDWHILE     (T(k) ≤ p) ∧ q
    WHILE T(l) ≤ p DO l:= l+1 ENDWHILE     (T(l) > p) ∧ (T(k) ≤ p) ∧ q
    IF l < k THEN
      interchange (T(l),T(k))
      k:= k-1
      l:= l+1
    ENDIF
                                           q
  ENDWHILE                                 (k > l) ∧ q
  interchange (T(i),T(k))
  RETURN (k)        ((i ≤ l' < k) ⟹ (T(l') ≤ T(k))) ∧ (k' > k ⟹ (T(k') > T(k)))
END
```

where q is the assertion

$$\Big(p = T(i)\Big) \wedge \Big((i \leq l' < l) \implies (T(l') \leq p)\Big) \wedge \Big(k' > k \implies (T(k') > p)\Big)$$

note that q is a loop invariant for the loop WHILE $l \leq k$ ....

Now let $q(T, i, j)$ be the assertion $i \leq k \leq l \leq j \implies (T(k) \leq T(l))$, and let q' be the assertion $((i \leq l' < k) \implies (T(l') \leq T(k))) \wedge (k' > k \implies (T(k') > T(k)))$; the following annotations show the partial correctness of **Quicksort**:

```
PROGRAM Quicksort
VAR i,j,k: integer
VAR T: integer list
BEGIN
  READ i,j,T
  IF i < j THEN
    pivot(T,i,j; k)              q'
    Quicksort(T, i, k-1)         q' ∧ q(T, i, k-1)
    Quicksort(T, k+1, j)         q' ∧ q(T, i, k-1) ∧ q(T, k+1, j)
  ENDIF                          q(T, i, j)
  PRINT T
END
```

INDEX

absorbing state, 268
absorption probability, 277
almost impossible event, 231
almost sure event, 231
antichain, 21
antihomomorphism
    (of Boolean algebras), 59
antisymmetric relation, 13
arity, 9
associative operation, 9
asymptotic power series
    expansion, 177
atomic formula, 83
average complexity of Quicksort, 284
average waiting time before
    absorption, 277

Bayes's rule for the
    probability of causes, 236
Bernoulli trials, 255
bijective mapping, 5
binary operation, 9
binomial coefficients, 108
binomial distribution, 255
Boolean algebra, 58
bootstrapping , 181
bound occurrence, 86

cardinality, 7
Cartesian product (of sets), 1
chain, 21
chain (of a graph), 188
characteristic function, 5
characteristic polynomial, 128
Chebyshev's inequality, 251
circuit, 188
clause (program, definite), 103
combination with repetition, 118
commutative operation, 9
complement, 2
complement of a relation, 11
complementary event, 228
complemented lattice, 33
complete finite-state automaton, 209
complete lattice, 27
complete recurrence, 125
complete transition system, 209
completeness (theorem), 76, 96
computation
    (of a transition system), 208
concatenation (of two languages), 39
conditional probability, 233
congruence, 15
connected component, 191
connected graph, 191
continued fraction, 289

continuous mapping, 28
convolution product, 150, 152
correlation coefficient (of a pair
    of random variables), 248
countable set, 8
co-variance (of a pair
    of random variables), 248
cycle, 147, 188

De Morgan's laws, 2, 33
deterministic finite-state
    automaton, 209
deterministic transition system, 209
diameter of a graph, 190
difference, 2
difference (of sequences), 143
direct product (of ordered sets), 21
directed graph, 183
disjoint subsets, 2
distributive lattice, 31
domain, 4

edge, 183
equivalence class, 14
equivalence relation, 14
Euler (path or circuit), 188
event, 228
exponential generating series, 157

fact, 103
factor ordering, 19
factor set, 15
finite index semi-congruence, 214
finite-state automaton, 209
first order formula, 83
fixed point, 29
formal power series, 150
formula (closed, ground), 89
formula (universal), 99
free monoid, 10

free occurrence, 86
function, 4

generating function
    (of a random variable), 253
generating series, 152
geometric distribution, 257
graph, 183
graph (of a Markov chain), 274
greatest element, 22
greatest lower bound, 23
ground term, 83, 98

Hamiltonian (path or circuit), 188
height of a binary tree, 51
Herbrand basis, 98
Herbrand model, 99
Herbrand structure, 98
Herbrand universe, 98
hierarchy, 177
Hoare's assertion method, 294
homogeneous recurrence, 125
homomorphism
    (of Boolean algebras), 59
homomorphism (of monoids), 205
Horn clause, 103
hypergeometric distribution, 257

impossible event, 228
independence
    (of two random variables), 244
independent events, 237
induction (first principle), 36
induction (second principle), 38
inductive definition of a mapping, 48
inductive definition of a subset, 40
injective mapping, 5
inorder traversal, 51
instance, 90
instance (ground), 90

Index 399

interpretation, 71
interpretation of a formula, 71
intersection, 2
intersection of relations, 11
interval, 22
invertible series, 153
irreducible Markov chain, 268
irreflexive relation, 13
isomorphic graphs, 185
iteration (of a language), 206

joint distribution
    (of two random variables), 242

k-combination, 108
k-permutation, 108
k-permutation with repetition, 118

language, 39
lattice, 27, 30
least element, 22
least upper bound, 23
left segment, 22
lexical variant, 92
linear extension, 20
linear homogeneous recurrence, 127
linear recurrence, 124
linear recurrences with
    parameters, 140
literal, 103
locally finite ordered set, 22
logical consequence, 73
loop, 186
lower bound, 22

mapping, 4
marginal distributions
    (of two random variables), 243
Markov chain, 263

Markov's inequality, 251
maximum, 22
mean (of a random variable), 245
minimal automaton, 217
minimum, 22
model, 96
monoid, 10, 204
monotone mapping, 20
monotonic mapping, 20
multinomial distribution, 256
multiset, 118
mutually exclusive events, 228

non-homogeneous linear
    recurrences, 134
non-homogeneous recurrence, 125
notation $O(g)$, 170
notation $o(g)$, 171
notation $\Omega(g)$, 170
notation $\theta(g)$, 171

occurrence, 84
one-to-one correspondence
    mapping, 5
one-to-one mapping, 5
onto mapping, 5
operation, 9
order relation, 17
ordered set, 20
ordering, 17
ordering (lexicographic), 18
ordering (prefix), 18
orientation, 185

palindromes, 213
partial correctness, 294
partial fraction expansion, 157
partial ordering, 18
partially ordered set, 20
partition, 3, 229

partition recurrence, 125, 141
Pascal's triangle, 108
path, 188
permutation, 107
persistent state
 (of a Markov chain), 269
planar graph, 196
Poisson distribution, 259
polynomial recurrence, 124
poset, 20
potential matrix
 (of a Markov chain), 271
predecessor (sequence), 143
prefixes (of a language), 210
prenex formula, 93
preorder, 18
preorder traversal, 52
principal part, 177
probability distribution, 230
probability distribution
 (of a random variable), 242
probability space, 230
product (of sequences), 143
product (of two languages), 206
product space, 240
product ($\star$ product of sequences), 143
proof, 73
propositional calculus, 70
propositional formula, 70
provable sequent, 73, 95

Quicksort, 281

random variable, 241
range, 4
reachable finite-state
 automaton, 216
recognizable language, 210
recurrence of degree k, 124

recurrent state
 (of a Markov chain), 269
reflexive relation, 13
regular expressions, 207
regular subset, 207
rooted tree, 199
rooted tree (ordered), 202

sample space, 227
sampling with replacement, 256
sampling without replacement, 257
satisfiable formula, 71, 89
satisfiable formula
 (in a structure), 89
semantical consequence, 72
semantics , 71
semi-congruence (right), 214
semi-group, 10
sentence, 89
sequent, 71
sequent true in I, 72
sequent valid, 72
set, 1
shift operation series, 153
simple graph, 186
Skolem functions, 102
Skolemization, 102
soundness (theorem), 76, 96
standard deviation
 (of a random variable), 247
state (of a Markov chain), 262
stochastic matrix, 265
strict ordering, 17
strongly connected graph, 191
structural induction (proof by), 43
structure, 86, 98
subformula, 84
subgraph, 186
submonoid, 205
substitution, 70

Index

substitution method, 141
summation factors (method), 141
summation (of a sequence), 143
sure event, 228
surjective mapping, 5
symmetric difference, 2
symmetric relation, 13

tautology, 71
term, 83
theory, 96
total correctness, 294
total ordering, 18
total probabilities rule, 236
trace (of a path), 208
transient state
    (of a Markov chain), 269
transition matrix, 265
transition probability, 265
transition system, 208
transitive relation, 13
tree, 197
trial, 227
tribe, 229

truth value, 71, 87

undetermined coefficients
    (method of), 139
undirected graph, 184
union, 2
union of relations, 11
unit, 9
universal induction principle, 53
universally valid formula, 89
unsatisfiable formula, 71
upper bound, 22

valid formula, 71, 89
valuation, 87
variance (of a random variable), 247
vertex, 183

weak law of large numbers, 252
well ordering, 25
well-founded ordering, 25

Prentice Hall International Series in Computer Science (*continued*)

PEYTON JONES, S.L. and LESTER, D., *Implementing Functional Languages*
POMBERGER, G., *Software Engineering and Modula 2*
POTTER, B., SINCLAIR, J., and TILL, D., *An Introduction to Formal Specification and Z*
ROSCOE, A.W. (ed.), *A Classical Mind: Essays in honour of C.A.R. Hoare*
ROZENBERG, G. and SALOMAA, A., *Cornerstones of Undecidability*
RYDEHEARD, D.E. and BURSTALL, R.M., *Computational Category Theory*
SHARP, R., *Principles of Protocol Design*
SLOMAN, M. and KRAMER, J., *Distributed Systems and Computer Networks*
SPIVEY, J.M., *An Introduction to Logic Programming*
SPIVEY, J.M., *The Z Notation: A reference manual* (2nd edn)
TENNENT, R.D., *Principles of Programming Languages*
TENNENT, R.D., *Semantics of Programming Languages*
WATT, D.A., *Programming Language Concepts and Paradigms*
WATT, D.A., *Programming Language Processors*
WATT, D.A., WICHMANN, B.A. and FINDLAY, W., *ADA: Language and methodology*
WATT, D.A., *Programming Language Syntax and Semantics*
WELSH, J. and ELDER, J., *Introduction to Modula 2*
WELSH, J. and ELDER, J., *Introduction to Pascal* (3rd edn)
WELSH, J. and HAY, A., *A Model Implementation of Standard Pascal*
WIKSTRÖM, A., *Functional Programming Using Standard ML*
WOODCOCK, J. and DAVIES, J., *Using Z: Specification, refinement, and proof*